Plasma Physics: Concepts and Applications

Plasma Physics: Concepts and Applications

Editor: Eddie Rocco

NY RESEARCH PRESS
New York

Published by NY Research Press
118-35 Queens Blvd., Suite 400,
Forest Hills, NY 11375, USA
www.nyresearchpress.com

Plasma Physics: Concepts and Applications
Edited by Eddie Rocco

International Standard Book Number: 978-1-63238-650-2 (Hardback)

Cataloging-in-Publication Data

Plasma physics : concepts and applications / edited by Eddie Rocco.
p. cm.
Includes bibliographical references and index.
ISBN 978-1-63238-650-2
1. Plasma (Ionized gases). 2. Physics. I. Rocco, Eddie.
QC718 .P53 2019
530.44--dc23

Contents

Preface

This book has been an outcome of determined endeavour from a group of educationists in the field. The primary objective was to involve a broad spectrum of professionals from diverse cultural background involved in the field for developing new researches. The book not only targets students but also scholars pursuing higher research for further enhancement of the theoretical and practical applications of the subject.

Plasma is a fundamental state of matter. It is generally produced by heating or by subjecting a neutral gas to a strong electromagnetic radiation thereby converting it into an ionized gas. In plasma physics, the bulk properties of plasma like magnetization, thermal properties, non-thermal properties as well as classification and production of commercial and industrial plasmas are studied. It has applications in diverse areas such as fusion technology, food processing, enhanced oil recovery and plasma medicine. Research in theoretical domains of plasma physics extends into the areas of plasma production in high energy interactions in nature, plasma stability, equilibria and confinement among others. This book unravels the recent studies in the field of theoretical and applied plasma physics. It also provides significant information of the technological advances and recent research frontiers to help develop a holistic understanding of this field. This book presents all aspects of this discipline in a comprehensive way and will be beneficial to researchers and students.

It was an honour to edit such a profound book and also a challenging task to compile and examine all the relevant data for accuracy and originality. I wish to acknowledge the efforts of the contributors for submitting such brilliant and diverse chapters in the field and for endlessly working for the completion of the book. Last, but not the least; I thank my family for being a constant source of support in all my research endeavours.

Editor

1

Dispersion relation and growth rate in two-stream thermal plasma-loaded free-electron laser with helical wiggler

S. Meydanloo · S. Saviz

Abstract Linear theory of the two-stream free-electron laser consisting of a relativistic electron beam transported along the axis of thermal plasma-loaded helical wiggler is proposed and investigated. The dispersion relation is derived employing linear fluid theory. The characteristics of the dispersion relation are analyzed by numerical solutions. The results show in that in the special values of the plasma temperature the growth rate is considerably enhanced. It is also shown that the growth rate after critical plasma density gradually decreases. Moreover, in the presence of the two-electron beam the growth rate of electrostatic mode is two times greater than that for electromagnetic mode.

Keywords Two-stream free-electron laser · Thermal plasma

Introduction

In the recent years, the free-electron laser (FEL) in the presence of the plasma has been widely discussed. Pei et al. [1] have proposed and examined cold plasma-loaded FEL. They have shown that the efficiency of the FEL considerably enhanced in the presence of the dense plasma. Pant et al. [2] illustrated that in the whistler mode FEL the growth rate considerably enhanced in the presence of the plasma. Serbto et al. [3], by making use of the collective description derived by Bonifacio et al. [4], discussed the characteristic of a gas-filled FEL and shown that the efficiency and power enhanced considerably. Zongjon Shi [5] investigated the cylindrical waveguide with a cold plasma-loaded FEL and shown that growth rate is enhanced at low frequency and it is decreased at high frequency. The electron trajectories and gain in the whistler mode FEL were investigated by Jafari et al. [6, 7]. The effects of the two-stream instability on the linear gain and growth rate of the two-stream FEL in different configuration were investigated by Saviz et al. [8–10]. They discussed the gain in two-stream FEL with planar wiggler and ion-channel guiding. Especially it is necessary to study the thermal plasma-loaded FEL with helical wiggler.

In the present paper, we investigated the effect of plasma density and temperature on the electrostatic and electromagnetic mode of the two-stream FEL. The research is not done until now. We show that the effect of growth rate occurs at the special range of the plasma temperature and plasma density. Moreover, we show that because of the presence of the two-stream instability, the growth rate of the electrostatic mode is two times greater than that of the electromagnetic mode.

The configuration of the present paper is as follows: The physical model is given in "Physical model". In "The general dispersion equation", we obtain the general dispersion equation using linear fluid theory. The numerical results and discussion are given in "The characteristic of dispersion equation" and conclusions are given in "Conclusion".

Physical model

The considered physical configuration is the idealized one-dimensional limit in which the two-stream relativistic electron beams are passing through the thermal plasma-

S. Meydanloo · S. Saviz (✉)
Plasma Physics Research Center, Science and Research Branch, Islamic Azad University, Tehran, Iran
e-mail: azarabadegan@gmail.com; shahrooz.saviz@srbiau.ac.ir

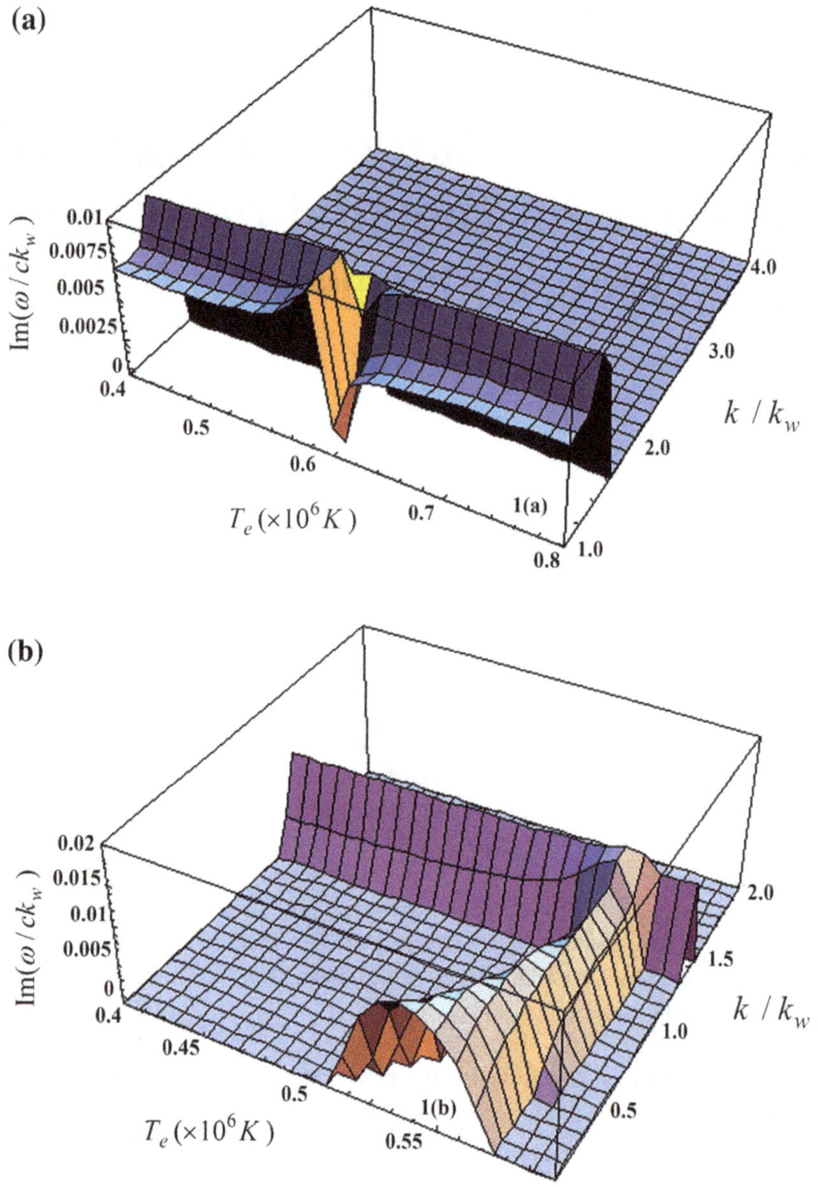

Fig. 1 The normalized growth rate $\mathrm{Im}(\omega/ck_w)$ of the electrostatic wave as a function of k/k_w and plasma electron temperature T_e **a** for the electrostatic mode, and **b** for the electromagnetic mode

loaded helical wiggler. We consider that the two-electron beams are moving along the z direction. The idealized helical wiggler magnetic field is described by the vector potential as follows:

$$A_w(z) = -A_w[\hat{e}_x \cos(k_w z) + \hat{e}_y \sin(k_w z)], \tag{1}$$

where, $A_w = B_w/k_w$, $\lambda_w = 2\pi/k_w$ and λ_w is the wiggler period, k_w is the wiggler wave number, and B_w denotes the amplitude of wiggler field. The vector potential of optical field is given as follows:

$$A_s(z,t) = (B_s/k_s)[\hat{e}_x \cos(k_s z - w_s t + \varphi_s(t)) - \hat{e}_y \sin(k_s z - w_s t + \varphi_s(t))], \tag{2}$$

where, k_s is optical wave number, B_s is the field strength and $\varphi_s(t)$ is initial of optical phase, which is usually looked as zero. The interaction among electron beams, background thermal plasma and optical field is described by the wave equation as follows:

$$(\partial^2/\partial z^2 - (1/c^2)\partial^2/\partial t^2)\vec{A} = -(4\pi/c)\vec{J}_\perp, \tag{3}$$

where the transverse current is as follows:

$$J_\perp = -\frac{en_{0p}}{m}\delta P_{p\perp} - \frac{en_{0b1}}{\gamma_{01}m}\delta P_{b\perp 1} - \frac{en_{0b2}}{\gamma_{02}m}\delta P_{b\perp 2} - \frac{eP_{0b\perp 1}}{\gamma_{01}m}\delta_{nb1} - \frac{eP_{0b\perp 2}}{\gamma_{02}m}\delta_{nb2}, \tag{4}$$

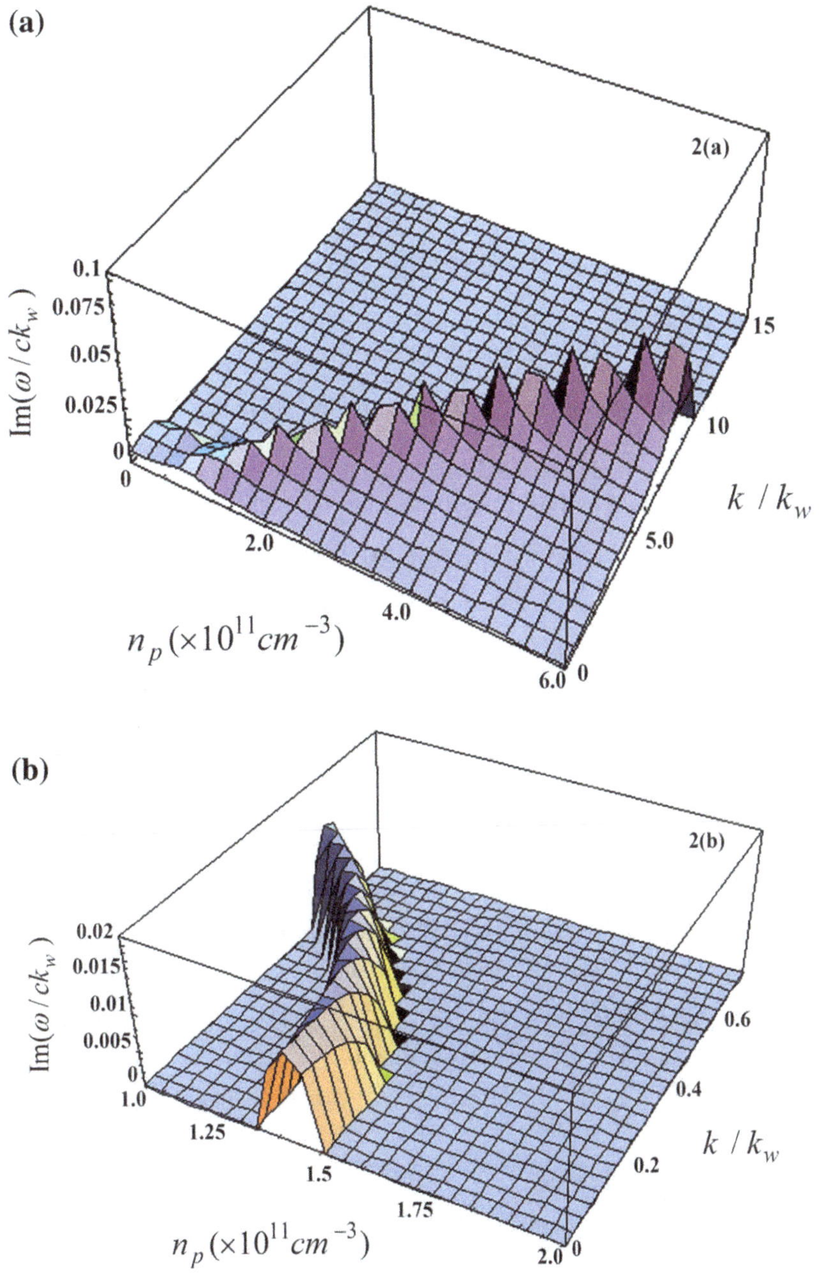

Fig. 2 The normalized growth rate $\mathrm{Im}(\omega/ck_w)$ of the electrostatic wave as a function of k/k_w and plasma density n_p **a** for the electrostatic mode, and **b** for the electromagnetic mode

In Eq. (3b), n_{0p} and n_{0b} are the plasma and electron beams density, respectively. The $P_{0b\perp1}$ and $P_{0b\perp2}$ are the two-beam transverse momentum, and γ_{01} and γ_{02} are relativistic factor for two beams. The electrostatic field is as follows:

$$\vec{E}_z(t) = \hat{e}_z E_z \sin[-k_1 z + w_1 t + \theta] = \hat{e}_z E_z \sin(\Psi + \delta) \quad (5)$$

where $\Psi = -k_1 z + w_1 t$ is the pondermotive wave phase, and δ is the mismatch phase of a three-wave interaction. The phase-matching condition is given by $w_1 = -w_s$ and $k_1 = -(k_s + k_w)$. In Eq. (5), E_z is the total longitudinal electric field produced by perturbed density of two-electron beams and background thermal plasma given by Poisson Equation:

$$\partial E_z/\partial z = -4\pi e(\delta n_{b1} + \delta n_{b2} + \delta n_p). \quad (6)$$

The general dispersion equation

The momentum transfer equation for the electron beams and plasma is obtained as follows:

$$d\vec{P}_b/dt = -e\left[\vec{E}_z + \vec{E}_s + V_b/c \times (\vec{B}_w + \vec{B}_s)\right], \quad (7)$$

By linearizing the Eq. (8) with $P_{b1,2} = P_{01,2} + \delta P_{b1,2}$ and $n_{b1,2} = n_{01,2} + \delta n_{b1,2,}$ the equilibrium and perturbed momentum of electron beam are obtained as follows:

$$P_{01,2} = e/(cA_{\rm w}) + \gamma_{01,2} m v_{01,2}, \quad \delta P_{b1,2} = (e/c)A_s. \tag{8}$$

Using the momentum equation, the following equations are obtained for the plasma:

$$d(m\delta\vec{v}_{\rm p\perp})/dt = (e/c)\partial/\partial t\vec{A}_{\rm s} - (e/c)\delta\vec{v}_{\rm pz} \times \vec{B}_{\rm w}, \tag{9}$$

$$d(m\delta\vec{v}_{\rm pz})/dt = -eE_z - (e/c)\delta\vec{v}_{\rm p\perp} \times \vec{B}_{\rm w} - (3k_b T_e/n_{0p} m_{\rm e})(\partial\delta n_{\rm p}/\partial z), \tag{10}$$

where $T_{\rm e}$ is the plasma electron temperature and $k_{\rm b}$ is the Boltzman constant.

After some straightforward algebra, the dispersion relation is obtained by Eqs. (3)–(10) as follows:

$$\begin{aligned}
&(w_{\rm s}^2 - c^2k_{\rm s}^2 - \omega_{\rm p}^2 - (\omega_{\rm b1}^2/\gamma_{01}) - (\omega_{\rm b2}^2/\gamma_{02}))\{[(w_1 - k_1v_1)^2\\
&(w_1 - k_1v_2)^2 - (\omega_{\rm b1}^2/\gamma_{01}^3)(w_1 - k_1v_2)^2 - (\omega_{\rm b2}^2/\gamma_{02}^3)\\
&(w_1 - k_1v_1)^2](w_1^2 - \omega_{\rm ce}^2/2 - \omega_{\rm T}^2 - \omega_{\rm p}^2)\\
&- \omega_{\rm p}^2(w_{\rm s}^2 - c^2k_{\rm s}^2 - \omega_{\rm p}^2 - (\omega_{\rm b1}^2/\gamma_{01}) - (\omega_{\rm b2}^2/\gamma_{02}))\\
&\{[(w_1 - k_1v_1)^2(w_1 - k_1v_2)^2 - (\omega_{\rm b1}^2/\gamma_{01}^3)(w_1 - k_1v_2)^2\\
&- (\omega_{\rm b2}^2/\gamma_{02}^3)(w_1 - k_1v_1)^2](w_1^2 - \omega_{\rm ce}^2/2 - \omega_{\rm T}^2\\
&- \omega_{\rm p}^2) - \omega_{\rm p}^2 - 2k_1(\omega_{\rm p}^2\omega_{\rm b1}^2/\gamma_{01}^4k_{\rm w})(w_1 - k_1v_2)^2\\
&- 2k_1\omega_{\rm p}^2\omega_{\rm b2}^2/(k_{\rm w}\gamma_{02}^4)(w_1 - k_1v_1)^2 + \omega_{\rm p}^2[(w_1 - k_1v_1)^2\\
&(w_1 - k_1v_2)^2 - (\omega_{\rm b1}^2/\gamma_{01}^3)(w_1 - k_1v_2)^2 - (\omega_{\rm b2}^2/\gamma_{02}^3)\\
&(w_1 - k_1v_1)^2] + 2(\omega_{\rm b1}^2/\gamma_{01}^4)(\omega_{\rm b2}^2/\gamma_{01}^4)(k_1/k_{\rm w})^2\\
&(w_1^2 - \omega_{\rm ce}^2/2 - \omega_{\rm T}^2) - (k_1/k_{\rm w})^2(w_1^2 - \omega_{\rm ce}^2/2 - \omega_{\rm T}^2)\\
&[\omega_{\rm b1}^2\omega_{\rm b2}^2/(\gamma_{01}^5\gamma_{02}^3) + \omega_{\rm b1}^2\omega_{\rm b2}^2/(\gamma_{02}^5\gamma_{01}^3)]] = 0
\end{aligned} \tag{11}$$

where $\omega_{\rm p}^2 = 4\pi n_{0\rm p}e^2/m_{\rm e}$, $\omega_{\rm b1,2}^2 = 4\pi n_{0\rm b1,2}e^2/m_{\rm e}$, $\omega_{\rm T}^2 = 3k_{\rm b}T_{\rm e}k^2/m_{\rm e}$, $w_1 = w_{\rm s}$ and $k_1 = k_{\rm w} + k_{\rm s}$.

The characteristic of dispersion equation

In this section, we focus on the characteristic of the dispersion equation, Eq. (11), through the numerical analysis. In the numerical calculation, we have found two unstable growing roots: one for electrostatic wave and the other for electromagnetic wave. In the presence of the thermal plasma, we have illustrated the growth rate as a function of normalized wave number ($k/k_{\rm w}$) and plasma electron temperature ($T_{\rm e}$). Figure 1a shows the growth rate ($\mathrm{Im}\omega/(ck_{\rm w})$) of the electrostatic mode with respect to the plasma temperature, $T_{\rm e}$ (in the region $0.4 \times 10^7K < T_{\rm e} < 0.6 \times 10^7K$). In this figure, the parameters used are chosen to be $\gamma_1 = 1.25$, $\gamma_2 = 1.232$, $B_w = 2000G$, $n_p = 2.5 \times 10^{11}/cm^3$ and $n_{b1,2} = 4 \times 10^8/cm^3$. As shown in this figure, the growth rate rapidly increases with plasma temperature and reaches the maximum near the plasma optimum temperature, and decreases when the plasma temperature was above the optimum value. These interesting phenomena come from the interaction of two-stream instability coupling FEL instability. Figure 1b shows the growth rate ($\mathrm{Im}\omega/(ck_{\rm w})$) of the electromagnetic mode with respect to the plasma temperature, $T_{\rm e}$ (in the region $0.4 \times 10^7K < T_{\rm e} < 0.6 \times 10^7K$). As seen in Fig. 1b, the maximum values of wave number occur in the special range of plasma temperature (nearly $0.5 \times 10^7K < T_{\rm e} < 0.57 \times 10^7K$). The effect of plasma density on the growth rate of electrostatic and electromagnetic modes, for specified values of parameters and range of normalized $k(k/k_{\rm w})$, is given in

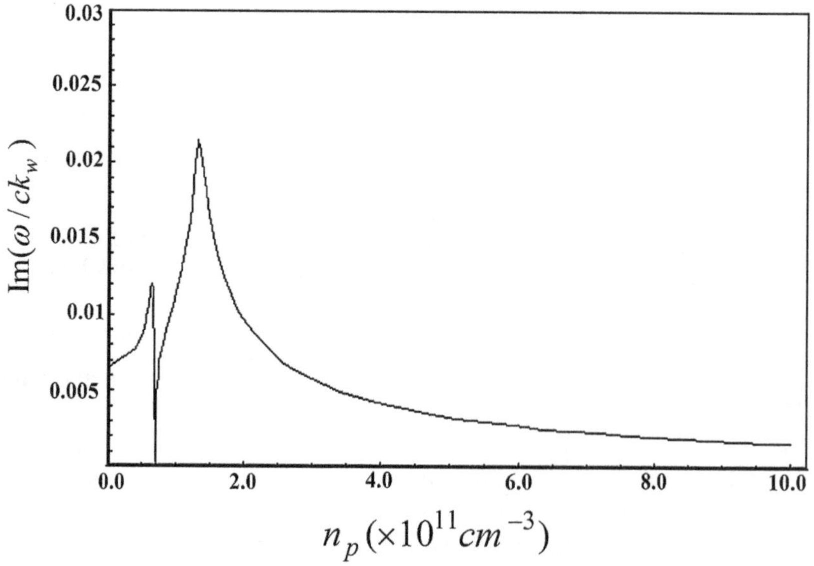

Fig. 3 The normalized growth rate $\mathrm{Im}(\omega/ck_{\rm w})$ of the electrostatic wave versus plasma density $n_{\rm p}$ for $k/k_{\rm w} = 1.0$

Fig. 2a, b, respectively. The chosen parameters are as: $\gamma_1 = 1.25$, $\gamma_2 = 1.232$, $B_w = 2000G$, $T_e = 5 \times 10^6 K$ and $n_{b1,2} = 4 \times 10^8/\text{cm}^3$. Figure 2a shows that at fixed values of wave number there is the critical plasma density after which the growth rate decreases gradually with plasma density. This phenomenon is because of the fact that the synchronism condition of electron beam and optical wave is destroyed at higher plasma density. The plasma effect on the electromagnetic mode is given in Fig. 2b. Figure 2b shows that increase in the plasma temperature causes the shift of wave number to the lower values. This figure also shows that the maximum growth rate of this instability occurs in the special range of plasma density corresponding to the given range of wave number. The growth rate of the electrostatic mode is two times greater than that of the electromagnetic mode. This difference is because of the fact that the two-stream instability affects the electrostatic mode and this difference is because of the two-stream instability. The growth rate as a function of the plasma density at $k/k_w = 1.0$ is shown in Fig. 3 which is the 2D plot of Fig. 2a and is for clarity.

Conclusion

In this work, we have investigated the instability of two-stream FELs that contain thermal plasma. The dispersion relation is obtained by employing the linear fluid theory and calculated numerically. The results show that the growth rate depends on the plasma temperature and plasma density. It was found that the growth rate in the special values of the plasma temperature enhances considerably and the growth rate after critical plasma density gradually decreases. In the plasma-loaded FEL [1–5], all authors use the single-electron beam but in the present work we use the two-electron beams. In the single electron beam case there is only the electromagnetic mode but as seen the results there is the new electrostatic mode that is two times stronger than electromagnetic mode. The electrostatic mode does not appear in single-stream FEL.

References

1. Pei, W.B., Chen, Y.S.: The effect of background plasma in the undulator on free electrom lasers. Int. J. Electronics **65**, 551 (1988)
2. Pant, K.K., Tripathi, V.K.: IEEE. Trans. Plasma Sci. **22**, 217 (1994)
3. Serbeto, A., Virginia Alves, M.: High-gain free-electron-laser amplifier with warm plasma background: linear analysis. IEEE. Trans. Plasma. Sci. **21**, 243 (1993)
4. Bonifacio, R., Pellegrini, C., Narducci, L.M.: Collective instabilities and high-gain regime in a free electron laser. Opt. Commun. **50**, 373 (1984)
5. Shi, Zongjun, Yang, Ziqiang, Liang, Zheng: Int. J. Infra. Mill. Waves. **24**, 1823 (2003)
6. Jafarinia, S., Jafari, S.: Investigation of the electron trajectories and gain regimes of the whistler pumped free-electron laser. Phys. Plasma. **20**, 043106 (2013)
7. Saviz, S., Jafari, S.: Two-stream whistler-pumped free-electron laser. J. Plasma Phys (2014). doi:10.1017/S0022377814000750
8. Saviz, S., Rezaei, Z., Farzin Aghamir, M.: Gain in two stream free electron laser with planar wiggler and ion-channel guiding. Phys. Plasma. **19**, 023115 (2012)
9. Mehdian, H., Saviz, S.: Lectron trajectory and growth rate in a two-stream electromagnetically pumped free electron laser and axial guide field. Phys. Plasma. **15**, 093103 (2008)
10. Mehdian, H., Saviz, S., Hasanbeigi, A.: Two-stream instability in free electron lasers with a planar wiggler and an axial guide magnetic field. Phys. Plasma. **15**, 043103 (2008)

2

Degradation of bromophenol blue molecule during argon plasma jet irradiation

Ziba Matinzadeh[1] · Farhad Shahgoli[2] · Hamed Abbasi[3] · Mahmood Ghoranneviss[1] · Mohammad Kazem Salem[1]

Abstract The aim of this paper is to study degradation of a bromophenol blue molecule (C19H10Br4O5S) using direct irradiation of cold atmospheric argon plasma jet. The pH of the bromophenol blue solution has been measured as well as its absorbance spectra and conductivity before and after the irradiation of non-thermal plasma jet in various time durations. The results indicated that the lengths of conjugated systems in the molecular structure of bromophenol blue decreased, and that the bromophenol blue solution was decolorized as a result of the decomposition of bromophenol blue. This result shows that non-thermal plasma jet irradiation is capable of decomposing, and can also be used for water purification.

Keywords Bromophenol blue · Cold atmospheric plasma · Conductivity · Degradation · Spectrophotometry · pH

Introduction

During the past few years, many studies have been carried out to find the effect of low-temperature plasma with a variety of remote and local plasmas in atmospheric pressure on biological and industrial fields such as different surface treatment [1–9], sterilization, and purification [10–12] processes using different electrical discharges including corona discharge, dielectric barrier discharge, micro-hollow cathode discharge, atmospheric pressure plasma jet, etc. One of the most important applications of plasma treatment is degradation of various molecules in water and other solutions. A number of papers have studied the generation of various plasma sources in water and also in contact with water [13–15]. Bruggeman et al. reviewed the atmospheric pressure non-thermal discharges in liquids and in contact with liquids [16]. Sugiarto et al. have investigated degradation of organic dyes by the pulsed discharge plasma in contaminated water in three discharge modes including streamer, spark, and spark–streamer mixed modes [17]. One of the most interesting topics, in recent years, has been studying degradation of different materials in aqueous solution such as degradation of methyl orange [18], methyl violet [19], phenol [20], methanol [21], diuron [22], 1-naphthylamine [23], pharmaceutical compound pentoxifylline [24], organophosphate pesticides [25], antibiotics [26], textile dyes [27], etc. In such studies, scientists are interested in investigating the effects of various parameters such as pH [28] and temperature [29] on the plasma-treatment process. The degradation process of some materials such as bromophenol blue during plasma irradiation has not been investigated; some studies concerned the degradation of bromophenol blue without using plasma irradiation [30–33]. On the other hand, the degradation process of some materials such as methyl blue has been well investigated. Some research concerned degradation and decolorization of methyl blue using different plasma sources such as dielectric barrier discharge [34, 35], radio frequency plasma [36], microwave discharge plasma [37], and corona discharge [38]. The aim of this paper is studying degradation of bromophenol blue molecules in water using

✉ Ziba Matinzadeh
z.matinzadeh@gmail.com

1 Plasma Physics Research Center, Science and Research Branch, Islamic Azad University, Tehran, Iran

2 Department of Energy Engineering and Physics, Amirkabir University of Technology, Tehran, Iran

3 Biomedical Laser and Optics Group, Department of Biomedical Engineering, University of Basel, Allschwil, Switzerland

capacitively coupled argon plasma jet irradiation. In the present study, atmospheric pressure plasma is irradiated to the solution of bromophenol blue to investigate the selected optical, chemical, and electrical properties of the solution after different treatment durations. The bromophenol blue solution was observed by spectrophotometry after 1, 5, and 10 min of irradiation, and they were compared with a solution without plasma treatment. Moreover, the pH and the conductivity behavior of the solution after different treatment durations of argon plasma were studied.

Materials and methods

Preparing bromophenol blue solution

Bromophenol blue (3,3′,5,5′-tetrabromophenolsulfonphthalein) powder from Merck company has been used to prepare the solution. Its formula is C19H10Br4O5S (CAS number: 115-39-9). Its molar mass equals 669.96 g/mol, and its bulk density equals 730 kg/m^3. The bromophenol solution was prepared by dissolving its powder in distilled water with a concentration of 10 mg/L.

Developing plasma jet

In this paper, the capacitively coupled argon plasma jet has been developed. The working gas was pure argon. The gas discharges were generated by a power supply with a frequency of 23 kHz and an applied voltage of 5.4 kV. In this plasma jet, there was a quartz tube (length 70 mm; inner diameter 9 mm; and outer diameter 14 mm) and a copper tube (inner diameter 7 mm; outer diameter 8.8 mm; and length 5 mm) which was used as a discharge electrode. Another copper tube (length 5 mm; inner diameter 14.1 mm; and outer diameter 15.5 mm) was used as a grounding electrode. The gas flow rate was 5 L/min, and the purity of argon gas was 99.999%. When the AC high voltage is applied, dielectric barrier discharge is induced in the glass tube between two electrodes, and the inflowing gas is excited and then released into the atmosphere. The plasma jet was directly irradiated onto 10 mL of this solution. The distance between the end of the quartz tube of plasma jet and the surface of the bromophenol blue solution was approximately 15 mm. Figure 1 shows a visual picture of the experimental setup.

The argon plasma jet was directly irradiated for 1–10 min onto 10 ml bromophenol blue solution in glass Petri dishes. The absorbance spectra of the bromophenol blue solution samples are obtained from spectrophotometer (Hach DR500), and the pH of the bromophenol blue solution samples is measured with pH meter (Metrohm 744), and the conductivity of samples is measured with conductometer (Metrohm 712). All measurements have been made in four various time durations: first, before irradiation of plasma jet; second, 1 min after irradiation of the plasma jet; third, 5 min after irradiation of the plasma jet; and finally, 10 min after irradiation of the plasma jet.

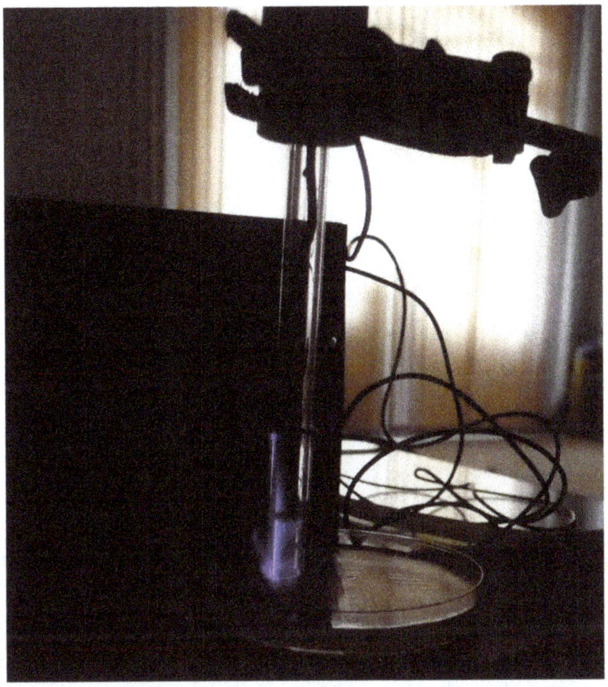

Fig. 1 A visual picture of the experimental setup

Results and discussion

Figure 2 shows the absorbance spectra of the four samples with different time durations. As shown in the figure, the absorbance of samples has decreased with the increasing irradiation duration.

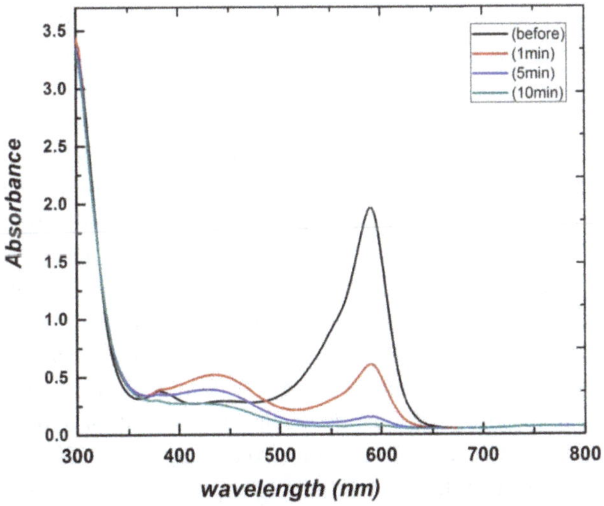

Fig. 2 The absorbance spectra of samples for different time durations

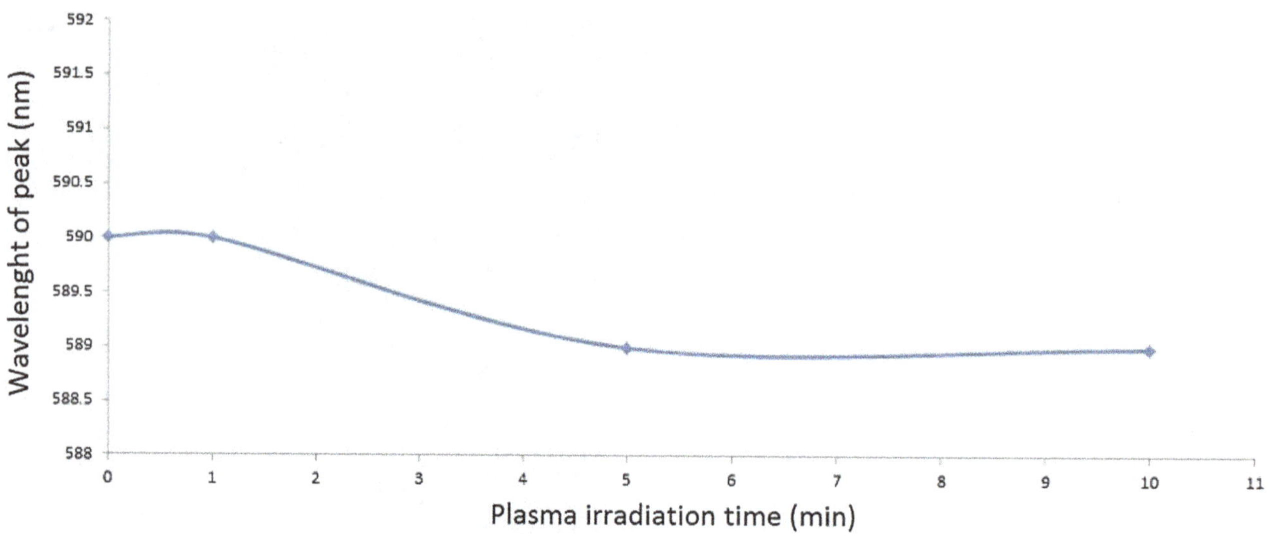

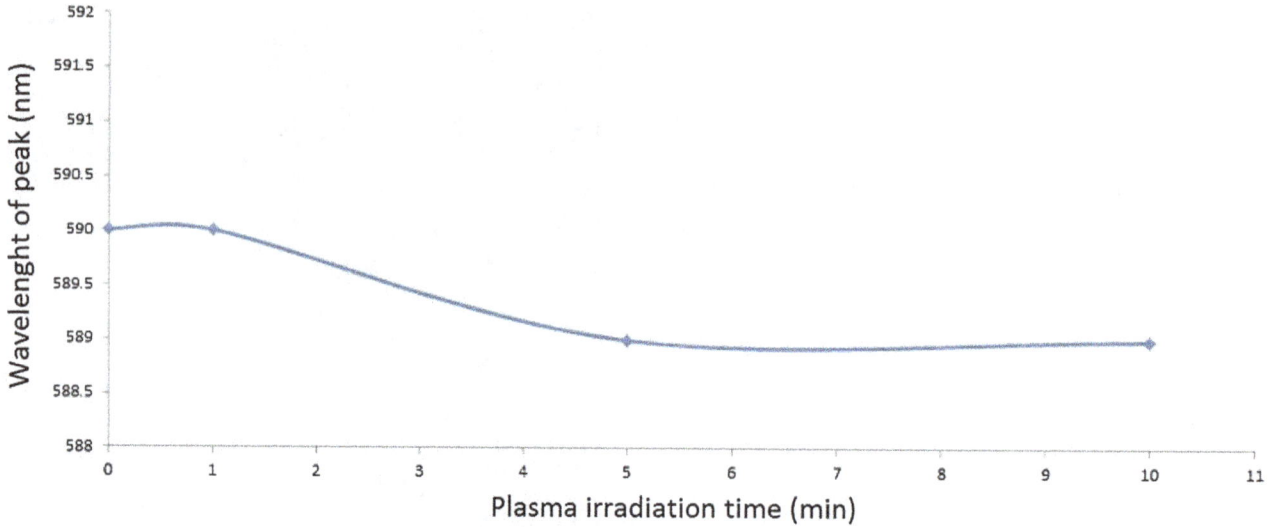

Fig. 3 The change in the wavelength of the peak with respect to plasma jet irradiation durations

This figure indicates the dependence of the absorbance spectrum of the bromophenol blue samples on the plasma jet irradiation duration. The peak wavelengths of all samples were observed around 590 nm.

Figure 3 shows a change in the peak wavelengths observed at approximately 590 nm with respect to plasma jet irradiation durations. The peaks shifted to the short wavelength region as the plasma jet irradiation time increased. It is known that the position of an absorbance peak is related to the lengths of conjugated systems in the molecular structure. A peak shifts to the long wavelength region when the lengths of conjugated systems increase, whereas it shifts to the short wavelength region when the lengths of that decrease [39]. The shift of peak wavelength to the short wavelengths indicates that the length of conjugated system in the bromophenol blue molecular structure has decreased due to plasma irradiation.

The variation of conductivity values during the reaction process has been shown in Fig. 4. The conductivity value of distilled water was measured about 10 ms/m and the conductivity value of the prepared bromophenol blue (BPB) solution before irradiation of plasma was measured about 3.2 ms/m. This distinction in conductivity values indicates that the addition of bromophenol blue powder to the distilled water resulted in a decrease in conductivity value of the solution; this is due to the fact that bromophenol blue powder increases the level of impurity. As it can be seen in the figure, the conductivity value has first decreased and then after 1 min increased. It can be hypothesized that during the first 1 min of plasma irradiation, plasma breaks the molecules of the solution into the neutral products and then after continuing the irradiation, the molecules in the solutions break up into the charged products. This behavior of BPB-related materials is proved by in situ resistance measurements of its ohmic response [40].

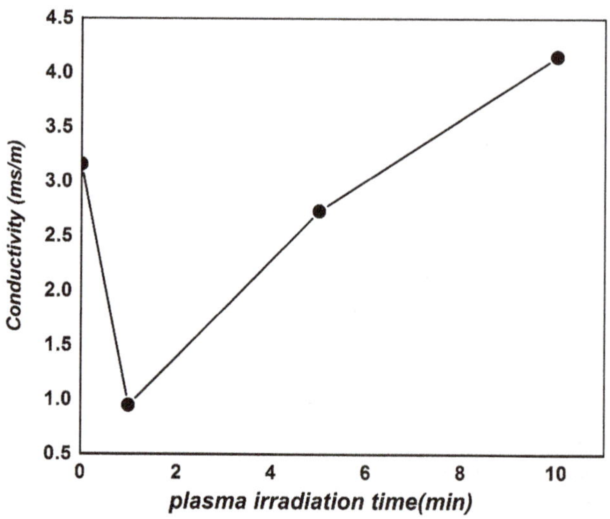

Fig. 4 The variation of conductivity values during the reaction process

As shown in Eq. (1), high-energy electrons (e^-) in the plasma jet collide with H_2O molecules in the solution to generate hydroxyl (OH) radicals and hydrogen (H) [41]:

$$H_2O + e^- \rightarrow OH^{\cdot} + H^{\cdot} + e^-. \quad (1)$$

Sanroman et al. suggested that bromophenol blue (BPB) molecule interacts with hydroxyl radical (OH) as below [31]:

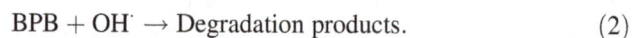

$$BPB + OH^{\cdot} \rightarrow \text{Degradation products}. \quad (2)$$

This argument is in agreement with our experiments, since hydroxyl (OH) radicals are used in Eq. (2) and hydrogen radicals (H) are left. There are some possible

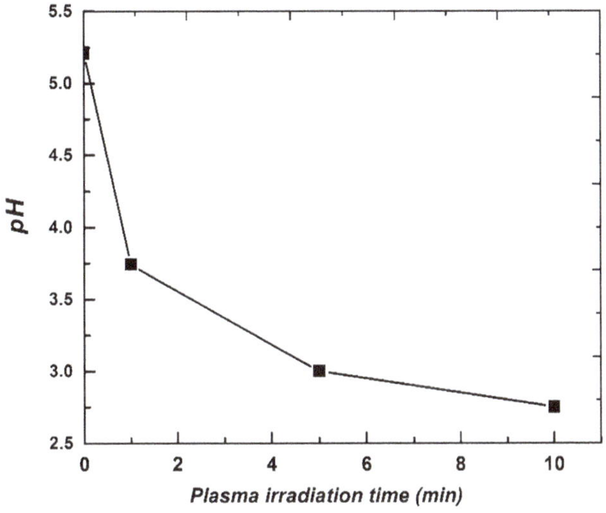

Fig. 5 The diagram of pH for the bromophenol blue solution plotted against plasma jet irradiation durations

degradation pathways. One can find the chemical mechanism and the possible degradation processes of bromophenol blue molecule in [31–33]. Figure 5 shows the diagram of pH for the bromophenol blue solutions plots against plasma jet irradiation durations. The decrease in pH value is due to the generation of hydrogen ions (H^+) in the solution. This result shows that the concentration of hydrogen ions in the solution increases as a result of plasma jet irradiation. This is in a good agreement with the results shown in Fig. 4, because, as shown in the figure, the conductivity value has increased with the increasing plasma irradiation duration and accordingly generating hydrogen ions (H^+) in the solution.

Figure 6 shows the color variation of different bromophenol blue solutions at different treatment durations. The solution was blue before irradiation, lightened after 5 min of irradiation, and became almost transparent after 10 min of irradiation.

The absorption and emission bands of material and as a consequence their colors depend on their molecular structure (energy levels). The colors of the solutions here are in a good agreement with their absorption spectra shown in Fig. 2. As is obvious from Fig. 2, before plasma treatment, the solution has a strong absorption in red region, medium absorption in green region and weak absorption in blue region; therefore, the main reflected/transmitted spectra must be in blue area, and it is in good agreement with the color of the solution before treatment. After 1 min of radiation, the solution has an equal absorption in whole regions, so it looks transparent. But after more irradiation of plasma, as it can be seen in the absorption spectra, due to the change in the molecular structure of the solution, it absorbs less light in red region in comparison with the solution with 1 min of plasma radiation, so it looks yellow

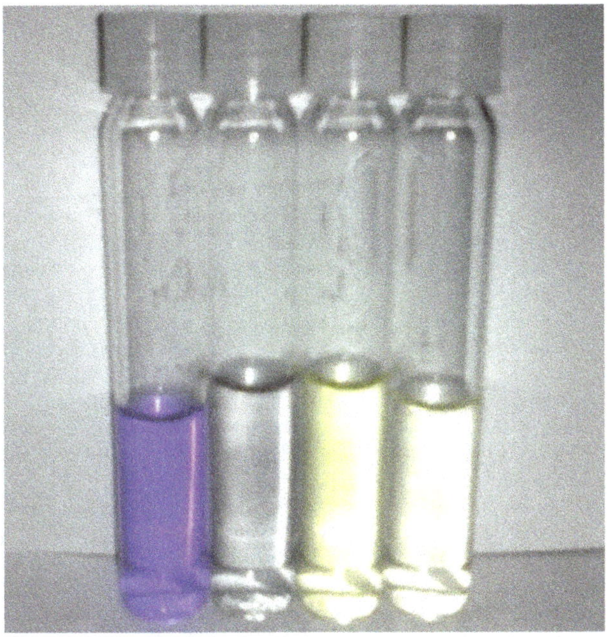

Fig. 6 The color variation of different bromophenol blue solutions at different treatment durations. *From left to right* without treatment, 1-min treatment, 5-min treatment, and 10 min treatment

(yellow color is the combination of blue absorption with red and green reflection/transmission).

Conclusion

The degradation process of bromophenol blue molecule by direct irradiation of non-thermal atmospheric argon plasma using a local plasma jet was experimentally studied. The absorbance spectra, the conductivity value, and the pH value of the bromophenol blue solutions have been measured before and after different plasma irradiation durations. The measured values were in good agreement and indicated that the lengths of conjugated systems in the molecular structure of bromophenol blue decreased, and that the bromophenol blue solution was decolorized as a result of plasma irradiation. Atmospheric cold plasmas have proved their potential to be effective in the treatment of aqueous solutions such as bromophenol blue.

Acknowledgements Hereby, the authors would like to express their deepest appreciation to Ms. Bahareh Abbasi and Dr. Ramin Rahmani for their guidance and persistent helps in editing the manuscript, and also Mr. Shahriar Mirpour for his scientific discussion.

References

1. Nijdam, S., van Veldhuizen, E., Bruggeman, P., Ebert, U.: An introduction to nonequilibrium plasmas at atmospheric pressure. In: Parvulescu, V.I., Magureanu, M., Lukes, P. (eds.) Plasma Chemistry and Catalysis in Gases and Liquids, pp. 1–44. Wiley-VCH Verlag & Co. KGaA, Weinheim, Germany (2012)
2. Attri, P., Arora, B., Choi, E.H.: Utility of plasma: a new road from physics to chemistry. RSC Adv. **3**(31), 12540–12567 (2013)
3. Nehra, V., Kumar, A., Dwivedi, H.: Atmospheric non-thermal plasma sources. Int. J. Eng. **2**(1), 53 (2008)
4. Tendero, C., Tixier, C., Tristant, P., Desmaison, J., Leprince, P.: Atmospheric pressure plasmas: a review. Spectrochim. Acta Part B **61**(1), 2–30 (2006)
5. Schutze, A., Jeong, J.Y., Babayan, S.E., Park, J., Selwyn, G.S., Hicks, R.F.: The atmospheric-pressure plasma jet: a review and comparison to other plasma sources. Plasma Sci. IEEE Trans. **26**(6), 1685–1694 (1998)
6. Laroussi, M., Akan, T.: Arc-free atmospheric pressure cold plasma jets: a review. Plasma Process. Polym. **4**(9), 777–788 (2007)
7. Konesky, G.: Dwell time considerations for large area Cold Plasma decontamination. In: SPIE Defense, Security, and Sensing, International Society for Optics and Photonics (2009)
8. Manolache, S., Jiang, H., Rowell, R.M., Denes, F.S.: Hydrophobic wood surfaces generated by non-equilibrium, atmospheric pressure (NEAPP) plasma-enhanced coating. Mol. Cryst. Liq. Cryst. **483**(1), 348–351 (2008)
9. Yun, T.K., Kim, J.H., Lee, D.K.: Enhancing the surface wettability of poly (ethylene terephthalate) film by atmospheric pressure plasma treatment with Ar and N_2 gas mixture. Mol. Cryst. Liq. Cryst. **586**(1), 188–195 (2013)
10. de Brito Benetoli, L.O., Cadorin, B.M., Baldissarelli, V.Z., Geremias, R., de Souza, I.G., Debacher, N.A.: Pyrite-enhanced methylene blue degradation in non-thermal plasma water treatment reactor. J. Hazard. Mater. **237**, 55–62 (2012)
11. Gao, J.: A novel technique for waste water treatment by contact glow-discharge electrolysis. Pak. J. Biol. Sci. **9**(2), 323–329 (2006)
12. Malik, M.A., Ghaffar, A., Malik, S.A.: Water purification by electrical discharges. Plasma Sources Sci. Technol. **10**(1), 82 (2001)
13. Foster, J., Sommers, B., Weatherford, B., Yee, B., Gupta, M.: Characterization of the evolution of underwater DBD plasma jet. Plasma Sources Sci. Technol. **20**(3), 034018 (2011)
14. Sato, M.: Environmental and biotechnological applications of high-voltage pulsed discharges in water. Plasma Sources Sci. Technol. **17**(2), 024021 (2008)
15. Locke, B., Sato, M., Sunka, P., Hoffmann, M., Chang, J.-S.: Electrohydraulic discharge and nonthermal plasma for water treatment. Ind. Eng. Chem. Res. **45**(3), 882–905 (2006)
16. Bruggeman, P., Leys, C.: Non-thermal plasmas in and in contact with liquids. J. Phys. D Appl. Phys. **42**(5), 053001 (2009)
17. Sugiarto, A.T., Ito, S., Ohshima, T., Sato, M., Skalny, J.D.: Oxidative decoloration of dyes by pulsed discharge plasma in water. J. Electrostat. **58**(1), 135–145 (2003)
18. Huang, F., Chen, L., Wang, H., Feng, T., Yan, Z.: Degradation of methyl orange by atmospheric DBD plasma: Analysis of the degradation effects and degradation path. J. Electrostat. **70**(1), 43–47 (2012)
19. Chen, G., Zhou, M., Chen, S., Chen, W.: The different effects of oxygen and air DBD plasma byproducts on the degradation of methyl violet 5BN. J. Hazard. Mater. **172**(2), 786–791 (2009)
20. Sato, M., Tokutake, T., Ohshima, T.; Sugiarto, A.T.: Aqueous phenol decomposition by pulsed discharge on water surface. In: Industry Applications Conference, 2005. Fortieth IAS Annual Meeting. Conference Record of the 2005, IEEE (2005)
21. Ma, Y., Chen, J., Yang, B., Yu, Q.: Degradation of high concentration methanol in aqueous solution by dielectric barrier discharge. Plasma Sci. IEEE Trans. **41**(7), 1716–1724 (2013)
22. Feng, J., Zheng, Z., Sun, Y., Luan, J., Wang, Z., Wang, L., Feng, J.: Degradation of diuron in aqueous solution by dielectric barrier discharge. J. Hazard. Mater. **154**(1), 1081–1089 (2008)
23. Gao, J., Yu, J., Lu, Q., Yang, W., Li, Y., Pu, L.: Plasma degradation of 1-naphthylamine by glow-discharge electrolysis. Pak. J. Biol. Sci. **7**(10), 1715–1720 (2004)
24. Magureanu, M., Piroi, D., Mandache, N.B., David, V., Medvedovici, A., Parvulescu, V.I.: Degradation of pharmaceutical compound pentoxifylline in water by non-thermal plasma treatment. Water Res. **44**(11), 3445–3453 (2010)
25. Hu, Y., Bai, Y., Yu, H., Zhang, C., Chen, J.: Degradation of selected organophosphate pesticides in wastewater by dielectric barrier discharge plasma. Bull. Environ. Contam. Toxicol. **91**(3), 314–319 (2013)
26. Magureanu, M., Piroi, D., Mandache, N., David, V., Medvedovici, A., Bradu, C., Parvulescu, V.: Degradation of antibiotics in water by non-thermal plasma treatment. Water Res. **45**(11), 3407–3416 (2011)
27. Tichonovas, M., Krugly, E., Racys, V., Hippler, R., Kauneliene, V., Stasiulaitiene, I., Martuzevicius, D.: Degradation of various textile dyes as wastewater pollutants under dielectric barrier discharge plasma treatment. Chem. Eng. J. **229**, 9–19 (2013)
28. Ikawa, S., Kitano, K., Hamaguchi, S.: Effects of pH on bacterial inactivation in aqueous solutions due to low-temperature atmospheric pressure plasma application. Plasma Process. Polym. **7**(1), 33–42 (2010)
29. Benetoli, L.O.D.B., Cadorin, B.M., Postiglione, C.D.S., Souza, I.G.D., Debacher, N.A.: Effect of temperature on methylene blue decolorization in aqueous medium in electrical discharge plasma reactor. J. Braz. Chem. Soc. **22**(9), 1669–1678 (2011)
30. Sanroman, M.A., Pazos, M., Ricart, M.T., Cameselle, C.: Electrochemical decolourisation of structurally different dyes. Chemosphere **57**(3), 233–239 (2004)
31. Salem, I.A.: Kinetics of the oxidative color removal and degradation of bromophenol blue with hydrogen peroxide catalyzed by copper (II)-supported alumina and zirconia. Appl. Catal. B **28**(3), 153–162 (2000)
32. Hong, J., Ta, N., Yang, S.G., Liu, Y.Z., Sun, C.: Microwave-assisted direct photolysis of bromophenol blue using electrodeless discharge lamps. Desalination **214**(1), 62–69 (2007)
33. Dlamini, L.N., Krause, R.W., Kulkarni, G.U., Durbach, S.H.: Photodegradation of bromophenol blue with fluorinated TiO_2 composite. Appl. Water Sci. **1**(1–2), 19–24 (2011)
34. Manoj Kumar Reddy, P., Rama Raju, B., Karuppiah, J., Linga Reddy, E., Subrahmanyam, C.: Degradation and mineralization of methylene blue by dielectric barrier discharge non-thermal plasma reactor. Chem. Eng. J. **217**, 41–47 (2012)
35. Huang, F., Chen, L., Wang, H., Yan, Z.: Analysis of the degradation mechanism of methylene blue by atmospheric pressure dielectric barrier discharge plasma. Chem. Eng. J. **162**(1), 250–256 (2010)
36. Miyamoto, I., Maehara, T., Miyaoka, H., Onishi, S., Mukasa, S., Toyota, H., Kuramoto, M., Nomura, S., Kawashima, A.: Effect of the temperature of water on the degradation of methylene blue by the generation of radio frequency plasma in water. J. Plasma Fusion Res. Ser. **8**, 0627–0631 (2009)
37. Wang, B., Sun, B., Zhu, X., Yan, Z., Liu, Y., Liu, H.: Degradation of methylene blue by microwave discharge plasma in liquid. Contrib. Plasma Phys. **53**(9), 697–702 (2013)
38. Magureanu, M., Piroi, D., Gherendi, F., Mandache, N.B., Parvulescu, V.: Decomposition of methylene blue in water by corona discharges. Plasma Chem. Plasma Process. **28**(6), 677–688 (2008)

39. Kuwahata, H., Kimura, K., Ohyama, R.-I.: Decolorization of methylene blue aqueous solution by atmospheric-pressure plasma jet. e J. Surface Sci. Nanotechnol. **8**, 381–383 (2010)
40. Ferreira, J., Girotto, E.M.: pH effects on the ohmic properties of bromophenol blue-doped polypyrrole film. J. Braz. Chem. Soc. **21**(2), 312–318 (2010)
41. Kuwahata, H., Mikami, I.: Generation of H_2O_2 in distilled water irradiated with atmospheric-pressure plasma jet. e J. Surface Sci. Nanotechnol. **11**, 113–115 (2013)

3

Negative and positive dust grain effect on the modulation instability of an intense laser propagating in a hot magnetoplasma

N. Sepehri Javan[1]

Abstract The modulation instability of intense circularly polarized laser beam in hot magnetized dusty plasma is studied. A nonlinear equation describing the interaction of laser with dusty plasma in the quasi-neutral approximation is derived. The effect of negative and positive dust grains on the laser modulation growth rate is studied. It is shown that the existence of positive dust grains instead of ions can substantially improve the modulation growth rate.

Keywords Dusty plasma · Laser · Modulation · Nonlinear ineration · Growth rate · Magnetoactive

Introduction

Dusty plasmas are frequently found in different places of the cosmic environment. They exist in planetary rings, comet comae and tails, and interplanetary and interstellar molecular clouds [1–7]. They can be revealed in the vicinity of aircrafts [6, 7] and in the controlled plasma fusion [8–10]. In some industrial applications of plasma such as plasma processing of materials, the formation of dusty plasma has also been observed [11–14]. They can also be created during laser ablation experiments [15, 16]. In addition, complex dusty plasmas form in the flame of a humble candle, in the zodiacal light, cloud-to-ground lightings, and volcanic eruptions. Recently [17], it has been suggested that the ball lightning is the dusty plasma medium and it is created during oxidation of nanoparticle networks in the normal lightning strike on soil. Furthermore, dusty plasmas can be produced and investigated in laboratories. Dust grains not only can be intentionally added into the plasma, but can also appear because of different mechanisms in some experiments. The existence of heavy and highly ionized dust grains gives some special and extraordinary properties to the dusty plasma, providing great motivations to investigate it theoretically and experimentally. One of these interests is the study of interaction of laser with dusty plasmas and its related linear and nonlinear effects. These effects include wave dissipation [18], modulation and filamentation instabilities [19–22], linear and nonlinear wave propagation [18, 23–31], parametric instabilities [32], self-focusing [18, 33], etc. Moreover, interaction of laser with dusty plasmas has some important industrial applications. For instance, by interaction of high power lasers with molecular or atomic clusters, during which dusty plasma is created, high-energy electrons can be produced by three processes, i.e., inner ionization, outer ionization, and Coulomb explosion [34–36]. In some experiments, lasers are used in order to study the dynamics of different phenomena in dusty plasmas which some recent experiments about investigation of different exotic phenomena can be found in [37–42].

Here, we focus on the MI of intense lasers in magnetized dusty plasmas. The MI represents a fundamental subject in the theory of nonlinear waves. MI exists due to the interplay between the nonlinearity and dispersion/diffraction effects. The ponderomotive force created by the electromagnetic wave (EMW) stimulates low-frequency perturbations of the electrons density; then, they interact with the primary high-frequency EMW in which the amplitude of the pump wave becomes modulated and the MI of the EMW occurs. The MI of laser beams in plasmas and dielectrics has been the subject of several publications [43–45]. The MI of strong EMWs in plasmas with arbitrary large amplitude was

✉ N. Sepehri Javan
sepehri_javan@uma.ac.ir

[1] Department of Physics, University of Mohaghegh Ardabili, PO Box 179, Ardabil, Iran

studied by Shukla et al. in 1987 [46]. Most of the early publications about MI considered one-dimensional models in which the laser beam was represented as a plane wave [47, 48]. The MI of a laser pulse in the cold nonmagnetized plasma has been considered by several authors [46, 49, 50]. The MI of a linearly polarized laser pulse propagating in the cold magnetized plasma was studied by Jha et al. in 2005 [51]. The MI of the right-hand elliptically laser pulse in cold magnetized plasma has been investigated by Chen et al. in 2011 [52]. Recently, the MI of an intense circularly polarized laser beam in the hot magnetized electron–positron and electron–ion (e–i) plasmas as studied by Sepehri Javan [53, 54]. Our recent work [55] has extended the MI of the circularly polarized laser beam propagating along an external magnetic field in the non-Maxwellian plasma. In this article we study the MI of an intense laser beam in the magnetized hot dusty plasma. In the quasi-neutral approximation and by using a relativistic fluid model, we consider the presence of both negative and positive dust grains and investigate the effect of such grains on the MI. The organization of the paper is as follows. In Sect. 2, the basic assumptions are presented and a nonlinear wave equation is derived for the laser amplitude evolutions. An analytic expression for the growth rate of MI is obtained in Sect. 3. In Sect. 4, a numerical study of the MI of circularly polarized laser beam in the magnetized electron–ion–positive dust–negative dust (e–i–d+–d–) plasma is presented. The concluding remarks are made in Sect. 5.

Deriving a nonlinear wave equation

Let us consider the propagation of a circularly polarized EMW in a hot magnetized four-component plasma which contains electron, ion, and positive and negative dust grains. Each type of plasma particle may have its own specific temperature. To determine the quantities related to the electrons, ions, and positive and negative dust grains, we use indices e, i, d + and d−, respectively. We take the external magnetic field parallel to the z axis, i.e., $\mathbf{B_0} = B_0\hat{\mathbf{e}}_\mathbf{z}$. To describe the nonlinear dynamics of the interaction of EMW with the dusty plasma, we define the electric and magnetic fields **E** and **B** through the vector and scalar potentials **A**, φ as:

$$\mathbf{E} = -\frac{1}{c}\frac{\partial \mathbf{A}}{\partial t} - \nabla\varphi, \quad \mathbf{B} = \nabla \times \mathbf{A}, \tag{1}$$

where c is the speed of light.

Using Eq. (1) in Maxwell equations, one can easily obtain:

$$\frac{1}{c^2}\frac{\partial^2 \mathbf{A}}{\partial t^2} - \nabla^2\mathbf{A} = \frac{4\pi}{c}\mathbf{J}, \tag{2}$$

where $\mathbf{J} = -n_e e\mathbf{v}_e$ is the current density of electrons, e, $\mathbf{v}_e$ and n_e are the density, velocity and charge of the electron, respectively. We ignore the translational velocity of the heavy ions and dust grains. Now, we can write the relativistic fluid momentum equation for electrons as:

$$\frac{\partial \mathbf{p}_e}{\partial t} + (\mathbf{v}_e.\nabla)\mathbf{p}_e = -e\left[\mathbf{E} + \frac{1}{c}\mathbf{v}_e \times (\mathbf{B} + \mathbf{B_0})\right] - \frac{1}{n_e}\nabla\Pi_e, \tag{3}$$

where $\mathbf{p}_e$ and Π_e are the momentum and pressure of the electron, respectively. Substituting Eq. (1) into Eq. (3) leads to

$$\frac{\partial \mathbf{p}_e}{\partial t} + \frac{1}{\gamma_e m_0}(\mathbf{p}_e\cdot\nabla)\mathbf{p}_e = \frac{e}{c}\frac{\partial \mathbf{A}}{\partial t} + e\nabla\varphi - \frac{e}{\gamma_e m_0 c}\mathbf{p}_e \times (\nabla\times\mathbf{A}) - \frac{\omega_c}{\gamma_e}\mathbf{p}_e \times \hat{\mathbf{e}}_\mathbf{z} - k_B T_e \nabla\ln(n_e), \tag{4}$$

where T_e is the temperature of electrons, m_0 the electron rest mass, $\gamma_e = \sqrt{1 + p_e^2/m_0^2c^2}$ the relativistic Lorentz factor, $\omega_c = eB_0/m_0c$ the electron cyclotron frequency and k_B the Boltzmann constant.

We consider the propagation of circularly polarized wave along the external magnetic field and write the vector potential of this wave as:

$$\mathbf{A} = \frac{1}{2}\tilde{A}(\hat{\mathbf{e}}_\mathbf{x} + i\sigma\hat{\mathbf{e}}_\mathbf{y})\exp(-i\omega_0 t + ik_0 z) + c.c., \tag{5}$$

where ω_0, k_0 are the frequency and wave number, respectively. $\sigma = +1, -1$ denotes the right- and left-hand circularly polarized wave, respectively, and also $\tilde{A}(z,t)$ is the slowly varying amplitude that satisfies the following condition:

$$\left|\frac{1}{\omega_0}\frac{\partial \tilde{A}}{\partial t}\right| << |\tilde{A}|. \tag{6}$$

Inserting Eq. (5) into Eq. (4), we can find that Eq. (4) is satisfied by [56–58]:

$$\bar{\mathbf{p}}_e = \frac{\bar{\mathbf{A}}}{1 - \sigma\alpha/\gamma_e}, \tag{7}$$

and together with

$$n_e = n_{0e}\exp\left[\frac{e\varphi}{k_B T_e} - \beta_e\left(\gamma_e - 1 - \frac{\sigma\alpha}{2\gamma_e^2}|\bar{p}_e|^2\right)\right], \tag{8}$$

where n_{0e} is the unperturbed density of electrons, $\bar{\mathbf{p}}_e = \mathbf{p}_e/m_0c$ is the normalized electron momentum, $\bar{\mathbf{A}} = e\mathbf{A}/m_0c^2$ is the normalized vector potential, $\alpha = \omega_c/\omega_0$, $\beta_e = c^2/v_{T_e}^2$ and $v_{T_e}^2 = k_\mathrm{B}T_e/m_0$ is the electron thermal velocity. For weakly relativistic laser intensity, when $|\bar{\mathbf{A}}|^2, |\bar{\mathbf{P}}_e|^2 << 1$ and $\gamma_e \approx 1 + \frac{1}{2}|\bar{\mathbf{P}}_e|^2$, we can simplify density of electrons as follows:

$$n_e = n_{0e}\exp\left[\frac{e\varphi}{k_B T_e} - \frac{\beta_e}{2}\frac{|\bar{\mathbf{A}}|^2}{(1-\sigma\alpha)}\right]. \tag{9}$$

We suppose that the ion and dust grains slow motions are non-relativistic and, by assuming an isothermal equation of state for these heavy particles, obtain the following expressions for number densities:

$$n_i = n_{0i}\exp\left(-\frac{e\varphi}{k_B T_i}\right), \tag{10}$$

$$n_{d+} = n_{0d+}\exp\left(-\frac{z_+ e\varphi}{k_B T_{d+}}\right), \tag{11}$$

$$n_{d-} = n_{0d-}\exp\left(\frac{z_- e\varphi}{k_B T_{d-}}\right), \tag{12}$$

where T_j, n_{0j}, z_+ and z_- are the temperature and unperturbed density of j-type particle, and order of ionization of positive and negative dust grains, respectively.

Expanding Eqs. (9)–(12) and using them in the quasi-neutral condition, i.e., $n_i + z_+ n_{d+} - n_e - z_- n_{d-} = 0$, yield the following result:

$$\Phi = \frac{e\varphi}{k_B T_e} = \frac{\mu\beta_e|\bar{\mathbf{A}}|^2}{2(1-\sigma\alpha)}, \tag{13}$$

where

$$\mu = \frac{1}{1 + \zeta_i\delta_i^{-1} + z_+^2\zeta_{d+}\delta_{d+}^{-1} + z_-^2\zeta_{d-}\delta_{d-}^{-1}}, \tag{14}$$

$$\zeta_j = \frac{n_{0j}}{n_{0e}}, \quad \delta_j = \frac{T_j}{T_e}, \quad j = i, d+, d-. \tag{15}$$

Substituting Eq. (13) into Eq. (9) results in the following expression for the electron density:

$$n_e = n_{0e}\exp\left[\frac{(\mu-1)\beta_e}{2(1-\sigma\alpha)}|\bar{\mathbf{A}}|^2\right]. \tag{16}$$

In physical units, from Eq. (7) we can obtain the following for the velocity of electrons:

$$\mathbf{v}_e = \frac{e}{m_{0e}c}\frac{\mathbf{A}}{\gamma_e - \sigma\alpha}. \tag{17}$$

Also, for electron Lorentz factor, we can approximately write:

$$\gamma_e \approx \sqrt{1 + \frac{|\bar{\mathbf{A}}|^2}{(1-\sigma\alpha)^2}}. \tag{18}$$

Now, taking Eqs. (16) and (17) into consideration, we derive the nonlinear current density as follows:

$$-\frac{4\pi}{c}\mathbf{J} = \frac{\omega_p^2}{c^2}\frac{\mathbf{A}}{(\gamma_e - \sigma\alpha)}\exp\left[\frac{(\mu-1)\beta_e}{2(1-\sigma\alpha)}|\bar{\mathbf{A}}|^2\right], \tag{19}$$

where $\omega_p = \sqrt{4\pi n_0 e^2/m_0}$ is the electron Langmuir frequency.

In the weakly relativistic regime of laser intensity, we can expand the nonlinear current density of Eq. (19) with respect to the normalized vector potential amplitude and save only the second orders of amplitude. In this case, substituting simplified current density, together with the vector potential in the form of Eq. (5) into Eq. (2) leads to the following equation for the EMW envelope evolutions:

$$\left(\nabla^2 - \frac{1}{c^2}\frac{\partial^2}{\partial t^2}\right)ae^{i(k_0 z - \omega_0 t)} = k_p^2\left[\frac{1}{1-\sigma\alpha} - |a|^2 N\right]ae^{i(k_0 z - \omega_0 t)}, \tag{20}$$

where

$$N = \frac{\omega_0^4}{2(\omega_0 - \sigma\omega_c)^4} + \frac{\omega_0^2}{2(\omega_0 - \sigma\omega_c)^2}(1-\mu)\beta_e, \tag{21}$$

and $a = e\tilde{A}/m_0c^2$, $k_p = \omega_p/c$ are the normalized amplitude of vector potential, and wave number of the plasma wave, respectively.

For e–i plasma when $\zeta_i = 1$ and $\zeta_{d+} = \zeta_{d-} = 0$, the nonlinear term reduces to

$$N = \frac{\omega_0^4}{2(\omega_0 - \sigma\omega_c)^4} + \frac{\omega_0^2}{2(\omega_0 - \sigma\omega_c)^2}\frac{\beta_e}{1 + T_e/T_i}, \tag{22}$$

which agrees with the results of Sepehri Javan and Nasirzadeh [56].

Derivation of nonlinear dispersion relation and MI

To derive the nonlinear dispersion relation, Eq. (18) is simplified in the new form:

$$\begin{aligned}&\frac{\partial^2 a}{\partial t^2} - c^2\frac{\partial^2 a}{\partial z^2} - 2i\omega_0\frac{\partial a}{\partial t} - 2ik_0c^2\frac{\partial a}{\partial z}\\ &+ \left[-\omega_0^2 + c^2k_0^2 + \omega_p^2\left(\frac{1}{1-\sigma\alpha} - N|a|^2\right)\right]a\\ &= 0.\end{aligned} \tag{23}$$

In the last term of Eq. (23), the coefficient of a is the nonlinear dispersion relation. In the absence of interaction between EMW and plasma, when amplitude is a real constant ($a = a_0$), we can derive the nonlinear dispersion relation for magnetoplasma with negative and positive dust grains as follows:

$$\begin{aligned}&c^2k_0^2 - \omega_0^2\\ &+ \omega_p^2\left[\frac{1}{1-\sigma\alpha} - a_0^2\left(\frac{\omega_0^4}{2(\omega_0 - \sigma\omega_c)^4} + \frac{\omega_0^2}{2(\omega_0 - \sigma\omega_c)^2}\frac{(\mu-1)}{(1-\sigma\alpha)}\beta_e\right)\right] = 0.\end{aligned} \tag{24}$$

In the linear limit (when $a^2 \to 0$), Eq. (24) can be reduced to the well-known linear dispersion relation of circularly polarized EMW in the magnetized plasma:

$$k_0 = \frac{\omega_0}{c}\left(1 - \frac{\omega_p^2}{\omega_0(\omega_0 - \sigma\omega_c)}\right)^{1/2}. \quad (25)$$

It is worth mentioning that in the linear approximation, there is no contribution for dust grains on the dispersion Eq. (25) because we have investigated the evolution of high-frequency EMWs where heavy ions and dust particles cannot respond to this high frequency. However, traces of dust particles can be found in the nonlinear dispersion Eq. (24) through bipolar diffusion caused by slow motion of particles under the influence of ponderomotive laser force and thermal collision force.

By considering the condition of slowly varying amplitude (Eq. 6) and assuming that ω_0 and k_0 satisfy the linear dispersion of Eqs. (23), (25) can be modified as:

$$i\left(\frac{\partial a}{\partial t} + v_g\frac{\partial a}{\partial z}\right) + \frac{c^2}{2\omega_0}\frac{\partial^2 a}{\partial z^2} + \frac{\omega_p^2}{2\omega_0}N|a|^2 \quad a = 0 \quad (26)$$

where $v_g = \frac{k_0 c^2}{\omega_0}$ is the group velocity. Using the following dimensionless variables $\tau = \frac{\omega_p^2}{\omega_0}t$, $U_g = \frac{\omega_0}{\omega_p}\frac{v_g}{c}$ and $\xi = \frac{\omega_p}{c}z + U_g\tau$, Eq. (26) can be written as:

$$i\frac{\partial a}{\partial \tau} + \frac{1}{2}\frac{\partial^2 a}{\partial \zeta^2} + D_{\mathrm{NL}}\, a = 0, \quad (27)$$

where $D_{\mathrm{NL}} = N|a|^2/2$. Equation (27) is the well-known nonlinear Schrödinger equation (NLSE). This equation is frequently met in different areas of theoretical physics, especially in nonlinear optics. The NLSE describes the propagation of waves in nonlinear media taking into account both the group velocity dispersion (second term) and the nonlinearity (third term). It is a classical field equation whose important applications are in the propagation of EMWs in nonlinear optical fibers and planar waveguides [59] and to Bose–Einstein condensates confined to highly anisotropic cigar-shaped traps, in the mean-field regime [60]. Additionally, it can be revealed in the studies of small-amplitude gravity waves on the surface of deep zero-viscosity water [59], Langmuir waves in hot plasmas [59], propagation of plane-diffracted wave beams in the focusing areas of the ionosphere [61] and propagation of Davydov's alpha-helix solitons, which are responsible for energy transport along molecular chains [62].

The MI for the right- and left-hand circularly polarized EMW can be obtained using the usual method introduced by Shukla et al. [46]. In this approach, we suppose:

$$a = (a_0 + a_1)\exp(i\Lambda\tau), \quad (28)$$

where a_0 is a real constant and $a_0 >> |a_1|$,

$$\Lambda \equiv D_{\mathrm{NL}}(a = a_0) = \frac{1}{2}a_0^2 N. \quad (29)$$

By substituting Eq. (28) into Eq. (27) and linearizing obtained equation with respect to a_1, we can achieve:

$$i\frac{\partial a_1}{\partial \tau} + \frac{1}{2}\frac{\partial^2 a_1}{\partial \zeta^2} + \frac{1}{2}a_0^2 N(a_1 + a_1^*) = 0. \quad (30)$$

Introducing $a_1 = X + iY$, inserting it into Eq. (30) and separating the real and imaginary parts of this equation yield:

$$\begin{cases} \dfrac{\partial X}{\partial \tau} + \dfrac{1}{2}\dfrac{\partial^2 Y}{\partial \zeta^2} = 0, \\ -\dfrac{\partial Y}{\partial \tau} + \dfrac{1}{2}\dfrac{\partial^2 X}{\partial \zeta^2} + a_0^2 N = 0. \end{cases} \quad (31)$$

We consider the following oscillational form for X and Y:

$$\begin{pmatrix} X \\ Y \end{pmatrix} = \begin{pmatrix} \tilde{X} \\ \tilde{Y} \end{pmatrix}\exp(-i\Omega\tau + iK\zeta), \quad (32)$$

where $\tilde{X}$ and $\tilde{Y}$ are real amplitudes, Ω is the modulation frequency normalized by ω_p^2/ω_0 and K is the modulation wave number normalized by ω_p/c. By substituting Eq. (32) into the set of Eq. (31), we can obtain the following nonlinear dispersion relation of MI:

$$\Omega^2 = -\frac{K^2}{2}\left[a_0^2 N - \frac{K^2}{2}\right]. \quad (33)$$

The temporal growth rate $\Gamma = -i\Omega$ can be extracted from Eq. (33) as below:

$$\Gamma = \frac{K}{\sqrt{2}}\left[a_0^2 N - \frac{K^2}{2}\right]^{1/2}. \quad (34)$$

The maximum growth rate of MI that occurs at $K = K_m = a_0 N^{1/2}$ is

$$\Gamma_{\max} = \frac{a_0^2}{2}N. \quad (35)$$

It may be useful to note that for e–i plasma, when $\zeta_i = 1$ and $\zeta_{d+} = \zeta_{d-} = 0$, Eq. (33) reduces to,

$$\Gamma = \frac{K}{\sqrt{2}}\left[\frac{a_0^2}{2(1-\sigma\alpha)^2}\left(\frac{\beta_e}{(1 + T_e/T_i)} + \frac{1}{(1-\sigma\alpha)^2}\right) - \frac{K^2}{2}\right]^{1/2}. \quad (36)$$

Numerical discussions

For numerical studies, in all the investigated cases, we suppose an Nd:YAG laser with frequency $\omega_0 = 1.88 \times 10^{15}\ \mathrm{s}^{-1}$ (that corresponds to the laser wave length $\lambda \approx 1\ \mu\mathrm{m}$) and $a_0 = 0.271$ (laser intensity $I \approx 10^{17}\ \mathrm{W/cm^2}$); also, we

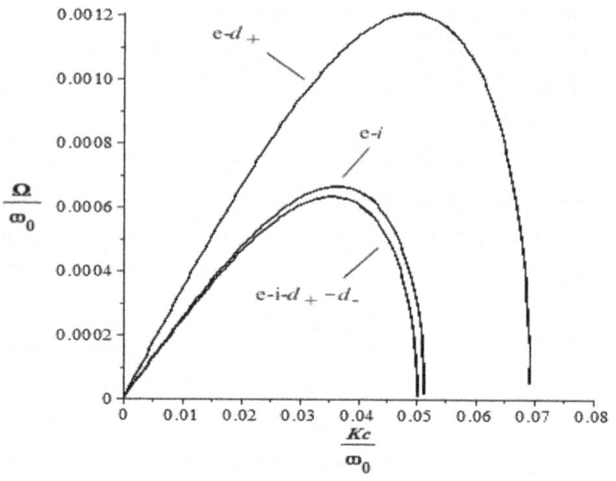

Fig. 1 Variations of the normalized MI growth rate with respect to the normalized modulation frequency for three different cases, e–i plasma with $\eta = \xi = 0$, e–d+ plasma with $\eta = 10^{-1}$, $\xi = 0$ and e–i–d+–d– plasma with $\eta = \xi = 5 \times 10^{-2}$, when $z_+ = z_- = 10$

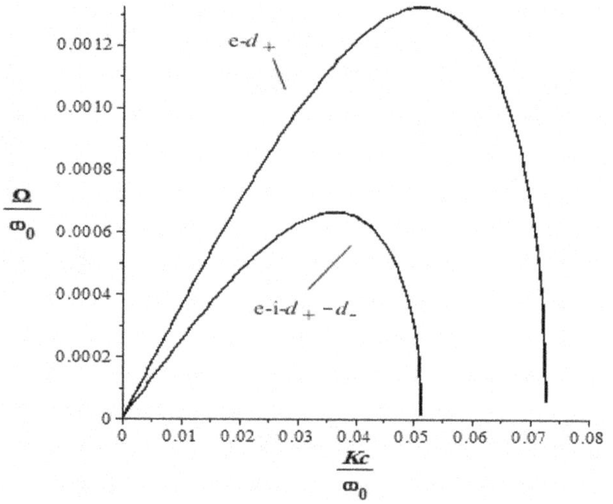

Fig. 2 Variations of the normalized MI growth rate with respect to the normalized modulation frequency for two different cases, e–d+ plasma with $\eta = 10^{-3}$, $\xi = 0$ and e–i–d+–d– plasma with $\eta = \xi = 5 \times 10^{-4}$, when $z_+ = z_- = 1000$

consider only the right-hand polarization laser in magnetized medium with $\alpha = 0.2$ and fix the temperature $T_j = 1$ keV for all the plasma components. For more clarification, we introduce two new parameters η and ξ as:

$$n_{0d+} = \eta n_0, \quad n_{0d-} = \xi n_0, \tag{37}$$

where we supposed $n_{0i} + z_+ n_{0d+} = n_{0e} + z_- n_{0d-} = n_0$ and set $n_0 = 10^{17}$ cm^{-3}.

Figure 1 shows variations of the normalized modulation growth rate Ω/ω_0 with respect to the normalized modulation wave number Kc/ω_0, when $z_+ = z_- = 10$. We have three different cases, in which $\eta = \xi = 0$ corresponds to the e–i plasma, $\eta = 10^{-1}$, $\xi = 0$ to the e–d+ plasma and $\eta = \xi = 5 \times 10^{-2}$ to the e–i–d+–d– plasma. We can see that adding positive dust grains instead of ions substantially increases the modulation growth rate, because localization of positive charges on the dust grains improves the ambipolar potential, which in turn leads to the sharpness of the density profile and consequently to more modulation. In the case of e–i–d+–d– plasma, decreasing the density of electrons and adding equivalent negative dust grains to the plasma results in the decrease in the laser modulation growth rate, because the nonlinear current is created by the motion of the electrons and decrease in the population of electrons leads to the decrease in the nonlinearity of the medium and consequently to the decrease in the growth rate. To investigate the effect of the order of ionization of dust grains on the spot size, in Fig. 2, we choose $z_+ = z_- = 1000$ for two different cases, the e-d + plasma with $\eta = 10^{-3}$, $\xi = 0$ and the e–i–d+–d– plasma with $\eta = \xi = 5 \times 10^{-5}$. We can see that the increase in the ionization order causes a small increase in the MI growth rate. It is worth mentioning that our numerical experiments show that an increase in the temperature causes a decrease in the modulation growth rate. In addition, magnetization of plasma enhances the modulation growth rate for the right-hand polarization and inversely reduces it for the left-hand one. These results are not new and have been investigated earlier [53, 54]; for brevity we do not provide them.

Conclusions

In this paper, we investigated the MI of a weakly relativistic laser propagating along an external magnetic field in the hot plasma containing positive and negative dust grains. The MI growth rate of the circularly polarized laser beam in the dusty plasma was obtained. It was found that adding the positive dust grains to plasma enhances the MI, but existence of the negative dust grains reduces it. Furthermore, the effect of the order of dust grain ionization on the MI was investigated and it was observed that its increase leads to the increase in the MI growth rate.

References

1. Goertz, C.K.: Dusty plasmas in the solar system. Rev. Geophys. **27**, 271 (1989)
2. Northrop, T.G.: Dusty plasmas. Phys. Scr. **45**, 475 (1992)
3. Tsytovich, V.N.: Dust plasma crystals, drops, and clouds. Usp. Fiz. Nauk **167**, 57 (1997). **[Phys. Usp. 40, 53 (1997)]**
4. Bliokh, P., Sinitsin, V., Yaroshenko, V.: Dusty and self-gravitational plasmas in space. Kluwer Acad. Publ, Dordrecht (1995)
5. Shukla, P.K., Mamun, A.A.: Introduction to dusty plasma physics. Institute of Physics Publishing, Bristol (2002)

6. Whipple, E.C.: Potentials of surfaces in space. Rep. Prog. Phys. **44**, 1197 (1981)
7. Robinson, P.A., Coakley, P.: Spacecraft charging-progress in the study of dielectrics and plasmas. IEEE Trans. Electr. Insul. **27**, 944 (1992)
8. Tsytovich, V.N., Winter, J.: On the role of dust in fusion devices. Usp. Fiz. Nauk **168**, 899 (1998). **[Phys. Usp. 41, 815 (1998)]**
9. Winter, J., Gebauer, G.: Dust in magnetic confinement fusion devices and its impact on plasma operation. J. Nucl. Mater. **266–269**, 228 (1999)
10. Winter, J.: Dust: a new challenge in nuclear fusion research? Phys. Plasmas **7**, 3862 (2000)
11. Jellum, G.M., Graves, D.B.: Particulates in aluminum sputtering discharges. J. Appl. Phys. **67**, 6490 (1990)
12. Anderson, H.M., Jairath, R., Mock, J.L.: Particulate generation in silane/ammonia rf discharges. J. Appl. Phys. **67**, 3999 (1990)
13. Shivatani, M., Fukuzawa, T., Watanabe, Y.: Formation processes of particulates in helium-diluted silane RF plasmas. IEEE Trans. Plasma Sci. **22**, 103 (1994)
14. Cui, C., Goree, J.: Fluctuations of the charge on a dust grain in a plasma. IEEE Trans. Plasma Sci. **22**, 151 (1994)
15. Trajanovic, Z., Senapati, L., Sharma, R.P., Venkatesan, T.: Stoichiometry and thickness variation of $YBa_2Cu_3O_{7-x}$ in off-axis pulsed laser deposition. Appl. Phys. Lett. **66**, 2418 (1995)
16. Fukushima, K., Kanka, Y., Badaye, M., Morishita, T.: Velocity distributions of ions in the ablation plume of a $Y_1Ba_2Cu_3O_x$ target. J. Appl. Phys. **77**, 5406 (1995)
17. Abrahamson, J., Dinniss, J.: Ball lightning caused by oxidation of nanoparticle networks from normal lightning strikes on soil. Nature **403**, 519 (2000)
18. Jana, M.R., Sen, A., Kaw, P.K.: Collective effects due to charge-fluctuation dynamics in a dusty plasma. Phys. Rev. E **48**, 3930 (1993)
19. Sambandan, G., Tripathi, V.K., Parashar, J., Bharuthram, R.: Nonlinear interaction of a high-power electromagnetic beam in a dusty plasma: two-dimensional effects. Phys. Plasmas **6**, 762 (1999)
20. Sharma, S.C., Gahlot, A., Sharma, R.P.: Effect of dust on an amplitude modulated electromagnetic beam in a plasma. Phys. Plasmas **15**, 043701 (2008)
21. Sodha, M.S., Mishra, S.K., Misra, S.: Nonlinear dependence of complex plasma parameters on applied electric field. Phys. Plasmas **18**, 023701 (2011)
22. El-Taibany, W.F., Kourakis, I., Wadati, M.: Low frequency localized wavepackets in dusty plasmas with opposite charge polarity dust components. Plasma Phys. Control. Fusion **50**, 074003 (2008)
23. Varma, R.K., Shukla, P.K., Krishan, V.: Electrostatic oscillations in the presence of grain-charge perturbations in dusty plasmas. Phys. Rev. E **47**, 3612 (1993)
24. Tripathi, K.D., Sharma, S.K.: Self-consistent charge dynamics in magnetized dusty plasmas: low-frequency electrostatic modes. Phys. Rev. E **53**, 1035 (1996)
25. Das, C., Janaki, M.S., Dasgupta, B.: Dust cyclotron instability in presence of dust charge fluctuations. Phys. Scr. T **75**, 216 (1998)
26. Dwivedi, C.B., Pandey, B.P.: Electrostatic shock wave in dusty plasmas. Phys. Plasmas **2**, 4134 (1995)
27. Popel, S.I., Yu, M.Y., Tsytovich, V.N.: Shock waves in plasmas containing variable-charge impurities. Phys. Plasmas **3**, 4313 (1996)
28. Paul, S.N., Mondal, K.K., Roychowdhury, A.: Effects of streaming and attachment coefficients of ions and electrons on the formation of soliton in a dusty plasma. Phys. Lett. A **257**, 165 (1999)
29. Janaki, M.S., Dasgupta, B.: Surface waves in a dusty plasma. Phys. Scr. **58**, 493 (1998)
30. Alam, M.K., Roy Chowdhury, A.: Surface wave propagation in a magnetized dusty plasma with charge fluctuation. Phys. Plasmas **6**, 3765 (1999)
31. Bhat, J.R., Pandey, B.P.: Self-consistent charge dynamics and collective modes in a dusty plasma. Phys. Rev. E **50**, 3980 (1994)
32. Burman, S., Paul, S.N., Roy Chowdhury, A.: Stimulated brillouin scattering in a magnetized dusty plasma with charge fluctuation. Phys. Plasmas **9**, 3752 (2002)
33. Mishra, S.K., Mishra, S., Sodha, M.S.: Self-focusing of a Gaussian electromagnetic beam in a complex plasma. Phys. Plasmas **18**, 043702 (2011)
34. Nantel, M., Ma, G., Gu, S., Cote, C.Y., Itatani, J., Umstadter, D.: Pressure ionization and line merging in strongly coupled plasmas produced by 100-fs laser pulses. Phys. Rev. Lett. **80**, 4442 (1998)
35. Springate, E., Hay, N., Tisch, J.W.G., Mason, M.B., Ditmire, T., Hutchinson, M.H.R., Marangos, J.P.: Explosion of atomic clusters irradiated by high-intensity laser pulses: scaling of ion energies with cluster and laser parameters. Phys. Rev. A **61**, 063201 (2000)
36. Last, I., Jortnera, J.: Electron and nuclear dynamics of molecular clusters in ultraintense laser fields. I. Extreme multielectron ionization. J. Chem. Phys. **120**, 1336 (2004)
37. Stoffels, E., Stoffels, W.W., Vender, D., Kroesen, G.M.W., de Hoog, F.J.: Laser-particulate interactions in a dusty RF plasma. IEEE Trans. Plasma Sci. **22**, 116 (1994)
38. Melzer, A.: Laser-experiments on particle interactions in strongly coupled dusty plasma crystals. Phys. Scr. **T89**, 33 (2001)
39. Nefedov, A.P., Petrov, O.F., Molotkov, V.I., Fortovs, V.E.: Formation of liquidlike and crystalline structures in dusty plasmas. JETP Lett. **72**, 218 (2000)
40. van de Wetering, F.M.J.H., Oosterbeek, W., Beckers, J., Nijdam, S., Gibert, T., Mikikian, M., Rabat, H., Kovačević, E., Berndt, J.: Interaction of nanosecond ultraviolet laser pulses with reactive dusty plasma. Appl. Phys. Lett. **108**, 213103 (2016)
41. Nosenko, V., Avinash, K., Goree, J., Liu, B.: Laser-excited mach cones in a dusty plasma crystal. Phys. Rev. E **62**, 4162 (2000)
42. van de Wetering, F.M.J.H., Oosterbeek, W., Beckers, J., Nijdam, S., Kovačević, E., Berndt, J.: Laser-induced incandescence applied to dusty plasmas. J. Phys. D Appl. Phys. **49**, 295206 (2016)
43. Esarey, E., Sprangle, P., Krall, J., Ting, A.: Overview of plasma-based accelerator concepts. IEEE Trans. Plasma Sci. **24**, 252 (1996)
44. Esarey, E., Sprangle, P., Krall, J., Ting, A.: Self-focusing and guiding of short laser pulses in ionizing gases and plasmas. IEEE J. Quantum Electron. **33**, 1879 (1997)
45. Shukla, P.K., Marklund, M., Eliasson, B.: Nonlinear dynamics of intense laser pulses in a pair plasma. Phys. Lett. A **324**, 193 (2004)
46. Shukla, P.K., Bharuthram, R.: Modulational instability of strong electromagnetic waves in plasmas. Phys. Rev. A **35**, 4889 (1987)
47. McKinstrie, C.J., Bingham, R.: Stimulated Raman forward scattering and the relativistic modulational instability of light waves in rarefied plasma. Phys. Fluids B **4**, 2626 (1992)
48. Sprangle, P., Esarey, E., Hafizi, B.: Intense laser pulse propagation and stability in partially stripped plasmas. Phys. Rev. Lett. **79**, 1046 (1997)
49. Shukla, P.K., Rao, N.N., Yu, M.Y., Tsintsadze, N.L.: Relativistic nonlinear effects in plasmas. Phys. Rep. **138**, 1 (1986)
50. Esarey, E., Kralland, J., Sprangle, P.: Envelope analysis of intense laser pulse self-modulation in plasmas. Phys. Rev. Lett. **72**, 2887 (1994)
51. Jha, P., Kumar, P., Raj, G., Upadhyaya, A.K.: Modulation instability of laser pulse in magnetized plasma. Phys. Plasmas **12**, 123104 (2005)
52. Chen, H.Y., Liu, S.Q., Li, X.Q.: Self-modulation instability of an intense laser beam in a magnetized pair plasma. Phys. Scr. **83**, 035502 (2011)
53. Sepehri Javan, N.: Modulation instability of an intense laser beam

in the hot magnetized electron-positron plasma in the quasi-neutral limit. Phys. Plasmas **19**, 122107 (2012)
54. Sepehri Javan, N.: Competition of circularly polarized laser modes in the modulation instability of hot magnetoplasma. Phys. Plasmas **20**, 012120 (2013)
55. Etemadpour, R., Sepehri, N.: Javan, Effect of super-thermal ions and electrons on the modulation instability of a circularly polarized laser pulse in magnetized plasma. Laser Part. Beams **33**, 265 (2015)
56. Sepehri Javan, N., Nasirzadeh, Zh: Self-focusing of circularly polarized laser pulse in the hot magnetized plasma in the quasi-neutral limit. Phys. Plasmas **19**, 112304 (2012)
57. Sepehri Javan, N., Rouhi Erdi, F.: Effect of dynamical non-neutrality on the modulational instability of laser propagating through hot magnetoplasma. Phys. Plasmas **22**, 062116 (2015)
58. Sepehri Javan, N., Hosseinpour Azad, M.: Thermal behavior change in the self-focusing of an intense laser beam in magnetized electron-ion-positron plasma. Beams **32**, 321 (2014)
59. Malomed, B.: Nonlinear Schrödinger equations. In: Scott, A. (ed.) Encyclopedia of nonlinear science, pp. 639–643. Routledge, New York (2005)
60. Pitaevskii, L., Stringari, S.: Bose-Einstein condensation. Clarendon, Oxford (2003)
61. Gurevich, A.V.: Nonlinear phenomena in the ionosphere. Springer, Berlin (1978)
62. Balakrishnan, R.: Soliton propagation in nonuniform media. Phys. Rev. A. **32**, 1144 (1985)

4

A study on dust acoustic traveling wave solutions and quasiperiodic route to chaos in nonthermal magnetoplasmas

Asit Saha[1] · Nikhil Pal[2] · Tapash Saha[1] · M. K. Ghorui[3] · Prasanta Chatterjee[2]

Abstract Bifurcations and chaotic behaviors of dust acoustic traveling waves in magnetoplasmas with nonthermal ions featuring Cairns–Tsallis distribution is investigated on the framework of the further modified Kadomtsev–Petviashili (FMKP) equation. The FMKP equation is derived employing the reductive perturbation technique (RPT). Bifurcations of dust acoustic traveling waves of the FMKP equation is presented. Using the bifurcation theory of planar dynamical systems, two new analytical traveling wave solutions for solitary and periodic waves are derived depending on the parameters α, α_1, q, l and U. Considering an external periodic perturbation, the chaotic behavior of dust acoustic traveling waves is investigated through quasiperiodic route to chaos. The parameter q significantly affects the chaotic behavior of the perturbed FMKP equation.

✉ Asit Saha
asit_saha123@rediffmail.com

Nikhil Pal
nikhilpal.math@gmail.com

Tapash Saha
tapashbrb@rediffmail.com

M. K. Ghorui
malaykr_ghorui@rediffmail.com

Prasanta Chatterjee
prasantachatterjee1@rediffmail.com

1 Department of Mathematics, Sikkim Manipal Institute of Technology, Sikkim Manipal University, Majitar, Rangpo, East Sikkim 737136, India

2 Department of Mathematics, Siksha-Bhavana, Visva-Bharati University, Santiniketan 731235, India

3 Department of Mathematics, B.B. College, Ushagram, Asansol 713303, India

Keywords Dusty plasma · Traveling wave · Chaotic behavior · Quasiperiodic route to chaos

Introduction

The physics of dusty plasmas is an important topic of growing research which has gained more and more interest over the last few decades not only from the academic point of view, but also from the view of its new aspects [1] in space and modern astrophysics, semiconductor technology, fusion devices, plasma chemistry, crystal physics, and biophysics. In 1989, Goertz [2] discussed collective effects in dusty plasmas which affect various waves, such as density waves in planetary rings and low-frequency plasma waves. The authors described briefly the possibility of charged grains forming a Coulomb lattice. Low temperature dusty plasmas is used in manufacturing of chips and material processing [3, 4] in industry, which is one of the greatest impacts on our everyday lives. Recently, a number of laboratory experiments [5–7] have demonstrated that highly ordered dust structures, i.e., dusty plasma crystals are formed when $\Gamma_c \geq 170$. Because of different types of dust charged grains in a plasma, a number of different wave modes are introduced, for example, dust acoustic mode [8], dust ion acoustic mode [9], dust lattice mode [10], Shukla–Varma mode [11], dust Berstain–Green–Kruskal mode [12] and dust drift mode [13]. Rao et al. [8] investigated the existence of a new extremely low-phase velocity dust acoustic waves (DAW) in an unmagnetized dusty plasma. Many experimental and theoretical observations performed by Angelo [14], Barkan et al. [15, 16], Nakamuro et al. [17] have confirmed the linear and nonlinear phenomena of both DAW and DIAW. Tomar et al. [18] studied the reflection of ion acoustic soliton in an inhomogeneous dusty plasma having two temperature electrons. Sabetkar and Dorranian [19] investigated the effect of obliqueness and external magnetic field on the characteristics of dust acoustic solitary waves in dusty plasma with two temperature nonthermal ions. El-Hanbaly et al. [20] studied the propagation of linear and nonlinear dust acoustic waves in a homogeneous unmagnetized, collisionless and dissipative dusty plasma consisted of extremely massive, micron-sized, negative dust grains. Tomar et al. [21] also investigated the evolution of solitons and their reflection and transmission in a plasma having negatively charged dust grains. Sabetkar and Dorranian [22] investigated the nonextensive effects on the characteristics of dust acoustic solitary waves in magnetized dusty plasma with two temperature isothermal ions. Dorranian and Sabetkar [23] studied the nonlinear dust acoustic solitary waves in a dusty plasma with two nonthermal ion species at different temperatures. The authors showed the effects of nonthermal coefficient, ions temperature, and ions number density on the amplitude and width of soliton in dusty plasma. Shahmansouri and Tribeche [24] investigated nonlinear dust acoustic (DA) shock waves in a nonextensive charge varying complex plasma and found that the influence of nonextensive particles and dust charge fluctuation affect the basic properties of the collisionless DA shock wave drastically. Shahmansouri and Mamun [25] carried out a theoretical investigation to study the basic properties of dust acoustic (DA) shock waves in a magnetized nonthermal dusty plasma containing cold viscous dust fluid, nonthermal ions, and nonthermal electrons. Shahmansouri and Borhanian [26] reported the nonlinear aspects of nonplanar dust acoustic (DA) solitary waves in an unmagnetized complex plasma comprising of cold dust grains, kappa-distributed ions as well as electrons.

There are some astrophysical and space plasmas environments containing particles with distribution functions which are quasi-Maxwellian up to the mean thermal velocities and present non-Maxwellian nonthermal tails when the particles gain high velocities and energies [27–29]. These types of plasmas are known as nonthermal plasmas which are observed in Mercury, in the solar wind, Saturn and in the Magnetospheres of the Earth [29, 30]. Tribeche et al. [31] generalized the model of Cairns et al. [32] and outlined a physically meaningful nonextensive nonthermal velocity distribution. They [31] studied the ion acoustic solitary waves in a plasma with nonthermal electrons featuring Tsallis distribution (Cairns–Tsallis). Recently, Williams and Kourakis [33] re-examined the Cairns–Tsallis model for ion acoustic solitons and concluded that the parameters q and α must be in the ranges $0 \leq \alpha < 0.25$ and $0.6 < q < 1$ subject to the physical cutoff imposed by the monotonicity condition $\alpha = \frac{(2q-1)}{4}$.

There are many important nonlinear dynamical systems in physics, chemistry and biology which clearly display different types of regular and chaotic behaviors depending upon the strength of control parameters, initial conditions, nature of external perturbation, and so on. Thus, to identify whether a given motion of a dynamical system is periodic or quasiperiodic or chaotic, one needs to perform quantitative measures in addition to the various qualitative features. Using numerical computations, some perturbed nonlinear evolution equations (Sine-Gordon, KdV and Schrodinger equations) have been investigated [34, 35]. But it is important to note that the presence of external perturbations introduces different dynamic behaviors like quasiperiodic behavior and chaotic behavior. Thus, addition of an external perturbation to a nonlinear integrable wave equation may provide quasiperiodic and chaotic motions. Considering an external perturbation, many authors have investigated chaos through different routes, such as, period doubling route [36] to chaos, quasiperiodic route [37] to chaos, crisis route [38] to chaos and intermittency route [39] to chaos.

Recently, Samanta et al. [40] studied bifurcations of dust ion acoustic traveling waves in a magnetized dusty plasma with a q-nonextensive electron velocity distribution using bifurcation theory of planar dynamical systems for the first time in the literature. Later on, a number works [41–45] on bifurcations of nonlinear waves in plasmas have been reported through perturbative and nonperturbative approaches. Saha and Chatterjee [46] studied propagation and interaction of dust acoustic multi-soliton in dusty plasmas with q-nonextensive electrons and ions. Very recently, Saha et al. [47] investigated the dynamic behavior of ion acoustic waves in electron–positron–ion magnetoplasmas with superthermal electrons and positrons in the framework of perturbed and non-perturbed Kadomtsev–Petviashili (KP) equations. Ghosh et al. [48] investigated the dynamic structures of ion acoustic waves in an unmagnetized plasma with q-nonextensive electrons and positrons applying the bifurcation theory of planar dynamical systems. Sahu et al. [49] studied the quasiperiodic behavior in quantum plasmas due to the presence of bohm potential. Zhen et al. [50] studied dynamic behavior of the quantum ZK equation in dense quantum magnetoplasma. But bifurcation and chaotic behaviors of nonlinear waves in plasmas on the framework of FMKP equation have not been reported to the best of our knowledge.

In this work, our aim is to investigate the bifurcation and chaotic behaviors of dust acoustic traveling waves in magnetoplasmas with nonthermal ions featuring Cairns–Tsallis distribution on the framework of FMKP equation using bifurcation theory of planar dynamical systems. We derive two new analytical solutions for solitary and periodic waves of the FMKP equation. Considering an external periodic perturbation, we study the chaotic behaviors of the perturbed FMKP equation through quasiperiodic route to chaos in the mentioned plasmas. In this case, we restrict the parameter ranges $0 \leq \alpha < 0.25$ and $0.6 < q < 1$ based on the study [33].

The remaining part of the paper is organized as follows. In the next section, we consider model equations and then derive the FMKP equation. Following this, we obtain a dynamical system of the FMKP equation after which bifurcations of phase portraits are obtained. In the subsequent section, two analytical traveling wave solutions of the FMKP equation are derived. Before the concluding section, we discuss the chaotic behavior of the perturbed FMKP equation. The study is concluded in the final section.

Basic equations

We consider a plasma model whose constituents are dynamic dust particles and nonthermal cold ions featuring Tsallis distribution in the presence of an external static magnetic field $M = \hat{x}M_0$ acting along the x-axis, where $\hat{x}$ is an unit vector along the x-axis. The normalized continuity, momentum and Poisson's equations are as follows:

$$\frac{\partial n}{\partial t} + \nabla.(n\tilde{U}) = 0, \tag{1}$$

$$\frac{\partial \tilde{U}}{\partial t} + (\tilde{U}.\nabla)\tilde{U} = \nabla\phi - \tilde{U} \times \hat{x}, \tag{2}$$

$$\nabla^2\phi = \alpha_1(n - n_i), \tag{3}$$

where $\alpha_1 = \frac{r^2}{\lambda^2}$, $r = \frac{C_s}{\Omega}$ is the dust gyroradius, $\lambda = \sqrt{T_i/4\pi e^2 n_0 z_{d0}}$ is the Debye length, $C_s = (T_i/m)^{1/2}$ is the dust acoustic velocity, $\Omega = \frac{eM_0}{mc}$ is the dust gyrofrequency, c is the speed of the light, m is the mass of dusts and z_d is the number of the charge residing on the dust grains, so that the charge of the dust $q_d = -ez_d$ with e is the elementary charge. ϕ is the plasma potential. n and $\tilde{U}$ denote number density and velocity of dust particles, respectively. We assume that the wave is propagating in the xy-plane. Here, n_{i0}, and n_0 are, respectively, the unperturbed number densities of ions and dust particles. The dust velocity $\tilde{U} = (u, v, w)$ is normalized to dust acoustic speed $C_s = \sqrt{\frac{T_i}{m}}$ and plasma potential ϕ is normalized to T_i/e. Space variables and time are normalized to the dust gyroradius r and inverse of the dust gyrofrequency Ω, respectively.

The nonextensive nonthermal velocity distribution [31] function is given by:

$$f_i(v_x) = C_{q,\alpha}\left(1 + \alpha\frac{v_x^4}{v_{ti}^4}\right)\left\{1 - (q-1)\frac{v_x^2}{2v_{ti}^2}\right\}^{\frac{1}{q-1}},$$

where $v_{ti} = (T_i/m_i)^{1/2}$ is the ion thermal velocity, T_i is the ion temperature, m_i is its mass, and $C_{q,\alpha}$ is the constant of normalization which is given by the following expressions:

$$C_{q,\alpha} = n_{i0}\sqrt{\frac{m_i}{2\pi T_i}}\frac{\Gamma(\frac{1}{1-q})(1-q)^{5/2}}{\Gamma(\frac{1}{1-q}-\frac{5}{2})[3\alpha + (\frac{1}{1-q}-\frac{3}{2})(\frac{1}{1-q}-\frac{5}{2})(1-q)^2]}$$

for $-1 < q < 1$,

and

$$C_{q,\alpha} = n_{i0}\sqrt{\frac{m_i}{2\pi T_i}}\frac{\Gamma(\frac{1}{q-1}+\frac{3}{2})(q-1)^{5/2}(\frac{1}{q-1}+\frac{3}{2})(\frac{1}{q-1}+\frac{5}{2})}{\Gamma(\frac{1}{q-1}+1)[3\alpha + (\frac{1}{q-1}+\frac{3}{2})(\frac{1}{q-1}+\frac{5}{2})(q-1)^2]}$$

for $q > 1$.

Here, α is a parameter determining the number of nonthermal ions present in the model, q stands for the strength of nonextensivity, and Γ is the standard Gamma function. For $q > 1$, the distribution function exhibits a thermal cutoff on the maximum value allowed for the velocity of the ions, given by

$$v_{\max} = \sqrt{\frac{2T_i}{m_i(q-1)}},$$

beyond which no probable states exist.

Integrating the nonthermal velocity distributed function $f_i(v_x)$ over all velocity space, one can obtain the ion density [31] as:

$$n_i = n_{i0}\left(1 - M\left(\frac{e\phi}{T_i}\right) + N\left(\frac{e\phi}{T_i}\right)^2\right)\left\{1 - (q-1)\left(\frac{e\phi}{T_i}\right)\right\}^{\frac{1}{q-1}+\frac{1}{2}},$$

where $M = -\frac{16\alpha q}{(5q-3)(3q-1)+12\alpha}$ and $N = \frac{16\alpha q(2q-1)}{(5q-3)(3q-1)+12\alpha}$.

In the limiting case, when $q \to 1$, the above ion density reduces to the nonthermal ion density of Cairns et al. [32] as

$$n_i = n_{i0}\left(1 + \frac{4\alpha}{1+3\alpha}\left(\frac{e\phi}{T_i}\right) + \frac{4\alpha}{1+3\alpha}\left(\frac{e\phi}{T_i}\right)^2\right) \times \exp\left(-\frac{e\phi}{T_i}\right),$$

and in the case, when $\alpha = 0$, the ion density reduces to the nonextensive ion density [51] as

$$n_i = n_{i0}\left\{1 - (q-1)\left(\frac{e\phi}{T_i}\right)\right\}^{\frac{1}{q-1}+\frac{1}{2}}.$$

The normalized ion number density [31] is given by

$$n_i = (1 - M\phi + N\phi^2)\{1 - (q-1)\phi\}^{\frac{1}{q-1}+\frac{1}{2}},$$

where $M = -\frac{16\alpha q}{(5q-3)(3q-1)+12\alpha}$ and $N = \frac{16\alpha q(2q-1)}{(5q-3)(3q-1)+12\alpha}$.

Equations (1)–(3) can be written in components form as:

$$\frac{\partial n}{\partial t}+\frac{\partial(nu)}{\partial x}+\frac{\partial(nv)}{\partial y}=0, \tag{4}$$

$$\frac{\partial u}{\partial t}+\left(u\frac{\partial}{\partial x}+v\frac{\partial}{\partial y}\right)u=\frac{\partial\phi}{\partial x}, \tag{5}$$

$$\frac{\partial v}{\partial t}+\left(u\frac{\partial}{\partial x}+v\frac{\partial}{\partial y}\right)v=\frac{\partial\phi}{\partial y}-w, \tag{6}$$

$$\frac{\partial w}{\partial t}+\left(u\frac{\partial}{\partial x}+v\frac{\partial}{\partial y}\right)w=v, \tag{7}$$

$$\left(\frac{\partial^2}{\partial x^2}+\frac{\partial^2}{\partial y^2}\right)\phi=\alpha_1\left[n-(1-M\phi+N\phi^2)\{1-(q-1)\}^{\frac{1}{q-1}+\frac{1}{2}}\right]. \tag{8}$$

Derivation of the FMKP equation

We employ the reductive perturbation technique (RPT) to derive the Kadomtsev–Petviashili(KP) equation. According to the RPT, the independent variables are stretched as:

$$\begin{cases} Y=\epsilon^2 y, \\ \eta=\epsilon(x-Vt), \\ \tau=\epsilon^3 t, \end{cases} \tag{9}$$

where V denotes the phase velocity of dust acoustic wave along the x-axis in magnetoplasmas with nonthermal ions featuring Tsallis distribution, and ϵ is a small parameter which characterizes the strength of the nonlinearity. The dependent variables in the above relations are expanded as:

$$\begin{cases} n=1+\epsilon^2 n_1+\epsilon^4 n_2+\cdots \\ u=\epsilon^2 u_1+\epsilon^4 u_2+\cdots \\ v=\epsilon^3 v_1+\epsilon^5 v_2+\cdots \\ w=\epsilon^3 w_1+\epsilon^5 w_2+\cdots \\ \phi=\epsilon^2\phi_1+\epsilon^4\phi_2+\cdots \end{cases} \tag{10}$$

Substituting the Eqs. (9)–(10) into the system of Eqs. (4)–(8) and equating the coefficient of lowest order of ϵ, one can obtain the phase velocity as

$$V^2=\frac{1}{(a+M)}, \tag{11}$$

where $a=\frac{q+1}{2}$.

Considering the coefficient of next order of ϵ, we obtain the KP equation as:

$$\frac{\partial}{\partial\eta}\left[\frac{\partial\phi_1}{\partial\tau}-A\phi_1\frac{\partial\phi_1}{\partial\eta}+B\frac{\partial^3\phi_1}{\partial\eta^3}\right]+C\frac{\partial^2\phi_1}{\partial Y^2}=0, \tag{12}$$

where $A=\frac{V}{2P}[3P^2-2Q]$, $B=\frac{V}{2P\alpha_1}$, $C=\frac{V}{2}$, with $P=a+M$, $b=\frac{(q+1)(3-q)}{8}$ and $Q=b+N+aM$.

The KP equation (12) depends on A which is a function of α and q. In Fig. 1, it is shown that A may be positive or negative depending on different values of q with fixed value of $\alpha=0.1$, but there is a critical point at which $A=0$, which can provide an infinite growth of the amplitude of the solitary wave solutions and periodic wave solutions of Eq. (12) which breaks down the validity of the RPT. In this case, q is called the critical parameter with critical value $q\simeq 0.8751$. Thus, the exact solutions of the Eq. (12) do not exist at the points which are very near to the critical values of the critical parameters. In this situation, the KP equation is unable to describe the nonlinear wave phenomena in this dusty plasma. So to describe the nonlinear wave features near or around or at $A=0$, we extend the study and want to obtain satisfactory solutions near and around the critical value. Therefore, we consider more higher order nonlinear equation to achieve the desired results.

We proceed for the modified Kadomtsev–Petviashili (MKP) equation by considering higher order coefficients of ϵ. We consider the same set of stretched coordinates but the previous expansions of the dependent variables are not valid. Therefore, we consider a set of new expansions of the dependent variables as follows:

$$\begin{cases} n=1+\epsilon n_1+\epsilon^2 n_2+\epsilon^3 n_3+\cdots \\ u=\epsilon u_1+\epsilon^2 u_2+\epsilon^3 u_3+\ldots \\ v=\epsilon^2 v_1+\epsilon^3 v_2+\epsilon^4 v_3+\ldots \\ w=\epsilon^2 w_1+\epsilon^4 w_2+\epsilon^6 w_3+\ldots \\ \phi=\epsilon\phi_1+\epsilon^2\phi_2+\epsilon^3\phi_3+\ldots. \end{cases} \tag{13}$$

Substituting the above expansions (13) along with the same stretched coordinates (9) into Eqs. (4)–(8) and equating the coefficients of different powers of ϵ and eliminating n_3, w_3 and ϕ_3, one can obtain the following equation:

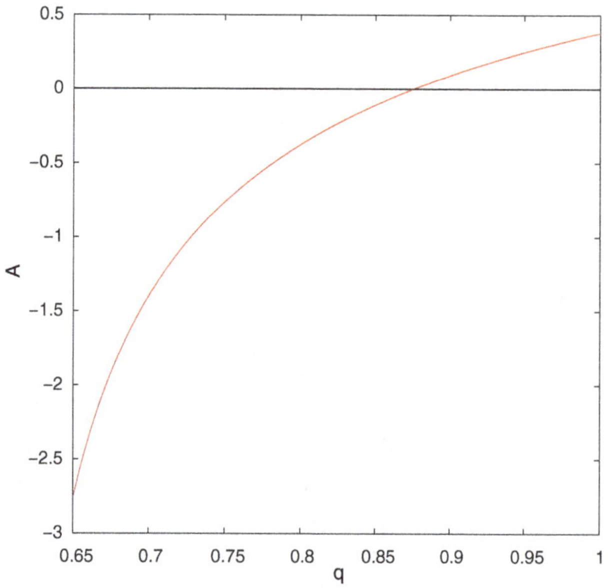

Fig. 1 A is a function of q when $\alpha=0.1$

$$\frac{\partial}{\partial \eta}\left[\frac{\partial \phi_1}{\partial \tau} - A\frac{\partial(\phi_1\phi_2)}{\partial \eta} - D\phi_1^2\frac{\partial \phi_1}{\partial \eta} + B\frac{\partial^3 \phi_1}{\partial \eta^3}\right] + C\frac{\partial^2 \phi_1}{\partial Y^2} = 0, \tag{14}$$

where the coefficients A, B and C are same as the coefficients of the KP equation and $D = \frac{3V}{2P}(R + 2P^3 - 3PQ)$ with $R = K + bM + aN$. It is clear that for the critical values of the parameters A may equal to zero and the Eq. (14) reduces to the following MKP equation:

$$\frac{\partial}{\partial \eta}\left[\frac{\partial \phi_1}{\partial \tau} - D\phi_1^2\frac{\partial \phi_1}{\partial \eta} + B\frac{\partial^3 \phi_1}{\partial \eta^3}\right] + C\frac{\partial^2 \phi_1}{\partial Y^2} = 0, \tag{15}$$

If A is at the same order of ϵ, but not zero, we derive the FMKP equation using the same stretched coordinates and same expansions as the MKP equation:

$$\frac{\partial}{\partial \eta}\left[\frac{\partial \phi_1}{\partial \tau} - A\phi_1\frac{\partial \phi_1}{\partial \eta} - D\phi_1^2\frac{\partial \phi_1}{\partial \eta} + B\frac{\partial^3 \phi_1}{\partial \eta^3}\right] + C\frac{\partial^2 \phi_1}{\partial Y^2} = 0. \tag{16}$$

Formation of dynamical system

To investigate all traveling wave solutions of the FMKP equation (16), we transform it to a dynamical system by introducing a new variable χ as follows:

$$\chi = (l\eta + mY - U\tau), \tag{17}$$

where l and m are the cosines of the angles made by wave propagation with η-axis and Y-axis, respectively. Here, U is the speed of dust acoustic traveling wave. Substituting $\psi(\chi) = \phi_1(\eta, Y, \tau)$ into the FMKP equation (16) and then integrating twice, the FMKP equation (16) takes the form

$$Bl^4\frac{d^2\psi}{d\chi^2} + (Cm^2 - lU)\psi - \frac{Al^2}{2}\psi^2 - \frac{Dl^2}{3}\psi^3 = 0. \tag{18}$$

Then, Eq. (18) can be written as the following dynamical system:

$$\begin{cases} \dfrac{d\psi}{d\chi} = z, \\ \dfrac{dz}{d\chi} = \dfrac{(lU - C(1-l^2) + \frac{Al^2}{2}\psi + \frac{Dl^2}{3}\psi^2)\psi}{Bl^4}. \end{cases} \tag{19}$$

The system (19) represents a planar Hamiltonian system with the following Hamiltonian function:

$$H(\psi, z) = \frac{z^2}{2} - \frac{1}{12Bl^4}(6(lU - C(1-l^2)) + 2Al^2\psi + Dl^2\psi^2)\psi^2. \tag{20}$$

The system (19) is a planar dynamical system with parameters α, α_1, q, l and U. It is interesting to note that the phase orbits defined by the vector fields of Eq. (19) determine all traveling wave solutions of the FMKP equation (16). Thus, we investigate bifurcations of phase portraits of Eq. (19) in the (ψ, z) phase plane as the parameters α, α_1, q, l and U are varied. In this case, we consider a physical system for which only bounded traveling wave solutions are meaningful. Therefore, our attention is to study only bounded traveling wave solutions of the FMKP equation (16). It is known that a solitary wave solution of Eq. (16) corresponds to a homoclinic orbit of Eq. (19). A periodic orbit of Eq. (19) corresponds to a periodic traveling wave solution of Eq. (16). The bifurcation theory of planar dynamical systems [52, 53] plays an important role in this study.

Phase plane analysis

In this section, we investigate the bifurcations of phase portraits of Eq. (19). When $AB\beta l \neq 0$ and $lU \neq C(1-l^2)$, then there are three equilibrium points at $E_0(\psi_0, 0)$, $E_1(\psi_1, 0)$ and $E_2(\psi_2, 0)$, where $\psi_0 = 0$, $\psi_1 = \frac{3}{2Dl^2}\{\frac{-Al^2}{2} + \sqrt{\frac{A^2l^4}{2} - \frac{4Dl^2}{3}(lU - C(1-l^2))}\}$ and $\psi_2 = \frac{3}{2Dl^2}\{\frac{-Al^2}{2} - \sqrt{\frac{A^2l^4}{2} - \frac{4Dl^2}{3}(lU - C(1-l^2))}\}$.

Let $M(\psi_i, 0)$ be the coefficient matrix of the linearized system of Eq. (19) at an equilibrium point $E_i(\psi_i, 0)$. Then, we have

$$J = \det M(\psi_i, 0) = \frac{(C(1-l^2) - lU)}{Bl^4} - \frac{1}{Bl^2}\{A\psi_i + D\psi_i^2\}. \tag{21}$$

By the theory of planar dynamical systems [52, 53], we know that the equilibrium point $E_i(\psi_i, 0)$ of the planar dynamical system (19) is a saddle point when $J < 0$ and the equilibrium point $E_i(\psi_i, 0)$ of the planar dynamical system (19) is a center when $J > 0$.

If $2lU > V(1-l^2)$, $3P^2 < 2Q$, $R + 2P^3 < 3PQ$, $\frac{5}{7} < q < 1$, $0 \leq \alpha < 0.25$, $0 < l < 1$, and $\alpha_1 > 0$, then the system (19) has three equilibrium points at $E_0(\psi_0, 0)$, $E_1(\psi_1, 0)$ and $E_2(\psi_2, 0)$, where $\psi_0 = 0$, $\psi_1 > 0$ and $\psi_2 < 0$. The equilibrium point $E_0(\psi_0, 0)$ is a saddle point, $E_1(\psi_1, 0)$ and $E_2(\psi_2, 0)$ are centers. There is a pair of homoclinic orbits at $E_0(\psi_0, 0)$ surrounding the centers $E_1(\psi_1, 0)$ and $E_2(\psi_2, 0)$ (see Fig. 2).

Using the above analysis, we have shown the phase portrait of Eq. (19) in Fig. 2 depending on some special values of the parameters α, α_1, q, l and U. It is seen that there is a pair of homoclinic orbits at the equilibrium point $E_0(\psi_0, 0)$ surrounding two centers at the equilibrium points $E_1(\psi_1, 0)$ and $E_2(\psi_2, 0)$ in Fig. 2. For these pair of homoclinic orbits of the dynamical system (19), the FMKP

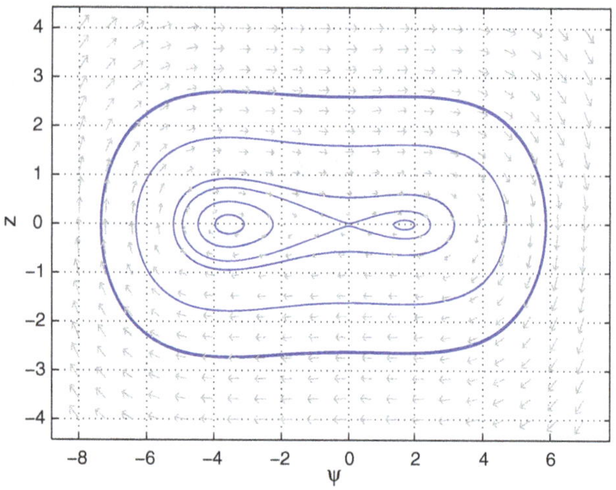

Fig. 2 Phase portrait of Eq. (19) for $l = 0.7, \alpha = 0.1, \alpha_1 = 0.1, q = 0.8$ and $U = 1$

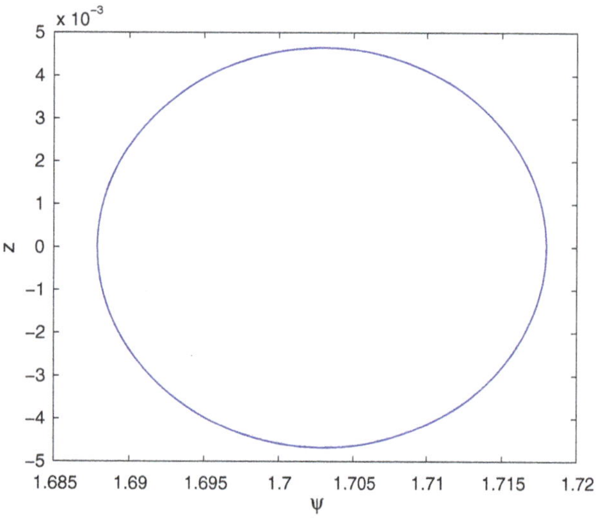

Fig. 3 Phase projection of system (19) with the same values of parameters as Fig. 2

equation has dust acoustic compressive and rarefactive solitary wave solutions.

In Fig. 3, we have presented one limit cycle about the center $E_1(\psi_1, 0)$ of the dynamical system (19) for $l = 0.7, \alpha = 0.1, \alpha_1 = 0.1, q = 0.8$ and $U = 1$. Corresponding to the limit cycles about the center $E_1(\psi_1, 0)$ of the dynamical system (19), we get a family of periodic wave solutions of the FMKP equation (16). In Fig. 4, we have presented the periodicity of Z based on system (19) with the same values of parameters as Fig. 2 and in Fig. 5, we have shown the periodicity of ψ based on system (19) with the same values of parameters as Fig. 2. We can obtain similar results in case of equilibrium point $E_2(\psi_2, 0)$.

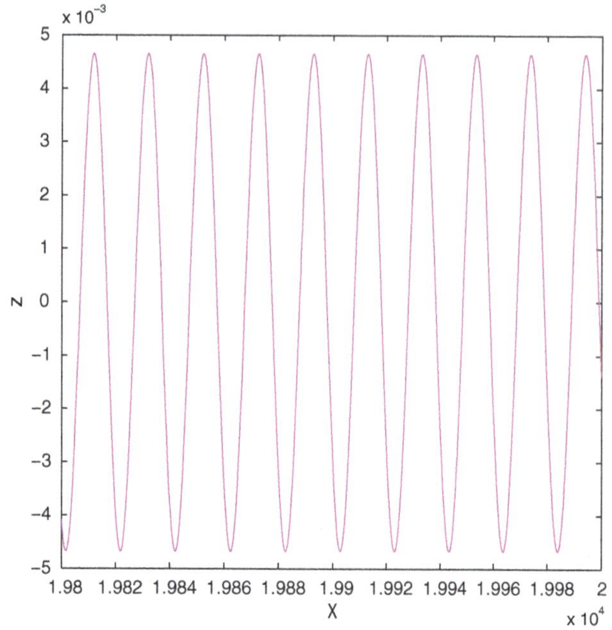

Fig. 4 Periodicity of Z based on system (19) with the same values of parameters as Fig. 2

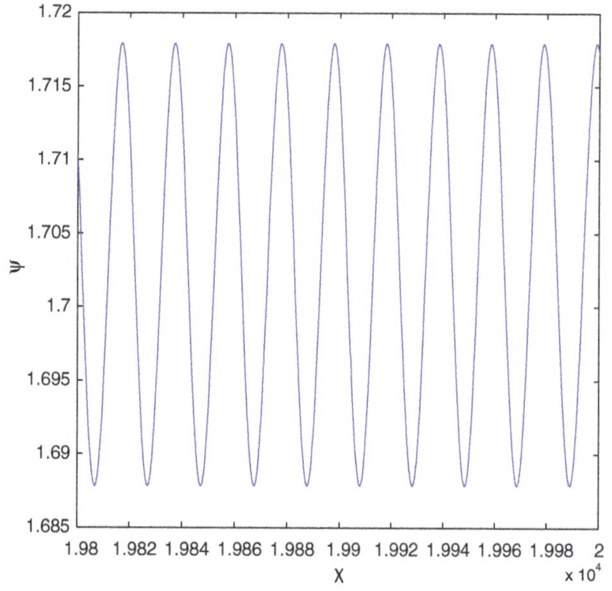

Fig. 5 Periodicity of ψ based on system (19) with the same values of parameters as Fig. 2

Analytical traveling wave solutions

In this section, using the planar dynamical system Eq. (19) and the Hamiltonian function Eq. (20), we derive analytical traveling wave solutions for solitary waves and periodic waves of the FMKP equation (16) depending on the parameters α, α_1, q, l and U. It should be noted that

$cn(\Omega_1\xi, k_1)$ is the Jacobian elliptic function [54] with the modulo k_1.

(1) Corresponding to the pair of homoclinic orbits at $E_0(\psi_0, 0)$ surrounding the centers $E_1(\psi_1, 0)$ and $E_2(\psi_2, 0)$ (see Fig. 2), the FMKP equation (16) has a pair of the solitary wave solutions (compressive and rarefactive types):

$$\psi(\chi) = \pm\frac{1}{\sqrt{2\left(1-\frac{b_1^2}{9a_1c_1}\right)\,\sin\left(2\sqrt{\frac{a_1}{c_1}}\chi\right)+\frac{b_1}{6a_1}}}, \quad (22)$$

where $a_1 = \frac{lU-C(1-l^2)}{Bl^4}$, $b_1 = \frac{A}{2Bl^2}$ and $c_1 = \frac{D}{3Bl^2}$.

(2) Corresponding to the family of periodic orbits about $E_2(\psi_2, 0)$ (see Fig. 2), the FMKP equation (16) has a family of the periodic traveling wave solutions:

$$\psi(\chi) = \frac{\alpha_2 B_1 + \beta_2 A_1 - (\alpha_2 B_1 - \beta_2 A_1)cn(\Omega_1\chi, k_1)}{B_1 + A_1 - (B_1 - A_1)cn(\Omega_1\chi, k_1)}, \quad (23)$$

where $A_1 = \alpha_2^2+\alpha_2\gamma_2+\delta_2, B_1 = \beta_2^2+\beta_2\gamma_2+\delta_2, \Omega_1 = \sqrt{-\frac{D}{6Bl^2}}$ and $k_1 = \frac{(\alpha_2-\beta_2)^2-(A_1-B_1)^2}{4A_1B_1}$ with $\alpha_2, \beta_2, \gamma_2$ and δ_2 are roots of the equation $h+\frac{1}{12Bl^4}(6(lU-C(1-l^2))+2Al^2\psi+Dl^2\psi^2)\psi^2 = -\frac{D}{12Bl^2}(\alpha_2-\psi)\quad(\psi-\beta_2)(\psi^2+\gamma_2\psi+\delta_2)$, satisfying $\alpha_2 > \beta_2$, and $\gamma_2^2-4\delta_2 < 0, h \in (h_2, 0)$, $h_2 = H(\psi_2, 0)$.

Sabetkar and Dorranian [22] investigated dust acoustic solitary waves (DASWs) in a magnetized four component dusty plasma and showed that due to electron nonextensivity, their dusty plasma model admitted positive potential as well as negative potential solitons. Dorranian and Sabetkar [23] also investigated the dust acoustic solitary waves in a dusty plasma on the frameworks of the KP and modified KP equations. The authors obtained the compressive and rarefactive solitary wave solutions in terms of sech$(\frac{\chi}{w})$ for some special values of the physical parameters. But in this work, we have obtained a new form of the compressive and rarefactive solitary wave solutions (22) and periodic wave solution (23) in terms of the Jacobean elliptic function. Thus, the dust acoustic compressive and rarefactive solitary waves of our work have been supported by the works [22, 23] reported in the literature.

Quasiperiodic route to chaos

In this section, we study the quasiperiodic and chaotic behaviors of the perturbed system given by:

$$\begin{cases} \dfrac{d\psi}{d\chi} = z, \\ \dfrac{dz}{d\chi} = \dfrac{\left(lU - C(1-l^2) + \dfrac{Al^2}{2}\psi + \dfrac{Dl^2}{3}\psi^2\right)\psi}{Bl^4} + f_0\,\cos(\omega\chi), \end{cases} \quad (24)$$

where $f_0\ \cos(\omega\chi)$ is an external periodic perturbation, f_0 is the strength of the periodic perturbation and ω is the frequency. It is to be noted that the difference between system (19) and system (24) is that only external periodic perturbation is added with system (24). Furthermore, existence of $f_0\cos(\omega\chi)$ in system (24) is a root that can turn system (19) into the chaotic state.

In Fig. 6, we have presented phase portrait of the perturbed system (24) for $l = 0.7, \alpha = 0.1, \alpha_1 = 0.1, q = 0.8, U = 1,\ f_0 = 0.02$ and $\omega = 1$ with initial condition $(\psi_0, z_0) = (1.72, 0.0001)$. It is found that the perturbed system (24) has quasiperiodic motion even if the external periodic perturbation is considered. Thus, a quasiperiodic motion of the system (24) is observed with incommensurable periodic motions and the trajectory in the phase space winds around torus filling its surface densely. In Figs. 7 and 8, we have presented the quasiperiodicity of Z and ψ, respectively, based on the system (24) with the same values of parameters as Fig. 6. If we increase strength of the periodic perturbation and consider $f_0 = 1$ with the same values of other parameters, then the perturbed system (24) shows chaotic motions. In Fig. 9, we have presented the phase portrait of the perturbed system (24) for

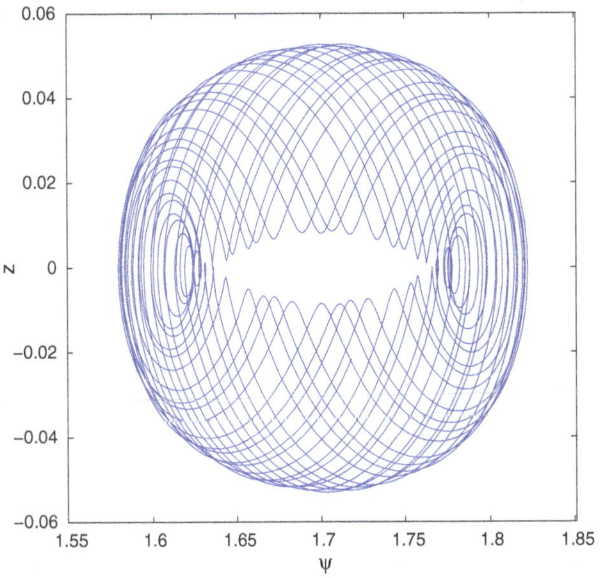

Fig. 6 Phase portrait of the perturbed system (24) for $l = 0.7, \alpha = 0.1, \alpha_1 = 0.1, q = 0.8, U = 1,\ f_0 = 0.02$ and $\omega = 1$ with initial condition $(\psi_0, z_0) = (1.72, 0.0001)$

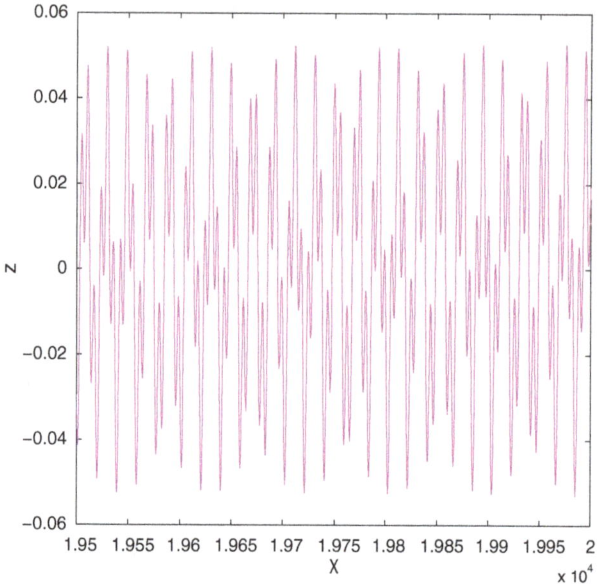

Fig. 7 Quasiperiodicity of Z based on system (24)

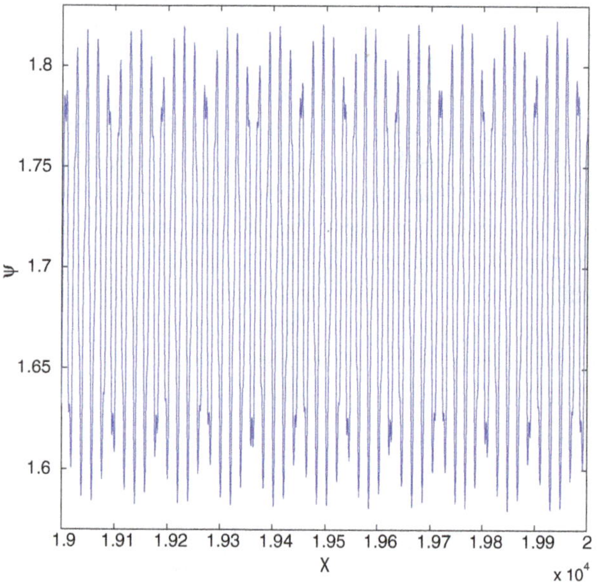

Fig. 8 Quasiperiodicity of ψ based on system (24)

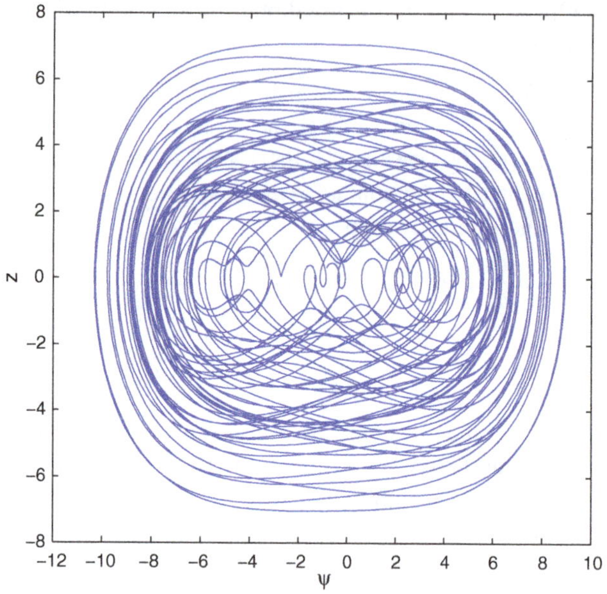

Fig. 9 Phase portrait of the perturbed system (24) for $l = 0.7, \alpha = 0.1, \alpha_1 = 0.1, q = 0.8, U = 1, f_0 = 1$ and $\omega = 1$ with initial condition $(\psi_0, z_0) = (1.72, 0.0001)$

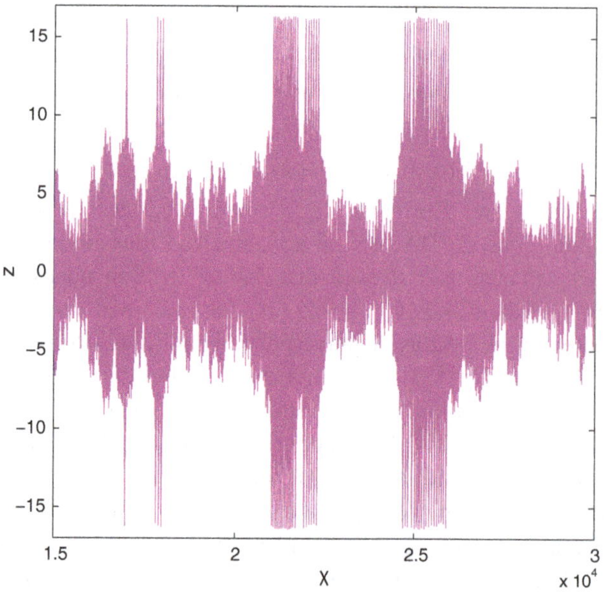

Fig. 10 Chaotic motions of Z based on system (24)

$l = 0.7, \alpha = 0.1, \alpha_1 = 0.1, q = 0.8, U = 1,$ $f_0 = 1$ and $\omega = 1$ with same initial condition as Fig. 6. In Figs. 10 and 11, we have presented the chaotic motions of Z and ψ, respectively, based on the system (24) with same values of parameters as Fig. 9. Thus, the developed chaotic motions occur (see Figs. 9, 10, 11) and the solutions ignore the periodic motions and represent random sequences of uncorrelated oscillations. Hence, the strength of the periodic perturbation plays a crucial role for the development of the quasiperiodic motion of the perturbed system (24) and transition from quasiperiodic motion to chaotic motion of the system (24). Thus, it is observed that the perturbed plasma system shows chaotic behavior through quasiperiodic route to chaos which is an important observation in this study.

Conclusions

In this paper, we have derived the FMKP equation for dust acoustic waves in magnetoplasmas with nonthermal ions featuring Cairns–Tsallis distribution. Applying the

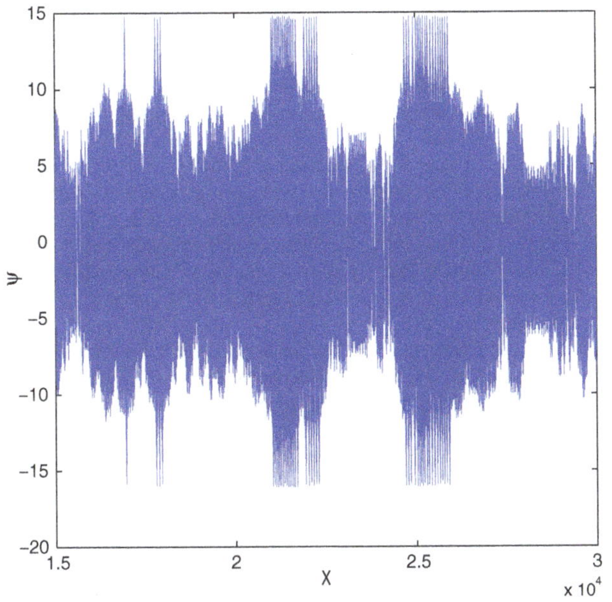

Fig. 11 Chaotic motions of ψ based on system (24)

bifurcation theory of planar dynamical systems to the FMKP equation, we have presented the existence of solitary and periodic traveling waves through phase plane analysis. Two new analytical solutions for the solitary waves (compressive and rarefactive) and periodic waves are obtained depending on parameters α, α_1, q, l and U. Considering an external periodic perturbation, the quasiperiodic and chaotic behaviors of dust acoustic waves are studied through numerical computations. The presence of the parameters q, α, and α_1 affects significantly on bifurcation of traveling wave solutions of the FMKP equation, the quasiperiodic and chaotic behaviors of the perturbed FMKP equation. It should be noted that for same set of values of parameters α, α_1, q, l and U, the unperturbed FMKP equation has solitary and periodic wave solutions, but the perturbed FMKP equation shows the quasiperiodic and chaotic behaviors based on the strength of the external periodic perturbation. It is also important to note that the dust acoustic waves of the perturbed FMKP equation represent the chaotic motions through quasiperiodic route to chaos.

Acknowledgments The authors are grateful to the reviewers for their useful comments and suggestions which helped to improve the paper.

References

1. Shukla, P.K., Mamun, A.A.: Introduction to Dusty Plasma Physics. Institute of Physics Publishing, Bristol (2002)
2. Goertz, C.K.: Dusty plasmas in the solar system. Rev. Geophys. **27**, 271 (1989)
3. Chen, F.F.: Industrial applications of low temperature plasma physics. Phys. Plasmas **2**, 2164 (1995)
4. Hopkins, M.B., Lawler, J.F.: Plasma diagnostics in industry. Plasmas Phys. Control. Fusion **42B**, 189 (2000)
5. Chu, J.H., Lin, I.: Direct observation of Coulomb crystals and liquids in strongly coupled rf dusty plasmas. Phys. Rev. Lett. **72**, 4009 (1994)
6. Thomas, H., Morfill, G.E., Dammel, V.: Plasma crystal: Coulomb crystallization in a dusty plasma. Phys Rev. Lett. **73**, 652 (1994)
7. Melzer, A., Trottenberg, T., Piel, A.: Experimental determination of the charge on dust particles forming Coulomb lattices. Phys. Lett. A **191**, 301 (1994)
8. Rao, N.N., Shukla, P.K., Yu, M.Y.: Dust-acoustic waves in dusty plasmas. Planet Space Sci. **38**, 543 (1990)
9. Kourakis, I., Shukla, P.K.: Lagrangian description of nonlinear dust-ion acoustic waves in dusty plasmas. Eur. Phys. J. D **30**, 97 (2004)
10. Melandso, F.: Lattice waves in dust plasma crystals. Phys. Plasmas **3**, 3890 (1996)
11. Shukla, P.K., Varma, R.K.: Convective cells in nonuniform dusty plasmas. Phys. Fluids B **5**, 236 (1993)
12. Tribeche, M., Zerguini, T.H.: Small amplitude Bernstein–Greene–Kruskal solitary waves in a thermal charge-varying dusty plasma. Phys. Plasmas **11**, 4115 (2004)
13. Shukla, P.K., Yu, M.Y., Bharuthram, R.: Linear and nonlinear dust drift waves. J. Geophys. Res. **96**, 21343 (1991)
14. D'Angelo, N.: Coulomb solids and low-frequency fluctuations in RF dusty plasmas. J. Phys. D **28**, 1009 (1995)
15. Barkan, A., Merlino, R.L., D'Angelo, N.: Laboratory observation of the dust-acoustic wave mode. Planet. Space Sci. **2**, 3563 (1995)
16. Barkan, A., Merlino, R.L., D'Angelo, N.: Experiments on ion-acoustic waves in dusty plasmas. Planet. Space Sci. **44**, 239 (1996)
17. Nakamura, Y., Bailing, H., Shukla, P.K.: Observation of ion-acoustic shocks in a dusty plasma. Phys. Rev. Lett. **83**, 1602 (1999)
18. Tomar, R., Malik, H.K., Dahiya, R.P.: Reflection of ion acoustic solitary waves in a dusty plasma with variable charge dust. J. Theor. Appl. Phys. **8**, 126 (2014)
19. Sabetkar, A., Dorranian, D.: Effect of obliqueness and external magnetic field on the characteristics of dust acoustic solitary waves in dusty plasma with two-temperature nonthermal ions. J. Theor. Appl. Phys. **9**, 141 (2015)
20. El-Hanbaly, A.M., El-Shewy, E.K., Sallah, M., Darweesh, H.F.: Linear and nonlinear analysis of dust acoustic waves in dissipative space dusty plasmas with trapped ions. J. Theor. Appl. Phys. **9**, 167 (2015)
21. Tomar, R., Bhatnagar, A., Malik, H.K., Dahiya, R.P.: Evolution of solitons and their reflection and transmission in a plasma having negatively charged dust grains. J. Theor. Appl. Phys. **8**, 138 (2014)
22. Sabetkar, A., Dorranian, D.: Non-extensive effects on the characteristics of dust-acoustic solitary waves in magnetized dusty plasma with two-temperature isothermal ions. J. Plasma Phys. **80**, 565 (2014)
23. Dorranian, D., Sabetkar, A.: Dust acoustic solitary waves in a dusty plasma with two kinds of nonthermal ions at different temperatures. Phys. Plasmas **19**, 013702 (2012)
24. Shahmansouri, M., Tribeche, M.: Nonextensive dust acoustic shock structures in complex plasmas. Astrophys. Space Sci. **346**, 165 (2013)
25. Shahmansouri, M., Mamun, A.A.: Dust-acoustic shock waves in a magnetized non-thermal dusty plasma. J. Plasma Phys. **80**, 593 (2014)
26. Shahmansouri, M., Borhanian, J.: Spherical Kadomtsev–Petviashvili solitons in a suprathermal complex plasma. Commun. Theor. Phys. **60**, 227 (2013)

27. Pierrard, V., Lemaire, J.: Lorentzian ion exosphere model. J. Geophys. Res. **101**, 7923 (1996)
28. Christon, S.P., Mitchell, D.G., Williams, D.J., Frank, L.A., Huang, C.Y., Eastman, T.E.: Energy spectra of plasma sheet ions and electrons from −50 eV/e to −1 MeV during plasma temperature transitions. J. Geophys. Res. **93**, 2562 (1988)
29. Maksimovic, M., Pierrard, V., Lemaire, J.F.: A kinetic model of the solar wind with Kappa distribution functions in the corona. Astron. Astrophys. **324**, 725 (1997)
30. Krimigis, S.M., Carbary, J.F., Keath, E.P., Armstrong, T.P., Lanzerotti, L.J., Gloeckler, G.: General characteristics of hot plasma and energetic particles in the Saturnian magnetosphere: Results from the Voyager spacecraft. J. Geophys. Res. **88**, 8871 (1983)
31. Tribeche, M., Amour, R., Shukla, P.K.: Ion acoustic solitary waves in a plasma with nonthermal electrons featuring Tsallis distribution. Phys. Rev. E **85**, 037401 (2012)
32. Cairns, R.A., Mamun, A.A., Bingham, R., Bostrom, R., Dendy, R.O., Nairn, C.M.C., Shukla, P.K.: Electrostatic solitary structures in non-thermal plasmas. Geophys. Res. Lett. **22**, 2709 (1995)
33. Williams, G., Kourakis, I.: Re-examining the Cairns–Tsallis model for ion acoustic solitons. Phys. Rev. E **88**, 023103 (2013)
34. Blyuss, K.B.: Chaotic behaviour of nonlinear waves and solitons of perturbed Korteweg–de Vries equation. Rep. Math. Phys. **46**, 47 (2000)
35. Moon, H.T.: Homoclinic crossings and pattern selection. Phys. Rev. Lett. **64**, 412 (1990)
36. Yorke, J.A., Alligood, K.T.: Period doubling cascades of attractors: a prerequisite for horseshoes. Commun. Math. Phys. **101**, 305 (1985)
37. Giglio, M., Musazzi, S., Perini, U.: Transition to chaotic behavior via a reproducible sequence of period-doubling bifurcations. Phys. Rev. Lett. **47**, 243 (1981)
38. Grebogi, C., Ott, E., Yorke, J.A.: Crises, sudden changes in chaotic attractors, and transient chaos. Phys. D **7**, 181 (1983)
39. Pomeau, Y., Manneville, P.: Intermittent transition to turbulence in dissipative dynamical systems. Commun. Math. Phys. **74**, 189 (1980)
40. Samanta, U.K., Saha, A., Chatterjee, P.: Bifurcations of dust ion acoustic travelling waves in a magnetized dusty plasma with a q-nonextensive electron velocity distribution. Phys. Plasma **20**, 022111 (2013)
41. Saha, A., Chatterjee, P.: Bifurcations of dust acoustic solitary waves and periodic waves in an unmagnetized plasma with nonextensive ions. Astrophys. Space Sci. **351**, 533 (2014)
42. Saha, A., Chatterjee, P.: Bifurcations of ion acoustic solitary waves and periodic waves in an unmagnetized plasma with kappa distributed multi-temperature electrons. Astrophys. Space Sci. **350**, 631 (2014)
43. Saha, A., Chatterjee, P.: Electron acoustic blow up solitary waves and periodic waves in an unmagnetized plasma with kappa distributed hot electrons. Astrophys. Space Sci. **353**, 163 (2014)
44. Saha, A., Chatterjee, P.: New analytical solutions for dust acoustic solitary and periodic waves in an unmagnetized dusty plasma with kappa distributed electrons and ions. Phys. Plasma **21**, 022111 (2014)
45. Saha, A., Chatterjee, P.: Dust ion acoustic travelling waves in the framework of a modified Kadomtsev-Petviashvili equation in a magnetized dusty plasma with superthermal electrons. Astrophys. Space Sci. **349**, 813 (2014)
46. Saha, A., Chatterjee, P.: Propagation and interaction of dust acoustic multi-soliton in dusty plasmas with q-nonextensive electrons and ions. Astrophys. Space Sci. **353**, 169 (2014)
47. Saha, A., Pal, N., Chatterjee, P.: Dynamic behavior of ion acoustic waves in electron–positron–ion magnetoplasmas with superthermal electrons and positrons. Phys. Plasma **21**, 102101 (2014)
48. Ghosh, U.N., Saha, A., Pal, N., Chatterjee, P.: Dynamic structures of nonlinear ion acoustic waves in a nonextensive electron-positron-ion plasma. J. Theor. Appl. Phys. **9**, 321 (2015)
49. Sahu, B., Poria, S., Roychoudhury, R.: Solitonic, quasi-periodic and periodic pattern of electron acoustic waves in quantum plasma. Astrophys. Space Sci. **341**, 567 (2012)
50. Zhen, H., Tian, B., Wang, Y., Zhong, H., Sun, W.: Dynamic behavior of the quantum Zakharov–Kuznetsov equations in dense quantum magnetoplasmas. Phys. Plasma **21**, 012304 (2014)
51. Tribeche, M., Djebarni, L., Amour, R.: Ion-acoustic solitary waves in a plasma with a q-nonextensive electron velocity distribution. Phys. Plasmas **17**, 042114 (2010)
52. Saha, A.: Bifurcation of travelling wave solutions for the generalized KP-MEW equations. Commun. Nonlinear Sci. Numer. Simul. **17**, 3539 (2012)
53. Guckenheimer, J., Holmes, P.J.: Nonlinear Oscillations. Dynamical Systems and Bifurcations of Vector Fields. Springer, New York (1983)
54. Byrd, P.F., Friedman, M.D.: Handbook of Elliptic Integrals for Engineer and Scientists. Springer, New York (1971)

5

Inactivation of *Aspergillus flavus* spores in a sealed package by cold plasma streamers

F. Sohbatzadeh[1,2] · S. Mirzanejhad[1,2] · H. Shokri[3] · M. Nikpour[1]

Abstract The main objective of this study is to investigate the inactivation efficacy of cold streamers in a sealed package on pathogenic fungi *Aspergillus flavus* (*A. flavus*) spores that artificially contaminated pistachio surface. To produce penetrating cold streamers, electric power supply was adapted to deposit adequate power into the package. The plasma streamers were generated by an alternating high voltage with carrier frequency of 12.5 kHz which was suppressed by a modulated pulsed signal at frequency of 110 Hz. The plasma exposition time was varied from 8 to 18 min to show the effect of the plasma treatment on fungal clearance while the electrode and sample remained at room temperature. This proved a positive effect of the cold streamers treatment on fungal clearance. Benefits of deactivation of fungal spores by streamers inside the package include no heating, short treatment time and adaptability to existing processes. Given its ability to ensure the safety and longevity of food products, this technology has great potential for utilization in food packaging and processing industry. In this study, moisture and pH changes of pistachio samples after plasma streamers treatment were also investigated.

Keywords *Aspergillus flavus* spore deactivation · Cold plasma streamers · Dielectric barrier discharge

✉ F. Sohbatzadeh
f.sohbat@umz.ac.ir

1 Atomic and Molecular Physics Department, Faculty of Basic Sciences, University of Mazandaran, Babolsar, Iran

2 Nano and Biotechnology Research Group, Faculty of Basic Sciences, University of Mazandaran, Babolsar, Iran

3 Faculty of Veterinary Medicine, Amol University of Special Modern Technologies, Amol, Iran

Introduction

Aspergillus flavus (*A. flavus*) is a saprotrophic and pathogenic fungus with a cosmopolitan distribution [22]. It is best known for its colonization of cereal grains and legumes. Post-harvest rot typically develops during harvest, storage, and/or transit. *A. flavus* infections can occur while hosts are still in the field (pre-harvest), but often show no symptoms (dormancy) until post-harvest storage and/or transport. In addition to causing pre-harvest and post-harvest infections, many strains produce significant quantities of toxic compounds known as mycotoxins, which are toxic to human and animals [22]. *A. flavus* is also an opportunistic human and animal pathogen causing aspergillosis in immune-compromised individuals [10].

Atmospheric pressure plasmas (APPs) system does not require vacuum systems and are a cost-effective and convenient alternative to low-pressure plasma systems [3]. APPs have the attractive feature of a non-equilibrium property by which hot electrons, cold neutral gas and ions can co-exist in the plasma state [14]. The energetic electrons lead to enhanced generation of reactive radicals and ions through collisions with the background neutral gas, while the gas temperature remains near the room temperature. As a result, APPs do not cause thermal damage to heat-sensitive biological systems such as cells and living tissue [9]. Plasma generators produce ultraviolet photons, charged particles and chemically active species. Dependent upon the working gases, the following species could be produced: negative and neutral atomic oxygen, ozone, singlet oxygen, superoxide, hydroxyl ions, and NO_x compounds. This led to the plasma sterilization technique being used as a simple, fast and toxic-free residue in the study of disinfection. Limitations in conventional sterilization methods have motivated the search for novel sterilization methods. In recent years, plasmas have various promising applications such as treatment of mammalian and cancerous cells [16], sterilization [8], bacteria inactivation [25], surface modification of Raw and Frit glazes [11], silicon rubber surface modification [26], protein destruction [6] blood coagulations [17], treatment of living tissue [7], wound care [21], teeth bleaching [19], and treatment of dental diseases [15].

Gas discharge in a gap between two electrodes covered by dielectric materials can be run in several distinct discharge modes: corona, glow and streamers [12, 18, 23]. In dielectric barrier discharge (DBD) scheme, corona and glow are widely employed in many applications because of its stability at high power. However, streamers are rarely employed for disinfection due to their relatively high temperature and damaging effect on the sample. If one can reduce the streamer temperature, they could be used in sterilization as well. In this regard, we developed a device for cold plasma streamers and showed that the cold streamers seem to be relatively simple and inexpensive as well as their non-toxic nature gives them the potential to be used in deactivation of micro-organisms.

The microdischarge streamers can be characterized as non-equilibrium weakly ionized plasma channels with futures similar to those of transient high-pressure glow discharges. They initiate when the breakdown electric field is reached. The current flow at the streamer location is terminated within a few ns after breakdown thus the temperature of the ions and neutral species remains close to the room temperature. The individual streamer properties depend on the gas properties, the pressure and the gap spacing. For a given industrial application of DBD, the large spacing between the electrodes is favored because of treatment restrictions. The advantage of the current apparatus is treating the large samples close to the room temperature while the common dielectric barrier glow discharges suppress the streamers by reducing the electrode gap distances.

The large gap spacing results in brighter, stronger, thicker and hotter streamers. Therefore, for large gaps, the temperature of the microdischarge streamer tends to grow. However, optimizing large gap spacing for a given application will prevent possible thermal damage of the target.

Elimination of fungus before toxin production must be the actual goal rather than the removal of toxins once produced. It is of great interest to develop novel, practical, and cost-effective methods or processes to reduce or if possible completely eliminate fungus before aflatoxins are proceed during storage. The contaminations of nuts by toxic fungal species, and consequently the presence of aflatoxins are unavoidable [1]. Low-pressure cold plasma (LPCP) using air and sulfur hexafluoride (SF6) was developed and tested for anti-fungal efficacy against *Aspergillus parasiticus* on various nut samples [2]. There are several works on spores deactivation by atmospheric pressure plasma. For example, Iseki et al. [13] used plasma of argon gas, Lim et al. [20] used Ar/O_2 plasma discharge and Uhm et al. [27] utilized argon gas plasma for spore deactivation. Recently, Connolly et al. [5] used Helium/air plasma at atmospheric pressure for antimicrobial efficacy against *E.coli* in a plastic package. However, deactivation of fungal spores inside a sealed package is very important and to our knowledge, it has not been investigated by cold plasma streamers, yet. It is one of the goals of this research to develop a room temperature instrument for *A. flavus* spores deactivation on pistachio surface inside a closed glass package without any feed gas. To lead the goal, we developed cold plasma streamer generator to change neutral air molecules that trapped in the package into reactive neutral species, positive- and negative-charged particles. In this regard, we adapted the electric power supply to produce cold streamers to penetrate into the package and cover the pistachio sample. We examined temperature, color, moisture and pH of raw pistachios before and after plasma exposure, too.

Materials and methods

Experimental setup

The setup of the DBD is mainly based on the principle of a dielectric barrier discharge and contains a high-voltage electrode, a ground electrode and a resistance. The discharge burned between two plane metal electrodes, both covered by plane Pyrex glass with 1 mm thickness. Diameters of circle metal electrodes were 44 mm and dimensions of plane Pyrex glass covering the electrodes were 100 × 75 mm. The distance between electrodes was 8 mm when plasma was used. The sample with its closed package was fixed in the middle of the discharge gap and was directly exposed to the plasma. The scheme of the experimental setup is shown in Fig. 1. The plasma was generated in open air by an alternating high voltage with carrier frequency of 12.5 kHz which was suppressed by a modulated pulsed signal at frequency of 110 Hz. Duty cycle of the applied voltage was 8.6 %. The electric power consumption was 2.49 W/cm^3 during the treatments. The full width at half maximum of each modulated signal pulse was 0.25 ms. The optical emission spectroscopy of the dielectric barrier discharge atmospheric cold plasma was carried out using a compact spectrometer (Solar Laser Systems, S-100) that records extremely wide optical range—from 190 to 1100 nm—with average resolution of about 1 nm. Figure 1 shows the schematic of the experimental setup to measure the optical emission spectra, the current and discharge voltage. An optical fiber was inserted in the

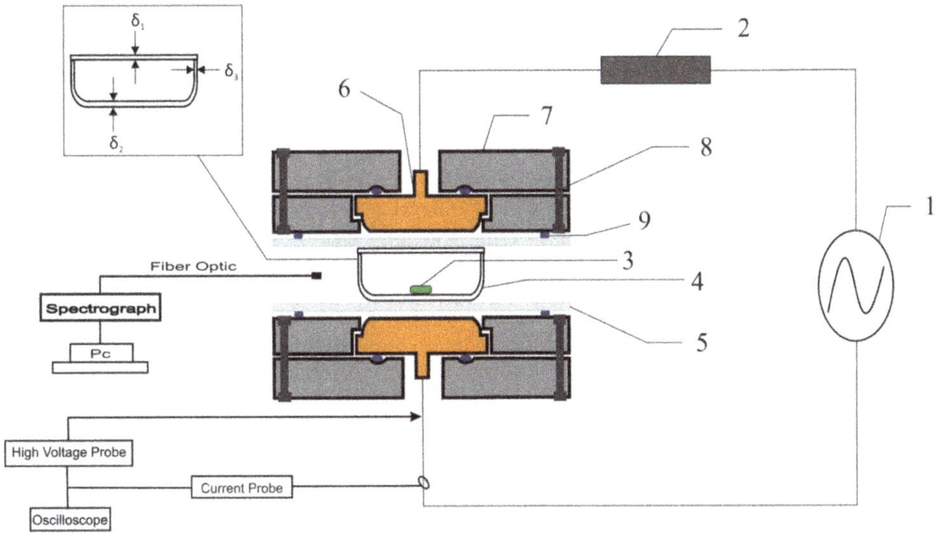

Fig. 1 Schematic setup of plasma streamer sterilization unit and the optical emission spectra measurements in the configuration. *1* Modulated high-voltage AC power supply, *2* resistance, *3* pistachio sample, *4* closed package, *5* dielectric, *6* electrode, *7* plexiglass sheet, *8* screw, *9* O-ring, $\delta_1 = \delta_2 = 2$ mm, $\delta_3 = 1$ mm

gap between two electrodes. The distance of the fiber end to the electrodes was 2 cm.

Voltage and current measurements were performed using a high-voltage probe (Tektronix P6015A) and a current monitor (Pearson 4100) with a digital oscilloscope (Tektronix DPO2012) (Fig. 1). Ozone concentration was measured outside the package by an ozone sensor (A-21ZX). It was about 8.5 ± 0.01 ppm.

Fungal strain

A. flavus (AF-1329) strain was chosen in this study. It was obtained from Fungal Collection of Faculty of Veterinary Medicine, Amol University of Special Modern Technologies, Amol, Iran.

Mycological examination before plasma exposure

Fungal strain was cultured on Sabouraud glucose agar (*Merck Co.,Darmstadt, Germany*) and incubated at 28 °C for 7 days. Conidia were harvested from 7-day-old cultures by pouring a sterile 0.1 % aqueous solution of Tween 80 onto the culture plates and scraping the plate surface with a bent glass rod to facilitate the release of conidia. The number of conidia in the suspension was adjusted to approximately 6×10^7 conidia/ml using a haemocytometer slide. Two microliters of the conidia suspension was inoculated on pistachio samples. The samples were dried under a laminar hood within an hour.

Plasma streamer treatment on contaminated pistachio samples

In this study, contaminated pistachio samples were prepared in closed package. The samples were analyzed before treatment. The samples were placed between the electrodes and then exposed to the plasma at 8, 10, 12, 14, 16 and 18 min. After plasma exposure, the treated samples were transferred for fungal analysis, each treatment was repeated thrice.

Mycological examination after plasma streamer exposure

Two hours after exposure, the dilution rates including 1:50, 1:500, 1:5000 and 1:50,000 were prepared in tubes containing 0.1 % peptone water solution. From each dilution, 100 µl of samples was spread on Sabouraud glucose agar, incubated at 28 °C and read visually after 7 days of incubation. The experiments were repeated three times. The number of *A. flavus* colonies was reported in each sample as colony forming units (CFUs).

$$10 \times \text{reverse dilution} \times \text{the number of fungal colonies} = \text{CFU/sample}$$

Measurement of pistachio moisture and pH

To investigate the effects of air plasma on moisture and pH of pistachio samples, we compared moisture and pH of uncontaminated pistachio samples with or without air plasma treatment. Moisture and pH analysis tests were carried out at two points: 0 and 18 min of plasma treatment.

Moisture of the samples was determined using the internationally approved method AACC 44-14A. This method determines moisture content in grain using a simple and acceptable formula: moisture content $= \frac{\text{mass change}}{\text{initial mass}} \times 100$. In this method, an oven is used. At first, powder of the sample is prepared, then 5.00 g of the pistachio powder is heated at 130 °C for 90 min on the oven. The net change in the mass is the moisture content of the sample. The experiment is repeated 5 times for each sample. To measure the sample pH, Iranian national standards number 37 was used. In this method, 10 grams of the pistachio powder is mixed with 100 CC deionized water, after settling the powder, the pH of the solution was measured by a pH meter.

Discussion

Non-thermal streamers initiated from the upper electrode, diffused into the sealed package and whelmed the pistachio sample, then finished at the bottom electrode. Modulated high voltage adapted adequate electric power transfer to the package. Therefore, negligible heat produced inside the package keeps the sample at room temperature during treatment. Since, the streamers are moderately high-density plasma and are rich in cold-reactive species, high-energy electrons and cold ions, the spore deactivation efficiency by our device seems to be satisfactory. On the other hand, modulated high voltage with low enough repetition rate let the streamers to be cold. Although the streamers are seen continuously, they actually are discrete in time. In this work, the duration time of each streamer is in the order of 10 μs as measured by current spikes using oscilloscope. As illustrated in Fig. 1, the electric field required for the electric discharge inside the package was about 1 kV/mm. So, according to the dielectric envelope of the package, electric discharge takes place due to charge accumulation on the dielectric surfaces. As shown in Fig. 2b, this process led to ignition of the trapped air molecules inside the package, in which cold streamers smudged the pistachio.

Surface removal of *A. flavus* from pistachio

Air plasma effect on *A. flavus* elimination was investigated at 8–18-min treatment periods. Figure 3 shows the inactivation kinetics of plasma gases of air in the case of fungal spores. The initial mold viable count on PDA plates was 1.2×10^5 CFU/sample in the contaminated control nut samples. Air plasma treatment at 8–16 min resulted about

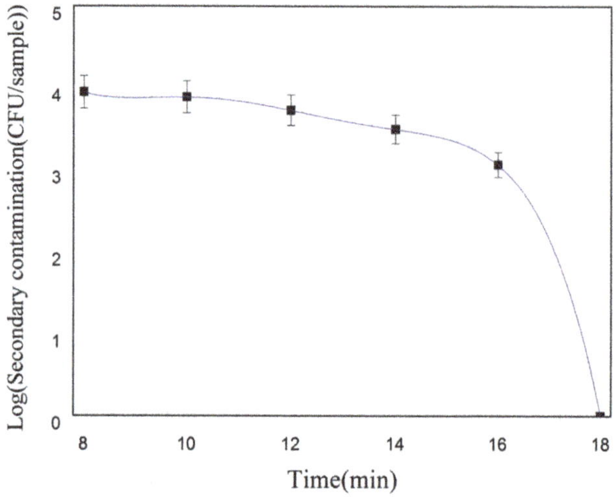

Fig. 3 Reduction effect of cold plasma streamers on *Aspergillus flavus*-contaminated pistachios

Table 1 The effect of cold plasma streamers on moisture and pH of pistachio samples

Sample status	Moisture (%)	pH
Without plasma treatment	3.73 ± 0.01	6.13 ± 0.01
Streamer treated (18 min)	2.62 ± 0.01	6.23 ± 0.01

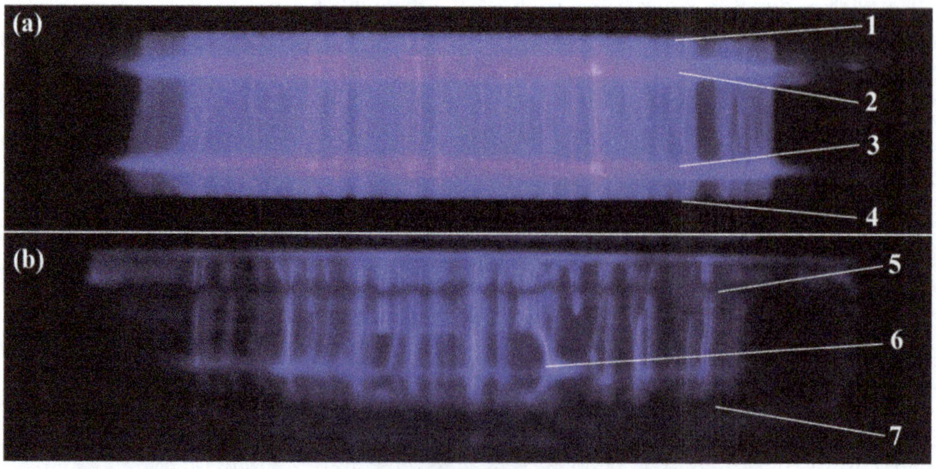

Fig. 2 **a**. Cold streamers before inserting the package into the plasma gap, *1* upper electrode, *2* upper dielectric barrier, *3* lower dielectric barrier, *4* lower electrode. **b** Cold streamers diffused in the closed package and whelmed the pistachio sample *5* upper sheet of package, *6* pistachio sample, *7* lower sheet of package

1-log reduction with a lag phase of *A. flavus* spores population. Figure 3 summarizes the results observed for the microbial-contaminated pistachio exposed to air plasma streamers by varying the plasma exposition time. It can be seen that the number of survival spores decreases with increase in the treatment time. As illustrated in Fig. 3, fungal contamination was completely removed after 18 min of plasma exposure.

Analysis of pH and moisture

No considerable changes in pH and moisture were observed after plasma treatment. Results showed that the plasma treatment of 18 min increased pH from 6.13 to 6.23 and decreased moisture from 3.73 to 2.62 % (Table 1). Figures 4 and 5 show the images of untreated and treated pistachio samples, respectively. It is seen that the treated sample is drier than the untreated one.

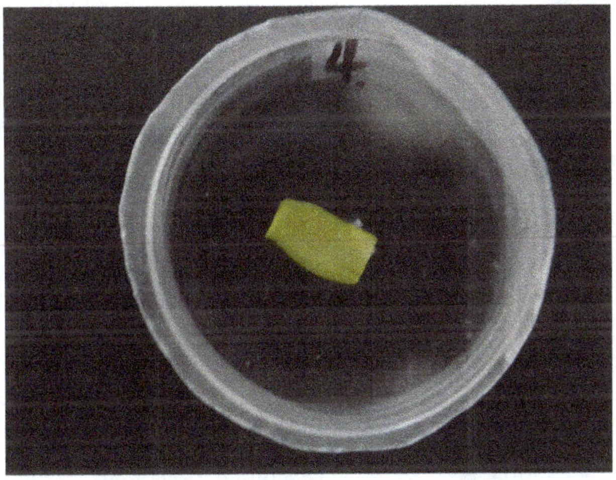

Fig. 4 Image of untreated pistachio cut inside a sealed package

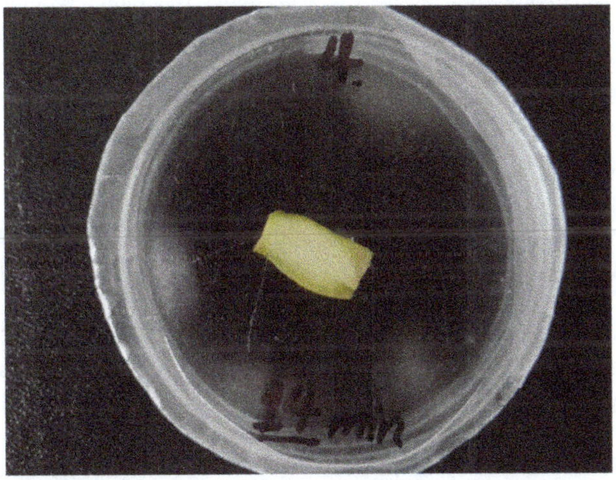

Fig. 5 Image of treated pistachio cut inside a sealed package for 14-min treatment by cold plasma streamers

Optical emission spectroscopy and V-I characteristics

Optical emission spectroscopy was used to measure the optical emission of the plasma streamers in air at atmospheric pressure. Emission spectrum of the DBD at 10.7 kV over the range of 200–1000 nm is presented in Fig. 6. The major peaks of active oxygen and nitrogen could be identified on light emission spectra. The reactive species can contribute to higher sterilization efficiency. On the other hand, the measurements revealed that the device produces less than 10 ppm ozone during the plasma treatments.

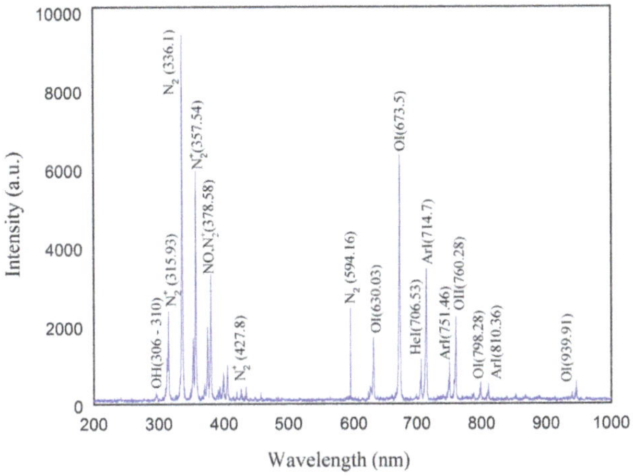

Fig. 6 Optical emission spectra of the cold plasma streamers

Table 2 Several important chemical reactive products in cold plasma streamers (Becker et al., 2005; Raizer and Allen 1997)

Reactive species	Reaction
Atomic oxygen	$O_2 + e \rightarrow O + O + e$
	$O_2 + e \rightarrow O + O^- + 2e$
	$h\upsilon(\lambda = 128nm) + O_2 \rightarrow 2O$
	$e + O_2 + O \rightarrow O_2^- + O$
	$e + O_2 \rightarrow e + O + O(^1D)$
Ozone	$O + O_2 + O_2 \rightarrow O_3 + O_2$
	$O + O_2 + M \rightarrow O_3 + (O_2 \text{ or } O)$
	$O + O_2 + N_2 \rightarrow O_3 + N_2$
Singlet oxygen	$O + O_3 \rightarrow O_2(^1\Delta_g) + O_2$
Oxygen negative ions	$e + O_2 + N_2 \rightarrow O_2^- + N_2$
	$e + O_2 + O_2 \rightarrow O_2^- + O_2$
	$O_2 + e \rightarrow O + O^- + 2e$
Hydroxyl	$O(^1D) + H_2O \rightarrow 2OH$
	$O + H_2O \rightarrow OH + O_2$
	$H_2O + e \rightarrow OH + H + e$

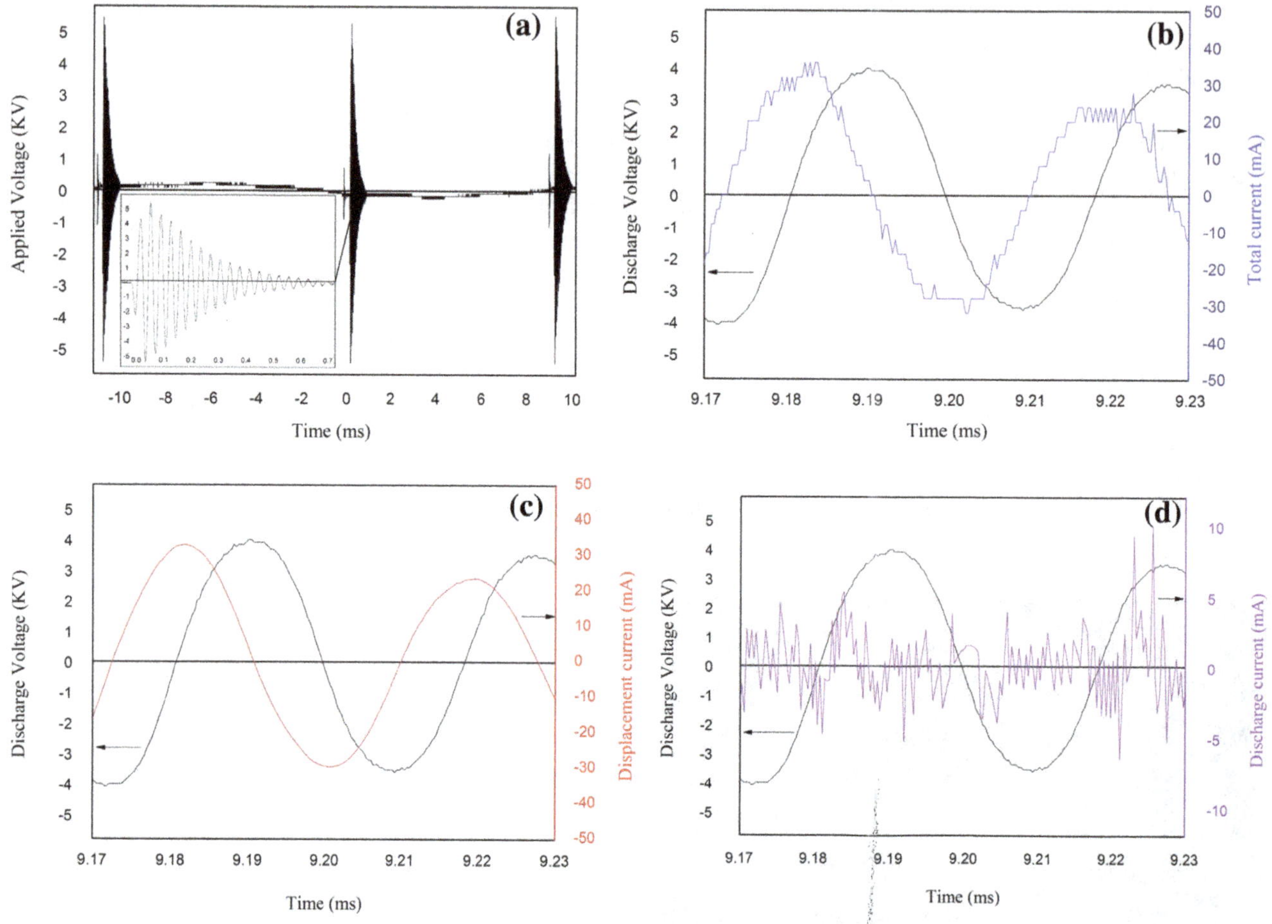

Fig. 7 Modulated voltage and current waveforms applied to the electrodes of the streamer generator, **a** discharge voltage. **b** A snap shot of the voltage and total current during 60 μs, **c** a snap shot of the displacement current evolution during 60 μs, **d** discharge current spikes versus time

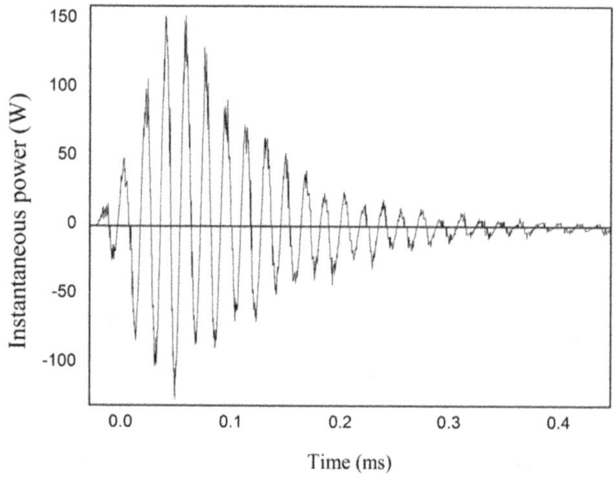

Fig. 8 Instantaneous power

Among the other species of plasma streamers, energetic electrons, atomic oxygen, singlet oxygen, ozone, negative oxygen molecule, and hydroxyl played the major role in deactivation process. The generation of atomic oxygen, singlet oxygen, ozone and hydroxyl takes place by the following reactions as indicated in Table 2 [3, 24].

Ozone is unstable, will degrade into molecular oxygen and monatomic oxygen, and the latter is another oxidant and some of the monatomic oxygen will combine with molecular oxygen, so they leave no harmful by-products at the end. Ozone, as a kind of gas, can pervade the whole space and leaves no dead angle. Furthermore, OH($A^2\sum^+ \rightarrow X^2\prod$) at 308 nm is the other discharge product present as the result of water vapor dissociation.

Potential targets in *fungal* cells include the cell wall, the plasma membrane, ribosomes, RNA, DNA, and structural and functional proteins. Cell destruction by plasma occurs if the levels of exposure exceed the capacity of the cell defense systems by affecting the many macromolecules within the cell including DNA, proteins and lipids. One of the most important mechanisms is DNA lesions by induction of breaks in DNA. Another mechanism is lipid peroxidation. Other mechanism is formation of the superoxide anion (O_2^-) in the oxidation/reduction (redox) reactions of mitochondria. Furthermore, the presence of humidity leads to the generation of OH, OH^+, and OH^-, which also have a noticeable disinfection effect.

Formation of such intense reactive species in plasma streamers is beneficial for spore deactivation applications. On the other hand, the energetic electrons inside a streamer play a key role in deactivation. They collide with the sample surface and damage the fungal cell walls. It is worth to note that the ratio of electrons to the neutral particles is about 10^{-6}. Such a low ionization degree together with modulated electric power supply guarantied deactivation of *A.flavus* spores on pistachio surface at room temperature by plasma streamers.

Figure 5 shows the typical voltage and current waveforms. It is seen that the duty cycle of the applied voltage is low enough for preventing heat production. In other words, it was 8.6 %. Therefore, the temperature of the electrodes and gas gap remained almost unchanged during 18-min plasma exposition. For clearance, a snapshot of carrier voltage wave and total current are illustrated in Fig. 7b. The DBD current includes displacement and discharge current. Figure 7c shows the displacement current after signal processing. Chaotic behavior of the streamers involved short current spikes in micro-seconds duration, randomly, which is shown in Fig. 7d; instantaneous power is shown in Fig. 8, too. The current spikes and instantaneous electric power are only shown during each modulated signal in Figs. 7 and 8. To identify averaged deposited electrical power to the plasma, we calculated mean power, P, using the following relation between instant voltage and current [4]:

$$P = \frac{\int_0^{\frac{T}{2}} U(t)I(t)\mathrm{d}(t)}{\int_0^{\frac{T}{2}} \mathrm{d}t} \quad (1)$$

The electric power consumption obtained was 2.49 W/cm^3 during the treatments.

Conclusions

The results suggested that the cold plasma streamers of air gas could be efficient in the disinfection of *A. flavus* spores in solid foods. The plasma streamers had no significant effect on the temperature, pH, moisture and color of the sample. The plasma streamer application with specifically chosen electric power supply and gases may offer a novel and efficient method for disinfections of dry food surfaces at atmospheric pressure. The research instruments developed in this work, guarantied deactivation of *A. flavus* spores on pistachio surface inside a closed package at room temperature. The disinfections of heat-sensitive food surfaces with plasma streamers open up an alternative novel route for the development of advanced technologies for the elimination of fungal contamination, provided that an ideal gaseous compound together with a proper electric power supply is employed. The modulation that we used in high-voltage wave form made the power supply to deposit less electric power into the glass package while deactivation of the *A. flavus* spores took place by reactive species, ions and energetic electrons.

References

1. Arrus, K., Blank, G., Abramson, D., Clear, R., Holley, R.A.: Aflatoxin production by *Aspergillus flavus* in Brazil nuts. J. Stored Prod. Res. **41**(5), 513–527 (2005)
2. Basaran, P., Basaran-Akgul, N., Oksuz, L.: Elimination of Aspergillus parasiticus from nut surface with low pressure cold plasma (LPCP) treatment. Food Microbiol. **25**(4), 626–632 (2008)
3. Becker, K.H., Kogelschatz, U., Schoenbach, K.H., Barker, R.J. (eds.): Non-Equilibrium Air Plasmas at Atmospheric Pressure. CRC Press (2004)
4. Cheng, C., et al.: Development of a new atmospheric pressure cold plasma jet generator and application in sterilization. Chin. Phys. **15**(7), 1544 (2006)
5. Connolly, J., Valdramidis, V.P., Byrne, E., Karatzas, K.A., Cullen, P.J., Keener, K.M., Mosnier, J.P.: Characterization and antimicrobial efficacy against *E. coli* of a helium/air plasma at atmospheric pressure created in a plastic package. J. Phys. D Appl. Phys. **46**, 035401 (2013)
6. Deng, X.T., Shi, J.J., Kong, M.G.: Protein destruction by a helium atmospheric pressure glow discharge: capability and mechanisms. J. Appl. Phys. **101**(7), 074701 (2007)
7. Dobrynin, D., Fridman, G., Friedman, G., Fridman, A.: Physical and biological mechanisms of direct plasma interaction with living tissue. New J. Phys. **11**(11), 115020 (2009)
8. Ehlbeck, J., Schnabel, U., Polak, M., Winter, J., Von Woedtke, T., Brandenburg, R., Weltmann, K.D.: Low temperature atmospheric pressure plasma sources for microbial decontamination. J. Phys. D Appl. Phys. **44**(1), 013002 (2011)
9. Fridman, G., Brooks, A.D., Balasubramanian, M., Fridman, A., Gutsol, A., Vasilets, V.N., Friedman, G.: Comparison of direct and indirect effects of non-thermal atmospheric-pressure plasma on bacteria. Plasma Process. Polym. **4**(4), 370–375 (2007)
10. Hedayati, M.T., Pasqualotto, A.C., Warn, P.A., Bowyer, P., Denning, D.W.: Aspergillus flavus: human pathogen, allergen and mycotoxin producer. Microbiology **153**(6), 1677–1692 (2007)
11. Ghasemi, M., Sohbatzadeh, F., Mirzanejhad, S.: Surface modification of Raw and Frit glazes by non-thermal helium plasma jet. J. Theor. Appl. Phys. **9**(3), 177–183 (2015)
12. Gherardi, N., Gouda, G., Gat, E., Ricard, A., Massines, F.: Transition from glow silent discharge to micro-discharges in nitrogen gas. Plasma Sources Sci. Technol. **9**(3), 340 (2000)
13. Iseki, S., Ohta, T., Aomatsu, A., Ito, M., Kano, H., Higashijima, Y., Hori, M.: Rapid inactivation of *Penicillium digitatum* spores using high-density nonequilibrium atmospheric pressure plasma. Appl. Phys. Lett. **96**(15), 153704 (2010)
14. Iza, F., Kim, G.J., Lee, S.M., Lee, J.K., Walsh, J.L., Zhang, Y.T., Kong, M.G.: Microplasmas: sources, particle kinetics, and biomedical applications. Plasma Process. Polym. **5**(4), 322–344 (2008)

15. Jiang, C., Chen, M.T., Gorur, A., Schaudinn, C., Jaramillo, D.E., Costerton, J.W., Gundersen, M.A.: Nanosecond pulsed plasma dental probe. Plasma Process. Polym. **6**(8), 479–483 (2009)
16. Kim, W., Woo, K.C., Kim, G.C., Kim, K.T.: Nonthermal-plasma-mediated animal cell death. J. Phys. D Appl. Phys. **44**(1), 013001 (2011)
17. Kuo, S.P., Tarasenko, O., Chang, J., Popovic, S., Chen, C.Y., Fan, H.W., Nikolic, M.: Contribution of a portable air plasma torch to rapid blood coagulation as a method of preventing bleeding. New J. Phys. **11**(11), 115016 (2009)
18. Laroussi, M., Kong, M.G., Morfill, G. (eds.): Plasma Medicine: Applications of Low-Temperature Gas Plasmas in Medicine and Biology. Cambridge University Press, Cambridge (2012)
19. Lee, H.W., Nam, S.H., Mohamed, A.A.H., Kim, G.C., Lee, J.K.: Atmospheric pressure plasma jet composed of three electrodes: application to tooth bleaching. Plasma Process. Polym. **7**(3–4), 274–280 (2010)
20. Lim, J.P., Uhm, H.S., Li, S.Z.: Influence of oxygen in atmospheric-pressure argon plasma jet on sterilization of Bacillus atrophaeous spores. Phys. Plasmas **14**, 093504 (2007)
21. Lloyd, G., Friedman, G., Jafri, S., Schultz, G., Fridman, A., Harding, K.: Gas plasma: medical uses and developments in wound care. Plasma Process. Polym. **7**(3–4), 194–211 (2010)
22. Machida, M., Gomi, K. (eds.): Aspergillus: Molecular Biology and Genomics. Caister Academic Press, NorfolK, UK (2010)
23. Rajasekaran, P., Mertmann, P., Bibinov, N., Wandke, D., Viöl, W., Awakowicz, P.: Filamentary and homogeneous modes of dielectric barrier discharge (DBD) in air: investigation through plasma characterization and simulation of surface irradiation. Plasma Process. Polym. **7**(8), 665–675 (2010)
24. Raizer, Y.P.: Gas Discharge Physics, 2nd edn. Springer, Berlin (1997)
25. Sohbatzadeh, F., Colagar, A.H., Mirzanejhad, S., Mahmodi, S.: *E. coli*, *P. aeruginosa*, and *B. cereus* bacteria sterilization using afterglow of non-thermal plasma at atmospheric pressure. Appl. Biochem. Biotechnol. **160**(7), 1978–1984 (2010)
26. Sohbatzadeh, F., Mirzanejhad, S., Ghasemi, M., Talebzadeh, M.: Characterization of a non-thermal plasma torch in streamer mode and its effect on polyvinyl chloride and silicone rubber surfaces. J. Electrostat. **71**(5), 875–881 (2013)
27. Uhm, H.S., Lim, J.P., Li, S.Z.: Sterilization of bacterial endospores by an atmospheric-pressure argon plasma jet. Appl. Phys. Lett. **90**(26), 261501 (2007)

6

Low-frequency shock waves in a magnetized superthermal dusty plasma

B. S. Chahal[1] · Yashika Ghai[1] · N. S. Saini[1]

Abstract The characteristics of low-frequency shocks in a magnetized dusty plasma comprising of negatively charged dust fluid, kappa-distributed electrons and ions have been investigated. Using the reductive perturbation method, the nonlinear Korteweg de–Vries–Burgers (KdV–B) equation which governs the dynamics of the dust acoustic (DA) shock waves is derived. The characteristics of shock structures are studied under the influence of various plasma parameters, viz. superthermality of ions, magnetic field, electron-to-dust-density ratio, kinematic viscosity, ion-to-electron-temperature ratio and obliqueness. The combined effects of these physical parameters significantly influence the characteristics of DA shock structures. It is observed that only negative potential shocks exist in a plasma environment comprising of dust fluid and superthermal electrons and ions such as that of Saturn's magnetosphere.

Keywords Dust acoustic · Shock waves · KdV–Burgers · Superthermal

✉ N. S. Saini
nssaini@yahoo.com

B. S. Chahal
chahal_bs@rediffmail.com

Yashika Ghai
yashu.gh92@gmail.com

1 Department of Physics, Guru Nanak Dev University, Amritsar, India

Introduction

The study of propagation properties of nonlinear solitary waves in different plasma environments (e.g., magnetized/unmagnetized multicomponent dusty/complex plasmas) is an important area of research in modern plasma physics. For the last many years, various types of nonlinear structures such as solitary waves, shocks, double layers and vortices have been studied theoretically as well as experimentally under different plasma conditions. When dispersion of the medium is completely balanced by nonlinearity, the dissipative effects can be negligible and solitary waves are formed. However, when the dissipative effects are more dominant than dispersion of medium, this leads to the formation of shock structures. The dissipation can be caused by fluid viscosity, inter-particle collision, Landau damping or can be due to nonadiabatic dust charge fluctuations. The dynamics of shock waves in a magnetized plasma are governed by KdV–Burgers equation whose solution has the form of shock structures. The shock structures play very important role in understanding astrophysical plasmas, e.g., supernova explosions [1], cosmic ray generation [2], bow shock region [3] and nonlinear dynamics of solar wind [4].

Over the last three decades, research activities in dusty plasmas have been witnessed in a number of studies with different velocity distributions of the charged particles [5, 6]. This is not only due to the omnipresence of dust but may also be due to the broad range applications of dusty plasmas in understanding the different collective processes in astrophysical and space environments [7–9], as well as in laboratory experiments [10]. The presence of dust particles (micron to submicron sized) is common in different plasma environments, e.g., magnetosphere of Earth, solar winds, ionosphere and planetary rings of Saturn [9, 11].

Charging of dust (positive or negative) depends upon the charging processes (e.g., photoionization, secondary electron emission, thermionic emission or by electron ion bombardments on surface of dust particles). The presence of charged dust in a plasma generates new eigenmodes such as dust acoustic (DA) waves [12], dust ion-acoustic DIA waves [13, 14], dust acoustic shock waves [15] and dust lattice waves [16].

A large number of investigations have been reported for the study of shock structures in plasmas both theoretically [15, 17–22] and experimentally [23–26]. Dust acoustic shock waves for two temperature ions in unmagnetized dusty plasma were investigated by Zhang and Wang [17]. Shock waves in an unmagnetized dusty plasma consisting of charged adiabatic dust fluid and Boltzmann-distributed electrons and ions have been studied by Rahman and Mamun [18] and only negative potential shocks were observed. It was also noticed that the effect of dust fluid temperature modifies the basic properties of shock structures. El-Hanbaly et al. [21] derived the modified KdV–Burgers equation in a dusty plasma consisting of Boltzmann-distributed electrons and trapped ions obeying vortex-like distribution and studied the shock solution in a dissipative space plasma. Recently, Tansim et al. [22] have studied dust acoustic shock waves in a four-component unmagnetized dusty plasma consisting of cold dust fluid, Boltzmann-distributed electrons and two temperature ions obeying q-nonextensive and Cairns distribution respectively. They have studied the combined effects of both nonextensive and nonthermal distribution of ions on the properties of dust acoustic shocks.

A number of observations and in situ measurements have confirmed the presence of excess superthermal population of charged particles in astrophysical and space plasmas, e.g., in the magnetospheres of Earth, Mercury, Saturn, Uranus and the solar wind [27–30]. Superthermal plasma behavior was also observed in various experimental plasma contexts, such as laser–matter interactions or plasma turbulence [31]. Such plasmas are best modeled by generalized Lorentzian (kappa) distribution rather than Maxwellian. The presence of excess superthermal charged particles population due to velocity space diffusion may lead to inverse power law distribution at a velocity much higher than the electron thermal speeds [32–36].

We shall use a three-dimensional κ-distribution function as [37]

$$f_k(v) = \frac{n_0}{(\pi\kappa\theta^2)^{3/2}} \frac{\Gamma(\kappa+1)}{\Gamma(\kappa-\frac{1}{2})} \left(1 + \frac{v^2}{\kappa\theta^2}\right)^{-(\kappa+1)}. \tag{1}$$

The effective thermal speed $\theta = (\frac{\kappa-3/2}{\kappa})^{1/2} v_{\text{th}}$ is related to thermal velocity $v_{\text{th}} = (2K_B T/m)^{1/2}$, m is species mass and T is the characteristic kinetic temperature. The spectral index κ is a measure of the slope of the energy spectrum of the superthermal particles forming the tail of the velocity distribution function. A small value of κ means particles are more superthermal and energy spectrum is hard. In the limit $\kappa \to \infty$, the kappa distribution approaches the Maxwellian limit. By integrating the distribution function given by Eq. (1) in the presence of electrostatic potential ϕ, one can obtain the number density of electrons/ions.

A number of investigations of solitary structures and shock waves in plasma with kappa-distributed electrons/ions have been reported by numerous authors in different kinds of plasma systems. Shahmansouri and Astaraki [38] investigated the properties of ion-acoustic solitary structures in a plasma system containing inertial ions and superthermal electrons and positrons which are assumed to be inertialess. Kundu et al. [39] studied the shock waves in unmagnetized dusty plasma where electrons are superthermally distributed. Burgers equation was derived using standard reductive perturbation method and it was concluded that shock waves are modified by the presence of positively and negatively charged dust particles and superthermal nature of kappa-distributed electrons/ions. Shahmansouri and Alinejad [40] studied the effect of superthermality of ions/electrons and dust charge fluctuation on dust acoustic shock waves.

It is a well-known fact that the propagation properties of electrostatic solitary structures can be modified under the influence of external magnetic field. Since most of the space and astrophysical plasmas are permeated by magnetic field, the effect of external magnetic field on electrostatic solitary waves in different kinds of plasma environments has been studied by various authors [41–53]. The obliquely propagating ion-acoustic (*IA*) solitons in a magnetized and weakly relativistic warm plasma have been investigated by deriving KdV equation and it was inferred that the soliton energy is lowered by stronger magnetic field and the solitons become narrower [41]. Furthermore, the effect of ion temperature and dust charging on characteristics of ion-acoustic (*IA*) solitons in a magnetized inhomogenous plasma has been reported by Malik [43] and Kumar et al. [44] respectively. The propagation properties of compressive solitons were analyzed by Malik and Malik [47] in a dusty plasma comprising of electrons, positrons and dust grains of either positive or negative charge. They derived the KdV equation and reported that the amplitude of a compressive soliton remains larger in case of positively charged dust grains as compared to dust grains having negative charge. Tomar et al. [48] studied the evolution and reflection of ion-acoustic solitary waves in an inhomogenous magnetized dusty plasma consisting of ions, electrons having two temperature and dust grains with varying charge. The evolution of solitons has also been studied by Tomar et al. [49] by deriving a modified KdV equation in a plasma having ions, two temperature electrons and negatively charged dust grains. Dust acoustic shock waves in magnetized homogenous dusty plasma have been investigated by Zhang and Xue [50]. In the past, ion-acoustic shock waves in magnetized plasma have been studied by Bains and Tribeche [51] and the effects of nonextensivity of electrons on the characteristics of shock structures have been analyzed. Shahmansouri and Mamun [52] studied the dust acoustic shock waves in magnetized nonthermal dusty plasma. Zaghbeer et al. [53] reported the study of dust acoustic shocks in magnetized dusty plasma with nonextensive electrons and ions. The KdV–Burgers equation was derived by using reductive perturbation method and found that DA shocks are significantly modified by the combined effects of dust fluid viscosity, external magnetic field and obliqueness.

Motivation of the present investigation is to study the combined influence of superthermality of charged particles, strength of magnetic field and other plasma parameters on the characteristics of shock structures in a homogenous dusty plasma. To the best of our knowledge, the properties of dust acoustic shocks in a magnetized dusty plasma with superthermal electrons and ions have not been investigated so far. In this study, our target is to investigate the effects of dust concentration, kinematic viscosity, superthermality of electrons and ions and the strength of ambient magnetic field on the characteristics of dust acoustic shocks. In the framework of κ-distribution, we shall develop a comprehensive formulation for three-dimensional dust acoustic shock waves projecting in a magnetized dusty plasma consisting of dust fluid, superthermal electrons and ions. The dust fluid viscosity will be responsible for dissipative effects. A nonlinear evolution equation known as KdV–Burgers equation is derived by using reductive perturbation method, and we shall analyze that the influence of various physical parameters modifies the width and amplitude of DA shocks due to dependence of nonlinear, dispersion and dissipative terms on such parameters.

The paper is structured as follows: In Sect. 2, the fluid model equations are described for DA shock waves. In Sect. 3, employing a reductive perturbation method, KdV–Burgers equation is derived and its solution is discussed. Sections 4 and 5 are devoted to parametric analysis and concluding remarks of present study respectively.

The fluid equations

We consider a plasma system consisting of negatively charged massive dust as inertial fluid, inertialess ions and electrons are considered to follow superthermal distribution. The dynamics of dust acoustic waves can be described by following set of normalized continuity, momentum and Poisson's equations as

$$\frac{\partial n_d}{\partial t} + \nabla \cdot (n_d \mathbf{u_d}) = 0, \tag{2}$$

$$\frac{\partial \mathbf{u_d}}{\partial t} + (\mathbf{u_d} \cdot \nabla)\mathbf{u_d} = \nabla\phi - (\mathbf{u_d} \times \mathbf{\Omega_d}) + \eta \nabla^2 \mathbf{u_d}, \tag{3}$$

$$\nabla^2 \phi = (n_d + \mu_e n_e - \mu_i n_i). \tag{4}$$

The equilibrium condition is $n_{i0} = n_{e0} + Z_d n_{d0}$, which reduces to $\mu_i = \mu_e + 1$, where $\mu_e = \frac{n_{e0}}{Z_{d0} n_{d0}}$ and $\mu_i = \frac{n_{i0}}{Z_{d0} n_{d0}}$. $\mathbf{u_d} = u\hat{x} + v\hat{y} + w\hat{z}$ where, u, v and w represent the dust particles velocities in x, y and z directions normalized by the dust acoustic speed $C_d = (Z_d T_i / m_d)^{1/2}$, ϕ represents the electrostatic potential normalized by T_i/e. The space and time coordinates are normalized by the dust Debye length $\lambda_{Dd} = (T_i / 4\pi e^2 Z_d n_{d0})^{1/2}$ and the inverse of the dust plasma frequency $\omega_{pd}^{-1} = (\frac{4\pi e^2 Z_d^2 n_{d0}}{m_d})^{-1/2}$, respectively. η is dust kinematic viscosity normalized by $\lambda_{Dd}^2 n_{d0} m_d \omega_{pd}$. The magnetic field is assumed to be along z-direction, i.e., $\mathbf{B} = B_0 \hat{\mathbf{k}}$ and $\mathbf{\Omega_d} = \frac{\omega_{cd}}{\omega_{pd}}$, where $\omega_{cd} = \frac{Z_d e \mathbf{B}}{m_d}$ is dust cyclotron frequency. The ion and electron densities obtained by integrating Eq. (1) are represented in normalized form as

$$\begin{aligned} n_e &= \left(1 - \frac{\sigma_i \phi}{\kappa_e - 3/2}\right)^{-\kappa_e + 1/2} \\ n_i &= \left(1 + \frac{\phi}{\kappa_i - 3/2}\right)^{-\kappa_i + 1/2}, \end{aligned} \tag{5}$$

where $\sigma_i = T_i/T_e$. The parameters κ_e and κ_i are significant in the present case and hold a physical meaning when $\kappa_s > 3/2$, where s=i, e for ions and electrons, respectively. In the limit $\kappa_e, \kappa_i \to \infty$, Eq. (5) agrees with the Maxwellian case expressions. The power law in Eq. (5) makes the Poisson's equation (4) intractable analytically. This difficulty is removed by assuming that any disturbance of the electrostatic potential is small (i.e., $\phi << 1$) in this regime, a Taylor expression of Eq. (4) around this parameter can be preferred and densities of electrons and ions are expressed as

$$n_j = 1 \pm c_{j1}\phi + c_{j2}\phi^2 \pm O(\phi^3) + \cdots, \tag{6}$$

where $j = e, i$ and lower sign (negative sign) is for ions. Truncating the expression at third order, Eq. (4) can be written as

$$\frac{\partial^2 \phi}{\partial x^2} + \frac{\partial^2 \phi}{\partial y^2} + \frac{\partial^2 \phi}{\partial z^2} = n_d - 1 + C_1\phi + C_2\phi^2 + C_3\phi^3, \tag{7}$$

the coefficients C_1 and C_2 are defined as

$$C_1 = \mu_e c_{e1} + \mu_i c_{i1}, \tag{8}$$

$$C_2 = \mu_e c_{e2} - \mu_i c_{i2}, \tag{9}$$

$$C_3 = \mu_e c_{e3} + \mu_i c_{i3}, \tag{10}$$

where $c_{e1} = \frac{\sigma_i(\kappa_e - 1/2)}{(\kappa_e - 3/2)}$, $c_{i1} = \frac{(\kappa_i - 1/2)}{(\kappa_i - 3/2)}$, $c_{e2} = \frac{\sigma_i^2(\kappa_e^2 - 1/4)}{2(\kappa_e - 3/2)^2}$, $c_{i2} = \frac{(\kappa_i^2 - 1/4)}{2(\kappa_i - 3/2)^2}$, $c_{e3} = \frac{\sigma_i{}^3(\kappa_e^2 - 1/4)(\kappa_e + 3/2)}{6(\kappa_e - 3/2)^3}$, $c_{i3} = \frac{(\kappa_i^2 - 1/4)(\kappa_i + 3/2)}{6(\kappa_i - 3/2)^3}$.

Derivation of KdV–Burgers equation

To investigate the dynamics of *DA* shock waves in a magnetized superthermal dusty plasma, we shall derive the KdV–B equation in the frame work of reductive perturbation method [54]. The stretched coordinates are introduced in the following form:

$$\xi = \epsilon^{1/2}(l_x x + l_y y + l_z z - V_0 t), \qquad \tau = \epsilon^{3/2} t \tag{11}$$

where ϵ, is a small parameter characterizing the nonlinearity. l_x, l_y and l_z are respectively the direction cosines of wave vector k along x-, y- and z-axis so that $l_x^2 + l_y^2 + l_z^2 = 1$. V_0 is the *DA* phase speed normalized by C_d. In a weak damping situation, the dust ion kinematic viscosity can be considered small but finite. This leads us to assume that

$$\eta \approx \epsilon^{1/2} \eta_0, \tag{12}$$

where η_0 is a finite parameter. The expansion of dependent variables n_d, u, v, w and ϕ around their equilibrium values in a power series is given as [55, 56]:

$$\begin{aligned} n_d &= 1 + \epsilon n_{d1} + \epsilon^2 n_{d2} + \cdots \\ u &= \epsilon^{3/2} u_1 + \epsilon^2 u_2 + \cdots \\ v &= \epsilon^{3/2} v_1 + \epsilon^2 v_2 + \cdots \\ w &= \epsilon w_1 + \epsilon^2 w_2 + \cdots \\ \phi &= \epsilon \phi_1 + \epsilon^2 \phi_2 + \cdots \end{aligned} \tag{13}$$

It is pertinent to mention here that the higher is the power of ϵ, the lower is the magnitude of perturbation of that physical quantity. It is a well-known fact that the solitons evolve in a homogenous plasma if the physical quantities have a slow time variation in comparison with their space variation, which leads to the balance between nonlinearity and dispersion in a given plasma medium. On the other hand, if the variation in nonlinearity is of the same order as dissipation in a given medium, then the nonlinear effects are balanced by dissipative effects and shock waves are formed. The dissipative effects are realized through kinematic viscosity of dust particles in a given plasma medium. The stretching in space and time variables is given by Eq. (11) where ξ is the space-like coordinate and τ is the time-like coordinate, whereas the stretching in kinematic viscosity is given by Eq. (12). The variation in other physical quantities such as density, velocity and potential is presented by Eq. (13). The Lorentz force acts only in the direction perpendicular to the ambient magnetic field, which is in the z- direction in present case. Hence, the magnitude of perturbation in v_x and v_y is taken to be lower than the perturbation in v_z. Using Eqs. (11)–(12) in Eqs. (2)–(4) and Eq. (7) and equating different powers of ϵ, the lowest order of ϵ leads to

$$n_{d1} = \frac{-l_z{}^2 \phi_1}{V_0{}^2}, \tag{14}$$

$$w_1 = -\frac{l_z \phi_1}{V_0}, \tag{15}$$

$$V_0 = \frac{l_z}{\sqrt{\mu_e c_{e1} + \mu_i c_{i1}}}. \tag{16}$$

On the other hand, the lowest order of ϵ, x- and y-components of momentum equation is reduced to

$$u_1 = -\frac{l_y}{\Omega_d}\frac{\partial \phi_1}{\partial \xi}, \qquad v_1 = \frac{l_x}{\Omega_d}\frac{\partial \phi_1}{\partial \xi}. \tag{17}$$

Next higher order in ϵ leads to the following set of evolution equations

$$\frac{\partial n_{d1}}{\partial \tau} - V_0 \frac{\partial n_{d2}}{\partial \xi} + \frac{\partial}{\partial \xi}(l_x u_2 + l_y v_2 + l_z w_2 + l_z n_{d1} w_1) = 0, \tag{18}$$

$$\frac{\partial w_1}{\partial \tau} - V_0 \frac{\partial w_2}{\partial \xi} + l_z w_1 \frac{\partial w_1}{\partial \xi} = l_z \frac{\partial \phi_2}{\partial \xi} + \eta_0 \frac{\partial^2 w_1}{\partial \xi^2}, \tag{19}$$

$$\frac{\partial^2 \phi_1}{\partial \xi^2} = C_1 \phi_2 + C_2 \phi^2 + n_{d2}, \tag{20}$$

$$u_2 = -\frac{V_0 l_x}{\Omega_d^2}\frac{\partial^2 \phi_1}{\partial \xi^2}, \tag{21}$$

$$v_2 = -\frac{V_0 l_y}{\Omega_d^2}\frac{\partial \phi_1}{\partial \xi^2}. \tag{22}$$

Using Eqs. (14)–(22) and eliminating the second-order perturbed quantities, we find the following KdV–Burgers equation

$$\frac{\partial \phi_1}{\partial \tau} + A \phi_1 \frac{\partial \phi_1}{\partial \xi} + B \frac{\partial^3 \phi_1}{\partial \xi^3} = C \frac{\partial^2 \phi_1}{\partial \xi^2}. \tag{23}$$

Equation (23) describes the nonlinear evolution of obliquely propagating *DA* shock waves in a magnetized superthermal plasma where the nonlinear, dispersion and dissipation coefficients A, B and C are respectively given by

$$A = -\left(\frac{3 l_z^2}{2 V_0} + \frac{V_0^3 C_2}{l_z^2}\right), \tag{24}$$

$$B = \left(1 + \frac{1 - l_z^2}{\Omega_d^2}\right)\frac{V_0^3}{2 l_z^2}, \tag{25}$$

$$C = \frac{\eta_0}{2}. \tag{26}$$

It can be observed easily that the obliqueness (l_z) and superthermality of ions/electrons (κ_i/κ_e) influence the nonlinear coefficient A and dispersion coefficient B. However, dissipation coefficient C depends only on kinematic viscosity η_0. The dispersion coefficient B is always positive and the nonlinear coefficient A remains negative for the given set of parameters. Thus, it is clear that only negative potential DA shocks are observed in the present study. If superthermality parameters $\kappa_e, \kappa_i \to \infty$, we get the same results as that of [52] for $\beta_e, \beta_i \to 0$ (i.e., the ions/electrons obey Maxwellian distribution).

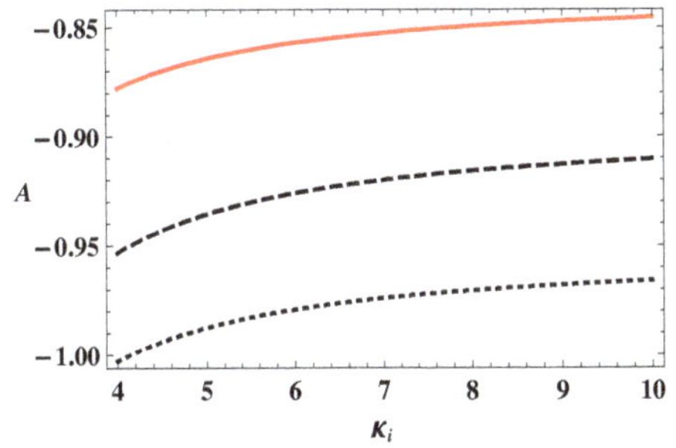

Fig. 1 (color online) Variation of nonlinear Coefficient A with κ_i at different obliqueness (via l_z) and electron-to-dust-density ratio (via μ_e). *Solid curve* is for $\mu_e = 0.2, \kappa_e = 3.5, l_z = 0.7, \sigma_i = 0.01$. *Dotted (Black) curve*: $l_z = 0.8$, *Dashed (Blue) curve*: $\mu_e = 0.3$

Solution of KdV–Burgers equation

To find the shock-like solution of KdV–B equation (23), we will adopt a reference frame moving with the shock speed. The spatial (ξ) and temporal (τ) coordinates are expressed as $\zeta = \alpha(\xi - V\tau)$ and $\tau = \tau$, here V is the shock speed and α^{-1} physically represents the shock width. The values of α account for spatially extended shock profile and vice versa. With new variables, transformed Eq. (23) reads as

$$-V\frac{d\phi_1}{d\zeta} + A\phi_1\frac{d\phi_1}{d\zeta} + B\alpha^2\frac{d^3\phi_1}{d\zeta^3} = C\alpha\frac{d^2\phi_1}{d\zeta^2}. \tag{27}$$

Integrating Eq. (27) once and using appropriate boundary conditions, $\phi_1 \to 0$, $\frac{d\phi_1}{d\zeta} \to 0$ and $\frac{d^2\phi_1}{d\zeta^2} \to 0$ for $\zeta \to \infty$, we obtain

$$-V\phi_1 + \frac{A}{2}(\phi_1)^2 + B\alpha^2\frac{d^2\phi_1}{d\zeta^2} = C\alpha\frac{d\phi_1}{d\zeta}. \tag{28}$$

Using hyperbolic tangent (tanh) method [57], we shall determine the analytical stationary solution of Eq. (23). Substituting $\phi_1 = U(Y)$, where, $Y = Tanh(\zeta)$, and $U(Y) = \Sigma_{n=0}^{\infty} a_n Y_n$, followed by balancing procedure, we get N = 2. Equating various powers of Y in Eq. (28), the shock solution of KdV–Burgers equation is determined as

$$\phi_1(\xi,\tau) = \phi_{max}\left(1 - \frac{1}{4}\left[1 + \tanh\left(\frac{\xi - V\tau}{W}\right)\right]^2\right), \tag{29}$$

where $\phi_{max} = 12C^2/25AB$ represents the shock amplitude, $W(= \alpha^{-1}) = 10B/C$ is the shock width and $V = 6C^2/25B$ is the shock speed.

Parametric analysis of dust acoustic shock structures

As obvious from the KdV–B equation (23), the nonlinear, dispersion and dissipation coefficients are functions of superthermality parameter of ions (κ_i) and electrons (κ_e), strength of magnetic field (via Ω_d), obliqueness (via l_z) and viscosity (via η_0). These parameters will effect the shock amplitude, width and velocity. Since the coefficient B is always positive, the nature of the shock structures depends upon the sign of the nonlinear coefficient (A). Positive sign of ($A > 0$) supports positive potential shocks, while negative sign ($A < 0$) corresponds to negative potential shocks. Figure 1 depicts the variation of nonlinear coefficient A with superthermality of ions (via κ_i) for different values of electron-to-dust density ratio (μ_e) and obliqueness (l_z). It is seen that for all chosen set of parameters, the coefficient A is always negative, and thus, only negative potential shock waves are formed. However, the absolute value of nonlinear coefficient A increases with increase in superthermality of ions (decrease in κ_i) and for a particular value of superthermality parameter the nonlinearity increases with increase in electron-to-dust density ratio. It is also observed that small obliqueness increases the nonlinearity in the system. Hence, it is concluded that absolute value of nonlinear coefficient A is enhanced with increase in superthermality of ions, electron-to-dust-density ratio and decrease in obliqueness. It is noteworthy to mention here that even in the case of dust acoustic solitary waves in an unmagnetized plasma with kappa-distributed electrons and ions, only negative potential nonlinear structures were reported if the dust is negatively charged [58].

To study the effect of various physical parameters on the profile of DA shock structures, we have studied numerically the stationary solutions of KdV–Burgers equation. Since nonlinear, dissipation and dispersion coefficients are dependent on the various physical parameters, it is important to trace the influence of such parameters on the characteristics of dust acoustic shock waves through the variations in various coefficients. We have displayed the numerical results in Figs. 2–8 depicting the influence of physical parameters on the profile of DA shocks governed by KdV–Burgers equation. The effect of superthermality on the speed and width of shock structures is depicted in

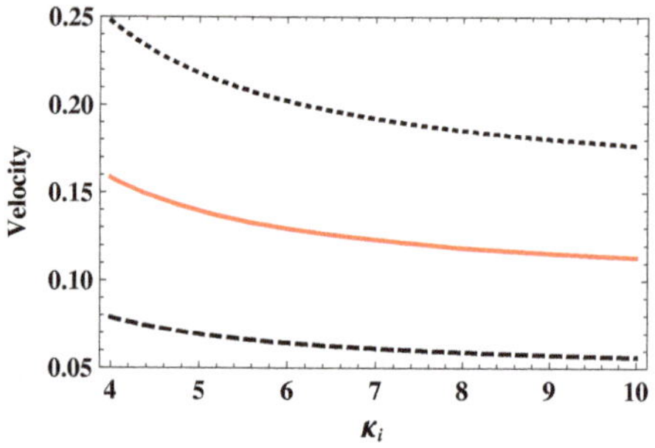

Fig. 2 (color online) Variation of velocity of shock waves with superthermality of ions at different values of magnetic field (via Ω_d) and kinematic viscosity (via η_0). *Solid curve* is for $\mu_e = 0.2$, $\kappa_e = 3.5, l_z = 0.7, \sigma = 0.01, \Omega_d = 1, \eta_0 = 0.8$. *Dotted (Black) curve*: $\eta_0 = 1$, *Dashed (Blue) curve*: $\Omega_d = 0.5$

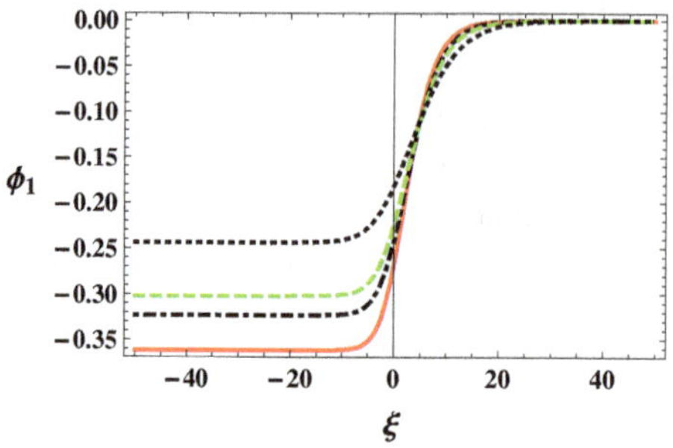

Fig. 4 (color online) Shock wave profile at different values of superthermality of ions (via κ_i) with values $\mu_e = 0.2, \kappa_e = 3.5$, $l_z = 0.7, \sigma = 0.01, \Omega_d = 1, \eta_0 = 0.8$. *Solid (Red) curve*: $\kappa_i = 4$, *Dot-dashed (Blue) curve*: $\kappa_i = 5$, *Dashed (Green) curve*: $\kappa_i = 6$, *Dotted (Black) curve*: $\kappa_i = 25$

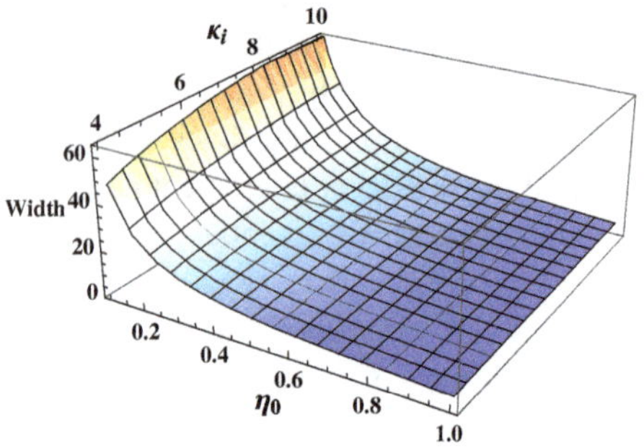

Fig. 3 (color online) 3D graph showing variation of width of shock structure with viscosity (via η_0) and superthermality of ions (via κ_i) for $\mu_e = 0.2, \kappa_e = 3.5, l_z = 0.7, \sigma = 0.01, \Omega_d = 1$

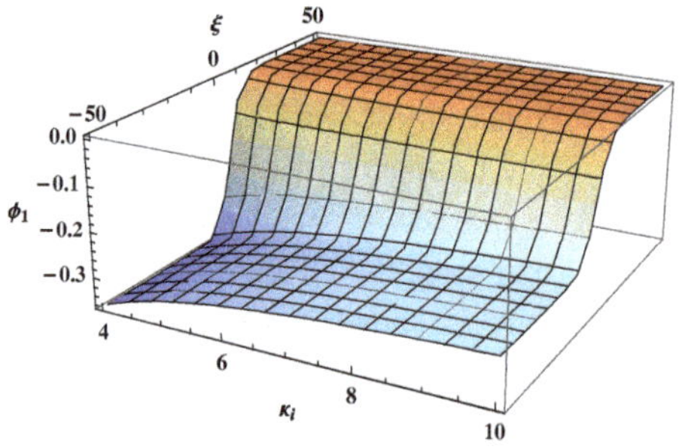

Fig. 5 (color online) Variation of 3D shock wave profile with superthermality of ions (via κ_i) for $\mu_e = 0.2, \kappa_e = 3.5, l_z = 0.7$, $\sigma = 0.01, \omega_d = 1, \eta_0 = 0.8$

Figs. 2 and 3. Figure 2 illustrates the variation of velocity of shocks with superthermality of ions for different values of magnetic field and viscosity. It is clear that the speed of shock waves is enhanced with increase in superthermality of ions (i.e., decrease in value of κ_i), but it has opposite effect on the width of the shock waves, i.e., the width of shock structures decreases with increase in superthermality of ions (see Fig. 3). The combined effects of superthermality of ions and kinematic viscosity on the width of shock structures affirm that a less viscous fluid is more attuned to superthermality parameter and support narrower shocks for more superthermal ions. The variation of shock profile with superthermality of ions is displayed in Fig. 4. It is observed that more superthermal ions lead to higher amplitude shock structures. It has also been asserted by Malik [43] for the case of ion-acoustic solitons that the amplitude of nonlinear structures increases with increase in temperature of ions. Overall, the amplitude and velocity of shock waves are enhanced with increase in superthermality (i.e., decrease in κ_i) and it is very sensitive for low range of κ_i. Thus, highly superthermal ions make the shocks taller, narrower and faster. This inference is further illustrated in 3D plot shown in Fig. 5, where more superthermal ions result in the formation of more abrupt shocks. Similar kind of variation of DA shock structures with superthermality of ions is observed in an unmagnetized electron-depleted dusty plasma environment containing two temperature superthermal ions [59].

Figure 6 presents the shock wave profile at different values of kinematic viscosity (via η_0). It is obvious from the graphs that higher values of kinematic viscosity support the shock structures with larger amplitude and smaller width. The results are in agreement as reported for the study of dust acoustic shock waves in a magnetized non-thermal dusty plasma [52]. The similar results are obtained for variation of magnetic field (via Ω_d) and ratio of electron to dust concentration (via μ_e), i.e., increase in Ω_d and μ_e makes the shocks more abrupt and faster as shown in Fig. 7 which are also consistent with the results reported by Shahmansouri and Mamun [52]. On the other hand, the decrease in obliqueness (increase in l_Z) results in shocks of smaller amplitude, whereas stronger is the magnetic field more spiky are the shock structures. This is due to the fact that the effect of magnetic field on shock structures is realized through dispersion coefficient B in KdV–Burgers equation. The value of dispersion coefficient decreases with increase in strength of magnetic field. Since the amplitude of shocks is inversely proportional to the dispersion coefficient, the shock structures become taller with

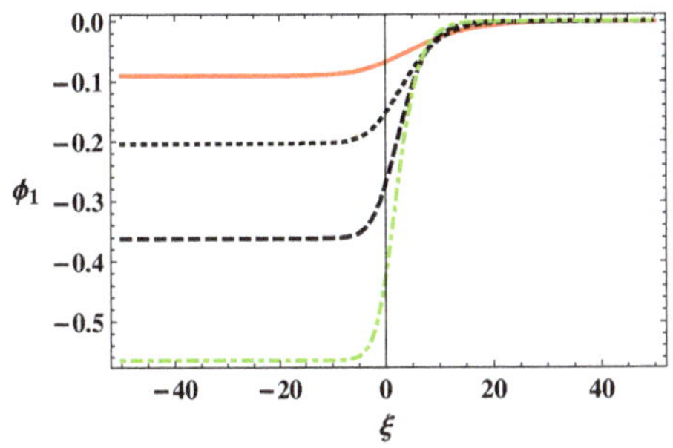

Fig. 6 (color online) Shock wave profile at different values of kinematic viscosity (via η_0) for $\mu_e = 0.2, \kappa_e = 3.5, \kappa_i = 4, l_z = 0.7, \sigma = 0.01, \Omega_d = 1$. *Solid (Red) curve*: $\eta_0 = 0.4$, *Dotted (Black) curve*: $\eta_0 = 0.6$, *Dashed (Blue) curve*: $\eta_0 = 0.8$, *Dotdashed (Green) curve*: $\eta_0 = 1$

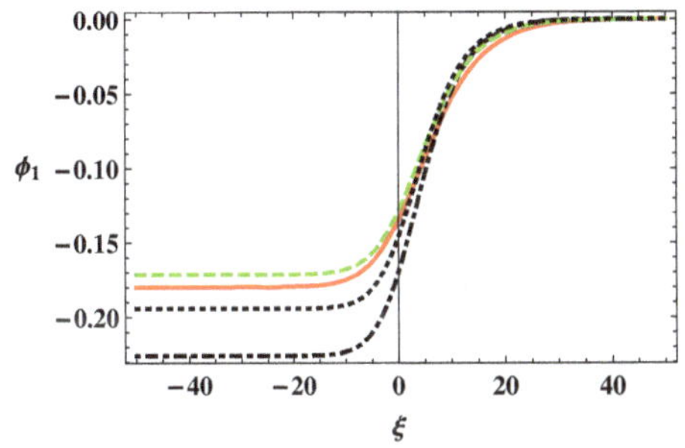

Fig. 7 (color online) Shock wave profile at different value of obliqueness (via l_z), electron-to-dust-density ratio (via μ_e) and magnetic field (via Ω_d). *Solid curve* is $\mu_e = 0.2, \kappa_e = 3.5, \kappa_i = 4, l_z = 0.7, \sigma = 0.01, \Omega_d = 0.5, \eta_0 = 0.8$. *Dotted (Black) curve*: $\mu_e = 0.4$, *Dashed (Green) curve*: $l_z = 0.8$, *Dotdashed (Blue) curve*: $\Omega_d = 0.6$

increase in strength of magnetic field. The opposite effect is observed by Malik [43] on the propagation properties of ion-acoustic (IA) solitons in a magnetized plasma where it is observed that the strength of magnetic field reduces the amplitude of ion-acoustic solitons. On the other hand, since the width of shock structures is directly proportional to the dispersion coefficient, the increase in strength of magnetic field tends to decrease the width of shocks structures and makes them narrower. Similar kind of behavior was observed by Kumar et al. [44]. Figure 8 illustrates the variation of shock structures with magnetic field in 3D. It is seen that magnetic field has a great influence on the shocks so formed such that the vanishing of magnetic field may even lead to disappearance of shock structures governed by the solution of KdV–Burgers equation.

Conclusions

We have examined the dust acoustic shocks in magnetized dusty plasma with superthermal electrons and ions. By employing reductive perturbation technique, KdV–Burgers

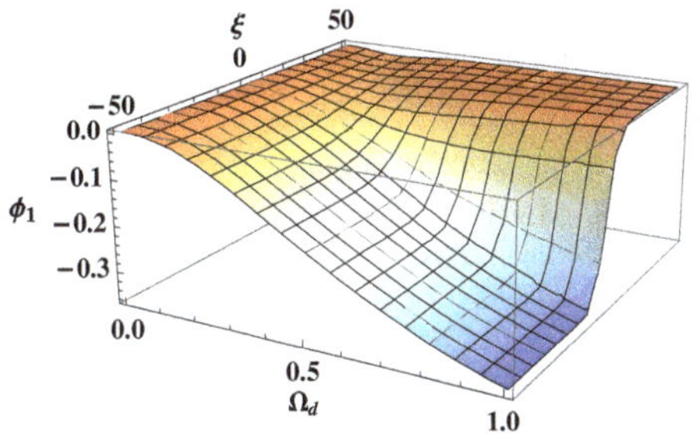

Fig. 8 (color online) Variation of 3D Shock wave profile with magnetic field (via Ω_d) for $\mu_e = 0.2, \kappa_e = 3.5, \kappa_i = 4, l_z = 0.7, \sigma = 0.01, \eta_0 = 0.8$

equation has been derived. The role of various plasma parameters (viz. superthermality of ions (via κ_i), electron-to-dust density ratio (via μ_e), dust kinematic viscosity of medium (via η_0), strength of magnetic field (via Ω_d) and obliqueness of magnetic field (via l_z)) on the characteristics of *DA* shock structures has been highlighted in the present study. Only negative potential shock structures are observed. The amplitude of shock structures increases with increase in superthermality of ions, dust concentration and increase in strength of magnetic field. On the other hand, width of shock structures decreases with increase in superthermality of ions.

For Cairns distribution (nonthermal distribution) in the presence of magnetic field, both polarities *DA* shock structures were obtained by Shahmansouri and Mamun [52], but under the influence of superthermality of charged particles and magnetic field, only negative potential *DA* shock structures are observed.

The findings of this investigation should be useful for wider understanding of the formation of dust acoustic shock structures in solar wind, Earth's magnetosphere, Earth's polar cap region and Saturn's rings where superthermal electrons/ions and dust particles are present along with an ambient magnetic field. These results may also be applicable in the laboratory experiments for laser–plasma interaction, where dust acoustic shocks may be observed in the presence of dust grains and superthermal particles.

Acknowledgements This work was supported by DRS-II (SAP) No.F/530/17/ DRS-II/2015(SAP-I) University Grants Commission (UGC), New Delhi, India. Y.G. gratefully acknowledges University Grants Commission (UGC) for providing scholarship under Basic Scientific Research (BSR) scheme.

References:

1. Kulsurd, R.M., Gunn, J.P., Ostriker, J.P., Gunn, J.E.: Acceleration of cosmic rays in supernova remnants. Phys. Rev. Lett. **28**, 636–639 (1972)

2. McClements, K.G., Dieckmann, M.E., Ynnerman, A., Chapman, S.G., Dendy, R.O.: Surfatron and Stochastic Acceleration of Electrons at Supernova Remnant Shocks. Phys. Rev. Lett. **87**, 255002 (2001). **(1-4)**
3. Bale, S.D., Balikhin, M.A., Horbury, T.S., Krasnoselskikh, V.V., Kucharek, H., Mobius, E., Walker, S.N., Balogh, A., Burgess, D., Lembege, B., Lucek, E.A., Scholer, M., Schwartz, S.J., Thomsen, M.F.: Quasi-perpendicular shock structure and processes. Space Sci. Rev. **118**, 161–203 (2005)
4. Lee, E., Parks, G.K., Wilbr, X., Lin, N.: Nonlinear Development of Shocklike Structure in the Solar Wind. Phys. Rev. Lett. **103**, 031101 (2009). **(1-4)**
5. Sabetkar, A., Dorranian, D.: Effect of obliqueness and external magnetic field on the characteristics of dust acoustic solitary waves in dusty plasma with two-temperature nonthermal ions. J. Theor. Appl. Phys. **9**, 141–150 (2015)
6. Shahmohammadi, N., Dorranian, D., Hakimipagouh, H.: Effect of superthermal electrons on the characteristics of dust acoustic solitary waves in a magnetized hot dusty plasma with dust charge fluctuation. Can. J. Phys. **93**, 344–352 (2015)
7. Goertz, C.K.: Dusty plasmas in the solar system. Rev. Geophys. **27**, 271–292 (1989)
8. Mendis, D.A., Rosenberg, M.: Cosmic Dusty Plasma. Annu. Rev. Astron. Astrophys. **32**, 419–463 (1994)
9. Shukla, P.K., Mamun, A.A.: Introduction to Dusty Plasma Physics. Institute of Physics, Bristol (2002)
10. Selwyn, G.S.: A Phenomenlogical Study of Particulates in Plasma Tools and Processes. Jpn. J. Appl. Phys. **32**, 3068–3073 (1998)
11. Veerhest, F.: Waves and instabilities in dusty space plasmas. Space Sci. Rev. **77**, 267–302 (1996)
12. Rao, N.N., Shukla, P.K., Yu, M.Y.: Dust- acoustic waves in dusty plasmas. Planet. Space Sci. **38**, 543–546 (1990)
13. Barkan, A., Merlino, R.L., D'Angelo, N.: Laboratory observation of the dust-acoustic wave mode. Phys. Plasmas. **2**, 3563–3565 (1995)
14. Shukla, P.K., Silin, V.P.: Dust ion-acoustic wave. Phys. Scripta. **45**, 508 (1992)
15. Melandsø, F., Shukla, P.K.: Theory of dust-acoustic shocks. Planet. Space Sci. **43**, 635–648 (1995)
16. Homann, A., Melzer, A., Peters, S., Piel, A.: Determination of the dust screening length by laser-excited lattice waves. Phys. Rev. E. **56**, 7138–7141 (1997)
17. Zang, L.P., Wang, J.K.: Nonplanar dust-acoustic shock waves for two-temperature ions in dusty plasma with dissipative effects and transverse perturbations. Phys. Plasmas. **13**, 022303 (2006)
18. Rahman, A., Mamun, A.A.: Shock Waves in an Adiabatic Dusty Plasma. Chin. J. Phys. **46**, 601–607 (2008)
19. Misra, A.P., Adhikari, N.C., Shukla, P.K.: Ion-acoustic solitary waves and shocks in a collisional dusty negative-ion plasma. Phys. Rev. Lett. E. **86**, 056406 (2012). **(1-10)**
20. Chatterjee, P., Ghosh, D.K., Sahu, B.: Planar and nonplanar ion acoustic shock waves with nonthermal electrons and positrons. Astrophys. Space Sci. **339**, 261–267 (2012)
21. El-Hanbaly, A.M., El-Shewy, E.K., Sallah, M., Darwees, H.F.: Linear and nonlinear analysis of dust acoustic waves in dissipative space dusty plasmas with trapped ions. J. Theor. Appl. Phys. **9**, 167–176 (2015)
22. Tasnim, I., Masud, M.M., Anowar, M.G.M., Mamun, A.A.: Dust-acoustic shockwaves in nonthermal dusty plasmas with two population ions. IEEE Trans. Plasma Sci. **43**, 2187–2194 (2015)
23. Heinrich, J.R., Kim, S.H., Merlino, R.L.: Laboratory Observations of Self-Excited Dust Acoustic Shocks. Phys. Rev. Lett. **103**, 115002 (2009). **(1–4)**
24. Merlino, R.L., Heinrich, J.R., Hyun, S.-H., Meyer, J.K.: Nonlinear dust acoustic waves and shocks. Phys. Plasmas **19**, 057301 (2012). **(1–7)**
25. Nakamura, Y., Bailung, H., Shukla, P.K.: Observation of Ion-Acoustic Shocks in a Dusty Plasma. Phys. Rev. Lett. **83**, 1602–1605 (1999)
26. Nakamura, Y., Bailung, H., Saitou, Y.: Observation of ion-acoustic shock waves undergoing Landau damping. Phys. Plasmas **11**, 3925–3931 (2004)
27. Maksimovic, M., Pierrard, V., Riley, P.: Ulysses electron distributions fitted with Kappa functions. Geophys. Res. Lett. **24**, 1151–1154 (1997)
28. Krimigis, M., Carbary, J.F., Keath, E.P., Armstrong, T.P., Lanzerotti, L.J., Gloeckler, G.: General characteristics of hot plasma and energetic particles in the Saturnian magnetosphere: results from Voyager spacecraft. J. Geophys. Res. **88**, 8871–8892 (1983)
29. Christon, S.P., Mitchell, D.G., Williams, D.J., Frank, I.A., Huang, C.Y., Eastman, T.E.: Energy spectra of plasma sheet ions and electrons from $\approx$50 eV/e to $\approx$1 MeV During Plasma Temperature Transitions. J. Geophys. Res. **93**, 2562–2572 (1988)
30. Pierrard, V., Lemaire, J.: Lorentzian ion exosphere model. J. Geophys. Res. **101**, 7923–7934 (1996)
31. Magni, S., Roman, H.E., Barni, R., Riccardi, C., Pierre, T.H., Guyomarc'h, D.: Statistical analysis of correlations and intermittency of a turbulent rotating column in a magnetoplasma device. Phys. Rev. E **72**, 026403 (2005). **(1–7)**
32. Formisano, V., Moreno, G., Palmiotto, F.: Solar Wind Interaction with the Earth's Magnetic Field: 1. Magnetosheath. J. Geophys. Res. **78**, 3714–3730 (1973)
33. Marsch, E., Mülhäuser, K.H., Schwenn, R., Rosenbauer, H., Phillip, W., Neubauer, F.M.: Solar wind protons: three-dimensional velocity distributions and derived plasma parameters measured between 0.3 and 1 AU. J. Goephys. Res. **87**, 52–72 (1982)
34. Leubner, M.P.: Fundamental issues on kappa-distributions in space plasmas and interplanetary proton distributions. Phys. Plasmas **11**, 1308–1316 (2004)
35. Lazar, M., Schlickeiser, R., Poedts, S., Tautz, R.C.: Counter-streaming magnetized plasmas with kappa distributions I. Parallel wave propagation. Mon. Not. R. Astron. Soc. **390**, 168–174 (2008)
36. Schippers, P., Blane, M., Andre, N., Damdouras, I., Lewis, G.R., Gilbert, L.K., Pearson, A.M., Karupp, N., Gumet, D.A., Coates, A.J., Krimigis, S.M., Young, D.A., Dougherry, M.K.: Multi-instrument analysis of electron populations in Saturn's magnetosphere. J. Geophys. Res. **113**, A07208 (2008). **(1–10)**
37. Hellberg, M.A., Mace, R.L., Baluku, T.K., Kourakis, I., Saini, N.S.: Comment on Mathematical and physical aspects of Kappa velocity distribution [Phys. Plasmas 14, 110702 (2007)]. Phys. Plasmas **16**, 094701 (2009). **(1–5)**
38. Shahmansouri, M., Astaraki, E.: Transverse perturbation on three-dimensional ion acoustic waves in electron–positron–ion plasma with high-energy tail electron and positron distribution. J. Theor. Appl. Phys. **8**, 189–201 (2014)
39. Kundu, S.K., Ghosh, D.K., Chatterjee, P., Das, B.: Shock waves in a dusty plasma with positive and negative dust, where electrons are superthermally distributed. Bulg. J. Phys. **38**, 409–419 (2011)
40. Shahmansouri, M., Alinejad, H.: Dust acoustic shock waves in a suprathermal dusty plasma with dust charge fluctuation. Astrophys. Space Sci. **343**, 257–263 (2013)
41. Malik, H.K.: Ion acoustic solitons in a weakly relativistic magnetized warm plasma. Phys. Rev. E **54**, 5 (1996)
42. Malik, R., Malik, H.K., Kaushik, S.C.: Soliton propagation in a moving electron-positron pair plasma having negatively charged dust grains. Phys. Plasmas **19**, 032107 (2012)
43. Malik, H.K.: Soliton reflection in magnetized plasma: Effect of ion temperature and nonisothermal electrons. Phys. Plasmas **15**, 072105 (2008)

44. Kumar, R., Malik, H.K., Singh, K.: Effect of dust charging and trapped electrons on nonlinear solitary structures in an inhomogeneous magnetized plasma. Phys. Plasmas **19**, 012114 (2012)
45. Malik, H.K., Singh, S., Dahiya, R.P.: Kadomtsev Petviashvili solitons in an inhomogenous plasmas with finite temperature drifting ions. Phys. Lett. A **195**, 369–372 (1994)
46. Malik, H.K., Tomar, R., Dahiya, R.P.: Conditions for reflection and transmission of an ion acoustic soliton in a dusty plasma with variable charge dust. Phys. Plasmas **21**, 072112 (2014)
47. Malik, R., Malik, H.K.: Compressive solitons in a moving e-p plasma under the effect of dust grains and an external magnetic field. J. Theor. Appl. Phys. **7**, 65 (2013)
48. Tomar, R., Malik, H.K., Dahiya, R.P.: Reflection of ion acoustic solitary waves in a dusty plasma with variable charge dust. J. Theor. Appl. Phys. **8**, 126 (2014)
49. Tomar, R., Bhatnagar, A.: Malik, H.K. and Dahiya, R.P: Evolution of solitons and their reflection and transmission in a plasma having negatively charged dust grains, J. Theor. Appl. Physics **8**, 138 (2014)
50. Zhang, L.P., Xue, J.K.: Shock wave in magnetized dusty plasmas with dust charging and nonthermal ion effects. Phys. Plasmas **12**, 042304 (2005). **(1–6)**
51. Bains, A.S., Tribeche, M.: Oblique shock dynamics in nonextensive magnetized plasma. Astrophys. Space Sci. **351**, 191–195 (2014)
52. Shahmansouri, M., Mamun, A.A.: Dust-acoustic shock waves in a magnetized non-thermal dusty plasma. J. Plasma Phys. **80**, 593–606 (2014)
53. Zaghbeer, S.K., Salah, H.H., Sheta, N.H., El-Shewy, E.K., Elgarayhi, A.: Dust acoustic shock waves in dusty plasma of opposite polarity with non-extensive electron and ion distributions. J. Plasma Phys. **80**, 517–528 (2014)
54. Washimi, H., Taniuti, T.: Propagation of ion-acoustic solitary waves of small amplitude. Phys. Rev. Lett. **17**, 996–998 (1966)
55. Kumar, R., Malik, H.K.: Nonlinear Solitary Structures in an Inhomogeneous Magnetized Plasma having Trapped Electrons and Dust Particles with Different Polarity. J. Phys. Soc. Japan **80**, 044502 (2011)
56. Malik, H.K., Kumar, R., Lonngren, K.E., Nishida, Y.: Collision of ion acoustic solitary waves in a magnetized plasma: Effect of dust grains and trapped electrons. Phys. Rev. E **92**, 063107 (2015)
57. Malfliet, W., Hereman, W.: The Tanh method: I. exact solutions of nonlinear evolution and wave equations. Phys. Scr. **54**, 563 (1996)
58. Baluku, T.K., Hellberg, M.A.: Dust acoustic solitons in plasmas with kappa-distributed electrons and/or ions. Phys. Plasmas **15**, 123705 (2008)
59. Ghai, Y., Saini, N.S.: Shock waves in dusty plasma with two temperature superthermal ions. Astrophys. and Space Sci. **58**, 362 (2017)

7

The effect of external magnetic field on the density distributions and electromagnetic fields in the interaction of high-intensity short laser pulse with collisionless underdense plasma

Masoomeh Mahmoodi-Darian[1] · Mehdi Ettehadi-Abari[2] · Mahsa Sedaghat[2]

Abstract Laser absorption in the interaction between ultra-intense femtosecond laser and solid density plasma is studied theoretically here in the intensity range $I\lambda^2 \simeq 10^{14}-10^{16}$ W cm^{-2} µm^2. The collisionless effect is found to be significant when the incident laser intensity is less than 10^{16} W cm^{-2} µm^2. In the current work, the propagation of a high-frequency electromagnetic wave, for underdense collisionless plasma in the presence of an external magnetic field is investigated. When a constant magnetic field parallel to the laser pulse propagation direction is applied, the electrons rotate along the magnetic field lines and generate the electromagnetic part in the wake with a nonzero group velocity. Here, by considering the ponderomotive force in attendance of the external magnetic field and assuming the isothermal collisionless plasma, the nonlinear permittivity of the plasma medium is obtained and the equation of electromagnetic wave propagation in plasma is solved. Here, by considering the effect of the ponderomotive force in isothermal collisionless magnetized plasma, it is shown that by increasing the laser pulse intensity, the electrons density profile leads to steepening and the electron bunches of plasma become narrower. Moreover, it is found that the wavelength of electric and magnetic field oscillations increases by increasing the external magnetic field and the density distribution of electrons also grows in comparison to the unmagnetized collisionless plasma.

✉ Masoomeh Mahmoodi-Darian
m.darian@kiau.ac.ir

1 Department of Physics, Karaj Branch, Islamic Azad University, Karaj, Iran

2 Physics Department and Laser Research Institute of Beheshti University, G.C., Evin, 19839 Tehran, Iran

Keywords Laser plasma interaction · Ponderomotive force · Underdense magnetized plasma · Nonrelativistic regime · Electrons density distribution

Introduction

Theoretical and experimental studies of electromagnetic wave propagation in magnetized plasmas are of key importance for a vast range of problems in space and laboratory physics [1–3]. The acceleration of electrons in the interaction of a high-intensity laser beam with plasma may have important applications in various domains such as laser particle acceleration, ion acceleration for fusion action, and the generation of intense and short-duration γ-ray sources for radiography [4–6]. Furthermore, the interaction of an ultra short high-intensity laser pulse with plasma without external dc magnetic field has been studied extensively [7–11]. When an external magnetic field is applied to the plasma, it is a medium capable to convert different initial energies to tunable coherent radiations. In 1976 it was shown that a linearly polarized laser beam interacting with hot magnetized plasma produces a radiative force along the electric field of laser beam. This magnetic radiation force at 5 MG is as large as the ponderomotive force for a 3-µm scale height [12]. In 2002, the generation of the huge azimuthal self-magnetic field of over 340 MG near the critical density surface was measured [13]. Two years later, this amount was scaled up to more than 700 MG by the same group in over dense plasma [14] while in underdense plasma fields in the order of 100 MG is presented [15]. The self-magnetic field in plasma is a function of laser intensity and Qiao et al. [16] showed, by their analytical model for Nd-glass laser at the intensity of 10^{20} W cm^{-2}, the obtained magnetic field of about 90 MG.

In studying the plasma parameters in an external magnetic field, the intensity should be in the same order where few megagauss magnetic field facilities for magnetized laser-plasma experiments are established [17, 18]. A method is also described for choosing experimental parameters in studies of high energy density physics relevant to fusion energy, as well as other applications by using megagauss magnetic fields [19]. The structure of these systems is based on the compact megagauss magnetic field generation in single-turn coils which created new frontiers for scientific experiments [20].

Extremely high azimuthal magnetic fields play an essential role in the particle transport, propagation of laser pulses, laser beam self-focusing, penetration of laser radiation into the overdense plasma, and the plasma electron and ion acceleration, where the first direct measurements of high-energy proton generation (up to 18 MeV) and propagation into a solid target during such intense laser plasma interactions were reported. Measurements of the deflection of these energetic protons were carried out which imply that magnetic fields in excess of 30 MG exist inside the target [21]. Although some valuable experimental works have been reported, the introduced analytical works did not properly explain the fundamental parameters of the magnetized collisionless plasma, including electric field oscillation, electron density distribution or the derived equations, being related to very special conditions [22, 23]. In the recent studying of plasma, such as ponderomotive acceleration of plasma in the applied magnetic field and improving of self-focusing in presence of the applied external magnetic fields, the range of magnetic field intensities are in the range of tens of megagauss [24, 25]. Based on the report of Gupta et al. [25], they did not have the effect of external magnetic field up to 20 MG in an underdense subrelativistic plasma and the effect is enhanced by increasing the magnetic field intensity up to 45 MG, where we have noticed an almost similar dependence of mentioned parameters to the magnetic field intensity in the present analytical work. Although some valuable experimental works are reported, the introduced analytical works did not properly explain the fundamental parameters of the magnetized plasma including electric and magnetic field oscillations or the derived equations being related to very special conditions [26–32].

Furthermore, the numerical analysis of high-power laser propagation in a magnetized plasma was considered by Druce et al. [33]. Here, a CO_2 laser with 10.6 μm wavelength interacts with magnetically confined plasma in the density range of $10^{17}-10^{18}$ cm^{-3}. Here, by using of the momentum and energy equations for the plasma species and assuming the dependence of T_e to the coordinates, the laser propagation in magnetized confined collisionless plasma is analyzed. In addition, in the previous work [34], we studied the nonlinear propagation of an intense laser pulse through an underdense magnetized plasma. Here, we considered the semi-infinite plasma in the region of $z \phi 0$ in attendance of a constant external magnetic field B_0 in y direction. The laser pulse is irradiated perpendicular into the plasma in z direction and the polarization of the incident wave is assumed to be linear. In this work it is obvious that the high azimuthally external magnetic field is considered [i.e., $\mathbf{k} = (0, 0, k)$, $\mathbf{B_0} = (0, B_0, 0)$, $\mathbf{E} = (E_x, 0, 0)$, $\mathbf{B} = (0, B_y, 0)$].

In the current theoretical study, by considering the circular symmetry case where the electric and magnetic fields of the laser are perpendicular to B_0 (external magnetic field) and with assumption an azimuthally polarized laser pulse $(E_\theta(z,t) = \hat{\theta}E_\theta(z)\exp(-i\omega t))$, where $\hat{\theta}$ is the unit vector in the azimuthal direction, we have studied the nonlinear structure of electromagnetic wave propagation of an ultra-short laser beam in underdense collisionless magnetized plasma. An external magnetic field is applied in the direction of laser beam propagation in homogeneous plasma. To achieve the nonlinear equation for the electric field in the plasma, we use the Maxwell equations and the equation of electrons motion while taking into account the average ponderomotive force per unit volume acting on the plasma electrons. Since the ion mass is much greater than the electron mass, we neglect the effect of the ponderomotive force on the ions. In order to imply the magnetic field in our formulation, we use the external plasma current density $\mathbf{J} = -4\pi n_e e\mathbf{V}$ in Ampere's law, in which e, n_e, and $\mathbf{V}$ are the electron charge, the electron density and the drift velocity of electrons. Furthermore, using the Maxwell and the transfer momentum equations, we found the dispersion relation of the transverse wave, in isothermal collisionless magnetized plasma. In our work, we show that in the presence of external magnetic field in the direction of laser beam propagation, the density of the electrons increases.

This work is organized as follows: after introduction the theoretical model and formulation are presented in Sect. II. Here, the basic equations and fundamental assumptions are given. In Sect. III, we have discussed the numerical method used in this work, and finally Sect. IV is devoted to conclusion.

The theoretical model and formulation

In order to formulate the nonlinear propagation of an intense laser pulse through an underdense magnetized collisionless isothermal plasma in nonrelativistic regime let us consider a semi-infinite plasma $(z > 0)$ in a constant external magnetic filed $\mathbf{B_0}$ in the z direction. In this model, we assume circular symmetry and a field $\mathbf{E}$ that is azimuthally polarized ($E_\theta(z,t) = \hat{\theta}E_\theta(z)\exp(-i\omega t)$) and it propagates in the $+z$ direction, where $\hat{\theta}$ is the unit vector in the azimuthal direction. In this case we should mention that polarization is one of the most important characteristics of laser radiation. While determining the polarization state of the beam, one can speak about type of polarization at the point of the beam cross section, homogeneity of ellipsometrical parameters over the beam cross section, and stability of polarization characteristics in time. From the nowadays viewpoint, the conventional types of polarization have substantial disadvantages. In the case of linear polarization, the parameters of the beam interaction with the matter depend upon the direction of polarization. In the case of circular polarization, these parameters are time averaged, i.e., not optimum from view point either of minimum losses or maximum absorption. The modes with inhomogeneous polarization, radial or azimuthal, are known in the laser resonator theory. In the case of radial (azimuthal) polarization, the direction of the electrical vector in the plane of the beam cross section is parallel (perpendicular) to the radial direction. Here, we consider a semi-infinite plasma in $z > 0$ region in a constant external magnetic filed $\mathbf{B_0}$ in the z direction. $z > 0$ region is taken to be filled with a homogeneous density profile of plasma with plasma–vacuum interface at $z = 0$. It should be noted that, in the problem of interaction of laser beams with plasma, if the vector **E** is parallel to the plane of incidence (radial polarization), the resonance absorption is maximum and if the vector **E** is perpendicular to the plane of incidence (azimuthal polarization), no resonance absorption occurs. In this case the electrons in plasma oscillate in the wave electrical field along lines of equal density without generating electrostatic fields. Such type of polarization that we used in this manuscript is useful for investigating ponderomotive forces affecting electron density profile [35–37]. As we see from Fig. 1, the electromagnetic wave enters normally into the plasma slab.

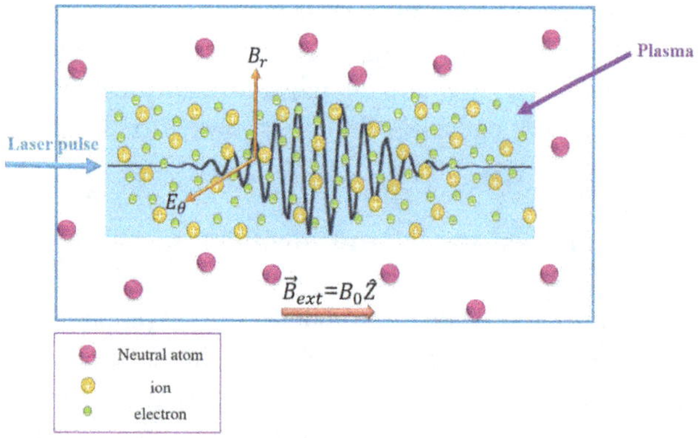

Fig. 1 Schematic view of the laser pulse propagation through the magnetized underdense plasma is shown. Direction of the applied uniform magnetic field B_0 is in z direction

To develop the wave equation for the oscillating electric and magnetic fields, one can start with Faraday's induction and Ampere's laws. By considering these Maxwell's equation, we have:

$$\nabla \times \mathbf{E} = -i\omega\mathbf{B}, \tag{1}$$

$$\nabla \times \mathbf{B} = \mu_0(\mathbf{J} - i\varepsilon_0\omega\mathbf{E}). \tag{2}$$

Now, if we assume that the laser electromagnetic transverse wave fields with frequency ω propagate through plasma along the z direction, the laser pulse electric field evolves as

$$\frac{d^2}{dz^2}E_\theta + k_0^2E_\theta + i\omega\mu_0J_\theta = 0. \tag{3}$$

Here, we consider the parameter $\mu_0\varepsilon_0\omega^2 = k_0^2 = \omega^2/c^2$ as the vacuum propagation constant. Now, to obtain the laser pulse electric field in plasma, we should use the appropriate electron current density J_θ in Eq. (3). As we know, the plasma electron current density equation is $J_\theta = -n_e eV_{e\theta}$ and for this purpose we should consider the equation of electrons motion in collisionless isothermal plasma as

$$m_e n_e \frac{d\mathbf{V_e}}{dt} = -en_e\left(\mathbf{E} + \frac{1}{c}\mathbf{V_e} \times \mathbf{B}\right) - \nabla P_e + \mathbf{F_{Pe}}, \tag{4}$$

where m_e, V_e, $P_e = n_eT_e$ and $\mathbf{F_{Pe}}$ are electron mass, electron velocity, pressure of electrons, and the average ponderomotive force defined by the laser pulse envelope. In this work, since the ion mass is much greater than the electron mass and $\omega \gg \omega_{pi}$, where $\omega_{pi} = \sqrt{\frac{4\pi n_i e^2}{m_i}}$, we neglect the ion motions. Then by writing Eq. (4) in the $\hat{r}$ and $\hat{\theta}$ directions, the v_r and v_θ components of electron velocity of magnetized collisionless isothermal plasma can be written as

$$v_r = \frac{ie}{m_e\omega}\frac{E_r - i\left(\frac{\omega_{ce}}{\omega}\right)E_\theta}{1 - \left(\frac{\omega_{ce}}{\omega}\right)^2}, \tag{5}$$

$$v_\theta = \frac{-ie}{m_e\omega}\frac{E_\theta - i\left(\frac{\omega_{ce}}{\omega}\right)E_r}{1 - \left(\frac{\omega_{ce}}{\omega}\right)^2}. \tag{6}$$

Here E_r and E_θ are radial and azimuthal electric field components, respectively. Now by considering an azimuthally polarized laser beam ($E_r = 0$), we reach to the following equations for v_r and v_θ components:

$$v_r = \frac{e}{m_e\omega}\left(\frac{\omega_{ce}}{\omega}\right)\frac{E_\theta}{1 - \left(\frac{\omega_{ce}}{\omega}\right)^2}, \tag{7}$$

$$v_\theta = \frac{-ie}{m_e\omega}\frac{E_\theta}{1 - \left(\frac{\omega_{ce}}{\omega}\right)^2}. \tag{8}$$

In the above equations $\omega_{ce} = eB_0/m_ec$ is the electrons cyclotron frequency due to the existence of the external magnetic field and ω is the laser pulse frequency. Here, the plasma electrons current density is obtained as

$$J_\theta = n_e e V_{e\theta} = \frac{i\varepsilon_0 \omega_{pe}^2 E_\theta}{\omega\left(1 - \frac{\omega_{ce}^2}{\omega^2}\right)}, \tag{9}$$

where $\omega_{pe0} = \sqrt{\frac{4\pi n_{e0} e^2}{m_e}}$ is the plasma electron frequency and n^e is the plasma density. Now, we consider the homogeneous collisionless magnetized isothermal plasma in the presence of the ponderomotive force due to the laser pulse. In this case, in the steady state, the ponderomotive force in attendance of the external magnetic field can be balanced with the electron pressure gradient force. Consequently, according to the momentum transfer Eq. (4) in the laser pulse propagation direction and assuming that the electron temperature T_e is independent of coordinates, we have

$$\frac{-n_e e^2}{2m_e\left(\omega^2 - \omega_{ce}^2\right)} \frac{\mathrm{d}E_\theta^2}{\mathrm{d}z} = T_e \frac{\mathrm{d}n_e}{\mathrm{d}z}, \tag{10}$$

where T_e is given in energy unit. Integrating Eq. (10) from n_{e0} to n_e, the electrons density becomes a function of laser pulse intensity as

$$n_e(z) = n_{e0} \exp\left(-\frac{e^2 E_\theta^2(z)}{m_e T_e\left(\omega^2 - \omega_{ce}^2\right)}\right). \tag{11}$$

It should be noted that, in the intermediate intensities 10^{14}–10^{16} W cm^{-2} μm^2 with laser pulse duration of order a few ns, it is convenient to assume that T_e is constant and the dominant spatial dependence comes from the electron density (n_e). This equation shows that the electron density is modified by the pondermotive force. Furthermore, by substituting Eq. (11) into Eq. (9) and using Eq. (3), the nonlinear equation for electric field propagation in collisionless plasma is obtained as

$$\frac{\mathrm{d}^2}{\mathrm{d}z^2}E_\theta + \left(\frac{\omega^2}{c^2}\right)\left\{1 - \left(\frac{\omega_{pe0}^2}{\omega^2\left(1 - \frac{\omega_{ce}^2}{\omega^2}\right)}\right) \times \exp\left(-\frac{e^2 E_\theta^2}{m_e T_e\left(\omega^2 - \omega_{ce}^2\right)}\right)\right\} E_\theta = 0. \tag{12}$$

In addition, the dielectric constant of a magnetized collisionless isothermal plasma can be found as follows:

$$\varepsilon - 1 - \frac{\omega_{pe0}^2 \exp\left(-\frac{e^2 E_\theta^2}{m_e T_e\left(\omega^2 - \omega_{ce}^2\right)}\right)}{\left(\omega^2 - \omega_{ce}^2\right)}. \tag{13}$$

As we see, it should be noted that the electromagnetic wave equation coupled with the equations of momentum transfer, particle conservation and energy in their stationary form and they are solved for obtaining the dielectric permittivity. Now, with having the dielectric permittivity variations, the electric field propagation through the plasma along the Z direction is obtained as Eq. (3). In addition, with having the electric field variation according to the Eq. (3), we can reach the quiver velocities of electrons Eqs. (5) and (6) and their variations in plasma. Here, the velocities depend on the electrons density, temperature parameters. In other words, as E_θ was obtained from Eq. (3), in which the current density J_θ depends on plasma electrons' density and temperature, we can conclude that the theta component of electric field is a function of temperature. Furthermore, as we know, for obtaining the ponderomotive force we should use the nonlinear theory (Ref. [22]). Here, it should be mentioned that the following equations, i.e., momentum transfer and the laser pulse electric field propagation in plasma have the directional dependence on the nonlinear ponderomotive force, and according to it, we conclude that all of them are nonlinear. Now, we analyze the nonlinear wave equation, dielectric permittivity, and the plasma electron density distribution in such plasmas. Substituting Eq. (11) into Eqs. (12) and (13), results in the inhomogeneous dielectric permittivity and the nonlinear equation for the electric field propagation in collisionless isothermal magnetized plasma:

$$\varepsilon = 1 - \left(\frac{\omega_{pe0}^2}{\left(\omega^2 - \omega_{ce}^2\right)}\right) \times \exp\left(-\frac{e^2 E_\theta^2(z)}{m_e T_e\left(\omega^2 - \omega_{ce}^2\right)}\right), \tag{14}$$

$$\frac{\mathrm{d}^2}{\mathrm{d}z^2}E_\theta + \left(\frac{\omega^2}{c^2}\right)\left\{1 - \left(\frac{\omega_{pe0}^2}{\left(\omega^2 - \omega_{ce}^2\right)}\right) \times \exp\left(-\frac{e^2 E_\theta^2(z)}{m_e T_e\left(\omega^2 - \omega_{ce}^2\right)}\right)\right\} E_\theta = 0. \tag{15}$$

Results and discussion

In sect. II, we investigated the theoretical model and formulated the nonlinear ponderomotive force effects on isothermal collisionless magnetized underdense plasma. From Eq. (15), it is clear that this equation is intensively nonlinear and does not have any analytical solution. Thus, we use the fourth order Runge–Kutta method to solve this equation numerically and find the mentioned changes inside plasma. In order to obtain the laser pulse electric field, and the electron density profiles, we introduce some dimensionless variables as follows:

$$a = \frac{eE_\theta}{m_e c_s \omega}, \quad \Omega_{pe} = \frac{\omega_{pe0}}{\omega},$$

$$\xi \frac{z\omega}{c}, \quad \omega_{ce0}\frac{\omega_{ce}}{\omega},$$

where $c_s = \left(\frac{T_e}{m_e}\right)^{1/2}$ is the sound velocity. Using these dimensionless parameters, Eqs. (11), (14), and (15) are written as follows:

$$\frac{n_e}{n_{e0}} = \exp\left(\frac{-a^2}{1-\omega_{ce0}^2}\right), \tag{16}$$

$$\varepsilon = 1 - \frac{\Omega_{pe}^2}{1-\omega_{ce0}^2}\exp\left(\frac{-a^2}{1-\omega_{ce0}^2}\right), \tag{17}$$

$$\frac{d^2a}{d\xi^2} + \left\{1 - \left(\frac{\Omega_{pe}^2}{1-\omega_{ce0}^2}\right) \times \exp\left(\frac{-a^2}{1-\omega_{ce0}^2}\right)\right\}a = 0. \tag{18}$$

Here, in the case of collisionless magnetized isothermal plasmas the following parameters of the laser beam and plasma have been chosen: $\frac{\omega_{pe0}}{\omega} = 0.6$ and $T_{e0} = 1$ KeV. As we know, in plasma when the condition of ($n_e < n_{\text{critical}}$) is satisfied the plasma is underdense and in the other case when this condition is not satisfied the plasma is overdense. Here in this work, because of the parameter $\frac{n_e}{n_{\text{critical}}}$ is proportional with $\omega_{pe}^2/\omega^2\left(\frac{n_e}{n_{\text{critical}}} = \frac{\omega_{pe}^2}{\omega^2}\right)$, the range of $\frac{n_e}{n_{\text{critical}}}$ is 0.36 and n_e is very small in comparison with n_{critical} and the plasma is underdense. In Fig. 2a, b, the effect of increasing laser intensity on the propagation of the electric field and the distribution of the electron density of magnetized collisionless plasma of this medium are shown.

In these figures the normalized external magnetic field is taken as $\omega_{ce}/\omega = 0.4$. Since laser intensity is proportional to the square of amplitude, the amplitude of the electric field in plasma is increased with increasing laser intensity. Furthermore, due to further decrease in electron density in the higher laser intensities, more decrease in the wavelength of fields takes place. The steepening of the electron density distribution is enhanced with an increase in laser pulse intensity. It is seen that when laser intensity is increased, the oscillations of the electron density become highly peaked and at the same time, their wavelengths tend to decrease. The physical reason of this effect is that since λ (the oscillation wavelength) is proportional to $\varepsilon^{-1/2}$, by increasing the dielectric constant the wavelength of electron's oscillation is decreased. It is noticeable that by increasing laser pulse intensity, the oscillations of the effective permittivity become highly peaked and the wavelength of these oscillations is decreased. Furthermore, we can see that by increasing laser intensity the oscillations of the effective permittivity are deviated from sinusoidal shape more intensively. Figure 3a, b shows the effect of the magnitude increment of the external magnetic field on the profiles of the electric field and the electron density distribution of this medium in underdense collisionless isothermal magnetized plasma in the situation of constant laser intensity. In the presence of the external magnetic field parallel to the laser pulse propagation direction, the electrons density distribution is increased in comparison to unmagnetized collisionless isothermal plasma. In this case the electron density distribution is increased more in

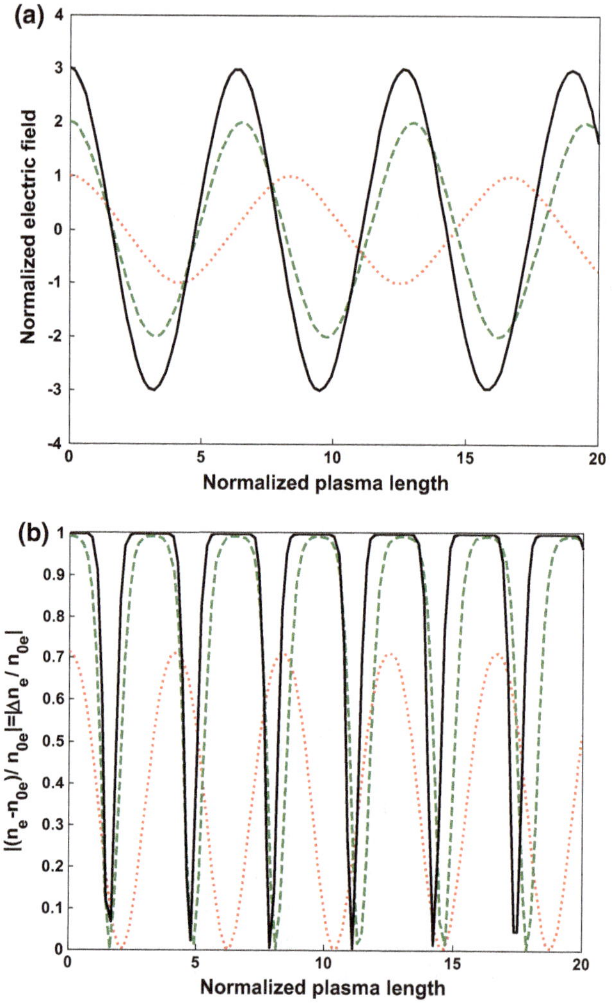

Fig. 2 The effect of increasing the laser pulse intensity in the case of the collisionless magnetized and isothermal plasma on the variations of the **a** normalized electric field, **b** normalized electrons density $\Delta n_e/n_{e0}$. The dimensionless laser pulse intensities are $a_0 = 1$ (*dotted line*), $a_0 = 2$ (*dotted-dashed line*), and $a_0 = 3$ (*solid line*). Electron temperature is $T_e = 1$ keV, the normalized cyclotron frequency is $\frac{\omega_{ce}}{\omega} = 0.4$, and the normalized plasma frequency is taken as $\frac{\omega_{pe0}}{\omega} = 0.6$

comparison with the electron density profile presented in the Ref. [34]. Here, the laser pulse should transfer more energy to the plasma electrons compared to unmagnetized plasma. It leads to an increase in the wavelength of the electric field. It is obvious that by increasing the external magnetic field, as a result of the increase of the electrons density distribution, the dielectric permittivity constant is decreased.

Summary and conclusion

In this paper, we have formulated the nonlinear propagation of an intense laser pulse through underdense magnetized collisionless isothermal plasma. From the results, one can see, in magnetized plasma, field wavelength is

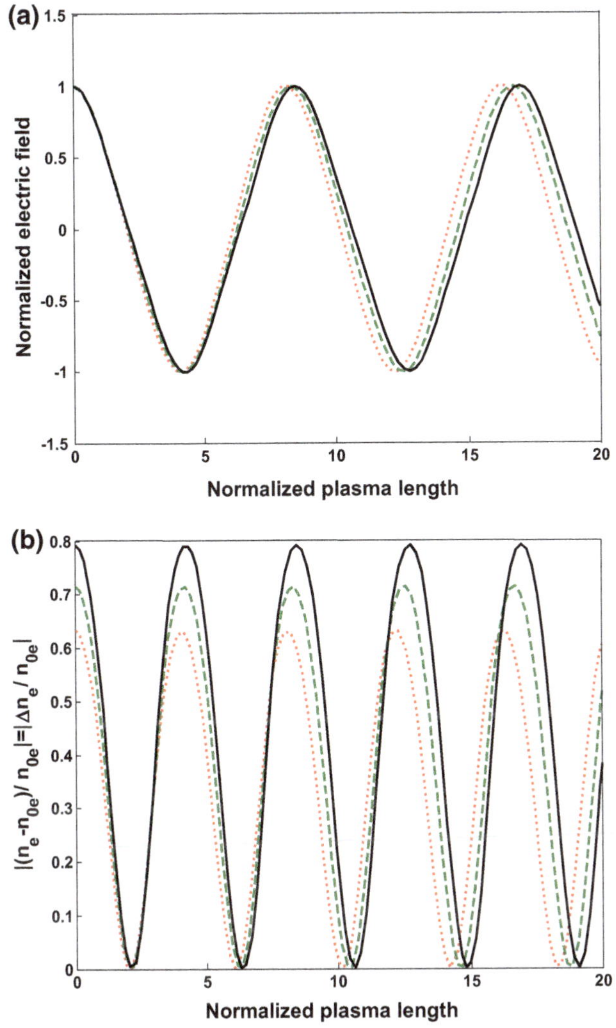

Fig. 3 The effect of increasing the external magnetic field in the case of the collisionless magnetized and isothermal plasma at the dimensionless laser pulse intensity of $a_0 = 1$ on the variations of **a**: normalized electric field, and **b** normalized electron density, $\Delta n_e/n_{e0}$ for different values of normalized cyclotron frequency $frac\omega_{ce}\omega = 0$ (*dotted line*), $\frac{\omega_{ce}}{\omega} = 0.4$ (*dotted-dashed line*), $\frac{\omega_{ce}}{\omega} = 0.6$ (*solid line*) is shown. Electron temperature is $T_e = 1$ keV, and the normalized plasma frequency is taken as $\frac{\omega_{pe0}}{\omega} = 0.6$

increased relative to the wavelength in unmagnetized plasma in the similar conditions. Also in the presence of the external magnetic field, the plasma electron density is increased and due to this effect, the effective permittivity of the mentioned plasma is decreased. Our numerical results are in good agreement with the mentioned work (R. L. Druce). Magnetized plasma plays as a capable medium to convert different inertial energies to tunable coherent radiation. This is because the electron density distribution is proportional to the plasma frequency and the magnitude of the external magnetic field at constant laser intensity. By adjusting the magnitude of the external magnetic field, the desired value of the electron density can be produced.

Acknowledgments This work was supported by Islamic Azad University, Karaj Branch. This work is a part of the project "The interaction of high intensity short laser pulse with the cold and magnetized plasma".

References

1. Ginzburg, V.L.: The Propagation of Electromagnetic Waves in Plasmas. Pergamon, Oxford (1970)
2. Chen, F.F.: Introduction to Plasma Physics and Controlled Fusion. Springer, Berlin (2010)
3. Mourou, G.A., Tajima, T., Bulanov, S.V.: Rev. Mod. Phys. **78**, 309 (2006)
4. Esarey, E., Goloviznin, R., Trans, I.E.E.E.: Plasma Sci. **24**, 252 (1996)
5. Cowan, T.E., Hunt, A.W., Phillips, T.W., Wilks, S.C., Perry, M.D., Brown, C., Fountain, W., Hatchett, S., Johnson, J., Key, M.H., Parnell, T., Pennington, D.M., Snavely, R.A., Takahashi, Y.: Rev. Sci. Instrum. **70**, 265 (1999)
6. Mangles, S.P.D., Thomas, A.G.R., Lundh, O., Landau, F., Kaluza, M.C., Persson, A., Wahistrom, C.G., Krusheinick, K., Najmudin, Z.: Phys. Plasmas **14**, 056702 (2007)
7. Gordienko, S., Pukhov, A.: Phys. Plasmas **12**, 043109 (2005)
8. Shokri, B., Niknam, A.R.: Phys. Plasmas **13**, 113110 (2006)
9. Davis, J., Petrov, G.M., Velikovich, A.L.: Phys. Plasmas **12**, 123102 (2009)
10. Wei, M.S., Mangles, S.P.D., Najmudin, Z., Walton, B., Gopal, A., Tatarakis, M., Krushelnic, K.: Phys. Phys. Rev. Lett. **93**, 155003 (2004)
11. Korovin, S.D., Kurkan, I.K., Loginov, S.V., Pegal, I.V., Polevin, S., Volkov, S.N., Zherlitsyn, A.: Laser Part. Beams **21**, 175 (2003)
12. Stamper, J.A., Bonder, S.E.: Phys. Rev. Lett. **37**, 435 (1976)
13. Tatarakis, M., Watts, I., Beg, F.N., Clark, E.L., Dangor, A.E., Gopal, A., Haines, M.G., Norreys, P.A., Wagner, U., Wei, M.-S., Zepf, M., Krushelnick, K.: Nature (London) **415**, 280 (2002)
14. Wagner, U., Tatarakis, M., Gopal, A., Beg, F.N., Clark, E.L., Dangor, A.E., Evanns, R.G., Haines, M.G., Norreys, P.A., Wei, M.S., Zeft, M., Krushelnick, K.: Phys. Rev. Lett. **70**, 026401 (2004)
15. Pukhov, A., meyer-ter-Vehn, J., Gopal, A.: Phys. Rev. Lett. **76**, 3975 (1996)
16. Qiao, B., Zhu, S., Zheng, C.Y., He, X.T.: Phys. Plasmas **12**, 053104 (2005)
17. Presura, R., Plechaty, C., Martinez, D., Bakeman, M.S., Laca, P.J., Haefner, C., Astanovit-skiy, A.L., Thompson, M., Trans, I.E.E.E.: Plasma Sci. **36**, 17 (2008)
18. Struve, K.W., Porter, J.L., Rovang, D.C.: Megagauss field generation for highenergy-density plasma science experiments, SANDIA report no. SAND2008-7015, (2008)
19. Siemon, R.E., Bauer, B.S., Awe, T.J., Angelova, M.A., Fuelling, S., Goodrich, T., Lindemuth, I.R., Makhin, V., Atchison, W.L., Faehl, R.J., Reinovsky, R.E., Turchi, P.J., Degnan, J.H., Ruden, E.L., Frese, M.H., Garanin, S.F., Mokhov, V.N.: J. Fusion Energy **27**, 235 (2008)
20. Portugall, O., Puhlmann, N., Mller, H.U., Barczewski, M., Stolpe, I., von Ortenberg, M.: J. Phys. D **32**, 2354 (1999)
21. Clark, E.L., et al.: Phys. Rev. Lett. **84**, 6703 (2000)
22. Eliezer, S.: The interaction of high power lasers with plasmas, Chap. III. IOP, Bristol (2002)
23. Liu, S.C.S., Tripathi, V.K.: Interaction of electromagnetic waves with electron beams and plasmas, Chap. VI. World Scientific, Singapore (1994)
24. Sharma, A., Tripathi, V.K.: Phys. Plasmas **16**, 043103 (2009)

25. Gupta, D.N., Hur, M.S., Suk, H.: Appl. Phys. Lett. **91**, 081505 (2007)
26. Yugami, N., Higashiguchi, T., Gao, H., Sakaii, S., Takahashi, K., Ito, H., Nishida, Y., Katsouleas, T.: Phys. Rev. Lett. **89**, 065003 (2002)
27. Dorranian, D., Starodubtsev, M., Kawakami, H., Ito, H., Yugami, N., Nishid, Y.: Phys. Rev. E **68**, 026409 (2003)
28. Dorranian, D., Ghoranneviss, M., Starodubtsev, M., Ito, H., Yugami, N., Nishida, Y.: Phys. Lett. A **331**, 77 (2004)
29. Tsung, F.S., Morales, G.J., Tonge, J.: Phys. Plasmas **14**, 042101 (2007)
30. Norreys, P.A., Beg, F.N., Sentoku, Y., Silva, L.O., Smith, R.A., Trines, R.M.G.M.: Phys. Plasmas **16**, 041002 (2009)
31. Davis, J., Petrov, G.M., Velikovich, A.L.: Phys. Plasmas **13**, 041002 (2009)
32. Pathak, V.B., Tripathi, V.K.: Phys. Plasmas **14**, 022105 (2007)
33. Druce, R.L., Kristiansen, M., Hagler, M.O.: Proceedings of the Indian Academy of Sciences - Section A **86**, 255–263 (1977)
34. Sadighi-Bonabi, R., Etehadi-Abari, M.: Phys. Plasmas **17**, 032101 (2010)
35. Nesterov, A.V., Nieziev, V.G.: J. Phys. D Appl. Phys. **33**, 1817–1822 (2000)
36. Niziev, V.G., Nesterov, A.V.: J. Phys. D Appl. Phys. **32**, 1455–1461 (1999)
37. Jordan, R.H., Hall, D.G.: Opt. Lett. **19**, 427–429 (1994)

8

Influence of ion-neutral collision parameters on dynamic structure of magnetized sheath during plasma immersion ion implantation

Mansour Khoram[1] · **Hamid Ghomi**[2]

Abstract A cold magnetized plasma sheath is considered to examine the gas pressure effect on the sheath dynamics. A fluid model is used to describe the plasma sheath dynamic. The governing fluid equations in the plasma are solved from plasma center to the target using the finite difference method and some convenient initial and boundary conditions at the plasma center and target. It is found that, the ion-neutral collision has significant effect on the dynamic characteristics of the high-voltage sheath in the plasma immersion ion implantation (PIII). It means that, the temporal profile of the ion dose on the target and sheath width are decreased by increasing the gas pressure. Also, the gas pressure substantially diminishes the temporal psychograph of ion incident angle on the target.

Keywords Dynamic sheath · Gas pressure · PIII · Magnetized sheath

Introduction

Plasma immersion ion implantation (PIII) [1, 2] has been shown to be an effective technique for semiconductor fabrication and material processing [3–6]. It emulates conventional ion-beam ion implantation (IBII) in a number of areas. For example, it has high sample throughput (high current density) and it is a paralleled processing technique in which the implantation time is independent of the wafer size.

The dynamic sheath model plays a very important role in PIII processes because it is used to predict process parameters and implantation results including the implant doses and energies. The optimum characteristics of plasma system can be determined from an appropriate and reasonable numerical analysis of the plasma-sheath dynamic. Unfortunately, accurate modeling and prediction of the ion implantation energy and dose in PIII is quite difficult because it is a complicated function of inter-related processing conditions such as plasma density, pulse duration, accelerating voltage, external magnetic field, ion mass, ion temperature, and ion charge state.

There has recently been an explosion of interest in the dynamic behavior of sheaths which is formed at an electrode biased with a pulsed negative voltage. This problem is of special interest in PIII technology, where ions are extracted from plasma, accelerated by a high potential drop in the sheath and injected into the surface layer of a material being treated.

In a simple model of PIII, the confined homogeneous plasma is brought in contiguity with a flat conducting electrode or target. A series of negative high-voltage pulses are applied to the electrode. As a result, electrons are repelled from the electrode and move back toward the plasma center leaving a positive ion sheath. This initial ion sheath, which is immediately formed after applying the negative voltage on the electrode, is called ion matrix sheath [7–9]. The matrix sheath is created on a time scale of the order of $1/\omega_p$ (where ω_p is the electron plasma frequency). At a characteristic time scale of the order of $1/\Omega_p$ (where Ω_p is the ion plasma frequency), the ions begin to move towards the electrode. Therefore, the sheath edge propagates into the plasma and a rarefactive ion

✉ Mansour Khoram
m.khoramabadi@srbiau.ac.ir

1 Department of Physics, Borujerd Branch, Islamic Azad University, Borujerd, Iran

2 Laser and Plasma Research Institute, Shahid Beheshti University, Evin, Tehran 1983963113, Iran

distribution begins to propagate into the sheath. Thickness of the time-dependent collisionless sheath is given by [10];

$$d(t) = d_0\left(\frac{2}{3}\Omega_p t + 1\right)^{1/3}, \tag{1}$$

where, $d_0 = \sqrt{2\varepsilon_0 V_T / en_0}$ is the thickness of the ion matrix sheath, n_0 is the plasma density, V_T is the applied negative voltage to the target, ε_0 is the free space permittivity and e is the unit of electric charge.

Navab Safa et al. [11] have investigated the sheath dynamics and implantation profiles during the PIII process on a long step shaped target in the presence of an external DC magnetic field and showed that the magnetic field inclination angle strongly affects the ion implanted dose and ion energy on the target surfaces. Minghao et al. [12] have analyzed the ion temperature and collision parameter effects on the plasma-sheath structure in the RF warm plasma-sheath. They have solved plasma fluid equations from plasma center to the wall (biased with a low-voltage sinusoidal source) and presented their results in time-averaged form. It has been examined [13, 14] the ion temperature and collision effects on the Bohm criteria in a multiply ionized and electronegative warm plasma sheath, respectively, and using these Bohm criterions it has been investigated [15, 16] the ion temperature and ion-neutral collision effects on the steady state plasma-sheath structure and sheath formation in warm magnetized plasmas.

In this paper, we will study the PIII process for different gas pressure to investigate the influence of ion-neutral collision on the dynamic behavior of a pulsed collisional magnetized sheath. Since the ion incident angle has a significant effect on the ion penetration at the target, we have used an external magnetic field in our model to be able to control the ion incident angle. Here, we use a fluid model to investigate the time evolution of a sheath expanding into the plasma. The model equations are solved through an implicit finite difference scheme.

After introduction, in "Fluid model and basic equations", a collisional magnetized plasma is formulated and the proper initial and boundary conditions are nominated. In "Numerical results and discussion", the numerical results and their explanation are presented. A summary and conclusions are presented in "Summary and conclusions".

Fluid model and basic equations

The implantation geometry is schematically displayed in Fig. 1. A collisional magnetized plasma limited at one side by a conducting flat target is considered to discuss the formation of dynamic sheath. The position of the target is taken at $x = 0$. We assume at the moment $t = 0$, the plasma which is located at the $x > 0$ is filled with stationary singly charged ions and the charge neutrality condition $n_i = n_e = n_0$ is fulfilled there. At the moment $t = 0^+$, the bias on the target is switched on from zero to a negative bias $V_T(t)$, drawing ions to the target and repelling electrons to the plasma. Under this configuration, the non-neutral plasma so-called plasma sheath is formed around the target.

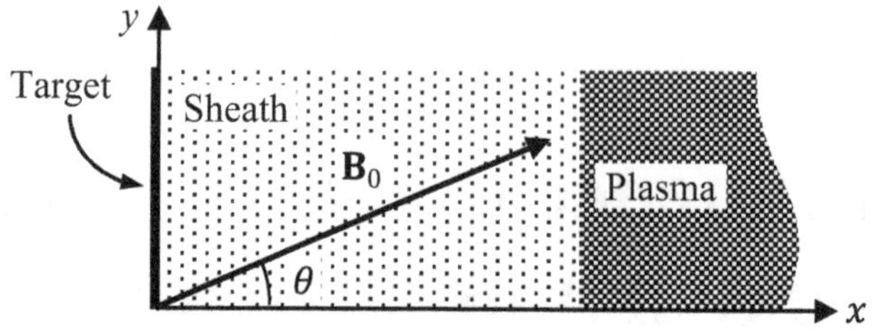

Fig. 1 A magnetized plasma-sheath as the simulation zone

On the other hand, far from the plasma sheath, there is a cold ($T_i = 0$) plasma with electron temperature T_e where the electric potential is $V = 0$. The fluid model is used to formulate the charged particles as electric fluids. The ion density and ion velocity distributions, as well as the electric potential distribution in the plasma-sheath evolve self-consistently in the fluid equations frame. The continuity and motion equations of the ions supplemented by the collisional term are:

$$\frac{\partial n_i}{\partial t} + \nabla \cdot (n_i \mathbf{v}) = 0, \tag{2}$$

$$mn_i\left(\frac{\partial}{\partial t} + \mathbf{v}\cdot\nabla\right)\mathbf{v} = en_i(\mathbf{E} + \mathbf{v}\times\mathbf{B}) - mn_i\nu\mathbf{v}, \tag{3}$$

where $\mathbf{E} = -\nabla V$ (with V as the electric potential) is the electric field, m, $\mathbf{v}$ and ν are the mass, velocity and collision frequency of ion, respectively.

Electrons in the electrostatic well of the target are in the thermal equilibrium, and their density n_e obeys the Boltzmann relation

$$n_e = n_0 \exp\left(\frac{eV}{k_B T_e}\right), \tag{4}$$

where k_B is the constant of Boltzmann. Also, the electric potential and ion and electron densities satisfy the Poisson's equation as follows:

$$\nabla^2 V = -\frac{e}{\varepsilon_0}(n_i - n_e). \tag{5}$$

On the basis of the above equations and the ion-neutral collision frequency $\nu = n_g \sigma_s (v/c_s)^{\beta} v$ (where $n_g = P_g / k_B T_g$, P_g and T_g are the density, pressure and temperature of neutral gas, respectively, σ_s is the ion-neutral collision cross section in the sound velocity c_s and β so-called the collision power parameter is a real number between -1 and 0 that $\beta = -1$ introduces the constant collision frequency, while $\beta = 0$ presents the constant collision mean free path), one can describe the dynamic structure of plasma sheath around target.

For the sake of convenience, it is efficient to introduce some normalized parameters and variables,

$$N_i = \frac{n_i}{n_0}, \quad N_e = \frac{n_e}{n_0}, \quad \mathbf{X} = \frac{\mathbf{x}}{\lambda_D}, \quad \mathbf{\Omega}_c = \frac{e\mathbf{B}}{m}$$

$$\Omega_p = \sqrt{\frac{n_0 e^2}{m\varepsilon_0}}, \quad \alpha = \frac{\lambda_D}{\lambda} = \lambda_D n_g \sigma_s, \quad \mathbf{\Omega} = \frac{\mathbf{\Omega}_c}{\Omega_p},$$

$$\tau = \Omega_p t, \quad \phi = \frac{eV}{k_B T_e}$$

$$\lambda_D = \sqrt{\frac{\varepsilon_0 k_B T_e}{n_0 e^2}}, \quad c_s = \sqrt{\frac{k_B T_e}{m}} = \lambda_D \Omega_p,$$

$$u_j = \frac{v_j}{c_s} (j = x, y, z)$$

where $\mathbf{\Omega}_c$ is the ion cyclotron frequency and $\mathbf{\Omega}$ is the ion cyclotron frequency to ion plasma frequency ratio. Indeed, $\mathbf{\Omega}_c$ introduces a vector paralleled to the magnetic field $\mathbf{B} = B_0(\cos\theta_0\hat{\mathbf{x}} + \sin\theta_0\hat{\mathbf{y}})$ with the magnitude eB_0/m. τ is the normalized time by $1/\Omega_p$, ϕ is the electric potential normalized by k_BT_e/e, α is the ion-neutral collision parameter, $\lambda = 1/n_g\sigma_s$ is the mean free path of ion collision, $\mathbf{u}$ is the ion velocity normalized by the ion sound velocity c_s, N_i and N_e are the ion and electron density, respectively, normalized by n_0, and $\mathbf{X}$ is the space coordinate normalized by Debye length λ_D. Since the target is planar, $\nabla \to (\partial/\partial x)\hat{\mathbf{x}}$ and the normalized one-dimensional form of Eqs. (2)–(5) will be as follows:

$$\frac{\partial N_i}{\partial \tau} + \frac{\partial (N_i u_x)}{\partial X} = 0, \tag{6}$$

$$\left(\frac{\partial}{\partial \tau} + u_x \frac{\partial}{\partial X}\right)\mathbf{u} = -\frac{\partial \phi}{\partial X}\hat{\mathbf{x}} + \mathbf{u} \times \mathbf{\Omega} - \alpha u^{\beta+1}\mathbf{u} \tag{7}$$

$$N_e = \exp(\phi) \tag{8}$$

$$\frac{\partial^2 \phi}{\partial X^2} = N_e - N_i. \tag{9}$$

Using the definition $\mathbf{\Omega} = eB_0(\cos\theta_0\hat{\mathbf{x}} + \sin\theta_0\hat{\mathbf{y}})/m\Omega_p = \Omega_0(\cos\theta_0\hat{\mathbf{x}} + \sin\theta_0\hat{\mathbf{y}})$, one can rewrite the vector Eq. (7) in the scalar form to find,

$$\left(\frac{\partial}{\partial \tau} + u_x \frac{\partial}{\partial X}\right)u_x = -\frac{\partial \phi}{\partial X} - \Omega_0 u_z \sin\theta_0 - \alpha u^{\beta+1} u_x \tag{10}$$

$$\left(\frac{\partial}{\partial \tau} + u_x \frac{\partial}{\partial X}\right)u_y = \Omega_0 u_z \cos\theta_0 - \alpha u^{\beta+1} u_y \tag{11}$$

$$\left(\frac{\partial}{\partial \tau} + u_x \frac{\partial}{\partial X}\right)u_z = \Omega_0 u_x \sin\theta_0 - \Omega_0 u_y \cos\theta_0 - \alpha u^{\beta+1} u_z. \tag{12}$$

Equations (6) and (8)–(12) make a complete set of equations describing the dynamic structure of plasma sheath. We solve the equations in a region from plasma to target supplied by a negative high voltage $V_T(t) = V_P[1 - \exp(-t/t_r)]$ in which, V_P and t_r are the voltage amplitude and rise time of pulse, respectively.

In order to examine the dynamic structure of plasma sheath and investigate the influence of plasma parameters on the sheath structure, the complete set of equations are numerically solved using a second-order finite difference scheme in space and a first-order finite difference scheme in time. After discretization of equations in full implicit finite difference scheme, some linearization is required. In order to linearize the equations in time, we used Taylor's expansion approximation; $f(t + \Delta t) \approx f(t) + (\partial f(t)/\partial t)\Delta t$.

To solve the Eqs. (6) and (8)–(12), some proper initial and boundary conditions are necessary. The boundary conditions at the sheath edge and on the target for the time interval $0 < \tau < \tau_p$ are:

$$\phi(0,\tau) = \phi_T(\tau), \phi(L,\tau) = 0,$$
$$u_x(L,\tau) = u_y(L,\tau) = u_z(L,\tau) = 0, N_i(L,\tau) = 1,$$

where $\tau_p = \Omega_p t_p$ is the normalized simulation time or normalized pulse duration, $\phi_T = eV_T/k_BT_e$ is the normalized voltage of target and L is the normalized length of the simulation area that introduces the location of a point in the plasma, sufficiently far from the sheath edge. Also, the initial conditions for the space interval $0 < X < L$ are:

$$\phi(X,0) = 0, u_x(X,0) = u_y(X,0) = u_z(X,0) = 0,$$
$$N_i(X,0) = 1.$$

Numerical results and discussion

The normalized equations can now be solved assuming some constant parameters. Atomic nitrogen is used as the ion that should be implanted on the target surface. Nitrogen implantation of steel surfaces by PIII has shown to improve mechanical properties and corrosion performance [17, 18]. Nitrogen plasma density is assumed $n_o = 5 \times 10^{14}$ m^{-3} ($\Omega_p = 7.8843 \times 10^6$Hz), electron temperature $T_e = 1$ eV, magnetic field amplitude $B_0 = 0.5$T ($\Omega_c = 3.44 \times 10^6$Hz) and the magnetic field deviation angle $\theta_0 = \pi/3$. According to the assuming data, one can find $\lambda_D = 332.85$ μm and $c_s = 2624.3$m/s. We choose $\sigma_s = 4 \times 10^{-19}$m^2 as the ion-neutral collision cross section and alter the gas pressure $P_g = k_BT_gn_g$ and for the both values of collision power parameter $\beta = 0$ and -1 to examine their effect on time evolution of the sheath dynamic structure.

To study the dynamic structure of sheath, we have numerically solved the complete set of equations assuming $\tau_r = \Omega_p t_r = 0.1\tau_p$, $\phi_p = eV_p/k_BT_e = -10000$, $\Delta\tau = \Omega_p\Delta t = 4 \times 10^{-3}$ as the normalized time steps, $n_t = 40{,}000$ as the total number of time steps, $\Delta X =$

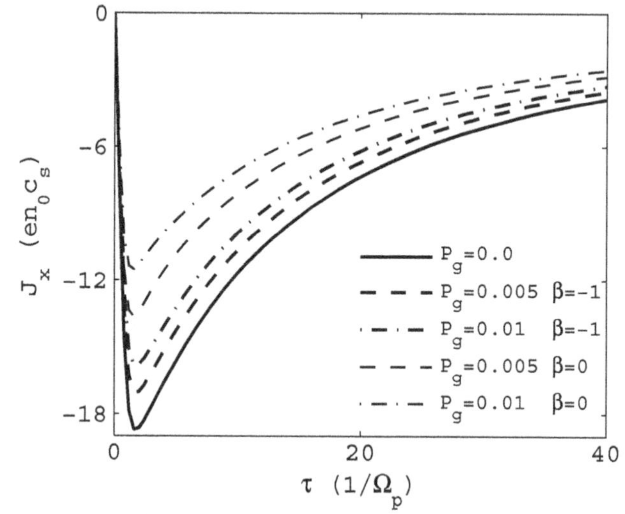

Fig. 2 Temporal variations of the ion current density J_x normalized by $J_0 = en_0c_s = 0.2099$ A/m^2 as a function of the gas pressure P_g and collision power parameter β. The constant parameters in the simulation are; $\phi_p = -10{,}000$, $\Delta\tau = 4 \times 10^{-3}$, $n_t = 40{,}000$, $\Delta X = 0.5$, $n_x = 1400$, $\tau_p = 160$, $1/\Omega_p = 0.1268\ \mu$s and $L = 700$

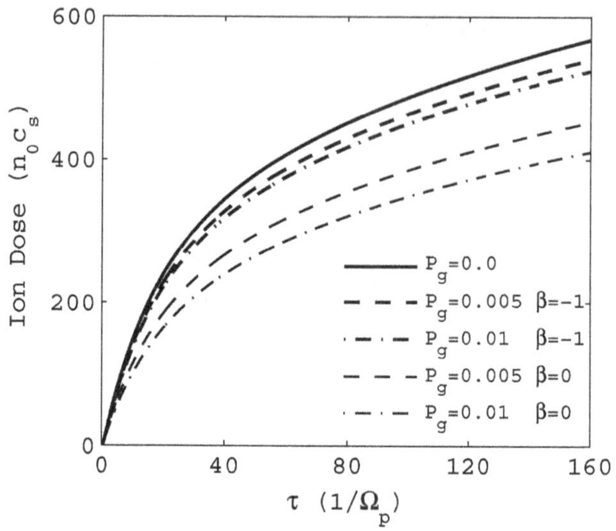

Fig. 3 Temporal variations of the ion dose on the target normalized by $n_0c_s = 1.3121 \times 10^{18}$ 1/m^{2}s as a function of the ion collision parameters. The constant parameters are the same as Fig. 2

$\Delta x/\lambda_D = 0.5$ as the normalized space steps and $n_x = 1400$ as the total number of space steps. Using the time and space data, one can find the pulse duration $t_p = n_t\Delta t = 20.293\ \mu$s ($\tau_p = 160$) and the simulation space $L\lambda_D = n_x\Delta x = 2.33$ cm ($L = 700$). The variable parameters are the gas pressure and collision power parameter which include the values $P_g = 0$, 0.05 and 0.1 Pa and $\beta = 0$ and -1, respectively. The simulation results are shown in Figs. 2, 3, 4 and 5.

Temporal variations of ion current density $J_x = en_iu_x$ perpendicular to the target surface is shown in Fig. 2 as a function of gas pressure and collision power parameter. It can be seen that the ion current density on the target increases from zero, reaches to a maximum value, decreases and saturates to a constant value. This figure shows that the both of gas pressure and collision power parameter reduce the maximum and saturation values of ion current density. This is because the ion collision term in the ion motion Eq. (3) decelerates the ion velocity toward the target.

Figure 3 displays the temporal profile of ion dose on the target as a function of gas pressure and collision power parameter. Since the ion dose on the target is defined by $D_x = \int_0^t J_x dt$ and is directly related to J_x, it has the same dependency on the ion collision parameters such as the ion current density. As it can be seen in Fig. 3, the both collision parameters P_g and β, reduce the ion dose on the target. According to Fig. 3, one can find the necessary time to achieve the required dose for specified ion collision parameters.

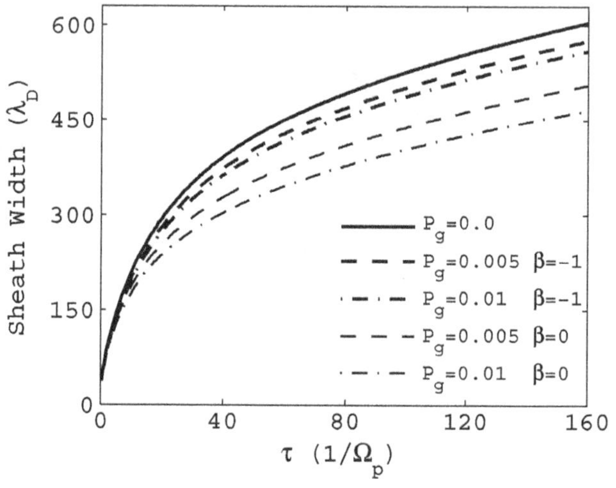

Fig. 4 Temporal variations of the sheath width normalized by Debye length $\lambda_D = 0.3328$ mm as a function of ion collision parameters P_g and β. The constant parameters are the same as Fig. 2

We have defined the sheath edge as a point between target and plasma in which the normalized electric potential reduces to $\phi = -0.01$. The distance between target and sheath edge is called the sheath width that has a critical role in PIII. Figure 4 exhibits temporal variations of sheath width normalized by Debye length $\lambda_D = 0.3328$mm as a function of the ion collision parameters. It is clear that the sheath width is a descending function of the both collision parameters P_g and β.

Ion incident angle is defined as the angle between ion velocity and normal vector on the target surface. It is then calculated via; $\mathrm{Arctan}\left(\sqrt{u_y^2 + u_z^2}/u_x\right)$. The vertical

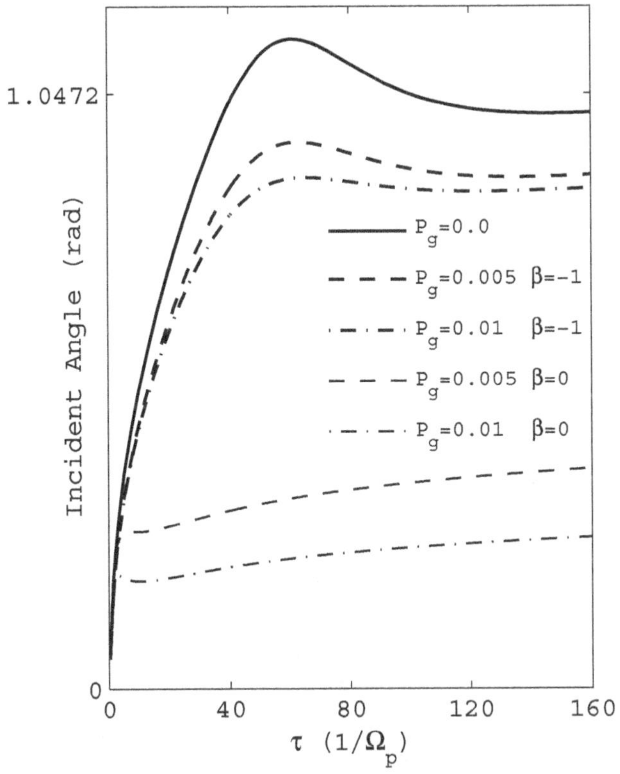

Fig. 5 Temporal variations of ion incident angle on the target as a function of ion collision parameters P_g and β. The constant parameters are the same as Fig. 2

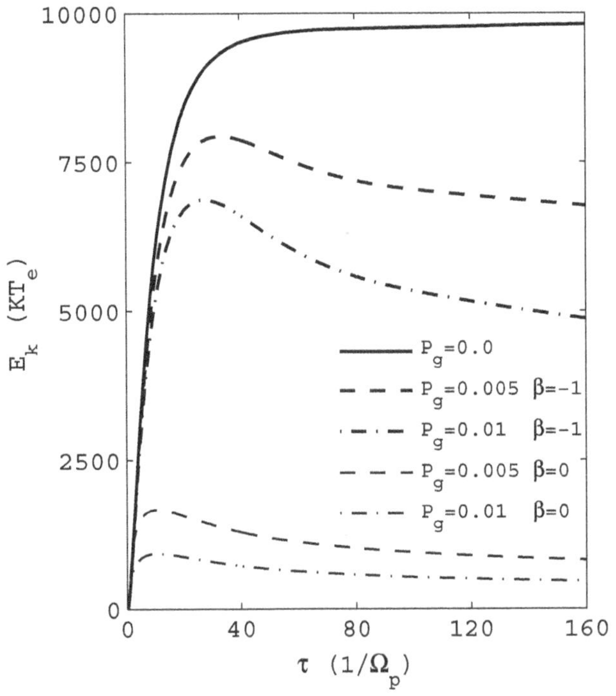

Fig. 6 Temporal variations of ion impact energy on the target normalized by $k_BT_e = 1$ eV as a function of the ion collision parameters P_g and β. The constant parameters are the same as Fig. 2

incident of the ion on the target increases the ion penetration while oblique incident of ion on the target grows up the sputtering of the target surface.

Temporal variations of ion incident angle are depicted in Fig. 5. This figure shows that the ion incident angle at the target surface increases from zero and after some fluctuations caused by the magnetic field is saturated to a constant value. In the absence of collision ($P_g = 0$), the constant incident angle is approximated to the magnetic field deviation angle $\theta_0 = \pi/3 = 1.0472\,\text{rad}$. It means the ion velocity on the target is paralleled to the external magnetic field. As a result, the magnetic force and fluctuations caused by it will be removed as soon as the ion velocity is constantly paralleled with the magnetic field. According to Fig. 5, the gas pressure reduces the saturated ion incident angle and diminishes the magnetic field effects.

Also, Fig. 5 shows that the ion incident angle is strongly reduced by increasing the collision power parameter from -1 (for constant collision frequency) to 0 (for constant collision mean free path). Indeed, further increasing the collision power parameter and the gas pressure reduces the ion incident angle. In other words, increasing the collision parameters turn the oblique incident to the perpendicular incident.

Time evolution of ion impact energy on the target surface defined by $E_k = m_i\left(v_{ix}^2 + v_{iy}^2 + v_{iz}^2\right)/2 = k_BT_e\left(u_x^2 + u_y^2 + u_z^2\right)/2$ is displayed in Fig. 6 as a function of the ion collision parameters P_g and β. This figure shows that the ion impact energy on the target surface increases from zero and is eventually saturated to a constant value by time. The saturation value of ion energy on the target is reduced by gas pressure and collision power parameter. As one can see, the ions never gain the full energy of bias voltage even in the absence of collision. Since ion transit time across the sheath is large (on the order of $1/\Omega_p$) and sheath expands during ion flight across the sheath, both the shape and magnitude of the potential barrier vary, so that the implanted ions do not get the full bias voltage. In other words, displacement current across the expanding sheath leads to an increase in the ion implanted current and causes a decrease in the implanted ion energy with respect to the stationary sheath with the same parameters [19, 20]. According to Fig. 6, one can anticipate the ion implanting energy for specified gas pressure and its proper collision model.

Summary and conclusions

In order to investigate the influence of the gas pressure and ion collision model on the dynamic behavior of plasma sheath in the PIII process, we have used the fluid model of plasma and numerically solved time-dependent equations of a pulsed magnetized plasma through an implicit finite difference scheme. Temporal variations of normalized current density, dose, incident angle and impact energy of ions on the target surface as well as sheath width were presented in the high-voltage sheath.

Calculations show that temporal profile of the ion current density and ion dose in the plasma-sheath decrease by increasing the gas pressure and collision power parameter. Sheath width is a descending function of the both collision parameters. Temporal profile of the ion incident angle on the target begins to increase from zero and after some fluctuations caused by magnetic field is saturated to a constant value. The constant value of incident angle is the same as the magnetic field deviation angle $\theta_0 = \pi/3$ in the absence of collision. It means that the magnetic force and fluctuations arising from it will be removed over time. The collision parameters including gas pressure and collision power parameter strongly decrease the saturated ion incident angle.

Also, temporal profile of the ion impact energy on the target surface shows a growing up from zero and saturating to a constant value over time. The constant value is significantly reduced by gas pressure and collision power parameter. The profile shows that in the dynamic sheath, unlike the static sheath, the ions never gain the full energy of bias voltage. This is because the dynamic sheath expands during ion flight across the sheath and both the shape and magnitude of the potential barrier vary. Using the temporal profile of the ion impact energy one can anticipate the ion implanting energy for specified collision parameters.

Acknowledgments This study was supported by Islamic Azad University, Borujerd Branch, Iran. The authors would like to acknowledge staffs of the university.

References

1. Conrad, J.R., Radtke, J., Dodd, R.A., Worzala, F.J.: Plasma source ion implantation technique for surface modification of materials. Appl. Phys. **62**, 4591 (1987)
2. Tendys, J., Donnelly, I.J., Kenny, M.J., Pollock, J.T.A.: Plasma immersion ion implantation using plasmas generated by radio frequency technique. Appl. Phys. Lett. **53**, 2143–2145 (1988)
3. Qin, S., McGruer, N., Chan, C., Warner, K.: Plasma immersion ion implantation doping using a microwave multipolar bucket plasma. IEEE Trans. Electron Devices **39**(10), 2354–2358 (1992)
4. Qin, S., Ghan, C.: Plasma Immersion Ion implantation doping experiments for microelectronics. J. Vac. Sci. Technol. B **12**(2), 962–968 (1994)
5. Cheung, N.W.: Plasma immersion ion implantation for ULSI procrssing. Nucl. Instrum. Meth. Phys. Res. B **55**, 811–820 (1991)
6. Pico, C.A., Lieberman, M.A., Cheung, N.W.: PMOS integrated circuit fabrication using BF3 plasma immersion ion implantation. J. Electron. Mater. **21**, 75–79 (1992)
7. Sander, K.F.: Theory of a thick dynamic positive-ion sheath. J. Plasma Phys. **3**, 353–370 (1969)
8. Lieberman, M.A., Lichtenberg, A.J.: Principles of plasma discharges and materials processing. Wiley, New York (1994)
9. Chu, P.K., Qin, S., Chan, C., Cheung, N.W., Larson, L.A.: Plasma immersion ion implantation—a fledgling technique for semiconductor processing. Mat. Sci. Eng. R **17**, 207–280 (1996)
10. Lieberman, M.A.: Model of plasma immersion ion implantation. J. Appl. Phys. **66**, 2926–2929 (1989)
11. Navab Safa, N., Ghomi, H., Khoramabadi, M., Ghasemi, S., Niknam, A.R.: External magnetic field effect on the sheath dynamics and implantation profiles in the vicinity of a long step shaped target in plasma immersion ion implantation. Vacuum **101**, 354–359 (2014)
12. Minghao, L., Yu, Z., Wanyu, D., Jinyuan, L., Xiaogang, W.: Effects of ion temperature on collisionless and collisional rf sheath. Plasma Sci. Technol. **8**(5), 544 (2006)
13. Khoramabadi, M., Ghomi, H., Shukla, P.K.: The Bohm-sheath criterion in plasmas containing electrons and multiply charged ions. J. Plasma Phys. **79**, 267 (2012)
14. Ghomi, H., Khoramabadi, M., Shukla, P.K., Ghorannevis, M.: Plasma sheath criterion in thermal electronegative plasmas. J. Appl. Phys. **108**, 063302 (2010)
15. Ghomi, H., Khoramabadi, M.: Influence of ion temperature on plasma sheath transition. J. Plasma Phys. **76**, 247 (2010)
16. Khoramabadi, M., Ghomi, H., Shukla, P.K.: Numerical investigation of the ion temperature effects on magnetized dc plasma sheath. J. Plasma Phys. **109**, 073307 (2011)
17. Tian, X.B., Chu, P.K.: Electrochemical corrosion properties of AISI304 steel treated by low-temperature plasma immersion ion implantation. Scr. Mater. **43**, 417–422 (2000)
18. Tian, X.B., Wei, C.B., Yang, S.Q., Fu, R.K.Y., Chu, P.K.: Corrosion resistance improvement of magnesium alloy using nitrogen plasma ion implantation. Surf. Coat. Technol. **198**, 454–458 (2005)
19. Kostov, K.G., Barroso, J.J., Ueda, M.: Two dimensional computer simulation of plasma immersion ion implantation. Braz. J. Phys. **34**, 1689–1695 (2004)
20. Wood, B.P.: Displacement current and multiple pulse effects in plasma source ion implantation. J. Appl. Phys. **73**, 4770–4778 (1993)

9

Space charge formation and Bohm's criterion in the edge of thermal electronegative plasma

Kiomars Yasserian[1] · Morteza Aslaninejad[2]

Abstract The collisional electronegative plasma space charge is investigated in the presence of the thermal positive ions. The Boltzmann distribution is assumed for electrons and negative ions and fluid equations are used to treat the accelerated positive ion through the sheath region. The influence of the positive ion temperature on the profile of the space charge is obtained for different negative ion concentration and negative ion temperature for collisionless and collisional cases. It is shown that the position of the space charge peak is independent of positive ion temperature while its amplitude depends on the positive ion temperature. The presence of the negative ion leads to damping of the space charge amplitude. In addition the thermal effect of the positive ion on the kinetic energy of the ion extracted from an ion source is studied in difference of collisionality and electronegativity. It is shown that, in the presence of thermal positive ion, the influence of the negative ion temperature on the sheath characteristics disappears. It is observed that in the presence of the hot positive ion, the two-fold feature of the space charge starts at higher values of negative ion temperature which is more pronounced in collisional case. Finally, the influences of the positive and negative ion temperature, as well as the electronegativity and collisionality on the net electric current are studied.

Keywords Plasma sheath · Ion source · Thermal plasma · Electronegative plasma

✉ Kiomars Yasserian
kiomars.yaserian@kiau.ac.ir

[1] Department of Physics, Karaj Branch, Islamic Azad University, Karaj, Iran

[2] Institute for Research in Fundamental Sciences (IPM), School of Particles and Accelerators, P.O. Box 19395-5531, Tehran, Iran

Introduction

Understanding sheath formation between plasma and an absorbing wall or electrode is necessary for many plasma devices as well as plasma processing. The transition layer between plasma and conducting wall can be separated into two sub regions as presheath and sheath. The presheath which is close to the plasma core determines the ion flux and the sheath determines the energy of the ions. In ion sources and for extracted ions from the core of ion source, it is important to predict the kinetic energy and the current of the ions through plasma. The formation of a sheath adjacent to the surrounding wall of the plasma is necessary if the wall is at a different potential from the quasi-neutral plasma. The potential of plasma surrounding wall is negative with respect to the plasma potential and therefore, the number of negative and positive species in the sheath region is different and consequently a space charge forms. Indeed, the positive space charge is responsible to maintain the plasma potential. Understanding the space charge formation and its mechanism is necessary for plasma surface interaction as well as plasma diagnostics such as of Langmuir probe [1–7].

In the sheath region, which is in contact to the wall, because of the absence of the ionizing collisions, the positive ion flux velocity remains constant. In the presheath region which is located between sheath and center of discharge, there are inelastic collisions such as ionizing collisions which leads to flux of the ions and increases toward the wall. The size of the sheath is very small in comparison with the presheath and the typical extension of the sheath is given by electron Debye length λ_D. Therefore, in the sheath region, the inelastic collisions can be ignored. In other word, for the asymptotic case $\frac{\lambda_D}{L} \to 0$, the ionizing collisions can be ignored where L is the plasma dimension which is usually in the order of ionizing length ([2] and references therein). However, the presence of elastic collisions between the accelerated ions and the neutral atom in the background sheath region can significantly influence the plasma-sheath characteristics. The collision frequency depends on the working pressure and degree of ionization of the plasma. Many papers have investigated the sheath and presheath regions in the presence of such collisions [8–11].

In addition, in most cases of interest such as plasma fusion device or plasma spray, the temperature of positive ions reaching the electrode cannot be ignored and its temperature should be included in the governing equations. The Refs. [12–15] consider the transition layer by including the ion temperature and obtaining the space charge characteristics in the sheath-presheath region.

Furthermore, the presence of the negative ions results some new features in plasma-sheath characteristics. The negative ions in plasmas have the positive effect of improving the performance of dry etching. An example is the cancelation of positive charge accumulated on the wafer. It is known that the accumulation of positive charge can induce serious damage and abnormal etching shapes [15–18]. Moreover, the sheath formation near the negatively biased surface of electrode that interfaces an insulator can act as lens, which focuses the ions to distinct location on the electrode surface [19, 20].

In our previous study, we focused on the influence of the positive ion collisions on the electronegative plasma sheath in the presence of cold positive ions [21]. For many plasma applications in industries and laboratories, the temperature of the positive ion cannot be ignored and the temperature of negative and positive ions is comparable. Therefore, we have to consider the temperature of the positive ion. Here, we concentrate on the collisional space-charge sheath in the presence of negative ion by including the temperature of the positive ions. Moreover, the net current passing through the sheath and positive ion kinetic energy in different conditions are obtained and the influences of the negative ion as well as the positive ion temperature on the sheath characteristic are investigated. The outline of the paper is as follows. In Sect. 2, a plasma-sheath model is introduced using the fluid equations for positive ions and Boltzmann distribution for negative species and the initial boundary conditions are introduced. The results of the model are presented in two subsections of Sect. 3 for collisionless and collisional cases. The conclusion remarks are given in Sect. 4.

The sheath model

In our model, because of high mobility, electrons are assumed to be thermalized and obey the Boltzmann distribution. In addition, it is assumed that the negative ions are isothermal. Therefore, the distribution of electrons and negative ions can be written as

$$n_e = n_{e0} \exp\left(\frac{e\phi}{T_e}\right) \tag{1}$$

$$n_- = n_{-0} \exp\left(\frac{e\phi}{T_-}\right) \tag{2}$$

where, ϕ is the electrostatic potential, T_e and T_- are the electron and negative ion temperatures and n_{e0} , n_{-0}, n_{i0} being the electron, negative ion and positive ion densities, at the beginning of the sheath.

In writing the above equations we assume that each negative ion has one electron charge. The quasi-neutrality condition in the electronegative discharge at the sheath edge expresses that

$$n_{i0} - n_{-0} - n_{e0} = 0 \tag{3}$$

where the n_{i0} is positive ion density at the sheath edge.

To treat the cold positive ions, the continuity and momentum transfer equations are employed.

$$\nabla \cdot (n_i V_i) = 0 \tag{4}$$

$$m_i n_i (V_i \cdot \nabla) V_i = -e n_i \nabla\phi - \nabla P_i - m_i n_i \nu V_i. \tag{5}$$

where the hypothesis of a constant positive ion flux in the sheath region has been employed in Eq. (4). Here m_i and V_i are the mass and flow velocity of positive ions,ν is the positive ion-neutral collision frequency, n_i is the positive ion density and P_i is the positive ion pressure. To close the set of fluid equations, one should introduce a state equation. In the current study we assume that $P_i = n_i kT$, where T is the positive ion temperature and k is the Boltzmann constant.

In our model, a slab geometry is considered along with the assumption that all parameters depend on coordinate z, directed from the sheath edge at $z = 0$, towards the wall and the z axis is normal to the wall. Thus, the continuity and momentum transfer equations can be written as

$$n_i V_{iz} = n_{i0} V_{iz0} \tag{6}$$

$$m_i V_{iz} \frac{dV_{iz}}{dz} = -e\frac{d\phi}{dz} - \frac{1}{n_i}\frac{dP_i}{dz} - m_i \nu V_{iz} \tag{7}$$

And the Poisson's equation becomes

$$\frac{d^2\phi}{dz^2} = -\frac{e}{\varepsilon_0}(n_i - n_e - n_-) \tag{8}$$

where ε_0 is vacuum permittivity. It is useful to normalize the quantities used in this model by the following dimensionless parameters.

$$\delta_- = \frac{n_{-0}}{n_{i0}};\quad \eta = -\frac{e\phi}{T_e};\quad \gamma = \frac{T_e}{T_-};\quad \xi = \frac{z}{\lambda_{De}};$$

$$\lambda_{De} = \sqrt{\frac{\varepsilon_0 T_e}{n_{e0} e^2}};\quad N_i = \frac{n_i}{n_{i0}};\quad N_e = \frac{n_e}{n_{e0}};\quad N_- = \frac{n_-}{n_{-0}};$$

$$u = \frac{V}{c};\quad M = \frac{V_{iz0}}{c};\quad c = \sqrt{\frac{T_e}{m_i}};\quad \alpha = \lambda_{De}\frac{\nu}{c}$$

where we have scaled z to Debye length, which is appropriate for the sheath region.

Including the Poisson's equation, a closed set of coupled differential equations is obtained, which describes the sheath structure. Vanishing the right hand side of Eq. (8) implies a presheath solution, not considered here.

$$N_e = \exp(-\eta) \tag{9}$$

$$N_- = \exp(-\gamma\eta) \tag{10}$$

$$N_i u_z = M \tag{11}$$

$$u_z \frac{du_z}{d\xi} = \frac{d\eta}{d\xi} - \alpha u_z - \frac{T}{N_i}\frac{dN_i}{d\xi} \tag{12}$$

The above equation can be written as

$$\frac{du_z}{d\xi} = \frac{u_z}{u_z^2 - T}\left(\frac{d\eta}{d\xi} - \alpha u_z\right) \tag{13}$$

Finally, the poison's equation becomes

$$\frac{d^2\eta}{d\xi^2} = \frac{\left(\frac{M}{u_z} - \delta_- \exp(-\gamma\eta) - (1-\delta_-)\exp(-\eta)\right)}{(1-\delta_-)} \tag{14}$$

According to the Ref. [22], by the Sagdeev potential and some algebra it can be shown that the ion velocity should satisfy the below inequality to enter the sheath

$$M \geq \left[\sqrt{\frac{\eta'_0 + T(1-\delta(1-\gamma))\eta'_0}{(1-\delta(1-\gamma))\eta'_0} + \frac{\alpha^2}{4[1-\delta(1-\gamma)]^2\eta'^2_0}} - \frac{\alpha}{2\eta'_0[1-\delta(1-\gamma)]}\right] \tag{15}$$

The inequality 15, states that the required flux velocity of positive ions at the sheath edge strongly depends on the number of negative ions and their temperature. For $\gamma = 1$, in which the two negative species are indistinguishable, we have the usual Bohm criterion. We have chosen the minimum value of the Eq. (15) for initial boundary condition of the ion velocity.

In the next section we attempt to solve the foregoing sets of equations according to the boundary conditions and acquire numerical results, considering the generalized Bohm criterion.

Results and discussion

Numerical solution in the collisionless case

The examples of the calculated space-charge in a pure electropositive plasma in the sheath region are shown in Fig. 1 where $\xi = 0$ coincides with plasma-sheath interface.

In the vicinity of the plasma-sheath edge, there is a peak which is the main characteristic of the sheath region. In the case of hot positive ion, the ions can penetrate more into the sheath and therefore, the profile of the space-charge amplitude increases. According to Eq. (4), the flux of the positive ions is an invariant quantity, hence by accelerating those ions through the sheath, the number density decreases. In addition, as the plasma potential is positive with respect to the plasma surrounding wall, the negative species repel from the wall. Therefore, the space-charge becomes low far from the wall.

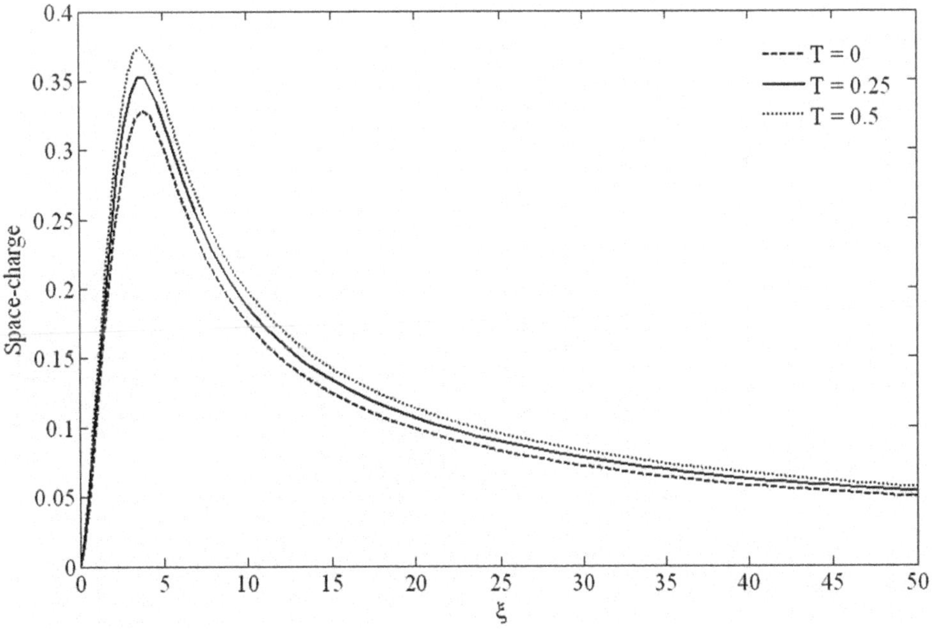

Fig. 1 The space-charge profile in the absence of collisions and negative ions for three values of positive ion temperature

As pointed out in the introduction, adding the negative ions in the plasma leads to great variation in the plasma-sheath characteristics. In Fig. 2, the profiles of the space-charge are drawn for a wide range of electronegativity without considering the thermal effects of the positive ions. From this figure, it can be seen that, adding the negative ions results in damping the profiles of the space-charge profile and furthermore, due to the attraction of the positive ions by low temperature negative ions, the main peak of the space-charge moves towards the plasma-sheath edge. It can be observed that in high and low electronegativity, the peak takes higher values with respect to the intermediate electronegativity. In addition, in high electronegativity, the width of the main peak becomes narrower. However, as shown in Fig. 3, in the case of thermal positive ions, raising the temperature of the positive ions leads to the enhancement of the space-charge. The interesting point is that the thermal effect of the positive ions is more pronounced in high electronegativity. In the case of the low concentration of the negative ions, increasing the positive ion temperature has no great influence on the increase of amplitude of the space-charge peak compared with the high electronegativity case. In addition, contrary to the case of cold positive ion, it is observed that the space-charge peak in high electronegativity is higher with respect to the low electronegativity.

The competition between the positive and negative ion temperature can determine the sheath characteristics and a twofold feature appears. In Fig. 4, the space-charge profiles are presented for different values of negative ion temperature and in the case of the collisionless cold positive ion condition. Far from the plasma wall, increase of γ leads to the decrease of the space charge. However, one can observe that for low enough negative ion temperature, twofold behavior of the space-charge emerges. This result is similar to the results of the Ref. [23] which has investigated the non-thermal electronegative plasma sheath. According to Fig. 5, by increasing the positive ion

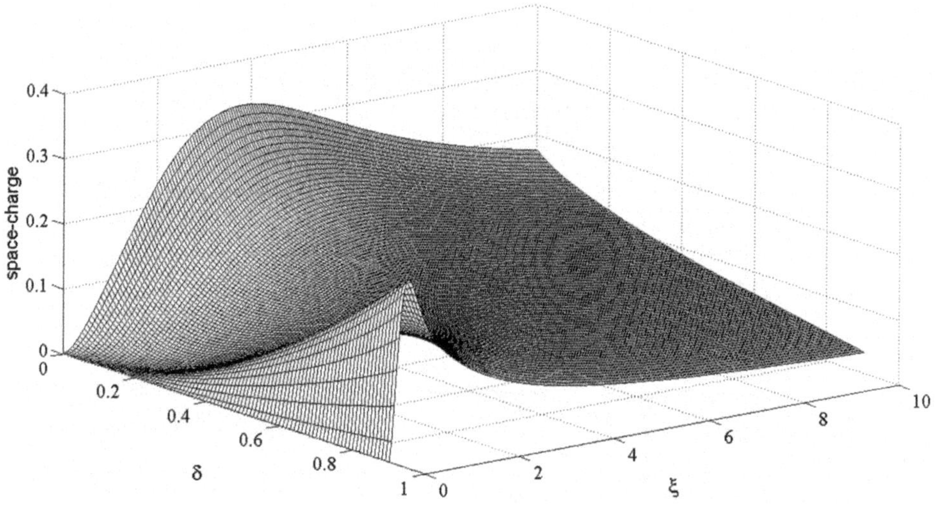

Fig. 2 The space-charge profile for different values of electronegativity in collisionless case for cold positive ions and $\gamma = 10$

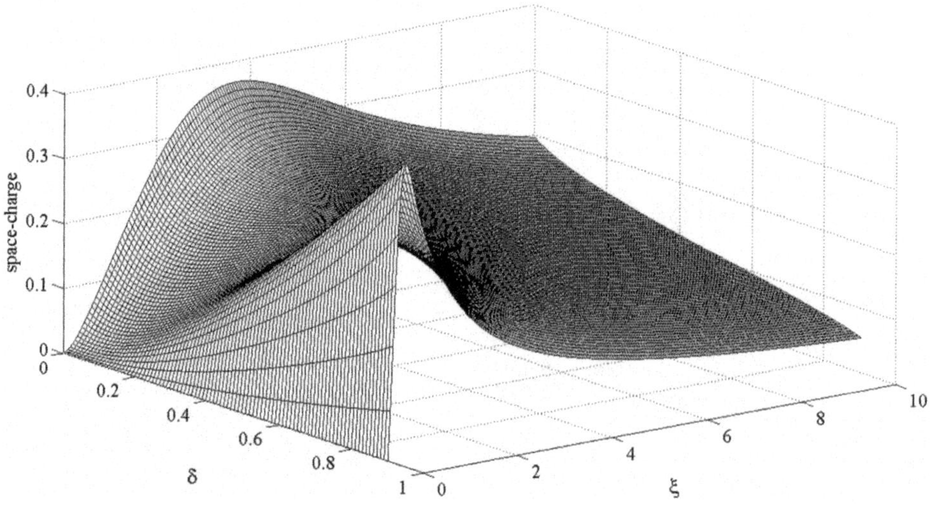

Fig. 3 The space-charge profile for different values of electronegativity in collisionless case for thermal positive ions and $\gamma = 10$

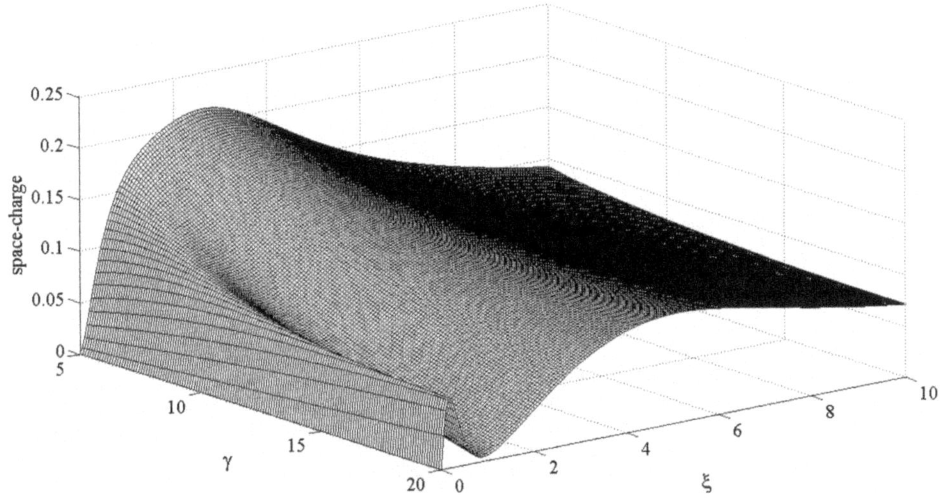

Fig. 4 The 3D space-charge profile for different values of negative ion temperature in the presence of cold positive ions and $\delta = 0.5$ in collisionless case

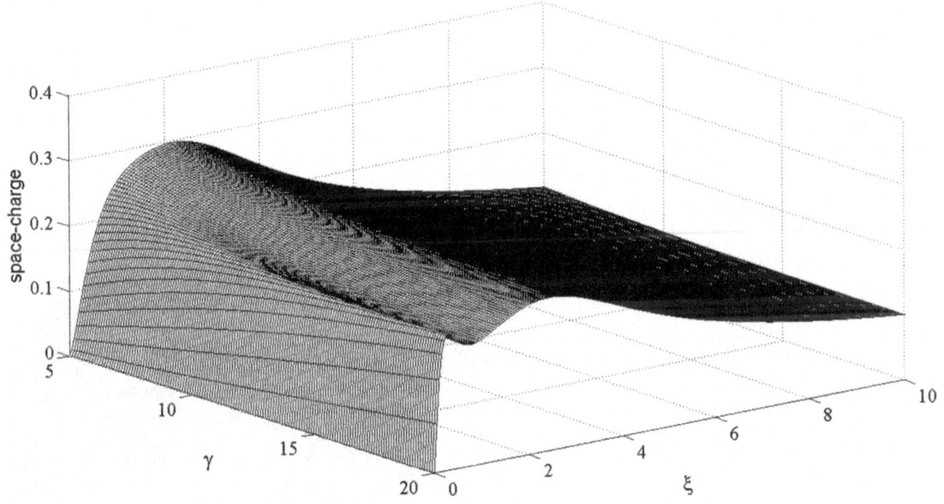

Fig. 5 The 3D space-charge profile for different values of negative ion temperature in the presence of thermal positive ions $T = 0.4$ and $\delta = 0.5$ in collisionless case

temperature, the difference between peak and dip of the space charge reduces and far from the sheath edge, the influence of negative ion temperature on the amount of the space-charge becomes negligible. In addition, in the presence of the hot positive ions, the space charge becomes stronger near the sheath edge.

The sheath solution in the collisional case

Now, we focus on the influence of the positive ion temperature on the sheath space-charge in the presence of collisions between positive ions and the neutral background. In Fig. 6, the space-charge profiles are presented for three values of positive ion temperature in the absence of negative ions.

By comparing Fig. 6 with the results of the collisionless case presented in Fig. 1, it can be deduced that apart from the positive ion temperature, the collisions lead to higher amplitude for the space-charge peak and does not change its position significantly.

Figures 7 and 8 show the dependence of the space-charge on the negative ion concentration in the presence of collisions for cold and hot positive ions cases. In the presence of collisions, as illustrated previously, the peak amplitude increases and in high electronegativity, the peaks become shrunk slightly for both cold and hot positive ions cases.

The collisions between the positive ions and neutral background result in decrease of positive ion velocity;

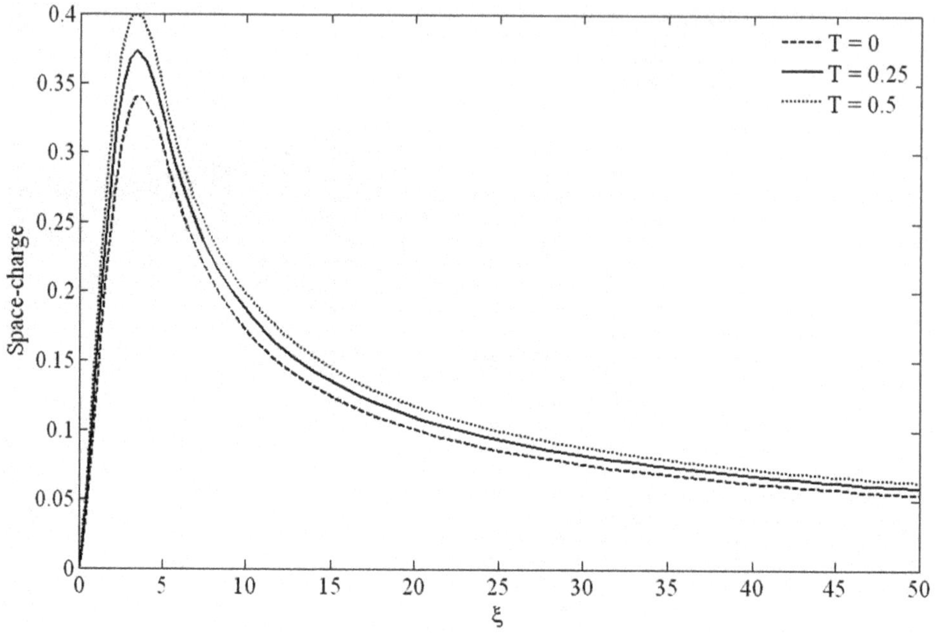

Fig. 6 The space-charge profile in the absence of negative ions for three values of positive ion temperature in collisional case $\alpha = 0.25$

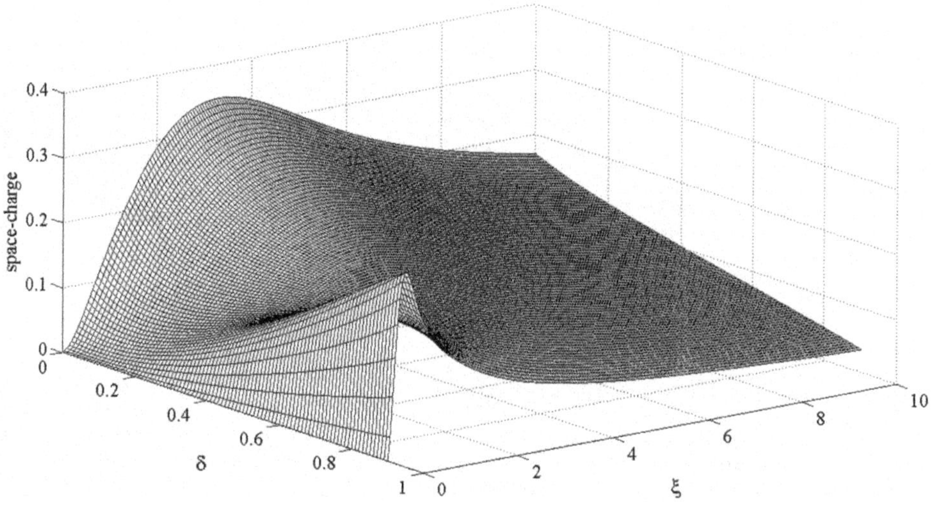

Fig. 7 The space-charge profile for different values of electronegativity in collisional case for cold positive ions $\alpha = 0.25$ and $\gamma = 10$

consequently raises the space-charge which is confirmed by comparison of Fig. 9 with Fig. 4.

As shown in Fig. 10 increasing the positive ion temperature, the ions can penetrate more in the sheath region and result in increasing the space charge profiles, which is more pronounced in the vicinity of the sheath edge. Decreasing the negative ion temperature leads to the damping of space charge which can be a consequence of attraction of the positive ion by the cold negative ions. However, it can be seen from Fig. 10, the temperature of the positive ions decrease the dependency of the space charge damping on the negative ion temperature.

The interesting point is that, in the presence of hot positive ions, the twofold feature starts at lower values of γ which is more pronounced in collisional case.

The kinetic energy of the accelerated ion and net current passing through the sheath

As mentioned in the introduction, one of the important quantities for plasma processing is the kinetic energy of the accelerated ion reaching the cathode. In Fig. 11, the kinetic energies of the positive ions reaching the wall versus the collision frequency are presented.

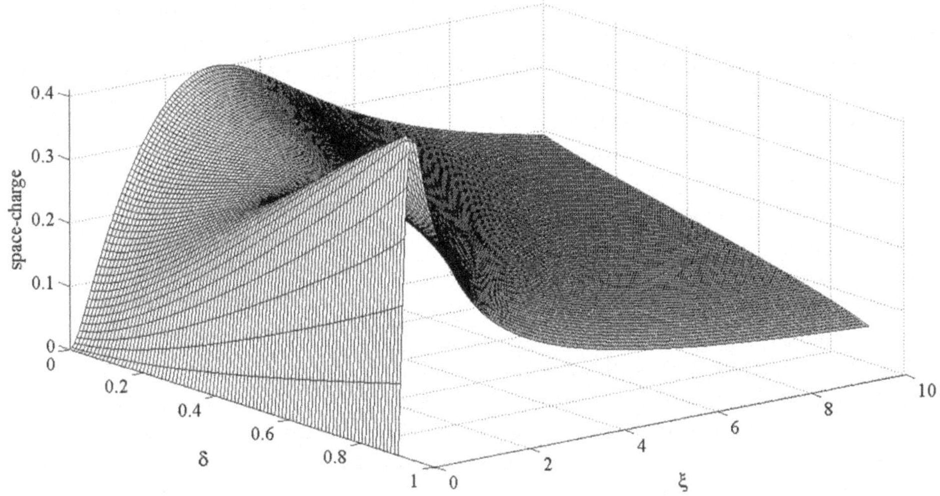

Fig. 8 The space-charge profile for different values of electronegativity in collisional case for thermal positive ions $T = 0.4$, $\alpha = 0.25$ and $\gamma = 10$

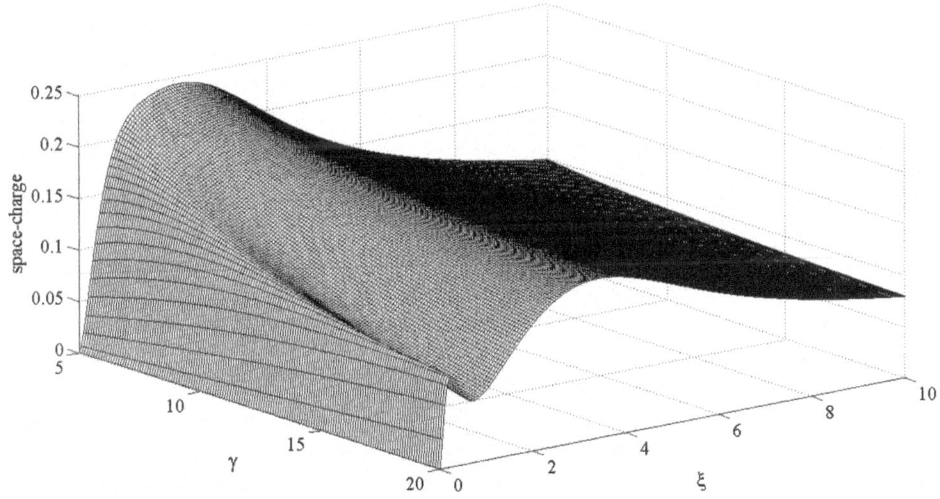

Fig. 9 The 3D space-charge profile for different values of negative ion temperature in the presence of cold positive ions and $\delta = 0.5$ in collisional case

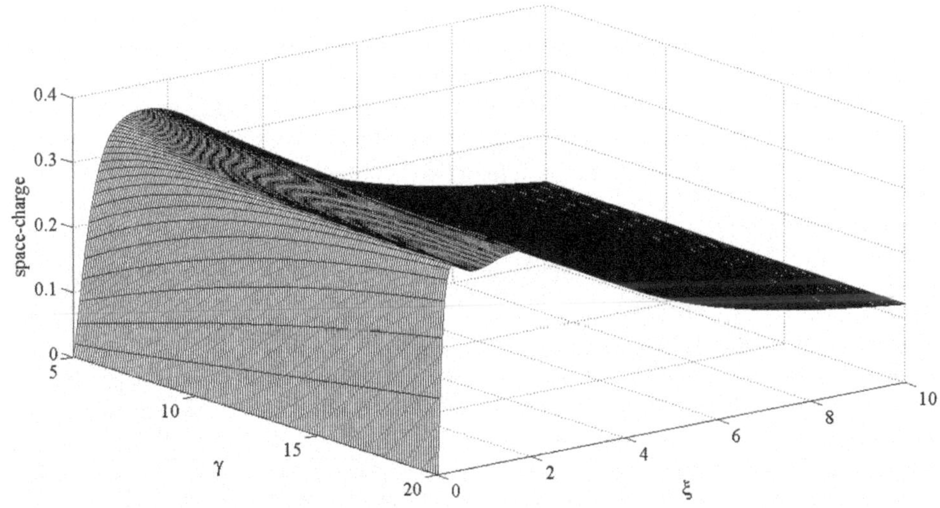

Fig. 10 The 3D space-charge profile for different values of negative ion temperature in the presence of thermal positive ions $T = 0.5$ and $\delta = 0.5$ in collisional case

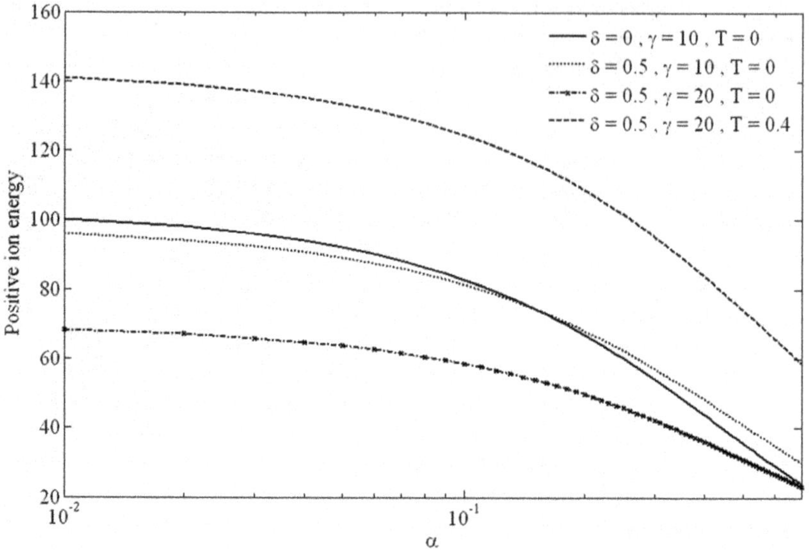

Fig. 11 The dependence of kinetic energy of positive ions on collision frequency

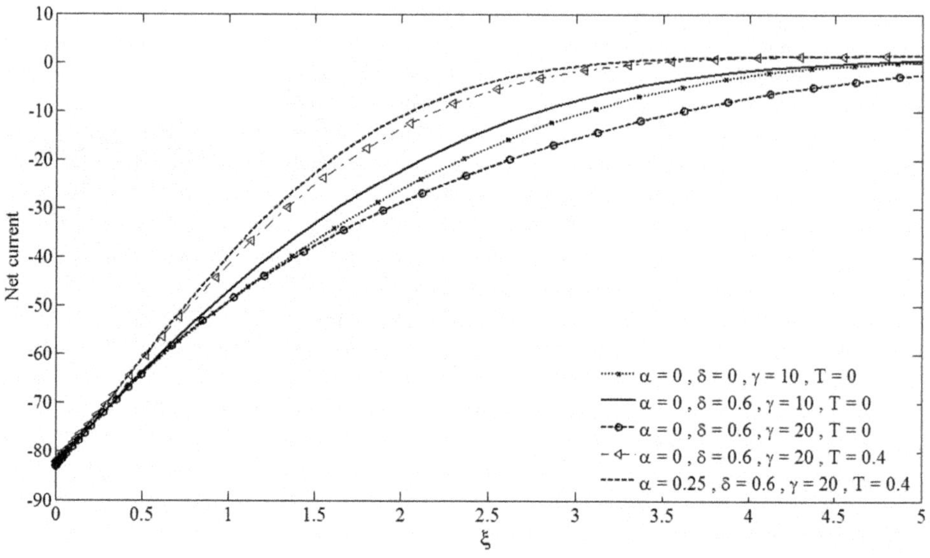

Fig. 12 The profile of net electric current across the sheath

Cases considered are for electropositive and electronegative plasmas for two values of γ in the presence of cold and hot positive ion. As can be seen, for all conditions, the collisions lead to a reduction in the positive ion kinetic energy. However, presence of the negative ions gives rise to more reduction in low collisionality while in high collision frequency, the kinetic energy of the positive ions increases. By decreasing the negative ion temperature, due to attraction by massive negative ions, the positive ions decelerate and, therefore, the kinetic energy reduces and finally by including the positive ion pressure in the force balance Eq. (5), the kinetic energy increases.

Now, we focus on the net current reaching the surrounding wall. In the current case, the net current can be defined as:

$$J_n = J_i - J_e - J_- = en_i V_{iz} - \frac{1}{4} en_e V_{the} - \frac{1}{4} en_- V_{th-} \quad (16)$$

where $V_{the} = \sqrt{\frac{8T_e}{\pi m_e}}$ and $V_{th-} = \sqrt{\frac{8T_-}{\pi m_-}}$ are the electron and negative ion thermal speeds, m_e and m_- are the masses of electron and negative ion, respectively. The thermal effects of the positive ions in the electric current are included in the first term of the left hand side of Eq. (17). The dimensionless net current scaled by $en_{e0}c_s$ [21] is

$$j_n = \frac{1}{1-\delta}\left(N_i u_z - (1-\delta)\sqrt{\frac{m_i}{2\pi m_e}} N_e - \delta\sqrt{\frac{m_i}{2\pi\gamma m_-}} N_-\right) \quad (17)$$

The oxygen and argon gasses are considered as the source of negative and positive atoms in the plasma.

In Fig. 12, we show the net electric current passing through the space-charge region in different conditions. It can be seen that the sheath thickness is sensitive to the presence of the negative ion concentration which leads to decrease of the sheath thickness. Decreasing the negative ion temperature leads to increments in the net current through the sheath and broadening the sheath thickness and furthermore, by increasing the positive ion temperature, sheath thickness decreases. In addition, the collisions result in decreasing the net current and finally, the presence of hot positive ion leads to decrease of the sheath thickness.

Conclusion

The characteristic features of space-charge region for thermal electronegative plasma are investigated in both collisionless and collisional cases by the set of fluid equations. The Bohm criterion for the positive ions entering the sheath is extracted using the Sagdeev potential method. In addition the profiles of the space-charge in the presence of the hot and cold positive ions are obtained for different electronegativity and negative ion temperature. It is shown that, increasing the positive ion temperature leads to enhancement of the space-charge amount and the influence of the thermal positive ion on the space-charge peak is more pronounced in high electronegativity. However, it is observed that the positive ion temperature has no influence on the peak position. Raising the positive ion temperature causes the effect of the negative ion temperature far from the sheath edge disappears. Moreover, the collisions does not change the peak position and twofold feature of the space-charge and in the presence of the hot positive ions, the twofold feature starts at lower values of relative temperature of electron to negative ion. In addition, the positive ion temperature leads to a decrease in the net current as well as the sheath thickness.

Acknowledgments The support of Karaj branch, Islamic Azad University is gratefully acknowledged for the present work.

References

1. Chodura, R.: Phys. Fluid **25**, 1628–1838 (1982)
2. Riemann, K.-U.: J. Phys. D Appl. Phys. **24**, 493–518 (1991)
3. Kim, G.H., Hershkowitz, N., Diebold, D.A., Cho, M.H.: Phys. Plasmas **2**, 3222–3233 (1995)
4. Riemann, K.-U.: Phys. Plasmas **4**, 4158 (1997)
5. Hromadka, J., Ibehej, T., Hrach, R.: Phys. Scr. **T161**, 014068 (2015)
6. Yasserian, K., Aslaninejad, M., Ghoranneviss, M.: Phys. Plasmas **16**, 023504 (2009)
7. Yasserian, K., Aslaninejad, M.: Phys. Plasmas **17**, 023501 (2010)
8. Franklin, R.N.: J. Phys. D Appl. Phys. **38**, 3412–3416 (2005)
9. Yasserian, K., Aslaninejad, M., Borghei, M., Eshghabadi, M.: J. Theor. Appl. Phys. **4**, 26 (2010)
10. Khoram, M., Ghomi, H.: J. Theor. Appl. Phys. **10**, 41 (2016)
11. Yasserian, K., Aslaninejad, M., Ghoranneviss, M., Aghamir, F.M.: J. Phys. D **41**, 105215 (2008)
12. Li. M, Vyvoda, M. A., Dew, S. K., Brett, M. J.: IEEE Trans. Plasma Sci. **28**, 248 (2000)
13. Minghao, L., Yu, Z., Wanyu, D., Jinyuan, L., Xiaogang, W.: Plasma Sci. Technol. **8**, 544 (2006)
14. Ghomi, H., Khorramabadi, M.: J. Plasma Phys. **76**(part 2), 247 (2010)
15. Araghi, F., Dorranian, D.: J. Theor. Appl. Phys **7**, 41 (2013)
16. Sheridan, T.E.: J. Phys. D Appl. Phys. **32**, 1761–1767 (1999)
17. Sheridan, T.E., Braithwaite, N., Boswell, R.W.: Phys. Plasmas **6**, 4375–4381 (1999)
18. Fernandez Palop, J.I., Ballesteros, J., Colomer, V., Hernandez, M.A., Morales Crespo, R.: J. Appl. Phys. **91**, 2587–2593 (2002)
19. Stamate, E., Sugai, H.: Phys. Rev. E **72**, 036407 (2005)
20. Stamate, E., Sugai, H.: Phys. Rev. Lett. **94**, 125004 (2005)
21. Yasserian, K., Aslaninejad, M.: Eur. Phys. J. D **67**, 161 (2013)
22. Ghomi, H., Khoramabadi, M., Shukla, P.K., Ghoranneviss, M.: J. Appl. Phys. **108**, 063302 (2010)
23. Yasserian, K., Aslaninejad, M.: Phys. Plasmas **19**, 073507 (2012)

10

High-frequency instabilities in an explosion-generated relativistic plasma

O. P. Malik · Sukhmander Singh · Hitendra K. Malik · A. Kumar

Abstract A realistic problem of an explosion-generated relativistic plasma is talked about with respect to the instabilities developed in such systems. For this, the dispersion equation is derived analytically and solved numerically for typical values of physical quantities. Our calculations reveal that two types of instabilities occur in the said plasma if the dust particles and relativistic effects of ions and electrons are considered. Both types of the instabilities are high-frequency instabilities, which carry growth rates of different magnitudes. In view of the magnitudes, the instability having higher/lower growth rate is called as high-frequency higher/lower growth rate instability. The relativistic effects of ions support the growth of these instabilities, whereas those of electrons suppress the growth of the instabilities. The waves propagating with larger phase velocity are found to grow at higher rates. There exists a critical value of the drift velocity of dust particles, above which another instability starts growing but with significantly lower growth rate.

O. P. Malik
Department of ECE, Al-Falah University, Dhauj, Faridabad, Haryana, India

S. Singh
Motilal Nehru College, South Campus, Delhi University, New Delhi 110 021, India

H. K. Malik (✉)
PWAPA Laboratory, Department of Physics, Indian Institute of Technology Delhi, New Delhi 110 016, India
e-mail: hkmalik@hotmail.com

A. Kumar
Department of Applied Sciences, Al-Falah University, Dhauj, Faridabad, Haryana, India

Keywords Explosion-generated relativistic plasma · Dust particles · High frequency instabilities · Dispersion equation · Growth rate

Introduction and motivation

Gamma rays are produced in a nuclear explosion, which interact with air molecules through a process called Compton effect, and electrons are scattered at high energies that ionize the atmosphere, hence, generating a powerful electrical field. A large-scale electromagnetic pulse (EMP) effect can be produced by a single nuclear explosion exploded high in the atmosphere. This effect is known as High Altitude EMP (HEMP) effect, which is harmless to people as it radiates outward, but can overload computer circuitry and damage it much more swiftly with effects similar to a lightning strike. A similar but smaller scale EMP effect can be created through explosion by nonnuclear devices having powerful batteries or reactive chemicals. This method is called high power microwave (HPM) effect. Due to generation of intense microwaves, these effects are also very damaging to electronics, but within a much smaller area. Depending upon the amount of energy that is coupled to the target, the microwave weapon systems have the ability to produce graduated effects in the target electronics [1]. Actually the electronic components, especially the integrated circuits, microelectronics, and components found in modern electronic systems, are very sensitive to the microwave emissions [2–4].

The HPM sources have been investigated by several researchers [2, 5, 6] as potential weapons for a variety of battle, damage, and terrorist applications in view of very short pulse (about 100 ps) released by an electromagnetic weapon and the damage they can cause. On the other hand, beams of ions and electrons are created in an explosion-generated plasma (EGP), which is a source of free energy. This energy can be transferred to waves generated by the oscillations of plasma species in the explosion. Then, these waves can evolve into different types of instabilities including explosive instabilities [7–10]. The explosive instabilities are of practical importance, as these seem to offer a mechanism for rapid dissipation of coherent wave energy into thermal motion and, hence, may be effective for plasma heating [7, 8]. The simplest wave coupling process that can exhibit explosive character is the coupling of three waves with fixed phases [11–13]. The phenomenon of a wave triplet was first described by Cairns [14] using the kinetic equation, in which all the three waves grow simultaneously. Considering a relativistic ion beam and a nonisothermal plasma, Fainshtein and Chernova [15] have investigated high-frequency instabilities and the high power electromagnetic radiation generation from the development of an explosive. Relativistic plasmas have also been explored for the instabilities and nonlinear waves that retain their shapes during propagation [16–21].

In the plasma generated by an explosion, the electrons may not follow the Boltzmann distribution and, hence, their finite mass needs to be taken into account. Moreover, the relativistic effects of plasma species should be included to realize the exact behavior of the waves or instabilities. Hence, in this article, we consider the relativistic effects of the electrons and ions. Also, we take into account the dust particles which are always present in most of such plasmas and whose charge may fluctuate due to currents flowing into the dust and emissions taking place [22, 23]. Further, we consider the temperature of electrons (T_e) to be higher than that of the ions (T_i) and dust particles (T_d), i.e., $T_e \gg T_i \gg T_d$, in view of a strongly nonisothermal plasma [24–27].

Basic fluid equations and dispersion equation

We consider that the plasma consists of electrons (density n_e, mass m), ions (density n_i, mass M), and dust particles (density n_d, mass m_d) having uniform mass and charge ($Z_d e$). If the unperturbed density of these species is n_{j0}, where j refers to e for electrons, i for ions, and d for dust particles, then the quasineutrality condition reads $n_{e0} = n_{i0} + \alpha Z_d n_{d0}$, where α is $+1$ for the positively charged dust particles and -1 for the negatively charged dust particles. If $\vec{v}_i$ ($\vec{u}_e$) is the ion (electron) fluid velocity along with their unperturbed values as v_0 and u_0, then the one-dimensional continuity and momentum equations for the ion, electron and dust fluids can be written as:

$$\frac{\partial}{\partial t}(\gamma_i n_i) + \frac{\partial}{\partial x}(\gamma_i n_i v_i) = 0, \tag{1}$$

$$\frac{\partial}{\partial t}(\gamma_e n_e) + \frac{\partial}{\partial x}(\gamma_e n_e u_e) = 0, \tag{2}$$

$$\frac{\partial n_d}{\partial t} + \frac{\partial}{\partial x}(n_d v_d) = 0, \tag{3}$$

$$\frac{\partial}{\partial t}(\gamma_e u_e) + u_e \frac{\partial}{\partial x}(\gamma_e u_e) + \frac{M}{m n_e}\frac{\partial n_e}{\partial x} - \frac{M}{m}\frac{\partial \varphi}{\partial x} = 0, \tag{4}$$

$$\frac{\partial}{\partial t}(\gamma_i v_i) + v_i \frac{\partial}{\partial x}(\gamma_i v_i) + \frac{\sigma}{n_i}\frac{\partial n_i}{\partial x} + \frac{\partial \varphi}{\partial x} = 0, \tag{5}$$

$$\frac{\partial v_d}{\partial t} + v_d \frac{\partial v_d}{\partial x} + \delta \frac{M}{n_{d0} m_d}\frac{\partial n_d}{\partial x} + \frac{\alpha M}{m_d}\frac{\partial \varphi}{\partial x} = 0. \tag{6}$$

Here v_d is the dust fluid velocity along with its unperturbed value as v_{d0}.

The system of equations can be closed with the following Poisson's equation:

$$\frac{\partial^2 \varphi}{\partial x^2} - n_e + n_i + \alpha n_d Z_d = 0 \tag{7}$$

In the above equations, the densities are normalized by a background density n_0, potential φ by T_e/e, time t by the inverse of frequency $\omega_{pi} = \sqrt{e^2 n_0/\varepsilon_0 M}$, space x by the Debye length $\lambda_{De} = \sqrt{\varepsilon_0 T_e/e^2 n_0}$, and velocities v_i, u_i and v_d by the ion acoustic speed $C_s = \sqrt{T_e/M}$. The ion- (dust-)-to-electron temperature ratio $T_i/T_e(T_d/T_e)$ is taken as $\sigma(\delta)$. Also $\gamma_e = \left(1 - \frac{u_0^2}{c^2}\right)^{-\frac{1}{2}}$ and $\gamma_i = \left(1 - \frac{v_0^2}{c^2}\right)^{-\frac{1}{2}}$ are the relativistic factors for the electrons and ions, respectively.

We consider the variation of perturbed quantities as $\psi_1 \sim \exp(i\omega t - ikx)$, where $\psi_1 \equiv n_{i1}, n_{e1}, n_{d1}, \vec{v}_{i1}, \vec{u}_{e1}, \vec{v}_{d1}, \vec{\varphi}_1$, ω is the frequency of oscillations and k is the wave number. Then, such variations are used in the basic fluid equations while employing the normal mode analysis. This way linear analysis yields:

$$n_{i1} = \frac{k^2 n_{i0}\varphi_1}{\gamma_i(\omega - kv_0)^2 - \sigma k^2}, \quad n_{e1} = \frac{k^2 \varphi_1}{\gamma_e \frac{m}{M}(\omega - ku_0)^2 - k^2},$$

$$n_{d1} = \frac{k^2 \alpha Z_d n_{d0}\varphi_1}{\frac{m_d}{M}(\omega - kv_{d0})^2 - k^2\delta}.$$

The use of these expressions in the Poisson's Eq. (7) gives rise to:

$$\frac{k^2\varphi_1}{\frac{m}{M}\gamma_e(\omega - ku_0)^2 - k^2} - \frac{k^2 n_{i0}\varphi_1}{\gamma_i(\omega - kv_0)^2 - \sigma k^2} - \frac{k^2\alpha^2 Z_d^2 n_{d0}\varphi_1}{\frac{m_d}{M}(\omega - kv_{d0})^2 - k^2\delta} + k^2\varphi_1 = 0 \tag{8}$$

After simplifying the above equation, we get the following dispersion equation:

$$\frac{m m_d}{M^2}\gamma_e\gamma_i\omega^6 - b_0\omega^5 + \omega^4 b_1 + \omega^3 b_2 + \omega^2 b_3 + \omega b_4 + a_1 a_2 a_3 + a_2 a_4 + a_1 a_4(n_{i0} - a_2) = 0. \tag{9}$$

Here

$$b_0 = \frac{m_\mathrm{d}}{M}\left(a_{11} + \frac{2m}{M}\gamma_i\gamma_\mathrm{e}k\upsilon_{\mathrm{d}0}\right),$$
$$b_1 = a_{14} + \frac{m_\mathrm{d}}{M}\left(\gamma_i - \gamma_\mathrm{e}\frac{n_{i0}m}{M}\right) - a_3\frac{m}{M}\gamma_i\gamma_\mathrm{e},$$
$$b_2 = a_3a_{11} - a_{15} + n_{i0}a_8 - a_5,$$
$$b_3 = a_3a_{12} - a_{16} + a_6 + n_{i0}a_9,$$
$$b_4 = a_{17} - a_7 + a_3a_{13} + n_{i0}a_{10}, a_1 = k^2\left(1 - \frac{m}{M}\gamma_\mathrm{e}u_0^2\right)$$
$$a_2 = k^2(\gamma_i\upsilon_0^2 - \sigma),\ a_3 = \alpha^2Z_\mathrm{d}^2n_{\mathrm{d}0},$$
$$a_4 = k^2\left(\frac{m_\mathrm{d}}{M}\upsilon_{\mathrm{d}0}^2 - \delta\right).$$

Together with $a_5 = \gamma_i\frac{m_\mathrm{d}}{M}k(\upsilon_0 + \upsilon_{\mathrm{d}0})$, $a_6 = \gamma_i\left[\frac{m_\mathrm{d}}{M}\left(4k^2\upsilon_0\upsilon_{\mathrm{d}0} + \frac{a_2}{\gamma_i}\right) + a_4\right]$, $a_7 = 2k\left(\gamma_i\upsilon_0a_4 + \frac{m_\mathrm{d}}{M}a_2\upsilon_{\mathrm{d}0}\right)$, $a_8 = \frac{2\gamma_e mm_\mathrm{d}}{M^2}k(u_0 + \upsilon_{\mathrm{d}0})$, $a_9 = \frac{m_\mathrm{d}}{M}\left(a_1 - \frac{4m}{M}\gamma_\mathrm{e}k^2u_0\upsilon_{\mathrm{d}0}\right) -\gamma_\mathrm{e}\frac{m}{M}a_4$, $a_{10} = \frac{2ku_0m}{M}a_4\gamma_\mathrm{e} - \frac{2m_\mathrm{d}}{M}k\upsilon_{\mathrm{d}0}a_1$, $a_{11} = \frac{2\gamma_\mathrm{e}\gamma_im}{M^2}k(u_0 + \upsilon_0)$, $a_{12} = \gamma_i\left(a_1 - \frac{4m}{M}\gamma_\mathrm{e}k^2u_0\upsilon_0\right) - \gamma_\mathrm{e}\frac{m}{M}a_2$, $a_{13} = 2k\left(\frac{m}{M}\gamma_\mathrm{e}a_2u_0 - \gamma_ia_1\upsilon_0\right)$,$a_{14} = \gamma_\mathrm{e}\gamma_i\frac{m}{M}a_4 - \frac{m_\mathrm{d}}{M}(a_{12} - 2ka_{11}\upsilon_{\mathrm{d}0})$, $a_{15} = \frac{m_\mathrm{d}}{M}(a_{13} - 2a_{12}k\upsilon_{\mathrm{d}0}) + a_4a_{11}$, $a_{16} = \frac{m_d}{M}(a_1a_2 - 2a_{13}k\upsilon_{d0}) + a_4a_{12}$, $a_{17} = \frac{2m_\mathrm{d}}{M}a_1a_2k\upsilon_{\mathrm{d}0} - a_4a_{13}$.

Numerical solution to dispersion equation and the results

We solve Eq. (9) numerically by giving typical values to various parameters in view of the explosion-generated relativistic plasma (EGRP). Hence $k = 1$, $n_{i0} = 1.1$, $M = 23.38 \times 10^{-27}$ kg, $m_\mathrm{d} = 10^{-20}$ kg, $\upsilon_{\mathrm{d}0} = 1$, $u_0 = 4{,}000$, $\upsilon_0 = 2{,}000, n_{\mathrm{nd}0} = 0.001, Z_\mathrm{d} = 100, T_i = 10\,\mathrm{eV}, T_\mathrm{d} = 1\,\mathrm{eV}$ and $T_\mathrm{e} = 50$ eV [28–32]. Our calculations reveal that two types of instabilities occur in the said plasma if the dust particles and relativistic effects are considered. Both types of the instabilities are high-frequency instabilities but carry growth rates of different magnitudes. In view of the magnitudes, the instability having higher growth rate γ_1 is taken to correspond to the high-frequency higher growth rate instability, which we call as HFHGI, and the instability having lower growth rate γ_2 is taken to correspond to the high-frequency lower growth rate instability, which we call as HFLGI. In addition to these instabilities, a propagating mode is also found to occur.

Figure 1 shows the variation of growth rates γ_1 and γ_2 along with their real frequencies $\omega_{\mathrm{R}1}$ and $\omega_{\mathrm{R}2}$ with normalized wave number k. Here one obtains that the growth rates attain different peaks corresponding to different values of k. The peak corresponding to HFHGI appears towards higher side of k compared to the case of HFLGI. However, the growth rate γ_2 is found to almost saturate for the longer values of k. Since the free energy needs to be coupled with the wave to cause the wave to grow, it appears that the maximum energy transfer takes place to the oscillations of a particular wavelength only. Hence, in our case, the growth rate of HFHGI is a maximum at a particular value of k. However, another type of instability (HFLGI) may attain the maximum value at much larger value of k. This can be confirmed if we plot the graph for much larger values of k.

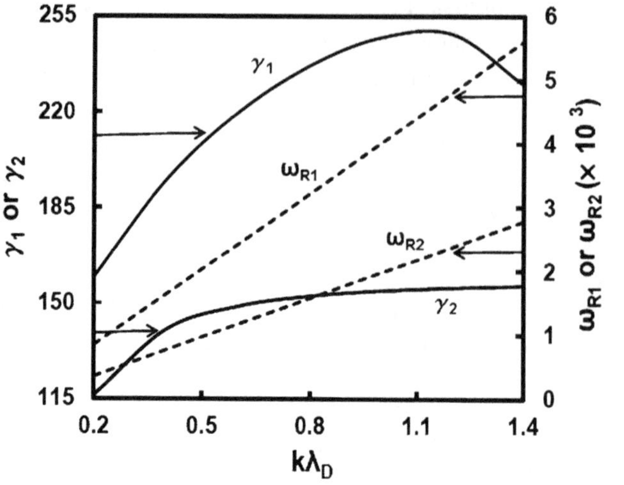

Fig. 1 Variation of growth rates γ_1 and γ_2 along with their real frequencies ω_{R1} and ω_{R2} with normalized wave number k, when $n_{i0} = 1.1$, $M = 23.38 \times 10^{-27}$ kg, $m_\mathrm{d} = 10^{-20}$ kg, $\upsilon_{\mathrm{d}0} = 1$, $u_0 = 4{,}000$, $\upsilon_0 = 2{,}000$, $n_{\mathrm{nd}0} = 0.001$, $Z_\mathrm{d} = 100$, $T_i = 10$ eV, $T_\mathrm{d} = 1$ eV and $T_\mathrm{e} = 50$ eV

Another interesting result is that both these instabilities propagate at constant velocities. The HFHGI propagates faster than the HFLGI. Hence, it can also be said that the wave propagating at higher speed evolves into an instability having larger growth rate. This is the similar result as obtained recently in a nonrelativistic EGP [33]. Consistent to Ref. 33, the present plasma also supports a propagating mode which has a constant velocity (Fig. 2).

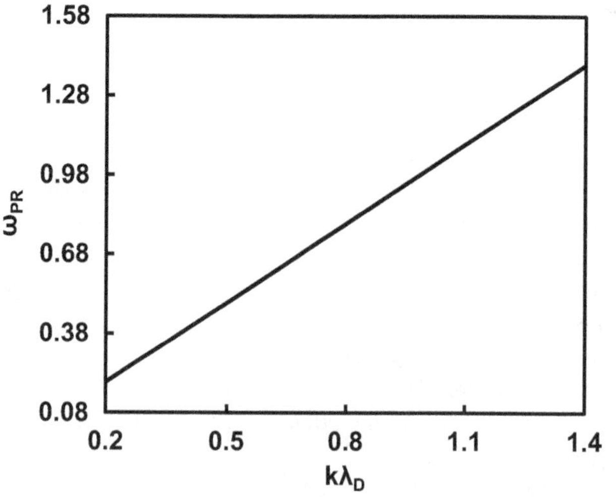

Fig. 2 Variation of frequency ω_{PR} of the propagating mode with normalized wave number, when the parameters are taken the same as in Fig. 1

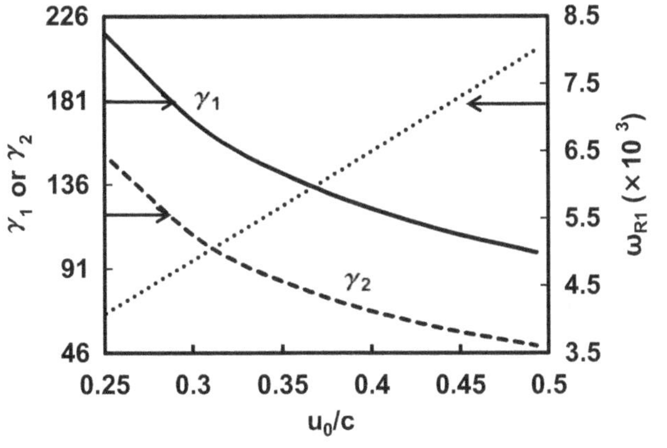

Fig. 3 Variation of growth rates γ_1 and γ_2 along with the real frequency ω_{R1} corresponding to HFHGI with relativistic speed of electrons, when $k = 1$ and the other parameters are the same as in Fig. 1

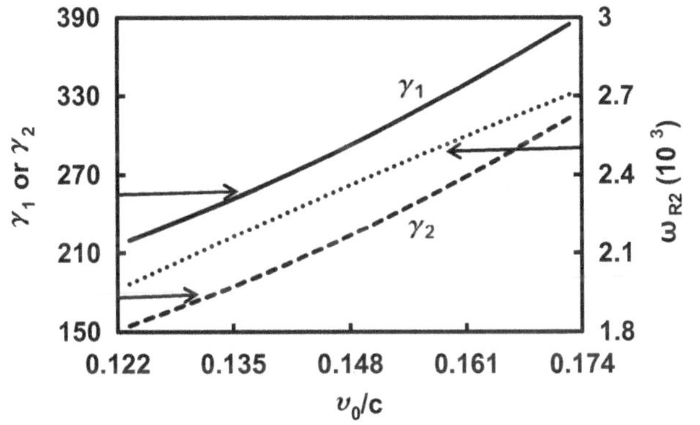

Fig. 4 Variation of growth rates γ_1 and γ_2 along with the real frequency ω_{R2} corresponding to HFLGI with relativistic speed of ions, when $k = 1$ and the other parameters are the same as in Fig. 1

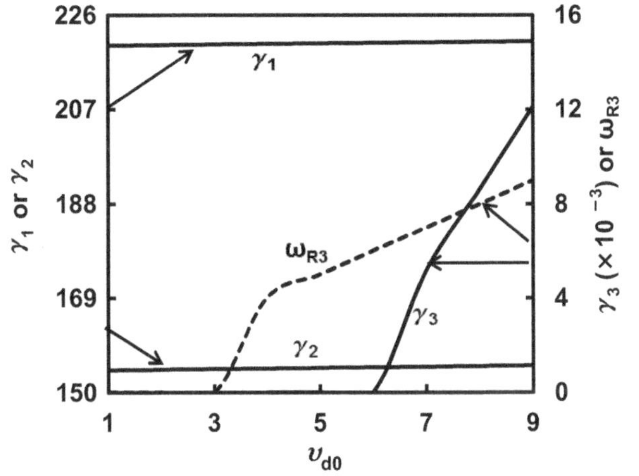

Fig. 5 Weak dependence of growth rates γ_1 and γ_2 on dust drift velocity, when $k = 1$ and other parameters are the same as given in Fig. 1

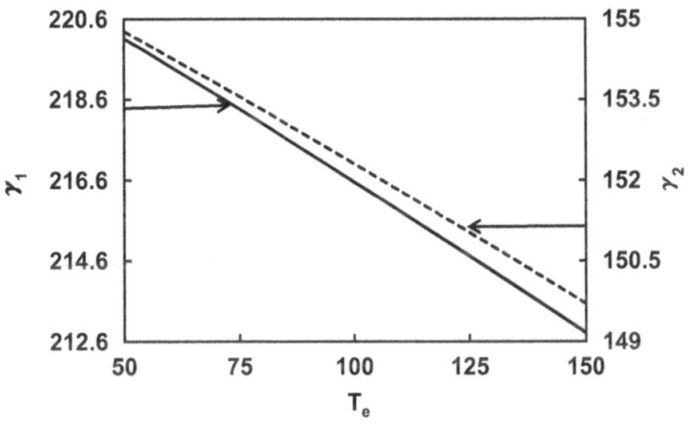

Fig. 6 Variation of growth rates γ_1 and γ_2 with electron temperature, when $k = 1$ and the other parameters are the same as in Fig. 1

Figures 3 and 4 show the effect of relativistic speeds of electrons and ions on the growth rates γ_1 and γ_2 along with the real frequencies ω_{R1} and ω_{R2}. Both types of the instabilities are observed to behave oppositely with the speeds v_0 and u_0. The growth rates γ_1 and γ_2 are reduced with the higher speed of the electrons, whereas these are enhanced for the higher speed of the ions. It means that the relativistic effects of the ions support the instability to grow, whereas due to the relativistic effect of the electrons the instabilities are suppressed. However, the propagation speed of the instabilities is enhanced in both the cases and this effect is linear in nature. Consistent to this observation in an EGRP, other investigators have also found the same effect of relativistic speeds of ions and electrons on the phase velocity of linear ion acoustic waves [16–21, 31].

With respect to the contribution of dust particles to these instabilities, we focus on Fig. 5 and observe that the growths of the instabilities remain unaltered by the drift velocity of the dust particles. However, an interesting feature of the dust grain effect is that a new type of instability develops in the system when the dust drift velocity exceeds a critical value; for example, $v_{d0} > 6$ in the present case. But the magnitude of the growth rate of this instability is 4 order less than the ones of HFHGI and HFLGI. On the other hand, the propagation speed of this new type of instability does not show linear dependence on the dust grain drift velocity.

Since this instability is the low-frequency instability, it means that this is related to the wave generated by the oscillations of the dust particles. Since the free energy needs to be coupled with the wave for its growth, it is plausible that the free energy is coupled with such oscillations when the dust particles attain certain value of the drift velocity. Hence, we get a critical value of the drift velocity of the dust particles for the evolution of this new type of the instability.

In Fig. 6, we study the effect of electron temperature on the growth rates γ_1 and γ_2. Here, the electron temperature is found to reduce the growth rates of both the instabilities. This result is contrary to the observation made in an explosion-generated nonrelativistic plasma [33]. Hence, it appears that due to the relativistic effect of the ions and electrons, the thermal motions of the electrons start suppressing the growth of the instabilities. On the other hand, Keidar and Beilis [34] had also observed similar effect of the electron temperature on the growth rate of dissipative instability.

The present calculations are for the realistic situation where a relativistic plasma is generated in an explosion. Hence, we have included the relativistic effects of the speeds of the ions and electrons. However, in view of heavy mass of the dust particles, this is justified to consider them as nonrelativistic but to drift with a finite speed. Our earlier article [33] depicts an approximate situation, where we considered only the initial drifts of the plasma species. In view of the nonrelativistic situation, we had also neglected the temperature of the dust species. However, the finite dust temperature is taken into account in the current calculations. Moreover, the instabilities talked about here correspond to the family of two-stream instabilities since there is a difference in the speeds of all the plasma species.

Finally, we mention that we did more calculations to examine the possibility of wave triplets in the present relativistic plasma. These calculations reveal that the wave triplet can evolve in the plasma only when the dust particles attain negative drift velocity. Since this is not physically acceptable, it can be said that the wave triplets are not possible in an EGRP. In support of this statement, we also understood that the waves are not found to grow for the negative values of k. It means that the possibility of negative energy wave is also ruled out in the plasma where the ions and electrons attain relativistic speeds.

Conclusions

We have investigated high-frequency instabilities in an EGRP under the effect of relativistic speeds of the ions and electrons. The dust particles were also taken into account with their initial nonrelativistic drifts. It was observed that two types of the instabilities generally evolve in such systems. However, a third type of instability appears in the system when the dust particles' drift velocity exceeds a critical value. The electron temperature was found to suppress the instabilities, whereas the ion temperature does not affect their growths. Both the instabilities were observed to be constant velocity waves. Consistent to the case of explosion-generated nonrelativistic plasma, the present plasma also supports a wave, which does not grow under the relativistic effects of the ions and electrons.

Acknowledgments CSIR, Government of India is thankfully acknowledged for the financial support.

References

1. Bäckström, M.G., Lövstrand, K.G.: Susceptibility of electronic systems to high-power microwaves: summary of test experience. IEEE Trans. Electromag. Compat. **46**, 396 (2004)
2. Giri, D.V., Tesche, F.M.: Classification of intentional electromagnetic environments. IEEE Trans. Electromag. Compat. **46**, 323 (2004)
3. Wik, M.W., Radasky, W.A.: Development of high-power electromagnetic (HPEM) standards. IEEE Trans. Electromag. Compat. **46**, 439 (2004)
4. Radasky W. A., Progress in IEC SC 77C high-power electromagnetics publications in 2009, In: Proceedings of 5th Asia-Pacific Conference on Environmental Electromagnetics (CEEM 2009), 5 (2009)
5. Efanov, V.: Gigawatt all solid state nano- and pico-second pulse generators for radar applications. In: Proceedings 14th IEEE Int. Pulsed Power Conf., Dallas, TX (2003)
6. Staines, G.: Compact sources for tactical RF weapon applications (Diehl), in Proc. AMEREM, Annapolis (2002)
7. Aamodt, R.E., Sloan, M.L.: Nonlinear interactions of positive and negative energy waves. Phys. Fluids **11**, 2218 (1968)
8. Nejoh, Y.N.: Modulational instability of relativistic ion-acoustic waves in a plasma with trapped electrons. IEEE Trans. Plasma Sci. **20**, 80 (1992)
9. Malik, H.K., Singh, S.: Resistive instability in a Hall plasma discharge under ionization effect. Phys. Plasmas **20**, 052115 (2013)
10. Singh, S., Malik, H.K., Nishida, Y.: High frequency electromagnetic resistive instability in a Hall thruster under the effect of ionization. Phys. Plasmas **20**, 102109 (2013)
11. Wilhelmson, H., Weiland, J.: Coherent non-linear interaction of waves in plasmas. Pergamon Press, Oxford (1977)
12. Wilhelmson, H.: On the explosive instabilities of waves in plasmas with special regard to dissipation and phase effects. Phys. Scr. **7**, 209 (1973)
13. Wilhelmson, H., Stenflo, I., Engelmann, F.: Explosive instabilities in the well defined phase description. J. Math. Phys. **11**, 1738 (1970)
14. Cairns, R.A.: The role of negative energy waves in some instabilities of parallel flows. J. Fluid Meek **92**, 1 (1979)
15. Fainshtein, S.M., Chernova, E.A.: Generation of high-power electromagnetic radiation from the development of explosive and high-frequency instabilities in a system consisting of a relativistic ion beam and a nonisothermal plasma. JETP **84**, 442 (1996)
16. Malik, H.K.: Ion acoustic solitons in a weakly relativistic magnetized warm plasma. Phys. Rev. E **54**, 5844 (1996)
17. Malik, H.K.: Ion acoustic solitons in a relativistic warm plasma with density gradient. IEEE Trans. Plasma Sci. **23**, 813 (1995)
18. Malik, H.K., Singh, K.: Small amplitude soliton propagation in a weakly relativistic two-fluid magnetized plasma: electron inertia contribution. IEEE Trans. Plasma Sci. **33**, 1995 (2005)
19. Singh, K., Kumar, V., Malik, H.K.: Electron Inertia Contribution to Soliton Evolution in an Inhomogeneous Weakly Relativistic Two-fluid Plasma. Phys. Plasmas **12**, 072302 (2005)
20. Malik, R., Malik, H.K., Kaushik, S.C.: Propagating and growing modes in a pair-ion plasma. Indian J. Phys. **85**, 1887 (2011)
21. Malik, R., Malik, H. K.: Instability in a weakly relativistic electron-positron plasma under the effect of dust grains, laser and plasma accelerator workshop , Taj Fort Aquada, Goa, India (2013)
22. Tomar, R., Malik, H.K., Dahiya, R.P.: Reflection of ion acoustic solitary waves in a dusty plasma with variable charge dust. J. Theor. Appl. Phys. **8**, 126 (2014)
23. Malik, R., Malik, H.K.: Compressive solitons in a moving e-p

plasma under the effect of dust grains and an external magnetic field. J. Theor. Appl. Phys. **7**, 65 (2013)
24. Landau, L.D., Lifshits, E.M.: Fluid Mechanics. Addison-Wesley, New York (1959)
25. Coppi, B., Rosenbluth, M.N., Sudan, R.N.: Nonlinear interactions of positive and negative energy modes in rarefied plasmas(I). Ann. Phys. **55**, 207 (1969)
26. Moiseev, S.S., Oraevsky, V.N., Pungin, V.G.: Nonlinear instabilities in plasmas and hydrodynamics-Taylor & Francis (1999)
27. Fainshtein, S.M.: On the possibility of generation of high-power low-frequency radiation as a result of evolution of explosive instability in the flow–nonisothermal plasma system. Radiophys. Quantum Electron J. **54**, 193 (2011)
28. Luo, Q.-Z., D'Angelo, N., Merlino, R.L.: Experimental study of shock formation in a dusty plasma. Phys. Plasmas **6**, 3455 (1999)
29. Merlino, R.L., Barkan, A., Thompson, C., D'Angelo, N.: Laboratory studies of waves and instabilities in dusty plasmas. Phys. Plasmas **5**, 1607 (1998)
30. Shukla, P.K., Mamun, A.A.: Dust-acoustic shocks in a strongly coupled dusty plasma. IEEE Trans. Plasma Sci. **29**, 221 (2001)
31. Nejoh.,Y.N.: Double layers, spiky solitary waves, and explosive modes of relativistic ion acoustic waves propagating in a plasma. Phys. Fluids B 4, 2830 (1992)
32. Esipchuk, Y.V., Tilinin, G.N.: Drift instability in a Hall-current plasma accelerator. Sov. Phys. Tech. Phys. **21**, 417 (1976)
33. Malik, O.P., Singh, S., Malik, H.K., Kumar A.: Low and high frequency instabilities in an explosion- generated-plasma and possibility of wave triplet. J. Theor. Appl. Phys. (2014)
34. Keidar, M., Beilis, I.I.: Electron transport phenomena in plasma devices with E × B Drift. IEEE Trans. Plasma Sci **34**, 804 (2006)

11

Theoretical analysis of a thermal plasma-loaded relativistic traveling wave tube having corrugated slow wave structure with solid electron beam

Zahra Javadi · Shahrooz Saviz

Abstract A relativistic traveling wave tube with thermal plasma-filled corrugated waveguide is driven by a finite solid electron beam with the entire system immersed in a strong longitudinal magnetic field that magnetized plasma and electron beam. The dispersion relation for the relativistic traveling wave tube is obtained by linear fluid theory. The numerical results show that the growth rate decreases by increasing plasma temperature, waveguide radius, plasma density and electron beam energy. As show in this paper the effect of electron beam density and corrugation period is to increase growth rate.

Keywords RTWT · Solid electron beam · Thermal plasma · Dispersion relation · Growth rate

Introduction

Relativistic traveling wave tube (RTWT) is an important high-power microwave (HPM) apparatus which has been developed in the past 20 years [1–3]. Most of the TWT mathematical analysis has been done by John Robinson Pierce and his colleagues at Bell Labs [4–6] and then have been developed by Chu, Jackson and Freund [7–9]. In TWT, sinusoidal corrugated slow wave structure (SWS) is used to reduce the phase velocity of the electromagnetic wave to synchronize it with the electron beam velocity, so that a strong interaction between the two can take place [10, 11]. TWT is extensively applied in satellite and airborne communications, radar, particle accelerators, cyclotron resonance and electronic warfare system. The plasma injection to TWT has been studied recently which can increase the growth rate and improve the quality of transmission of electron beam. We investigate the effect of thermal plasma and electron beam on the growth rate [12–17].

In this paper, RTWT with magnetized thermal plasma-filled corrugated waveguide with solid electron beam is studied. The dispersion relation of corrugated waveguide is derived from a solution of the field equations. By numerical computation, the dispersion characteristics of RTWT are analyzed in detail with different cases of various geometric parameters of slow wave structure.

In "Physical model", the physical model of RTWT filled with thermal plasma is established in an infinite longitudinal magnetic field. In "Dispersion equation", the dispersion relation of RTWT is derived. In "Numerical result and discussion", the dispersion characteristics of the RTWT are analyzed by numerical computation.

Physical model

The analysis presented in this paper is based on the SWS shown in Fig. 1. The SWS is a sinusoidal cylindrical waveguide that consists of an axially symmetric.

$$R(z) = R_0 + h\ \cos(\kappa_0 z), \tag{1}$$

$$K_n = k_z + n\kappa_0, \tag{2}$$

$$\kappa_0 = \frac{2\pi}{z_0}, \tag{3}$$

Z. Javadi
Central Tehran Branch, Islamic Azad University, Tehran, Iran

S. Saviz (✉)
Plasma Physics Research Center, Science and Research Branch, Islamic Azad University, Tehran, Iran
e-mail: azarabadegan@gmail.com

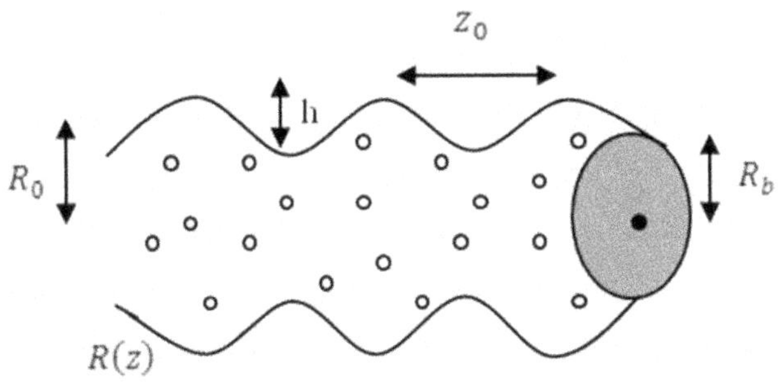

Fig. 1 Slow wave structure model of a plasma-filled relativistic traveling wave tube

Cylindrical waveguide, whose wall radius $R(z)$, varies sinusoidal according to the relation (1), h is the corrugation amplitude, $\kappa_0 = 2\pi/z_0$ is the corrugation wave number, and z_0 is the length of the corrugation period, R_0 is the waveguide radius and k_z is the axial number wave.

A finite solid relativistic electron beam with density n_b and radius R_b goes through the cylindrical waveguide, which is loaded completely with thermal, uniform and collisionless plasma of density n_p. The entire system is immersed in a strong, longitudinal magnetic field, which magnetizes both the beam and the plasma. Because of being anisotropic of dielectric constant it will be a tensor. In the beam-plasma case in a linearized scheme, the dielectric tensor in cylindrical coordinates may be given by:

$$[\varepsilon] = \varepsilon_0 \begin{bmatrix} \varepsilon_1 & \varepsilon_2 & 0 \\ -\varepsilon_2 & \varepsilon_1 & 0 \\ 0 & 0 & \varepsilon_3 \end{bmatrix}, \tag{4}$$

We assume that B_0 is very strong that $\varepsilon_1 = 1$ and $|\varepsilon_2|$ is negligibly small.

$$[\varepsilon] = \varepsilon_0 \times \begin{bmatrix} 1 & 0 & 0 \\ 0 & 1 & 0 \\ 0 & 0 & 1 - \dfrac{\omega_b^2}{\gamma^3(\omega - K_n v)^2 - 3K_n^2 v_{th_b}^2} - \dfrac{\omega_p^2}{\omega^2 - 3K_n^2 v_{th_p}^2} \end{bmatrix}, \tag{5}$$

$$\varepsilon_3 = 1 - \frac{\omega_b^2}{\gamma^3(\omega - K_n v)^2 - 3K_n^2 v_{th_b}^2} - \frac{\omega_p^2}{\omega^2 - 3K_n^2 v_{th_p}^2} \tag{6}$$

Here, $\omega_p = (\frac{n_p e^2}{m\varepsilon_0})^{1/2}$ is the plasma frequency, $\omega_b = (\frac{n_b e^2}{m\varepsilon_0})^{1/2}$ is the beam frequency, ω is the angular frequency of the electromagnetic wave, γ is the electron beam energy, $v_{th_p} = \sqrt{\frac{KT_p}{m}}$ is the thermal velocity of the plasma, $v_{th_b} = \sqrt{\frac{KT_b}{m}}$ is the thermal velocity of the beam, v is the velocity of the beam, T_b is the beam temperature, T_p is the plasma temperature and K is the Boltzmann constant.

Dispersion equation

In the above physical model, its Maxwell equations can be written as:

$$\nabla \times B = \frac{-i\omega}{c} \varepsilon . E, \tag{7}$$

$$\nabla \times E = \frac{i\omega}{c} B, \tag{8}$$

$$\nabla \cdot B = 0, \tag{9}$$

$$\nabla \cdot D = 0. \tag{10}$$

Suppose that every variable can be regarded as:

$$A = Ae^{i(K_n z - \omega t)}. \tag{11}$$

From Eqs. (7) and (8), we can obtain:

$$\nabla\nabla \cdot E - \nabla^2 E = \frac{\omega^2}{c^2} \varepsilon \cdot E \tag{12}$$

$$\nabla^2 E_z - [\nabla(\nabla \cdot E)]_z + \frac{\omega^2}{c^2} \varepsilon_3 E_z = 0, \tag{13}$$

Now Substituting Eq. (6) into Eq. (13), we have:

$$\nabla^2 E_z - [\nabla(\nabla \cdot E)]_z + \frac{\omega^2}{c^2} \times \left(1 - \frac{\omega_b^2}{\gamma^3(\omega - K_n v)^2 - 3K_n^2 v_{th_b}^2} - \frac{\omega_p^2}{\omega^2 - 3K_n^2 v_{th_p}^2}\right) E_z = 0 \tag{14}$$

From Eq. (10), we have:

$$\nabla \cdot \varepsilon_3 E_z = 0. \tag{15}$$

Substituting Eq. (6) into Eq. (15), we obtain:

$$\nabla \cdot E_z = \frac{\partial}{\partial z} E_z \left[\frac{\omega_b^2}{\gamma^3(\omega - K_n v)^2 - 3K_n^2 v_{th_b}^2} + \frac{\omega_p^2}{\omega^2 - 3K_n^2 v_{th_p}^2}\right]. \tag{16}$$

From Eqs. (13) and (16), the wave equation is obtained in the area of plasma beam as follows:

$$\left[\nabla_\perp^2 + \left(\frac{\omega^2}{c^2} - K_n^2\right) \times \left(1 - \frac{\omega_b^2}{\gamma^3(\omega - K_n)^2 - 3K_n^2 v_{th_b}^2} - \frac{\omega_p^2}{\omega^2 - 3K_n^2 v_{th_p}^2}\right)\right] E_z, \tag{17}$$

$$\nabla_\perp^2 = \nabla^2 - \frac{\partial^2}{\partial z^2}. \tag{18}$$

$$E_z(r, z) = E_{zn}(r) e^{i(K_n z - \omega t)}, \tag{19}$$

$$E_r(r, z) = \frac{iK_n}{\frac{\omega^2}{c^2} - K_n^2} e^{i(K_n z - \omega t)} \frac{dE_{zn}}{dr}. \tag{20}$$

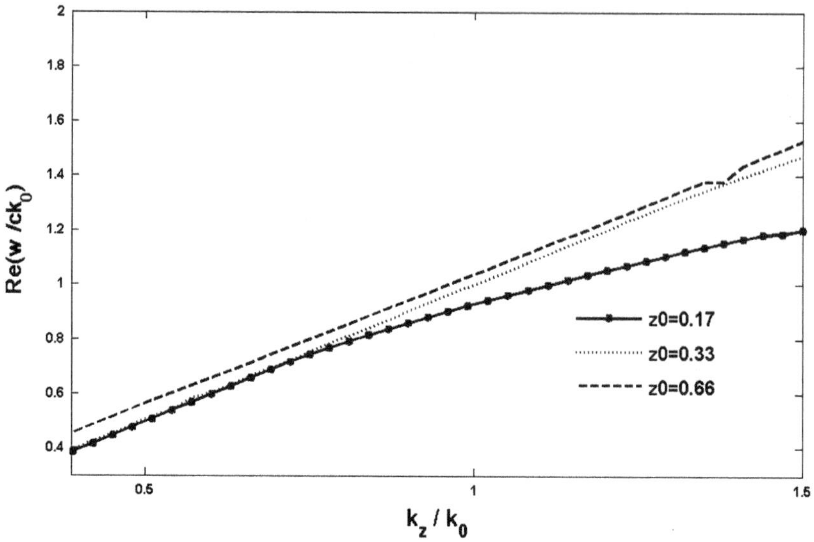

Fig. 2 Variation of normalized frequency Re $\left(\frac{\omega}{c\kappa_0}\right)$ with normalized wave number $\left(\frac{k_z}{\kappa_0}\right)$ for several values of the corrugation period. The chosen parameters are as follows: $n_p = 3 \times 10^{11}$, $R_0 = 1.60$, $h = 0.7$ and $T_p = 30 \times 10^8$

We investigate a ground state ($n = 0$) to solve the equation. With substituting Eq. (19) into Eq. (17), we have:

$$E_{zn} = \begin{cases} A_0 J_0(T_1 r) & 0 \le r \le R_b; \\ B_0 J_0(T_2 r) + C_0 N_0(T_2 r) & R_b \le r \le R(z); \end{cases} \quad (21)$$

$$T_1^2 = \left(\frac{\omega^2}{c^2} - K_n^2\right) \times \left(1 - \frac{\omega_b^2}{\gamma^3(\omega - K_n v)^2 - 3K_n^2 v_{th_b}^2} - \frac{\omega_p^2}{\omega^2 - 3K_n^2 v_{th_p}^2}\right), \quad (22)$$

$$T_2^2 = \left(\frac{\omega^2}{c^2} - K_n^2\right)\left(1 - \frac{\omega_p^2}{\omega^2 - 3K_n^2 v_{th_p}^2}\right). \quad (23)$$

The field components must satisfy the following continuity equations (first boundary condition):

$$E_z(r = R_b^-) = E_z(r = R_b^+), \quad (24)$$

$$E_r(r = R_b^-) = E_r(r = R_b^+). \quad (25)$$

As a result, the field components are obtained as follows:

$$B_0 = -\frac{\pi}{2} R_b s_0 A_0, \quad (26)$$

$$C_0 = -\frac{\pi}{2} R_b l_0 A_0, \quad (27)$$

where

$$s_0 = T_2 J_0(T_1 R_b) N_1(T_2 R_b) - T_1 J_1(T_1 R_b) N_0(T_2 R_b), \quad (28)$$

$$l_0 = T_1 J_0(T_2 R_b) J_1(T_1 R_b) - T_2 J_0(T_1 R_b) J_1(T_2 R_b). \quad (29)$$

At the perfectly conducting corrugated waveguide surface (Second boundary condition), the tangential electric field must be zero,

$$E_z(r = R(z)) + E_r(r = R(z))\frac{dR(z)}{dz} = 0. \quad (30)$$

With substituting Eqs. (19), (20) and (21) into Eq. (30), we investigate second boundary condition in ground state ($n = 0$) to achieve the dispersion equation.

$$B_0 J_0(T_2 R(z)) + C_0 N_0(T_2 R(z)) e^{i(K_0 z - \omega t)} + \frac{iK_0}{\frac{\omega^2}{c^2} - K_0^2} e^{i(K_0 z - \omega t)} \times \frac{d}{dR(z)}\left[B_0 J_0(T_2 R(z)) + C_0 N_0(T_2 R(z)) e^{i(K_0 z - \omega t)}\right] \times \frac{dR(z)}{dz} = 0, \quad (31)$$

Using the factorization of Eq. (31) and substituting B_0 and C_0, we obtain:

$$-e^{i(K_0 z - \omega t)} \frac{\pi}{2} R_b A_0 \left(1 + \frac{iK_0}{\frac{\omega^2}{c^2} - K_0^2}\frac{d}{dz}\right) \times [s_0 J_0(T_2 R(z)) + l_0 N_0(T_2 R(z))] = 0. \quad (32)$$

$$D \cdot A = D_n A_n = D_0 A_0 = 0, \quad (33)$$

"A" is a vector with element A_0 and "D" is a matrix with element D_0. With the help of derivative of Bessel functions, the dispersion relation can be obtained as, [18–22]

$$D_0 = \left(1 + \frac{iK_0}{\frac{\omega^2}{c^2} - K_0^2}\right) \times [T_2 h \kappa_0 (s_0 J_1 \sin(\kappa_0 z) + l_0 N_1 \sin(\kappa_0 z))] \quad (34)$$

Numerical result and discussion

The analysis of the dispersion relation is obtained by Eq. 34. First, we consider the dispersion analysis in the absence of the electron beam.

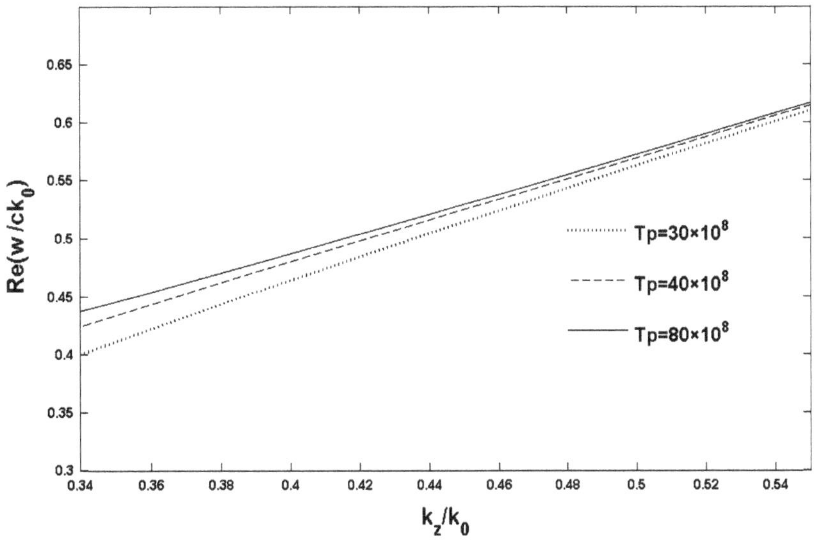

Fig. 3 Variation of normalized frequency $\mathrm{Re}\left(\frac{\omega}{c\kappa_0}\right)$ with normalized wave number $\left(\frac{k_z}{\kappa_0}\right)$ for several values of the plasma temperature. The chosen parameters are as follows: $n_p = 3 \times 10^{11}, R_0 = 1.60,\ h = 0.7,\ R_b = 0.7,\ z = 0.33$ and $z_0 = 0.66$

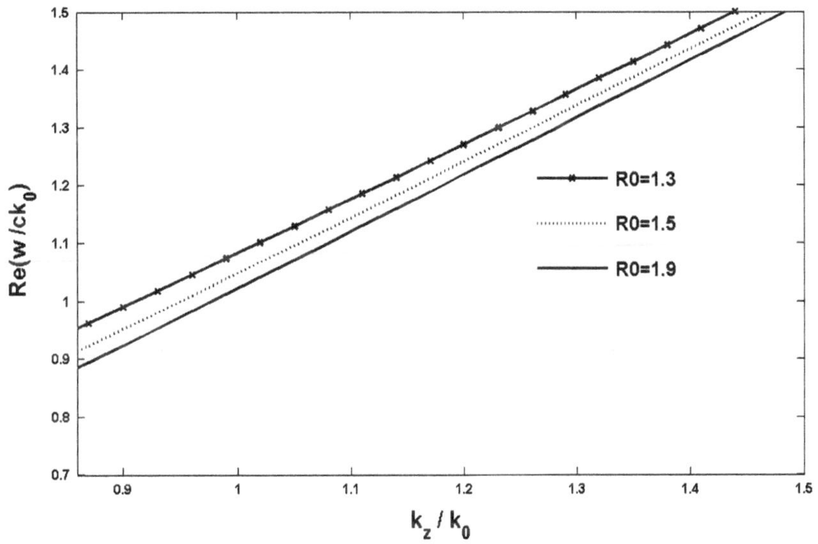

Fig. 4 Variation of normalized frequency $\mathrm{Re}\left(\frac{\omega}{c\kappa_0}\right)$ with normalized wave number $\left(\frac{k_z}{\kappa_0}\right)$ for several values of the waveguide radius. The chosen parameters are as follows: $n_p = 3 \times 10^{11},\ h = 0.7,\ R_b = 0.7, z = 0.33, T_p = 30 \times 10^8$ and $z_0 = 0.66$

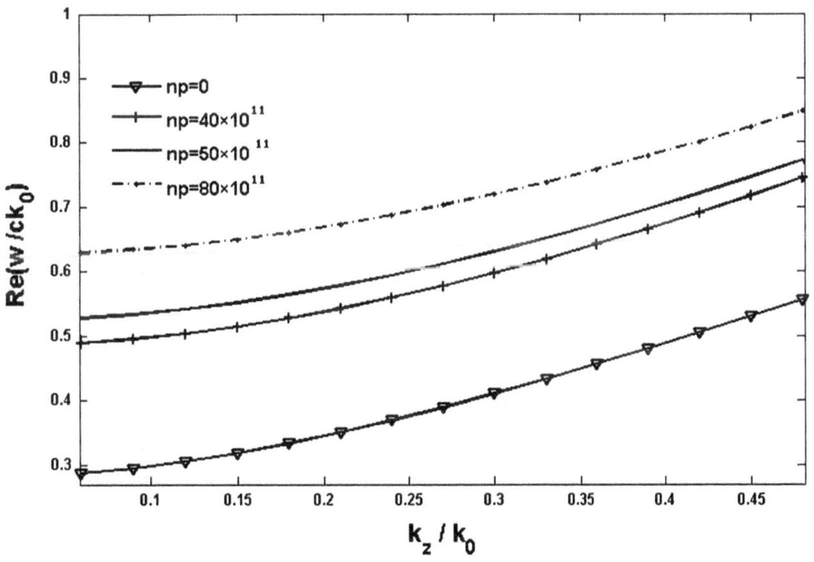

Fig. 5 Variation of normalized frequency $\mathrm{Re}\left(\frac{\omega}{c\kappa_0}\right)$ with normalized wave number $\left(\frac{k_z}{\kappa_0}\right)$ for several values of the plasma density. The chosen parameters are as follows: $R_0 = 1.60,\ h = 0.7, R_b = 0.7, z = 0.33,\ z_0 = 0.66$ and $T_p = 30 \times 10^8$

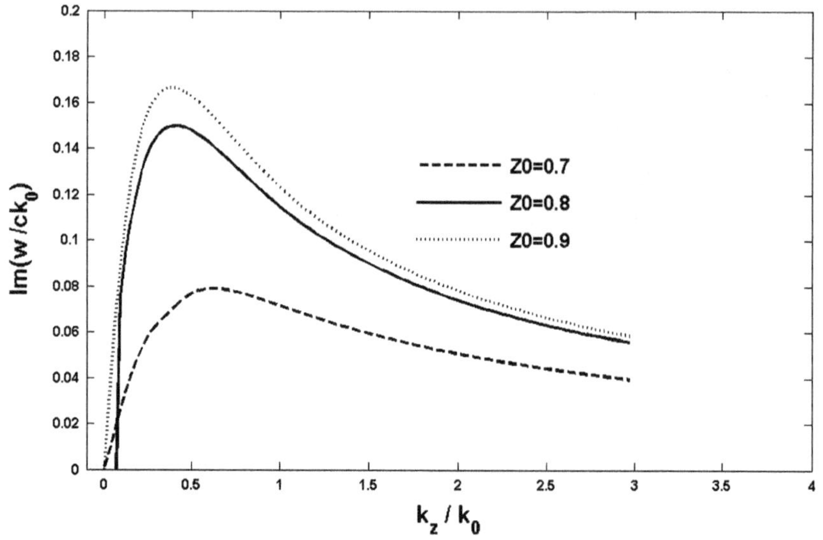

Fig. 6 Variation of the normalized growth rate $\mathrm{Im}\left(\frac{\omega}{c\kappa_0}\right)$ with normalized wave number $\left(\frac{k_z}{\kappa_0}\right)$ for several values of the corrugation period. The chosen parameters are as follows: $\gamma = 1.001, h = 0.7, R_0 = 1.6, R_b = 0.7,\ z = 0.33, n_b = 10 \times 10^{11}, n_p = 3 \times 10^{11}, T_p = 30 \times 10^8$ and $T_b = 20 \times 10^7$

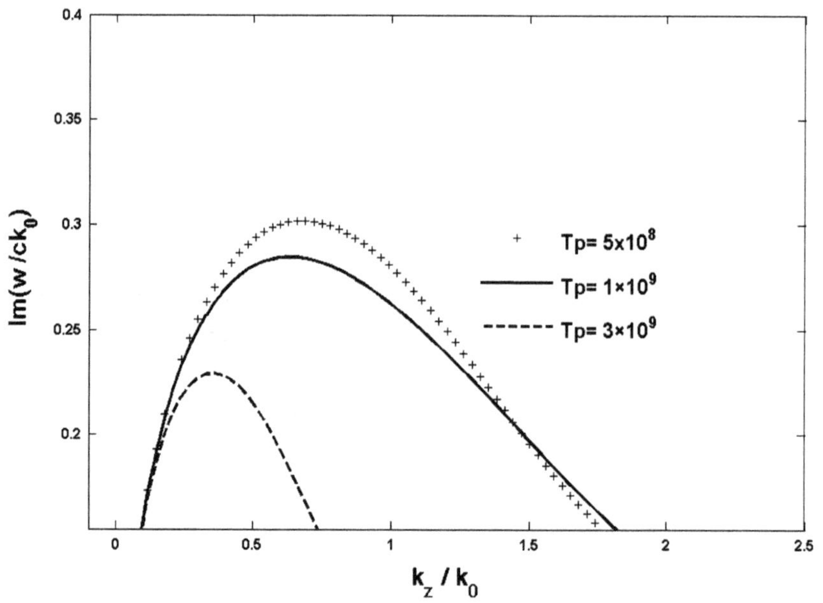

Fig. 7 Variation of the normalized growth rate $\mathrm{Im}\left(\frac{\omega}{c\kappa_0}\right)$ with normalized wave number $\left(\frac{k_z}{\kappa_0}\right)$ for several values of the plasma temperature. The chosen parameters are as follows: $n_p = 3 \times 10^{11}, h = 0.7, R_0 = 1.6,\ z = 0.33,\ \gamma = 1.001, z_0 = 0.66, n_b = 10 \times 10^{11}$ and $T_b = 20 \times 10^7$

Figure 2 shows the variation of normalized frequency $\mathrm{Re}\left(\frac{\omega}{c\kappa_0}\right)$ versus wave number $\left(\frac{k_z}{\kappa_0}\right)$ for several values of the corrugation period (z_0). As seen in this figure the effect of z_0 is to increase the frequency.

The effect of the plasma temperature on the frequency of the system as a function of k_z is shown in Fig. 3. This figure shows that the effect of plasma temperature increases the frequency. This effect is in good agreement with the thermal plasma dispersion relation $\omega^2 = \omega_p^2 + 3k_z^2 v_{th}^2$.

The effect of waveguide radius on the frequency of the wave as a function of k_z is shown in Fig. 4. As illustrated in this figure, the frequency decreases with increase in the waveguide radius. The phase velocity of the system decreases by increasing the radius. It may be cause to the wave exit from the resonant condition.

Variations of the frequency as a function of the wave number for different values of the plasma density are shown in Fig. 5. It is clear that from the Fig. 5, the effect of plasma considerably increases the frequency. This effect is in good agreement with the simple relation of the plasma $(\omega^2 = \omega_p^2 + 3k_z^2 v_{th}^2)$.

Now, we consider the analysis of the growth rate in the presence of the electron beam.

It is clear that from Fig. 6, the growth rate increases by increasing the corrugation period. The increase in the corrugation period helps the wave to be included in the resonant condition and finally increases the growth rate.

Fig. 8 Variation of the normalized growth rate $\mathrm{Im}\left(\frac{\omega}{c\kappa_0}\right)$ with normalized wave number $\left(\frac{k_z}{\kappa_0}\right)$ for several values of the waveguide radius. The chosen parameters are as follows: $n_p = 3 \times 10^{11}, h = 0.7,\ T_p = 30 \times 10^8, R_b = 0.7, z = 0.33, \gamma = 1.001, z_0 = 0.66, n_b = 10 \times 10^{11}$ and $T_b = 20 \times 10^7$

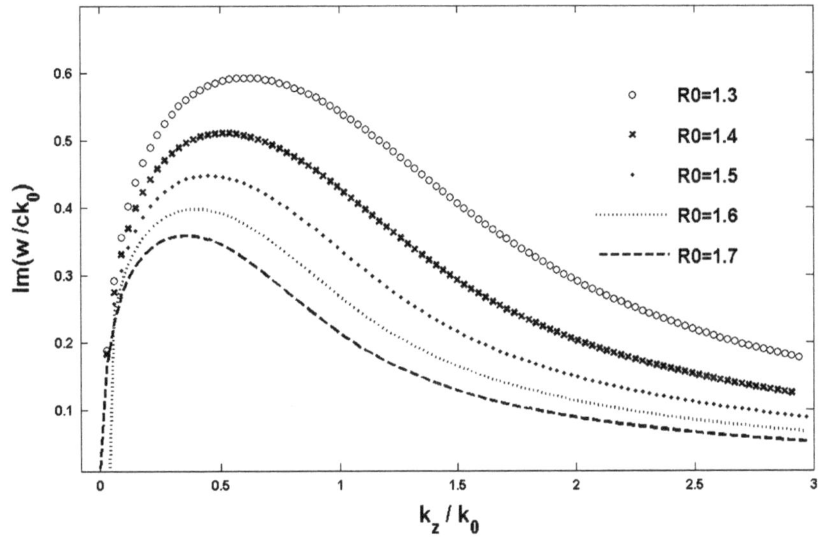

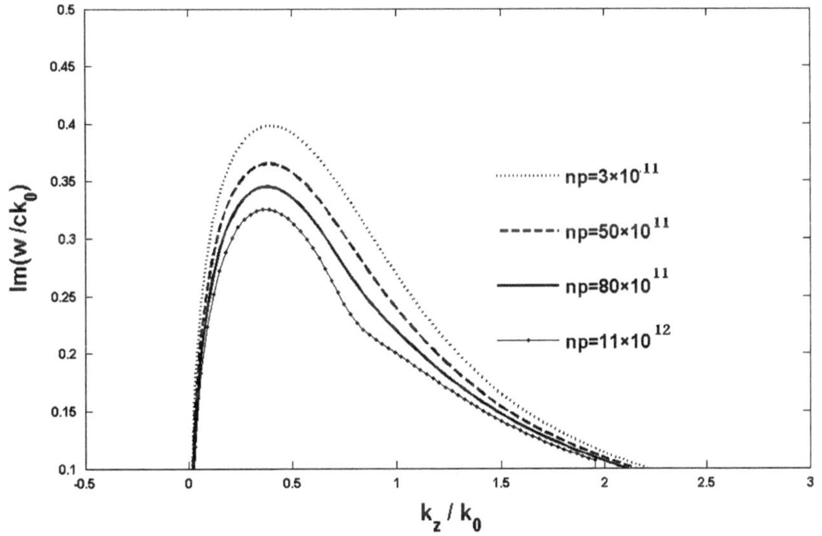

Fig. 9 Variation of the normalized growth rate $\mathrm{Im}\left(\frac{\omega}{c\kappa_0}\right)$ with normalized wave number $\left(\frac{k_z}{\kappa_0}\right)$ for several values of the plasma density. The chosen parameters are as follows: $R_0 = 1.6, h = 0.7, T_p = 30 \times 10^8, R_b = 0.7, z = 0.33, \gamma = 1.001, z_0 = 0.66, n_b = 10 \times 10^{11}$ and $T_b = 20 \times 10^7$

It is clear that from Fig. 7, the plasma temperature considerably decreases the growth rate. By increasing the plasma temperature, the electrons exit from the resonant condition and finally decrease the growth rate.

As seen in Fig. 8, the growth rate decreases by increasing the R_0. The waveguide radius has the important role in the determination of wave phase velocity. According to the Fig. 4, the phase velocity decreases by increasing the radius, and as seen in Fig. 8 the growth rate decreases by increasing the radius. By decreasing the phase velocity the wave exits from resonant condition.

The effect of the plasma density on the growth rate of the system as a function of the wave number is shown in Fig. 9. It is clear that in this frequency range the effect of plasma density is to decrease the growth rate of the system. According to the Fig. 5, the phase velocity increases by increasing the plasma density. By increasing the plasma density the wave exit from the resonant condition and finally the growth rate decreases.

It is clear that from Fig. 10, because of bunching effect, the increasing e-beam density increases the growth rate. As seen from Fig. 11, because of the synchronism condition, the effect of electron beam energy is to decrease the growth rate.

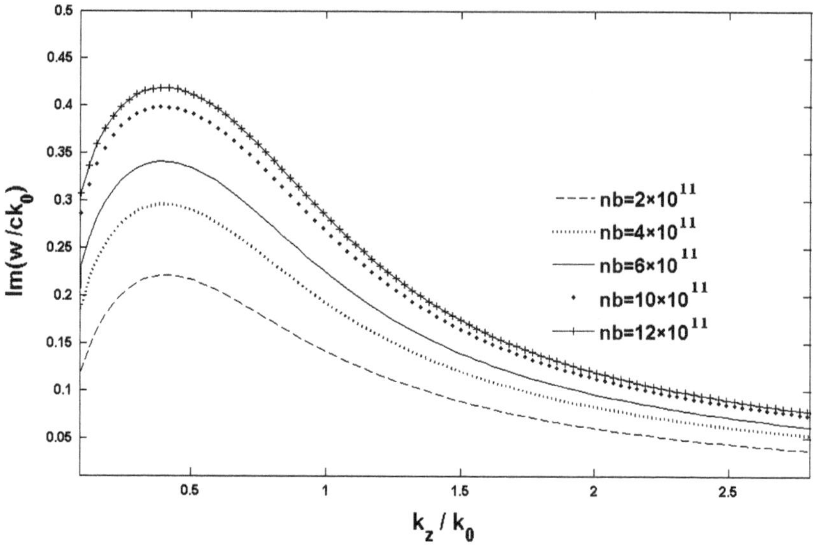

Fig. 10 Variation of the normalized growth rate $\mathrm{Im}\left(\frac{\omega}{c\kappa_0}\right)$ with normalized wave number $\left(\frac{k_z}{\kappa_0}\right)$ for several values of the electron beam density. The chosen parameters are as follows: $R_0 = 1.6, h = 0.7, z = 0.33$, $R_b = 0.7, \gamma = 1.001, z_0 = 0.66, n_p = 3 \times 10^{11}, T_b = 20 \times 10^7$ and $T_p = 30 \times 10^8$

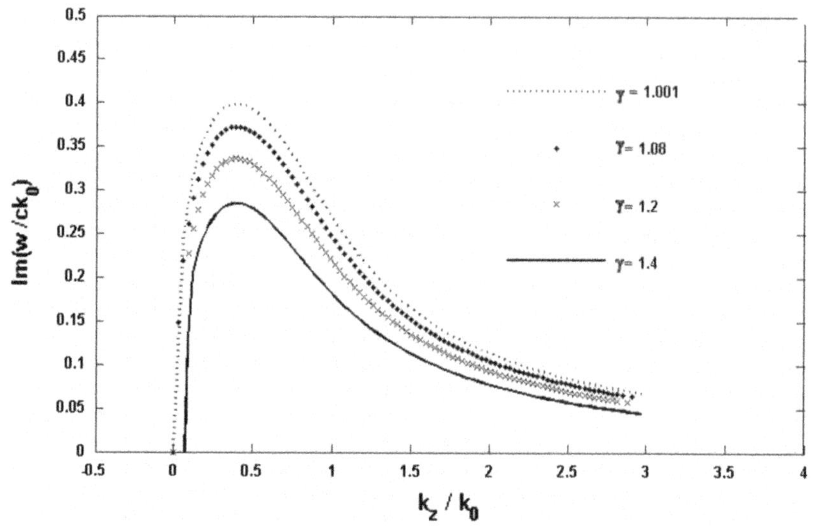

Fig. 11 Variation of the normalized growth rate $\mathrm{Im}\left(\frac{\omega}{c\kappa_0}\right)$ with normalized wave number $\left(\frac{k_z}{\kappa_0}\right)$ for several values of the electron beam energy. The chosen parameters are as follows: $R_0 = 1.6, h = 0.7, z_0 = 0.66$, $R_b = 0.7$, $z = 0.33$, $n_b = 10 \times 10^{11}$, $T_p = 30 \times 10^8$, $n_p = 3 \times 10^{11}$ and $T_b = 20 \times 10^7$

Conclusions

In this paper, useful results are obtained as:

1. In the absence of the electron beam, the frequency increases by increasing the length of the corrugation period, plasma temperature and plasma density.
2. The frequency decreases by increasing the waveguide radius in the absence of the electron beam.
3. In the presence of the electron beam, the increase in the corrugation period and e-beam density help the wave to be included in the resonant condition and finally increase the growth rate.
4. The growth rate decreases by increasing the plasma temperature, waveguide radius, plasma density and e-beam energy and decreases the phase velocity. By decreasing the phase velocity the wave exits from resonant condition.

References

1. Shiffler, D., Nation, J.A., Graham, S.K.: A high-power, traveling wave tube amplifier. IEEE Trans. Plasma Sci. **18**, 546 (1990)
2. Nusinovich, G.S., Carmel, Y., Antonsen, Jr.T.M.: IEEE Trans. Plasm. Sci. **26**, 628 (1998)
3. Kobayashi Jr, S., Antonsen, T.M., Nusinovich, G.S.: IEEE Trans. Plasma Sci. **26**, 669 (1998)
4. Pierce, J.R., Field, L.M.: Traveling-wave tubes. Proc. IRE **35**, 108 (1947)
5. Pierce, J.R.: Travelling-Wave Tubes. Van Nostrand Reinhold, New York (1950). (ch. 3)
6. Pierce, J.R.: Theory of the beam-type traveling wave tube. Proc. IRE **35**, 111 (1947)
7. Chu, L.J., Jackson, J.D.: Proc. IRE **36**, 853 (1948)
8. Freund, H.P., Ganguly, A.K.: Three-dimensional theory of the free electron laser in the collective regime. Phys. Rev. A **28**(6), 3438–3449 (1983)
9. Freund, H.P.: Multimode nonlinear analysis of free-electron laser, amplifiers in three dimensions. Phys. Rev. A **37**(9), 3371–3380 (1988)

10. Brillouin, L.: Wave Propagation in Periodic Structures. Dover, New York (1953)
11. Collin, R.E.: Field Theory of Guided Waves. McGraw-Hill, New York (1960)
12. Saviz, S.: The effect of beam and plasma parameters on the four modes of plasma-loaded traveling-wave tube with tape helix. J. Theor. Appl. Phys. **8**, 135 (2014)
13. Saviz, S., Salehizadeh, F.: Plasma effect in tape helix traveling-wave tube. J. Theor. Appl. Phys. **8**, 1 (2014)
14. Coaker, B., Challis, T.: Travelling wave tubes: modern devices and contemporary applications. Microw. J. **51**(10), 32 (2008)
15. Saviz, S., Shahi, F.: Analysis of axial electric field in thermal plasma-loaded helix traveling-wave tube with dielectric-loaded waveguide. IEEE Trans. Plasma Sci. **42**, 917 (2014)
16. Zavyalov, M.A., Mitin, L.A., Perevodchikov, V.I., Tskhai, V.N., Shapiro, A.L.: Powerful wideband amplifier based on hybrid plasma-cavity slow-wave structure. IEEE Trans. Plasma Sci. **22**, 600 (1994)
17. Saviz, S.: Plasma thermal effect on the growth rate of the helix traveling wave tube. IEEE Trans. Plasma Sci. **42**, 2023 (2014)
18. Miyamoto, K.: Parameter sensitivity of ITER type experimental tokamak reactor toward compactness. J. Plasma Fusion Res. **76**, 166 (2000)
19. Hong-Quan, X., Pu-Kun, L.: Theoretical analysis of a relativistic travelling wave tube filled with plasma. Chin. Phys. Soc. **16**(3), 766 (2007)
20. Chen, F.F.: Introduction to Plasma Physics and Controlled Fusion. Plenum Press, New York (1984)
21. Krall, N.A., Trivelpiece, A.W.: Principles of Plasma Physics. McGraw-Hill, New York (1973)
22. Goldston, R.J., Rutherford, P.H.: Introduction to Plasma Physics. Institute of Phyics Publishing, London (1995)

12

Proton driven plasma wakefield generation in a parabolic plasma channel

Y. Golian[1] · D. Dorranian[1]

Abstract An analytical model for the interaction of charged particle beams and plasma for a wakefield generation in a parabolic plasma channel is presented. In the suggested model, the plasma density profile has a minimum value on the propagation axis. A Gaussian proton beam is employed to excite the plasma wakefield in the channel. While previous works investigated on the simulation results and on the perturbation techniques in case of laser wakefield accelerations for a parabolic channel, we have carried out an analytical model and solved the accelerating field equation for proton beam in a parabolic plasma channel. The solution is expressed by Whittaker (hypergeometric) functions. Effects of plasma channel radius, proton bunch parameters and plasma parameters on the accelerating processes of proton driven plasma wakefield acceleration are studied. Results show that the higher accelerating fields could be generated in the PWFA scheme with modest reductions in the bunch size. Also, the modest increment in plasma channel radius is needed to obtain maximum accelerating gradient. In addition, the simulations of longitudinal and total radial wakefield in parabolic plasma channel are presented using LCODE. It is observed that the longitudinal wakefield generated by the bunch decreases with the distance behind the bunch while total radial wakefield increases with the distance behind the bunch.

✉ D. Dorranian
doran@srbiau.ac.ir

[1] Laser Laboratory, Plasma Physics Research Center, Science and Research Branch, Islamic Azad University, Tehran, Iran

Keywords Gaussian beam · LCODE · Parabolic plasma channel · Plasma wakefield accelerator · Whittaker functions

Introduction

Compared to the conventional radio frequency cavities, in plasma wakefield, the energy will be transferred from the driver bunch to the witness bunch, a way to achieve very high acceleration gradients [1–6]. The field in the plasma can be excited either by an ultra-intense, short laser pulse (or several pulses) [3, 7], or by a charged particle beam (an electron or proton beam) [8, 9]. Proton driven plasma wakefield acceleration has been recently proposed to attain electron energies of about TeV range [9–18]. Plasma wakefield generation in radially nonuniform plasma using charged particle beams [19] or in parabolic plasma channels using lasers [20–24] as drivers have been extensively studied. For the laser drivers, most of the studies are focused at parabolic channels with on-axis density minimum because of channeling properties of these channels, so that in a preformed plasma channel the laser pulse can propagate up to several Rayleigh lengths without substantial spreading. One of approaches in preventing of diffraction broadening of laser pulse and increasing of laser-plasma interaction distance is the guiding of the pulse in preformed plasma density channel [19, 25]. Therefore, generation of wakefields in a preformed channel becomes a subject of great interest.

In our recent work [26], we presented for the first time a thorough analytical model and solved the governing equations for the wakefield acceleration by proton beam in a parabolic density profile plasma, where the maximum plasma density is on the beam propagation axis along the *z*-

direction. It has a potential application for compensating the undesirable wave front curvature in the case of a nonlinear wakefield. But in this work, a plasma channel with parabolic density distribution was considered in which the minimum density of plasma is on the beam propagation axis and the behavior of wakefields is investigated in comparison of [26]. For laser drivers, a similar problem was addressed in [22], where a perturbative method was employed to obtain the components of wakefields. The governing equations and the solution for the longitudinal wakefield E_z is obtained analytically and the effects of the plasma channel radius, proton bunch parameters and wave number of plasma on the longitudinal electric field are investigated. In addition to analytical approach, the simulations of longitudinal and total radial wakefields in longitudinal and radial directions are presented using two-dimensional code LCODE [27].

The organization of the paper is as follows: "Introduction" is given in the next section followed by "Model description". "Results and discussions" are given before the last section. Finally, "Conclusion" is presented.

Model description

The accelerating model which we have used for proton driven plasma wakefield acceleration is the plasma channel model. The plasma has a variable density as:

$$n = n_0\left(1+\frac{r^2}{r_{\mathrm{pl}}^2}\right), \tag{1}$$

where r_{pl} is the plasma channel radius, and r is the radial coordinate. The minimum density of plasma (n_0) is on the beam propagation axis along the z-direction and $\frac{n_0 r^2}{r_{\mathrm{pl}}^2}$ represent radial variation in channel density. The density profile of proton beam is described as follows:

$$n_{\mathrm{b}} = n_{\mathrm{b0}} e^{\frac{-r^2}{2\sigma_r^2}} e^{\frac{-\xi^2}{2\sigma_z^2}}, \tag{2}$$

where $\xi = z - ct$, with c speed of light, $n_{\mathrm{b0}} = N/\left((2\pi)^{3/2}\sigma_r^2\sigma_z\right)$ N is the total number of protons in the bunch, and σ_r and σ_z are the beam radial and longitudinal characteristic lengths, respectively. Although the interaction of a dense $(n_{\mathrm{b}} > n_0)$, short proton beam (the wavelength of the plasma wave is of the order of the bunch length, $\sigma_z \sim \lambda_{\mathrm{p}}$) with a plasma is inherently nonlinear, we can use the linear theory which predicts the plasma response to either a short electron or proton beam [9, 28]. In principle, the linear regime is valid when the plasma density perturbation is small compared to the plasma density n_0. Since the density perturbation is typically of the order of the beam peak density n_{b0}, this criterion is equivalent to $n_{\mathrm{b0}}/n_0 \ll 1$. The simulations have shown that the predictions of the linear theory are good up to $n_{\mathrm{b0}}/n_0 \sim 1$ [29].

We note that charge density in the system can be expressed as $\rho = en_{\mathrm{b}} + en_{\mathrm{i}} - en_{\mathrm{e}} = en_{\mathrm{b}} + en_0 - e(n_0 + n_1)$. Indexes b, i and e denote the beam, ions and the electrons, respectively.

Using Maxwell's equations, the Newton's second law of motion for the plasma electrons, the equation of continuity for the plasma, and writing quantities in terms of an equilibrium and an oscillating term $E = E_0 + E_1$, $B = B_0 + B_1$, $v = v_0 + v_1$ and assuming that E_0, B_0 and v_0 are all equal to zero, we have:

$$\begin{aligned} &\nabla \cdot E_1 = \frac{1}{\varepsilon_0}(en_{\mathrm{b}} - en_1)\\ &\frac{\partial v_1}{\partial t} = -\frac{e}{m}E_1\\ &\frac{\partial n_1}{\partial t} + n\nabla \cdot v_1 = 0, \end{aligned} \tag{3}$$

where quantities indexed with 1 denote the perturbed parameters.

Using Eq. 3, one can arrive at,

$$\frac{\partial^2 n_1}{\partial t^2} + \omega_{\mathrm{p0}}^2\left(1+\frac{r^2}{r_{\mathrm{pl}}^2}\right)n_1 = \omega_{\mathrm{p0}}^2 n_{\mathrm{b}}\left(1+\frac{r^2}{r_{\mathrm{pl}}^2}\right), \tag{4}$$

where $\omega_{\mathrm{p0}} = \left(\frac{n_0 e^2}{m\varepsilon_0}\right)^{0.5}$ is the plasma frequency.

In Eq. 4, we use the change of variable $\xi = z - ct$, with c the speed of light. Time derivative $\partial/\partial t$ can be replaced by $-c\partial/\partial\xi$,

$$\frac{\partial^2 n_1}{\partial \xi^2} + k_{\mathrm{p0}}^2\left(1+\frac{r^2}{r_{\mathrm{pl}}^2}\right)n_1 = k_{\mathrm{p0}}^2 n_{\mathrm{b}}\left(1+\frac{r^2}{r_{\mathrm{pl}}^2}\right), \tag{5}$$

where k_{p0} is the plasma wave number. By substituting n_{b} in Eq. 5, we have:

$$\frac{\partial^2 n_1}{\partial \xi^2} + k_{\mathrm{p0}}^2\left(1+\frac{r^2}{r_{\mathrm{pl}}^2}\right)n_1 = k_{\mathrm{p0}}^2 n_{\mathrm{b0}} e^{\frac{-r^2}{2\sigma_r^2}} e^{\frac{-\xi^2}{2\sigma_z^2}}\left(1+\frac{r^2}{r_{\mathrm{pl}}^2}\right). \tag{6}$$

We can solve this differential equation using the Laplace transform of a system which is initially at rest,

$$\begin{aligned} n_1 = {}& \sqrt{2\pi}\sigma_z k_{\mathrm{p0}} n_{\mathrm{b0}}\left(1+\frac{r^2}{r_{\mathrm{pl}}^2}\right)^{0.5} e^{\frac{-r^2}{2\sigma_r^2}} e^{\frac{-k_{\mathrm{p0}}^2\left(1+\frac{r^2}{r_{\mathrm{pl}}^2}\right)\sigma_z^2}{2}}\\ &\times \sin\left(k_{\mathrm{p0}}\left(1+\frac{r^2}{r_{\mathrm{pl}}^2}\right)^{0.5}\xi\right). \end{aligned} \tag{7}$$

For an infinite plasma ($r_{pl} \to \infty$), this reduces to the expression derived for the perturbed density as in [30].

Also, we can solve the equations for the electric field as a function of the perturbed density. We note that the current density can be written as $J = J_b + J_e = J_b + (J_0 + J_1) = J_b + J_1$ in which J_b and J_e are beam and electron current density, respectively. J_0 and J_1 denote the unperturbed and the perturbed electron current density, respectively.

Using the Maxwell equations, we have:

$$\begin{aligned} &\nabla^2 E_1 - \nabla(\nabla \cdot E_1) = \frac{\partial}{\partial t}(\nabla \times B_1) \\ &\nabla \times B_1 = \mu_0 J_1 + \mu_0 J_b \hat{z} + \frac{1}{c^2}\frac{\partial E_1}{\partial t} \\ &\nabla^2 E_1 - \frac{\partial^2 E_1}{\partial \xi^2} = e n_0 \mu_0 c \left(1 + \frac{r^2}{r_{pl}^2}\right) \frac{\partial v_1}{\partial \xi} - \mu_0 e c^2 \frac{\partial n_b}{\partial \xi}\hat{z} \\ &+ \frac{e}{\varepsilon_0}\nabla n_b - \frac{e}{\varepsilon_0}\nabla n_1, \end{aligned} \tag{8}$$

where $\frac{\partial v_1}{\partial \xi} = \frac{e}{\mathrm{mc}} E_1$.

By substituting $\partial n_b/\partial \xi$, ∇n_b, ∇n_1 in Eq. 8 and separating the components we will have the longitudinal and radial wakefields as:

$$\begin{aligned} &\left[\nabla_r^2 - k_{p0}^2\left(1 + \frac{r^2}{r_{pl}^2}\right)\right] E_{1z}(r,\xi) = \frac{-e}{\varepsilon_0}\sqrt{2\pi}\sigma_z n_{b0} k_{p0}^2 \\ &\times \left(1 + \frac{r^2}{r_{pl}^2}\right) e^{\frac{-r^2}{2\sigma_r^2}} e^{\frac{-k_{p0}^2\left(1+\frac{r^2}{r_{pl}^2}\right)\sigma_z^2}{2}} \cos\left(k_{p0}\left(1 + \frac{r^2}{r_{pl}^2}\right)^{0.5} \zeta\right) \end{aligned} \tag{9}$$

and,

$$\begin{aligned} &\left[\nabla_r^2 - k_{p0}^2\left(1 + \frac{r^2}{r_{pl}^2}\right)\right] E_{1r}(r,\xi) = \frac{-e}{\varepsilon_0}\frac{r}{\sigma_r^2} n_{b0} e^{\frac{-r^2}{2\sigma_r^2}} e^{\frac{-\xi^2}{2\sigma_z^2}} \\ &- \frac{e}{\varepsilon_0}\sqrt{2\pi}\sigma_z n_{b0} k_{p0} e^{\frac{-r^2}{2\sigma_r^2}} e^{\frac{-k_{p0}^2\left(1+\frac{r^2}{r_{pl}^2}\right)\sigma_z^2}{2}} \times \sin\left(k_{p0}\left(1 + \frac{r^2}{r_{pl}^2}\right)^{0.5} \xi\right) \\ &\times \left[\frac{r}{r_{pl}^2}\left(1 + \frac{r^2}{r_{pl}^2}\right)^{-0.5} - \left(\frac{r}{\sigma_r^2} + \frac{k_{p0}^2\sigma_z^2 r}{r_{pl}^2}\right)\left(1 + \frac{r^2}{r_{pl}^2}\right)^{0.5}\right. \\ &\left. + \frac{k_{p0}\xi r}{r_{pl}^2}\cot\left(k_{p0}\left(1 + \frac{r^2}{r_{pl}^2}\right)^{0.5} \xi\right)\right] \end{aligned} \tag{10}$$

These equations could be compared with equations for a Gaussian beam in infinite plasma medium where ($r_{pl} \to \infty$). The equations reduce to their counterpart results expressed in [29, 31]. The solution for the homogeneous part of Eq. 9 with respect to the variable r for a finite parabolic plasma channel can be expressed by the Whittaker function:

$$\begin{aligned} E_{0z}(r) = &\frac{C_1}{r}\mathrm{whittakerM}\left(\frac{i k_{p0}^2}{4\left(-\frac{k_{p0}^2}{r_{pl}^2}\right)^{0.5}}, 0, i r^2\left(-\frac{k_{p0}^2}{r_{pl}^2}\right)^{0.5}\right) \\ &+ \frac{C_2}{r}\mathrm{whittakerW}\left(\frac{i k_{p0}^2}{4\left(-\frac{k_{p0}^2}{r_{pl}^2}\right)^{0.5}}, 0, i r^2\left(-\frac{k_{p0}^2}{r_{pl}^2}\right)^{0.5}\right) \end{aligned} \tag{11}$$

where C_1 and C_2 are constant and Whittaker M function is defined by

$$\mathrm{WhittakerM}\,(k,m,z) = e^{-z/2} z_1^{m+1/2} F_1\left(\frac{1}{2} + m - k, 1 + 2m; z\right),$$

where,

$${}_1F_1(a;b;z) = 1 + \frac{a}{b}z + \frac{a(a+1)}{b(b+1)}\frac{z^2}{2!} + \ldots = \sum_{k=0}^{\infty}\frac{(a)_k}{(b)_k}\frac{z^k}{k!}.$$

Whittaker W function is expressed as:

$$\mathrm{WhittakerW}\,(k,m,z) = e^{-z/2} z^{m+1/2} U\left(\frac{1}{2} + m - k, 1 + 2m; z\right),$$

where,

$$U(a,b,z) = z_2^{-a} F_0(a, 1 + a - b;; -z),$$

and $F_0(a,b;;z)$ is defined as a generalized hypergeometric function:

$${}_2F_0(a,b;;z) = \sum_{k=0}^{\infty}(a)_k(b)_k\frac{z^k}{k!}.$$

The detailed properties of the Whittaker M and Whittaker W functions can be found in [32]. Note that for infinite plasma ($r_{pl} \to \infty$), the solution takes the form of the well-known modified Bessel functions as given in [33]. Solution of the inhomogeneous equation can be expressed as the sum of the homogenous solution introduced above plus solution coming from the right hand side as:

$$
\begin{aligned}
E_{\mathrm{Inhom}-1z}(r,\xi) &= \frac{C_1}{r}\mathrm{whittakerM}\left(\frac{ik_{p0}^2}{4\left(-\frac{k_{p0}^2}{r_{pl}^2}\right)^{0.5}},0,ir^2\left(-\frac{k_{p0}^2}{r_{pl}^2}\right)^{0.5}\right)\\
&+\frac{C_2}{r}\mathrm{whittakerW}\left(\frac{ik_{p0}^2}{4\left(-\frac{k_{p0}^2}{r_{pl}^2}\right)^{0.5}},0,ir^2\left(-\frac{k_{p0}^2}{r_{pl}^2}\right)^{0.5}\right)\\
&+\frac{1}{r}\mathrm{whittakerW}\left(\frac{ik_{p0}^2}{4\left(-\frac{k_{p0}^2}{r_{pl}^2}\right)^{0.5}},0,ir^2\left(-\frac{k_{p0}^2}{r_{pl}^2}\right)^{0.5}\right)\\
&\times\int\left[\frac{-2r^2}{D}e^{-\frac{r_{pl}^2r^2+k_{p0}^2\sigma_r^2\sigma_z^2(r_{pl}^2+r^2)}{2r_{pl}^2\sigma_r^2}}Q\cos k_{p0}\left(\xi\left(1+\frac{r^2}{r_{pl}^2}\right)^{0.5}\right)(r^2+r_{pl}^2)\right]\\
&\times\left[\left(-\frac{k_{p0}^2}{r_{pl}^2}\right)^{0.5}\mathrm{whittakerM}\left(\frac{ik_{p0}^2}{4\left(-\frac{k_{p0}^2}{r_{pl}^2}\right)^{0.5}},0,ir^2\left(-\frac{k_{p0}^2}{r_{pl}^2}\right)^{0.5}\right)\right]dr\\
&+\frac{1}{r}\mathrm{whittakerM}\left(\frac{ik_{p0}^2}{4\left(-\frac{k_{p0}^2}{r_{pl}^2}\right)^{0.5}},0,ir^2\left(-\frac{k_{p0}^2}{r_{pl}^2}\right)^{0.5}\right)\times\int\left[\frac{2r^2}{D}\left(-\frac{k_{p0}^2}{r_{pl}^2}\right)^{0.5}\right]\\
&\times\left[e^{-\frac{r_{pl}^2r^2+k_{p0}^2\sigma_r^2\sigma_z^2(r_{pl}^2+r^2)}{2r_{pl}^2\sigma_r^2}}Q\cos k_{p0}\left(\xi\left(1+\frac{r^2}{r_{pl}^2}\right)^{0.5}\right)\right.\\
&\left.(r^2+r_{pl}^2)\mathrm{whittakerW}\left(\frac{ik_{p0}^2}{4\left(-\frac{k_{p0}^2}{r_{pl}^2}\right)^{0.5}},0,ir^2\left(-\frac{k_{p0}^2}{r_{pl}^2}\right)^{0.5}\right)\right]dr
\end{aligned}
\tag{12}
$$

where $Q=\frac{-e}{\varepsilon_0}\sqrt{2\pi}\sigma_z n_{b0}k_{p0}^2$ and D is defined as:

$$
\begin{aligned}
D &= ir_{pl}^2k_{p0}^2\mathrm{whittakerW}\left(\frac{ik_{p0}^2}{4\left(-\frac{k_{p0}^2}{r_{pl}^2}\right)^{0.5}},0,ir^2\left(-\frac{k_{p0}^2}{r_{pl}^2}\right)^{0.5}\right)\\
&\mathrm{whittakerM}\left(\frac{4\left(-\frac{k_{p0}^2}{r_{pl}^2}\right)^{0.5}+ik_{p0}^2}{4\left(-\frac{k_{p0}^2}{r_{pl}^2}\right)^{0.5}},0,ir^2\left(-\frac{k_{p0}^2}{r_{pl}^2}\right)^{0.5}\right)\\
&+4r_{pl}^2\left(-\frac{k_{p0}^2}{r_{pl}^2}\right)^{0.5}\mathrm{whittakerM}\left(\frac{ik_{p0}^2}{4\left(-\frac{k_{p0}^2}{r_{pl}^2}\right)^{0.5}},0,ir^2\left(-\frac{k_{p0}^2}{r_{pl}^2}\right)^{0.5}\right)\\
&\mathrm{whittakerW}\left(\frac{4\left(-\frac{k_{p0}^2}{r_{pl}^2}\right)^{0.5}+ik_{p0}^2}{4\left(-\frac{k_{p0}^2}{r_{pl}^2}\right)^{0.5}},0,ir^2\left(-\frac{k_{p0}^2}{r_{pl}^2}\right)^{0.5}\right)\\
&+2r_{pl}^2\left(-\frac{k_{p0}^2}{r_{pl}^2}\right)^{0.5}\mathrm{whittakerW}\left(\frac{ik_{p0}^2}{4\left(-\frac{k_{p0}^2}{r_{pl}^2}\right)^{0.5}},0,ir^2\left(-\frac{k_{p0}^2}{r_{pl}^2}\right)^{0.5}\right)\\
&\mathrm{whittakerM}\left(\frac{4\left(-\frac{k_{p0}^2}{r_{pl}^2}\right)^{0.5}+ik_{p0}^2}{4\left(-\frac{k_{p0}^2}{r_{pl}^2}\right)^{0.5}},0,ir^2\left(-\frac{k_{p0}^2}{r_{pl}^2}\right)^{0.5}\right).
\end{aligned}
$$

The two integrals in Eq. 12 over the plasma radius were solved by numerical integrations and $E_{\mathrm{Inhom}-1z}$ was obtained as a function of ξ and r.

It is to be noted here, however, that the limit of $\frac{\mathrm{WhittakerW}}{r}$ at $r=0$ approaches infinity. So the C_2 coefficient must be zero. The boundary conditions are those of a perfectly conducting tube of the radius $r_{\max}$.Thus, boundary conditions for the longitudinal electric field can be considered as $E_z(r=r_{pl})=0$ [34].

Results and discussion

Analytical results

To illustrate the behavior of wakefield in plasma channel, the parameters of beam and plasma are taken to be, $N=1\times10^{11}$, $r_{pl}=0.7\times10^{-3}\mathrm{m}=3.25\frac{c}{\omega_{p0}}$, $\sigma_r=0.43\times10^{-3}\mathrm{m}=2\frac{c}{\omega_{p0}}$, $\sigma_z=1\times10^{-4}\mathrm{m}=0.46\frac{c}{\omega_{p0}}$, $n_0=6\times10^{14}\mathrm{cm}^{-3}$, $\frac{c}{\omega_{p0}}=0.215\times10^{-3}\mathrm{m}$.

The variation of the longitudinal electric field in the longitudinal direction is presented in Fig. 1. The noticeable point in longitudinal wakefield (E_z) along the propagation axis is oscillatory behavior of wakefield and the acceleration gradient decreases with distance behind the bunch. Whereas in uniform plasma profile E_z in longitudinal direction changes uniformly.

Figure 2 shows the variations of longitudinal electric field in longitudinal direction at different plasma channel radii. To illustrate the exact modifications of longitudinal wakefield with channel radius, the small changes of channel radius are considered. The longitudinal wakefield are so sensitive to it.

As shown in this figure, by the modest increment in plasma channel radius, the acceleration gradient increases as the modest reduction in the size of proton bunch [35]. With further increases the plasma channel radius, the acceleration gradient tendency to constant level.

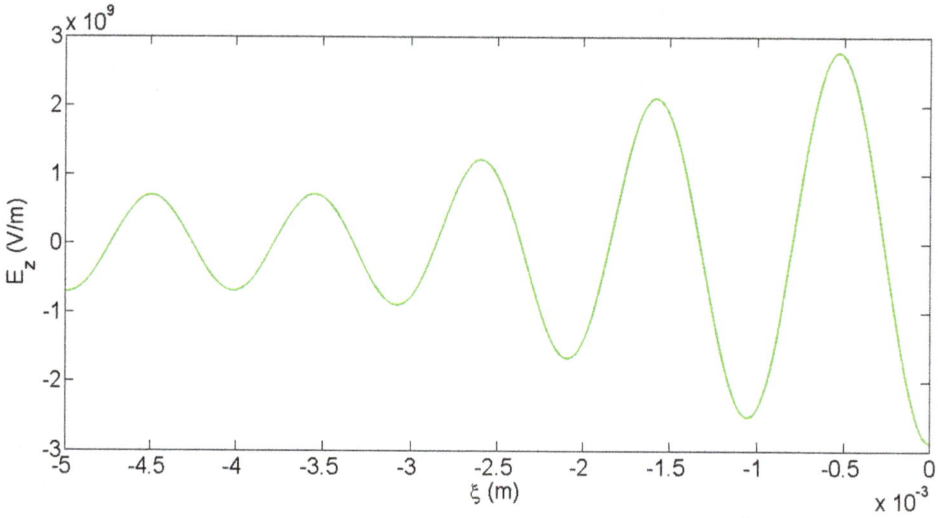

Fig. 1 The variation of longitudinal electric field in longitudinal direction for $r \approx 0$

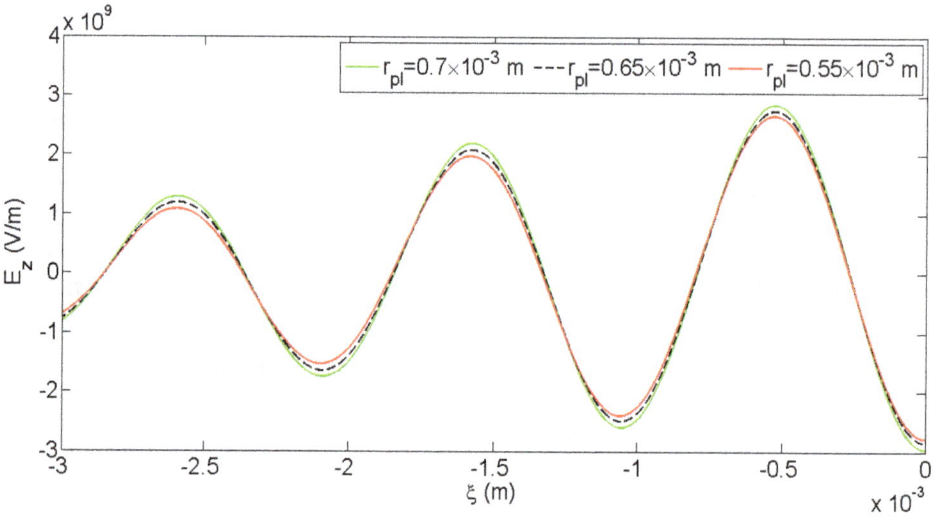

Fig. 2 The variations of longitudinal electric field in longitudinal direction at different plasma channel radii

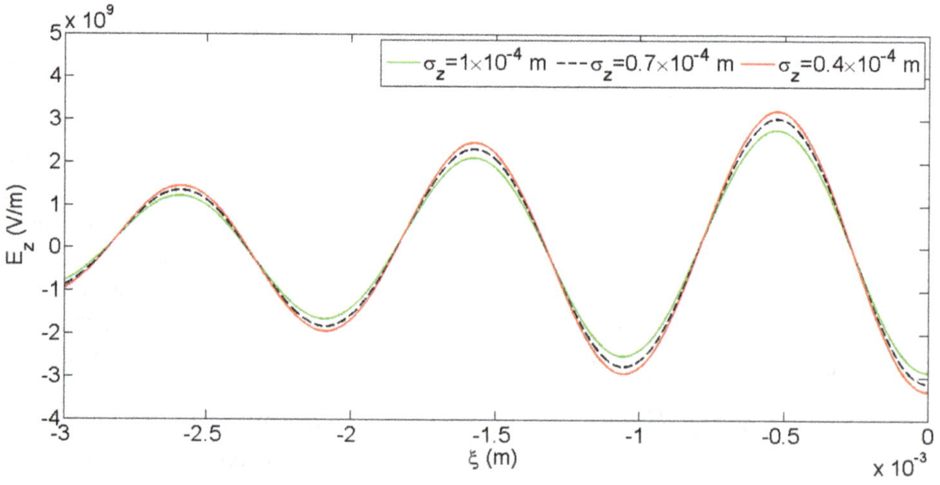

Fig. 3 The variation of the longitudinal electric field at different longitudinal characteristics of bunch

In Fig. 3, the variation of longitudinal electric field at different longitudinal size of the proton bunch is shown. The wakefield response is very sensitive to the bunch length. Also, longitudinal compression of the driver results in a shift of the optimum plasma density to higher value and to approximately proportional increase field amplitude

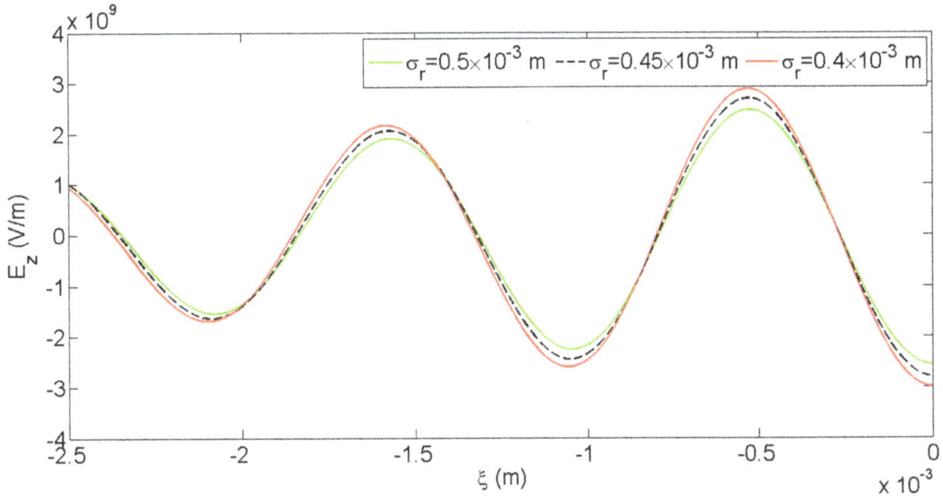

Fig. 4 The variations of the longitudinal electric field at different radial characteristics of the bunch

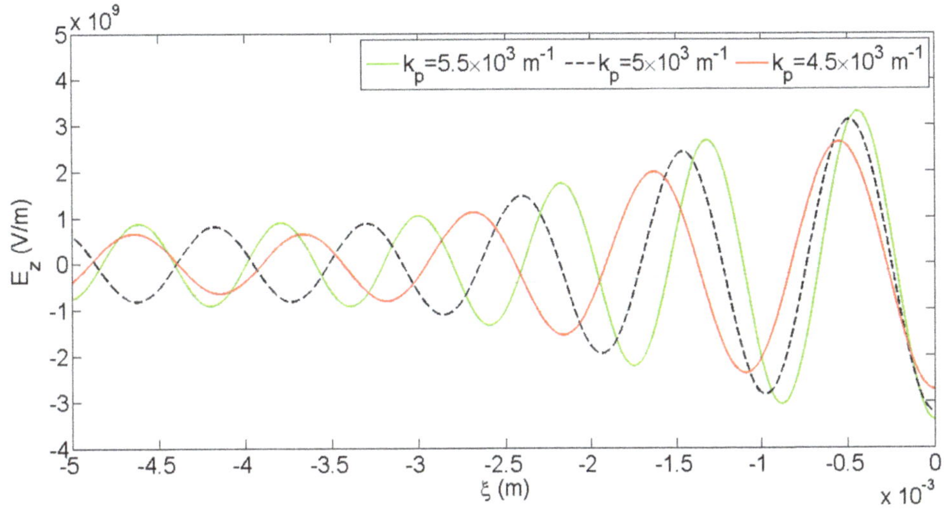

Fig. 5 The variation of longitudinal electric field at different wave numbers of plasma

[35]. Thus, the peak of accelerating field is controlled mainly by the driver length.

Also Fig. 4 shows the sensitivity of the longitudinal electric field to the radial characteristic of bunch. Similar to Fig. 3 by increasing the radial size of proton bunches, the generated wakefield becomes weaker. This is due to the aggregation of proton bunch [9, 28].

According to Figs. 3 and 4, it is expected that the dependence of wakefield on the number of beam particles scales nearly linearly.

Figure 5 shows the variations of longitudinal electric field at different wave numbers of the plasma. By increasing the wave number of the plasma, the oscillation of plasma electrons increases so the period of oscillations decreases. Also, by increasing the wave number of plasma, the longitudinal electric field will be amplified.

To create the wakefield efficiently, the driver must be focused to the transverse size $\sigma_r = k_p^{-1}$ and longitudinal size of bunch should be satisfy $\sigma_z \approx \sqrt{2k_p^{-1}}$. So, according to Figs. 3, 4, and 5, maximum acceleration gradient can be obtained by decreasing the size of bunch that is proportional to increment of wave number of plasma.

In Figs. 2, 3, 4, and 5, the parameters variations happen in nearby distances behind the bunch and it decreases far from the beam driver as the acceleration gradient is weaken.

Simulation results

It is instructive to compare the behavior of wakefields in simulation approach with the analytical results. For this purpose, two-dimensional fully electromagnetic relativistic code, LCODE [27] is used. The cylindrical coordinates and the comoving simulation window are utilized. In our simulations, kinetic model and plasma macroparticles is used. The complete description of LCODE is described in detail in Ref. [27].

The code allows an arbitrary initial transverse profile of the plasma density over the simulation window. So

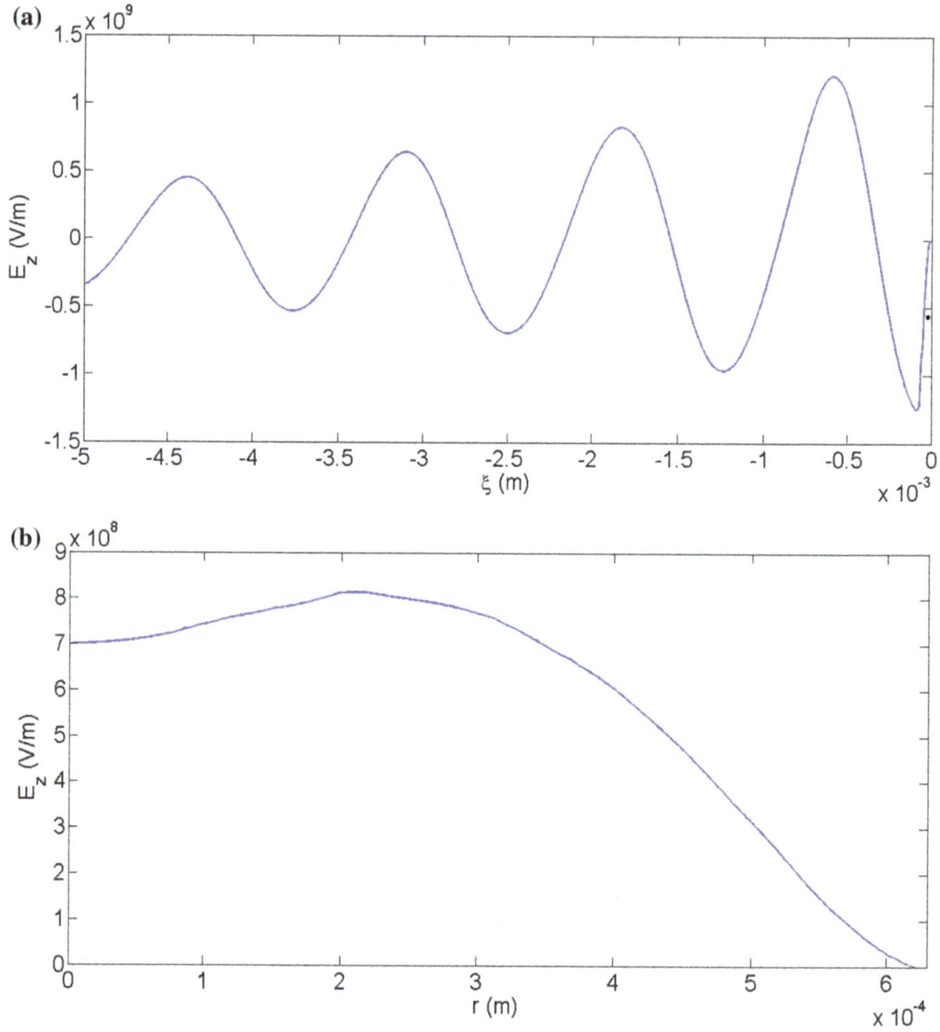

Fig. 6 **a** The variations of on axis longitudinal electric field in longitudinal direction and **b** in radial direction for $\xi = 0.43$ mm for the case in Fig. 6a

according to analytical approach, we used the distribution of parabolic plasma density. For this purpose, the new plasma.bin file generated according to our plasma profile and it imported as a parabolic plasma file.

The simulation window, in units of c/ω_p is 25 (in ξ) $\times$ 3 (in r). Both r and ξ steps are 0.01 c/ω_p where according to the analytical parameters $k_p = \frac{\omega_p}{c} = 4.65 \times 10^3$m. The beam is modeled by 4×10^7 macroparticles, and 15,000 plasma macroparticles are used. Also, initial distribution of beam particles in the transverse phase space is Gaussian and the beam current is calculated according to the parameters of the proton beam, $I_b(\xi) = \frac{\rho_b(0)\sigma_r^2}{2}$. The dependence of the beam current on the longitudinal coordinate is considered as cosine shape. The energy of proton bunch is about 1 TeV and the hydrogen plasma is used for the simulations in zero plasma temperature. The parameters of plasma and beam in simulation approach are similar to parameters in analytical approach as mentioned in previous section.

Figure 6 is presented the variations of on axis longitudinal electric field in the longitudinal and radial directions using LCODE. The behavior of longitudinal electric field in the longitudinal direction at $r = 0$ (Fig. 6a) is decreases with distance behind the bunch which has a good agreement with analytical results.

According to the Fig. 6b, the variations of longitudinal electric field in radial direction has little growth with radius and later scale down as r increases, and then go to zero. According to boundary condition of longitudinal wakefield in channel wall, $E_z(r = r_{pl}) = 0$, the decreasing treatment is expected.

The variations of total radial electric field $E_r - B_\Phi$ in longitudinal and radial directions are presented in Fig. 7. In Fig. 7a, the total radial electric field generated by the proton bunch increases with the distance behind the bunch. It is one of the significant points presented in this model. In Fig. 7b we have shown the variations of the total radial wakefield $E_r - B_\Phi$ in the radial direction. The radial electric field vanishes as $r = 0$ and it expected according to boundary condition in the origin, $E_r(r = 0) = 0$. Also, linear focusing caused by the plasma channel near the

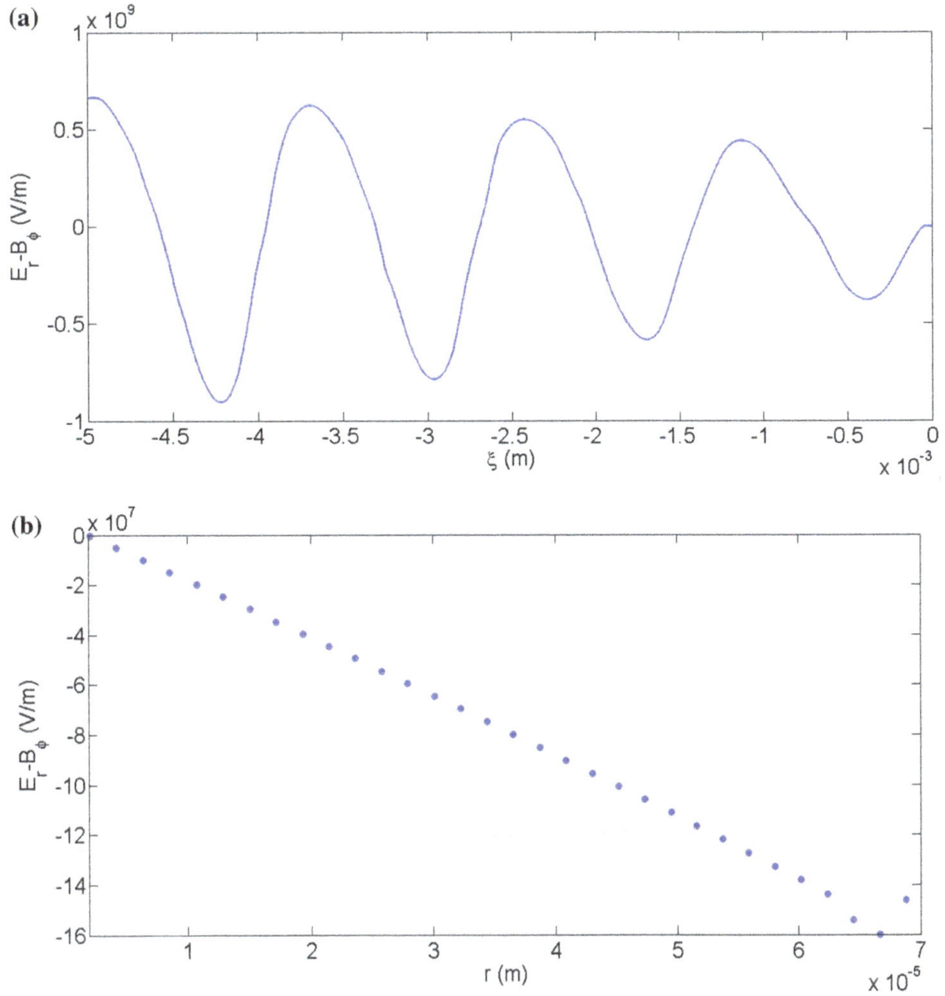

Fig. 7 **a** The variations of total radial electric field in longitudinal direction at $r = 0.2$ mm and **b** in radial direction at $\xi = 0.64$ mm

origin is obviously seen. Of course, the slope would change for different values of ξ [28]. The treatment of total radial wakefield in radial direction is good agreement with the response of wakefield to positron driver in [28] (or in the other word for positive drivers). It is seen for the case shown in the figure that the overall behavior of radial forces is focusing.

According to Figs. 6, and 7 and as an outstanding point, we can say that the wakefield energy that was initially stored in longitudinal oscillations near the axis is gradually transferred to the radial oscillations in the near-wall of channel.

Conclusion

A plasma channel with a parabolic density profile was considered as the proposed model for investigation of proton driven plasma wakefield acceleration. Plasma channel has the minimum density on the beam propagation axis. The governing equation in longitudinal direction was presented and the solution was expressed as Whittaker functions. The effects of plasma channel radius, proton bunch parameters and plasma parameter in proton driven plasma wakefield acceleration were studied. Effect of plasma channel on the acceleration of particles and laser beam has been investigated in several reports [15, 20–24]. According to analytical results, the strong dependence on bunch length suggests that far higher accelerating fields could be generated in the PWFA scheme with modest reductions in the bunch size. Also, the modest increment in plasma channel radius is needed to obtain maximum accelerating gradient. In addition, the simulations of wakefield generation by proton bunch in plasma channel are presented using LCODE. The behaviors of longitudinal and total radial wakefields in longitudinal and radial directions are investigated.

The longitudinal electric field in the longitudinal direction decreases with distance behind the bunch while the total radial electric field increases with the distance behind the bunch. These results are the significant points which are achieved in this suggested model. The longitudinal electric field scale down as r increases (in the wall), then go to zero.

The investigation of total radial electric field in radial direction showed that it vanishes as $r = 0$ and a linear focusing caused by the plasma channel near the origin is observed.

As another specific consequence, the wakefield energy that was initially stored in longitudinal oscillations near the axis is gradually transferred to the radial oscillations in the near-wall of channel.

References

1. Esarey, E., Schroeder, C.B., Leemans, W.P.: Physics of laser-driven plasma-based electron accelerators. Rev. Mod. Phys. **81**, 1129 (2009)
2. Katsouleas, K.: Plasma accelerators race to and beyond. Phys. Plasmas **13**, 055503 (2006)
3. Joshi, C.: The development of laser- and beam-driven plasma accelerators as an experimental field. Phys. Plasmas **14**, 055501 (2007)
4. Joshi, C., Blue, B., Clayton, C.E., Dodd, E., Huang, C., Marsh, K.A., Mori, W.B., Wang, S.: High energy density plasma science with an ultrarelativistic electron beam. Phys. Plasmas **9**, 1845 (2002)
5. Umstadter, D.: Review of physics and applications of relativistic plasmas driven by ultra-intense lasers. Phys. Plasmas **8**, 1774 (2001)
6. Leemans, W.P., Volfbeyn, P., Guo, K.Z., Chattopadhyay, S., Schroeder, C.B., Shadwick, B.A., Lee, P.B., Wurtele, J.S., Esarey, E.: Laser-driven plasma-based accelerators: Wakefield excitation, channel guiding, and laser triggered particle injection. Phys. Plasmas **5**, 1615 (1998)
7. Tajima, T., Dawson, J.M.: Laser electron accelerator. Phys. Rev. Lett. **43**, 267 (1979)
8. Chen, P., Dawson, J.M., Huff, R.W., Katsouleas, T.: Acceleration of electrons by the interaction of a bunched electron beam with a plasma. Phys. Rev. Lett. **54**, 693 (1985)
9. Caldwell, A., Lotov, K., Pukhov, A., Simon, F.: Proton-driven plasma-wakefield acceleration. Nat. Phys. **5**, 363 (2009)
10. Lotov, K.V.: Simulation of proton driven plasma wakefield acceleration. Phys. Rev. ST Accel. Beams **13**, 041301 (2010)
11. Kumar, N., Pukhov, A., Lotov, K.: Self-modulation instability of a long proton bunch in plasmas. Phys. Rev. Lett. **104**, 255003 (2010)
12. Lotov, K.V.: Controlled self-modulation of high energy beams in a plasma. Phys. Plasmas **18**, 024501 (2011)
13. Caldwell, A., Lotov, K.V.: Plasma wakefield acceleration with a modulated proton bunch. Phys. Plasmas **18**, 103101 (2011)
14. Lotov, K.V., Pukhov, A., Caldwell, A.: Effect of plasma inhomogeneity on plasma wakefield acceleration driven by long bunches. Phys. Plasmas **20**, 013102 (2013)
15. Yi, L., Shen, B., Lotov, K., Ji, L., Zhang, X., Wang, W., Zhao, X., Yu, Y., Xu, J., Wang, X., Shi, Y., Zhang, L., Xu, T., Xu, Z.: Scheme for proton-driven plasma-wakefield acceleration of positively charged particles in a hollow plasma channel. Phys. Rev. ST Accel. Beams **16**, 071301 (2013)
16. Assmann, R., et al.: Proton-driven plasma wakefield acceleration: a path to the future of high-energy particle physics. Plasma Phys. Control Fusion **56**, 084013 (2014)
17. Lotov, K.V., Sosedkin, A.P., Petrenko, A.V., Amorim, L.D., Vieira, J., Fonseca, R.A., Silva, L.O., Gschwendtner, E., Muggli, P.: Electron trapping and acceleration by the plasma wakefield of a self-modulating proton beam. Phys. Plasmas **21**, 123116 (2014)
18. Yi, L., Shen, B., Ji, L., Lotov, K., Sosedkin, A., Zhang, X., Wang, W., Xu, J., Shi, Y., Zhang, L., Xu, Z.: Positron acceleration in a hollow plasma channel up to TeV regime. Sci. Rep. **4**, 4171 (2014)
19. Khachatryan, A.G.: Excitation of nonlinear two-dimensional wake waves in radially-nonuniform plasma. Phys. Rev. E **60**, 6210 (1999)
20. Leemans, W., Siders, C.W., Esarey, E., Andreev, N., Shvets, G., Morti, W.B.: Plasma guiding and Wakefield generation for second generation experiments. IEEE Trans. Plasma Sci. **24**, 331 (1996)
21. Andreev, N.E., Chizhonkov, E.V., Frolov, A.A., Gorbunov, L.M.: On laser wakefield acceleration in plasma channels. Nucl. Instrum. Methods Phys. Res. **410**, 469 (1998)
22. Kim, J.U., Hafz, N., Suk, H.: Electron trapping and acceleration across a parabolic plasma density profile. Phys. Rev. E **69**, 026409 (2004)
23. Jha, P., Kumar, P., Upadhyaya, A.K., Raj, G.: Electric and magnetic wakefields in a plasma channel. Phys. Rev. ST Accel. Beams **8**, 071301 (2005)
24. Cormier-Michel, E., Esarey, E., Geddes, C.G.R., Schroeder, C.B., Paul, K., Mullowney, P.J., Cary, J.R., Leemans, W.P.: Control of focusing fields in laser-plasma accelerators using higher-order modes. Phys. Rev. ST Accel. Beams **14**, 031303 (2011)
25. Esarey, E., Sprangle, P., Krall, J., Ting, A.: Overview of plasma-based accelerator concepts. IEEE Trans. Plasma Sci. **24**, 252 (1996)
26. Golian, Y., Aslaninejad, M., Dorranian, D.: Whittaker functions in beam driven plasma wakefield acceleration for a plasma with a parabolic density profile. Phys. Plasmas **23**, 013109 (2016)
27. LCODE. http://www.inp.nsk.su/~lotov/lcode. Accessed 11 Mar 2016
28. Lee, S., Katsouleas, T., Hemker, R.G., Dodd, E.S., Mori, W.B.: Plasma-wakefield acceleration of a positron beam. Phys. Rev. E **64**, 045501(R) (2001)
29. Lu, W., Huang, C., Zhou, M.M., Mori, W.B., Katsouleas, T.: Limits of linear plasma wakefield theory for electron or positron beams. Phys. Plasmas **12**, 063101 (2005)
30. Holloway, J. A.: Simulating plasma Wakefield acceleration. M.S thesis, Imperial College London, Department of Physics (2010)
31. Katsouleas, T., Wilks, S., Chen, P., Dawson, J.M., Su, J.J.: Beam loading in plasma accelerators. Part Accel. **22**, 81 (1987)
32. Abramowitz, M., Stegun, I. A.: Handbook of mathematical functions with formulas, graphs, and mathematical tables. Dover, New York (1974)
33. Kallos, E.: Plasma Wakefield accelerators using multiple electron bunches. PhD thesis, University of Southern California (2008)
34. Lotov, K.V.: Fine wakefield structure in the blowout regime of plasma wakefield accelerators. Phys. Rev. ST Accel. Beams **6**, 061301 (2003)
35. Lee, S., Katsouleas, T., Hemker, R., Mori, W.B.: Simulations of a meter-long plasma wakefield accelerator. Phys. Rev. E **61**, 7014 (2000)

13

Water treatment by the AC gliding arc air plasma

Mehrnaz Gharagozalian[1] · Davoud Dorranian[1] · Mahmood Ghoranneviss[1]

Abstract In this study, the effects of gliding arc (G Arc) plasma system on the treatment of water have been investigated experimentally. An AC power supply of 15 kV potential difference at 50 Hz frequency was employed to generate plasma. Plasma density and temperature were measured using spectroscopic method. The water was contaminated with staphylococcus aureus (Gram-positive) and salmonella bacteria (Gram-negative), and Penicillium (mold fungus) individually. pH, hydrogen peroxide, and nitride contents of treated water were measured after plasma treatment. Decontamination of treated water was determined using colony counting method. Results indicate that G Arc plasma is a powerful and green tool to decontaminate water without producing any byproducts.

Keywords G Arc discharge system · Removing bacteria · Complete sterilization · Spectroscopy · Salmonella · Penicillium (mold fungus) · staphylococcus aureus

Introduction

There are different types of microorganisms which pollute water. For this reason, different physical and chemical methods are applied for removing these pollutants, such as the process of filtration, chlorination, UV irradiation, and ozonation, but these methods have several disadvantages. Therefore, researchers tried to employ new methods for inactivation of decontaminants in water [1]. Producing harmful byproducts in the nature and water as well as immunity of several microorganisms to such treatments are some disadvantages of convenient water treatment methods, which force researchers to look for new methods for treatment of polluted water. In the recent two decades, the use of plasma as an effective method for removing bacteria has been introduced and extensive research in this area has been done. Applying an electrical discharge in the water to sterilize is one of the new methods of decontamination, which has fixed many of the problems of the convenient methods [1].

It is more than two decades that plasma is used for removing the bacteria; however, sterilization with plasma systems is a new approach for removing all kinds of microorganisms. These systems on the contrary to other chemical methods are free of any biocides and toxic chemicals and remove all kinds of microorganisms in a short time in a very low temperature. Inactivation of microorganisms is relevant to the presence of various reactive plasma species, such as OH·, H·, O·, and HO_2 and molecular species such as H_2O_2, H_2, and O_2. Electric field, ultraviolet radiation, and shock waves are other parameters of plasma which may have different positive effects on the treatment of water [2–7].

In this experimental study, the G Arc plasma has been used in contact with water by inputting compressed air gas for inactivation of two kinds of bacteria and also one type of fungus. Staphylococcus aureus and Salmonella are Gram-positive and Gram-negative bacteria, respectively, and fungus is Penicillium (mold fungus). Staphylococcus is one of the toughest and most resistance spherical Gram-positive bacteria. The cell wall of these bacteria is very thick and grows large on a rich medium. Staphylococcus

✉ Davoud Dorranian
doran@srbiau.ac.ir

1 Biotechnology Lab., Plasma Physics Research Center, Science and Research Branch, Islamic Azad University, Tehran, Iran

aureus is responsible for a wide range of diseases such as mild skin infections (impetigo, folliculitis, etc.). Salmonella is a Gram- negative, non-spore-forming rod-shaped bacterium which can be transferred by eating or drinking contaminated food or water, in particular. Salmonella enters into the body through ingestion and easily makes its way through the stomach acid to its intestines and also is responsible for many diseases including vomiting and fever [8]. Both of these bacteria grow in water and may transfer to body from water. Recently, researches show that the fungal-like Penicillium spores grow better with water, especially in tap water, damp or water-damaged buildings, and in pool waters and create many problems. Damp or water-damaged building materials are in danger of fungal growth (mold growth) and endanger the health of building occupants and damages to the buildings. Other problems of fungal growth are blockage of water pipes, organoleptic deterioration, and pathogenic fungi [9, 10].

Experimental details

Plasma

The schematic and pictorial view of the plasma G Arc generator are shown in Fig. 1a, b. Air plasma was generated by a G Arc discharge setup. Electrodes were two half circular aluminum plates of 4 cm radius and 2 mm thickness. The gap between two electrodes was 2.5 cm in maximum case and 0.5 cm in minimum for producing plasma. There was 1 cm gap between the electrodes and the output part. The discharge was done using a 15 kV, AC power supply at 50 Hz frequency which was plugged into 220 V. The output voltage of power supply is shown in Fig. 2. Plasma was made with compressed air which was flown at 18 ml/min between electrodes. A container of water was placed on the stirrer. Using a sterile magnet in water, the output plasma could be covered and the plasma acted more effectively on the large area of the liquid. The volume of polluted water was 50 ml. Experiment was carried out at a room temperature of 24 °C. Plasma was diagnosed by a CCS200 Thorlabs spectrometer. Plasma spectrum is presented in Fig. 3.

Microbiological preparation

Microorganisms

Distilled water was polluted separately with two types of bacteria named staphylococcus aureus and Salmonella. Their features are different. Staphylococcus aureus is Gram-positive and Salmonella is Gram-negative. Water was also polluted by one type of fungus named Penicillium (mold fungus).

Suspension preparation

First of all, the bacteria were cultured on nutrient agar medium (Merck, Germany) and fungi were cultured on Sabouraud Dextrose Agar (SDA) (Merck, Germany). Then bacteria were incubated for 24 h at 36 °C and fungi were incubated for 1 week at 25 °C. For preparing the suspension of bacteria and fungi, 10 ml of distilled water was pipetted in a test tube and few colonies of bacteria and spores of fungi were added to it by loop until reaching a density of 0.5 McFarland standards which is equal to 1.5×10^8 CFU/mL. This suspension was used as a reference to adjust the turbidity of bacterial and fungal suspensions. Then 1–9 ml of suspension and distilled water were diluted for two times. Again 5 cc of polluted suspension were mixed with 45 cc distilled water for the last dilution. The final sample was 50 ml of solution with density of 1.5×10^5 CFU/ml polluted agent. The prepared bacterial solution was treated under G Arc plasma for 2–10 min. In the case of fungal pollution the time of treatment was 4–16 min. Then 100 μl of the controlled and treated samples in different mentioned times was cultured on said mediums. At last the mediums were incubated for 24 h at 36 °C and 1 week at 25 °C for bacteria and fungi, respectively.

The number of CFU which grew on the medium after incubating was counted by BZG 30 colony counter. Furthermore, the pH of water was measured using pH meter 8686 pH, AZ Instrument. SEM imaging was done employing AIS2100 together with gold coating by SC7620 device, hydrogen peroxide of treated water was measured using MQuant^{TM} Peroxidase Test Kit and nitrate of treated environment was measured by DR5000 Spectrophotometer.

Results and discussion

Spectroscopy of G Arc plasma

Spectroscopy was used to determine the density and temperature of G Arc air plasma in the wavelength range of 200–1100 nm. The air plasma spectrum was sketched through the origin software (Fig. 3). The spectrum shows some reactive species like N_2, O, H_α, H_β(486.1 nm), H_γ(410.2 nm) in plasma. Different species of hydrogen were generated in the collision of water molecules with electrons ($H_2O + e \rightarrow H + OH + e$). N_2 and O lines arise from air gases which were excited in the discharge process. Considering Fig. 3 there is no

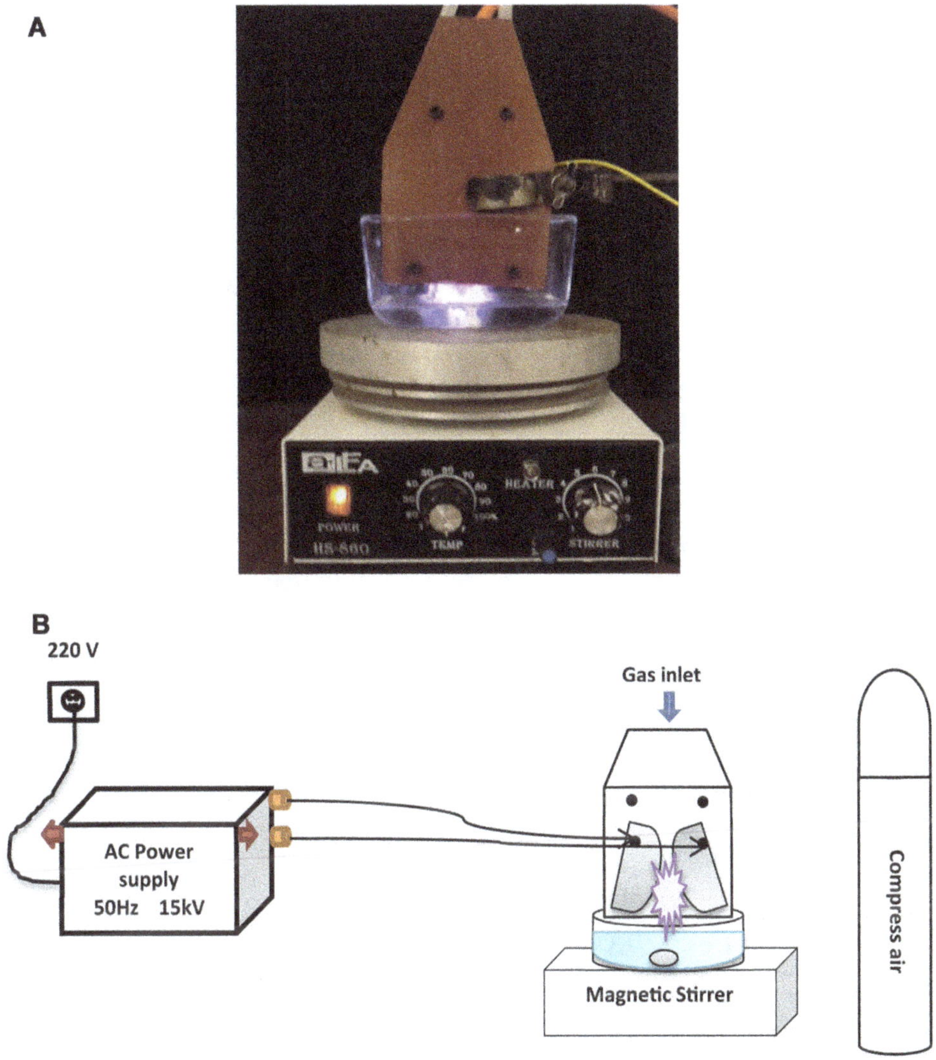

Fig. 1 The experimental setup of gliding arc plasma (**a**) with its schematic view (**b**)

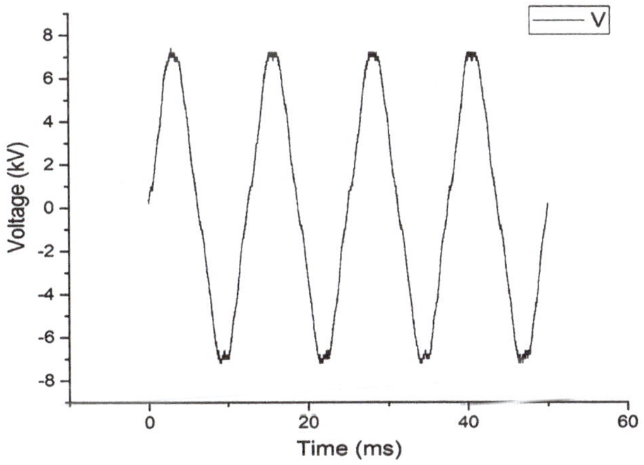

Fig. 2 Waveform of the power supply voltage with peak-to-peak-voltage 15 kV

emission in the interval of 200–300 nm wavelengths. Reactive oxygen species radiates at 777 nm. Generally, oxygen species is one of the most important materials for sterilization which has been mentioned in many reports on plasma sterilization. Atomic nitrogen radiates at 745 nm and N_2 as the other important reactive particle is detected in the range of 320–400 nm [11, 12]. Ultraviolet peaks are at 364.5, 388.5, 388.9, and 397 nm and the peaks of NO are at 296, 357.7, and 378.9 nm [13]. Such waves are radiated due to collisions between electron and neutral species. Ultraviolet radiation destroys organic components and dissolves hydrogen peroxide and ozone from each other to finally produce hydroxyls in plasma which removes the pollutants and increases the internal electrical energy efficiency.

When the water molecules are under the effect of electrical discharge, OH and H radicals are produced due to decomposition, ionization, rotational, and vibrational excitation reactions. Typical wavelength of OH radical emission is between 309 and 317.8 nm [13]. They are one of the most effective elements in the sterilization process because they can easily penetrate into the membrane of the

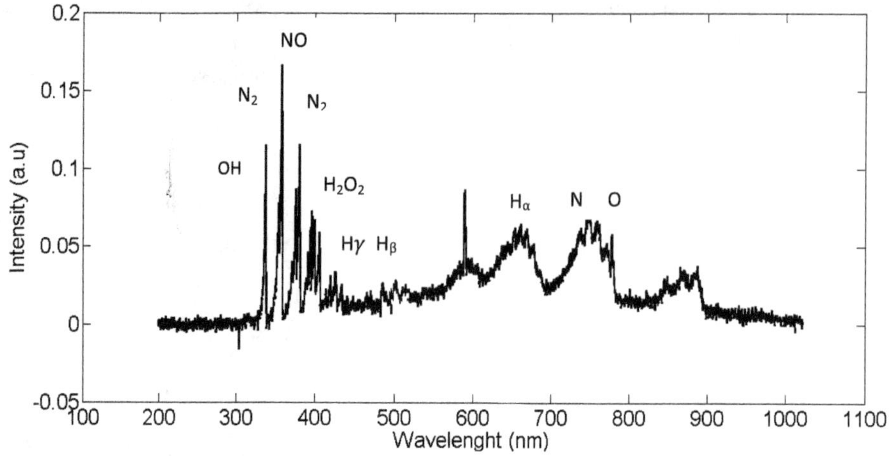

Fig. 3 The emission spectrum of G Arc plasma

fungus and bacteria to destroy them. Following reactions are the processes of OH and H production. The lifetime of hydroxyl is very short.

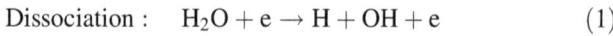

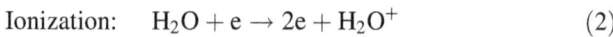

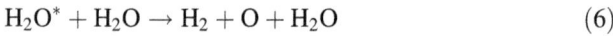

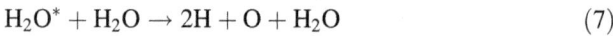

$$\text{Dissociation}: \quad H_2O + e \rightarrow H + OH + e \tag{1}$$

$$\text{Ionization}: \quad H_2O + e \rightarrow 2e + H_2O^+ \tag{2}$$

$$H_2O^+ + H_2O \rightarrow OH + H_3O^+ \tag{3}$$

$$\text{Rotational and vibrational excitation:} \quad H_2O + e \rightarrow H_2O^* + e \tag{4}$$

$$H_2O^* + H_2O \rightarrow H_2O + H + OH \tag{5}$$

$$H_2O^* + H_2O \rightarrow H_2 + O + H_2O \tag{6}$$

$$H_2O^* + H_2O \rightarrow 2H + O + H_2O \tag{7}$$

One of the interesting ways for analyzing the spectrum of plasma is measuring electron density, which may be obtained from H_α line in the spectrum. This type of atomic hydrogen is sensitive to stark broadening, which contains useful information such as electron density [14]. Stark broadening is obtained from the following equation:

$$\Delta\lambda_{\text{stark}} = 2.5 \times \alpha_{1/2} \times (n_e)^{2/3} \tag{8}$$

where $\alpha_{1/2}$ is a constant and n_e is the density of plasma electrons [15].

$\Delta\lambda_{\text{stark}}$ may be extracted from the equation

$$\Delta\lambda_L = \Delta\lambda_{\text{stark}} + \Delta\lambda_{\text{vanderwaals}} \tag{9}$$

where $\Delta\lambda_L$ is the width of H_α line. This line is shown by dots in Fig. 4, which is fitted by a Voigt curve using Origin software. $\Delta\lambda_{\text{vanderwaals}}$ is a kind of Lorentzian broadening which is obtained from $\Delta\lambda_{\text{vanderwaals}} = \frac{1.82}{T_g^{0.7}}$ (T_g is gas temperature) by taking van der Waals broadening, one can find $\Delta\lambda_{\text{stark}} = W = 31.6529$ and taking $\alpha_{1/2} = 2 \times 10^{-11}$ plasma density is found to be $5.03 \times 10^{16}\ 1/\text{cm}^3$. Regarding the point that $n_e \geq 10^{20}\text{m}^{-3}$ is the density of thermal plasma and $n_e \cong 10^{10}\text{m}^{-3}$ is the density of non-thermal plasma, the used G Arc plasma in this experiment is between thermal and non-thermal plasmas.

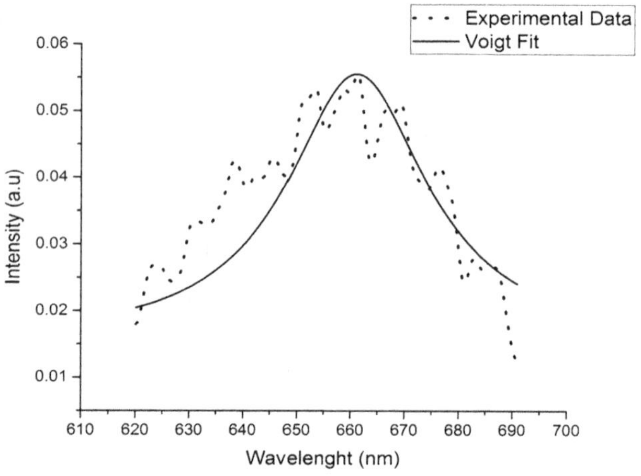

Fig. 4 Broadening Voigt for simulating the line of H_α

Plasma temperature is one of the considerable elements in sterilization which directly affects the biochemical reactions of bacteria [16]. Electron temperature of G Arc plasma was obtained by limiting the wavelength between 370 and 382 nm to SpecAir software for fitting which is shown in Fig. 5. Plasma temperature was calculated as follows:

$$T_e = 3000\,\text{K} = 0.2\,\text{eV}$$

During the experiment the temperature of water was checked. The maximum temperature of polluted water was 29 °C during the plasma treatment.

Results of colony counting

Results show that the numbers of CFU decreased proportionally with time under the treatment with the G Arc plasma. Number of CFU for Staphylococcus and Salmonella bacteria and Penicillium are shown in Figs. 6, 7, and 8 respectively. These data are plotted in Fig. 9. Colonies of bacteria were counted on nutrient agar after

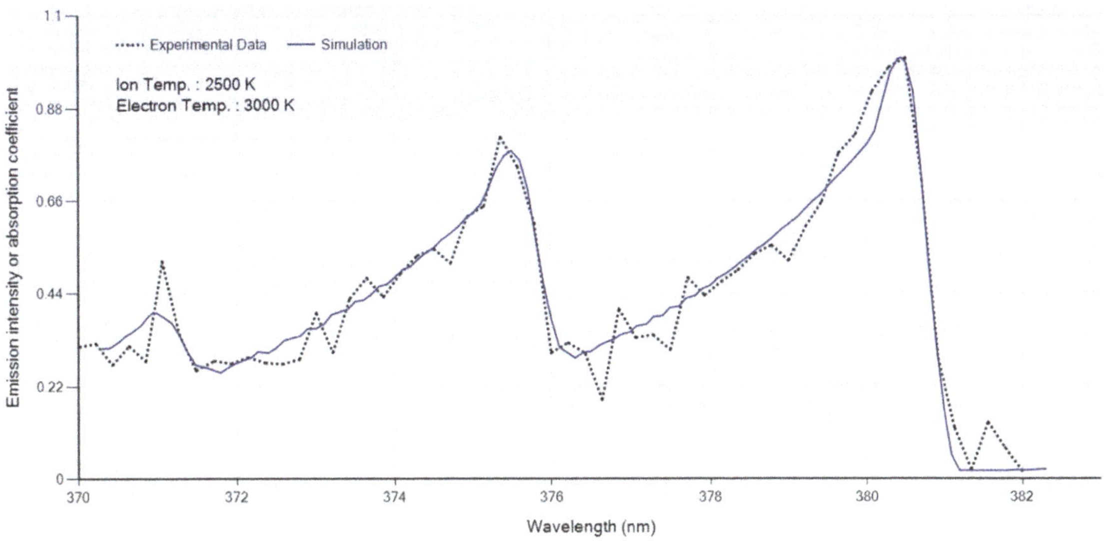

Fig. 5 Fitting for measuring ion and electron temperature of G Arc plasma using SpecAir software

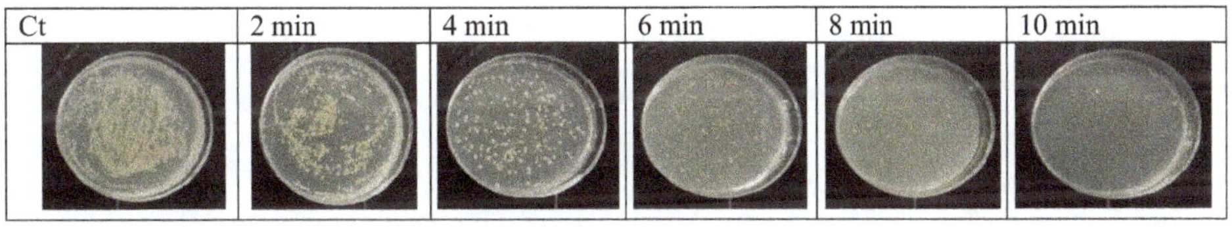

Fig. 6 Results of G Arc plasma on the Staphylococcus bacteria (Gram-positive) after 24-h incubation at 36 °C

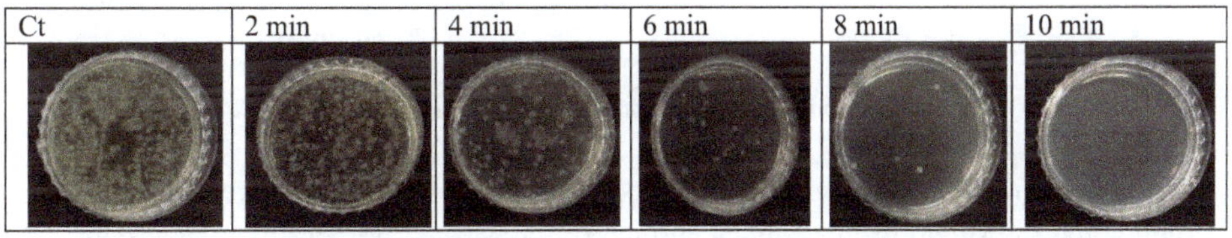

Fig. 7 Results of G Arc plasma on the Salmonella bacteria (Gram-negative) after 24-h incubation at 36 °C

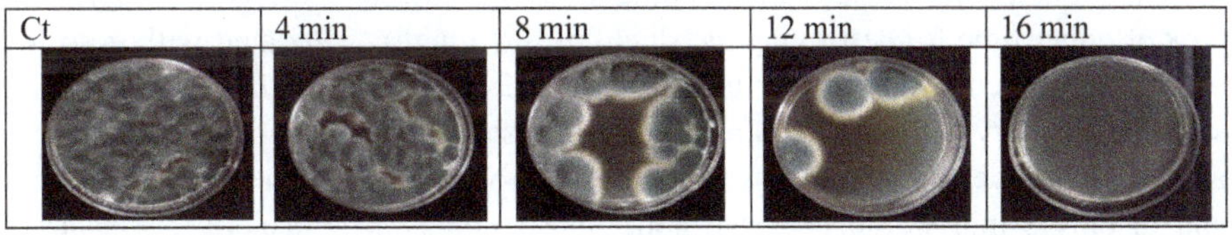

Fig. 8 Results of G Arc plasma on the Penicillium (mold fungus) after 1-week incubation at 25 °C

24-h incubation at 36 °C and colonies of Penicillium were counted on SDA after 1-week incubation at 25 °C. In this experiment, 10 min was enough for decontamination of bacteria but for the case of Penicillium, decontamination was completed in 16 min. In the case of Salmonella, the survival curve consists of two lines while for the case of Staphylococcus the survival curve consists of three lines. The first line of survival curve with very soft slope introduces the destruction of the shield of bacteria by the UV photons of plasma [1, 17]. Figure 3 shows that in comparison with other references, the number of UV photons is very large in our G Arc plasma. In this case, the slope of the first line is steep. In the case of Gram-negative bacteria, because the bacteria shield is thinner than the Gram-positive one, the first and second lines of survival curve have almost the same slope. In the case of Gram-positive bacteria, because of their thicker shield, the slope of the first line is smaller. The second stage of decontamination occurred between minutes 2 and 4 of plasma exposure. At this stage the DNA and RNA of bacteria react chemically with reactive species of plasma. This stage occurred faster for the Gram-positive bacteria. Finally, in 10 min decontamination was completed for both bacteria.

Among these three microorganisms, the time for decontamination of Penicillium (mold fungus) was longer than others. The survival curve of this fungus consists of two stages. It is more similar to the survival curve of Salmonella which is a Gram-negative bacterium. First

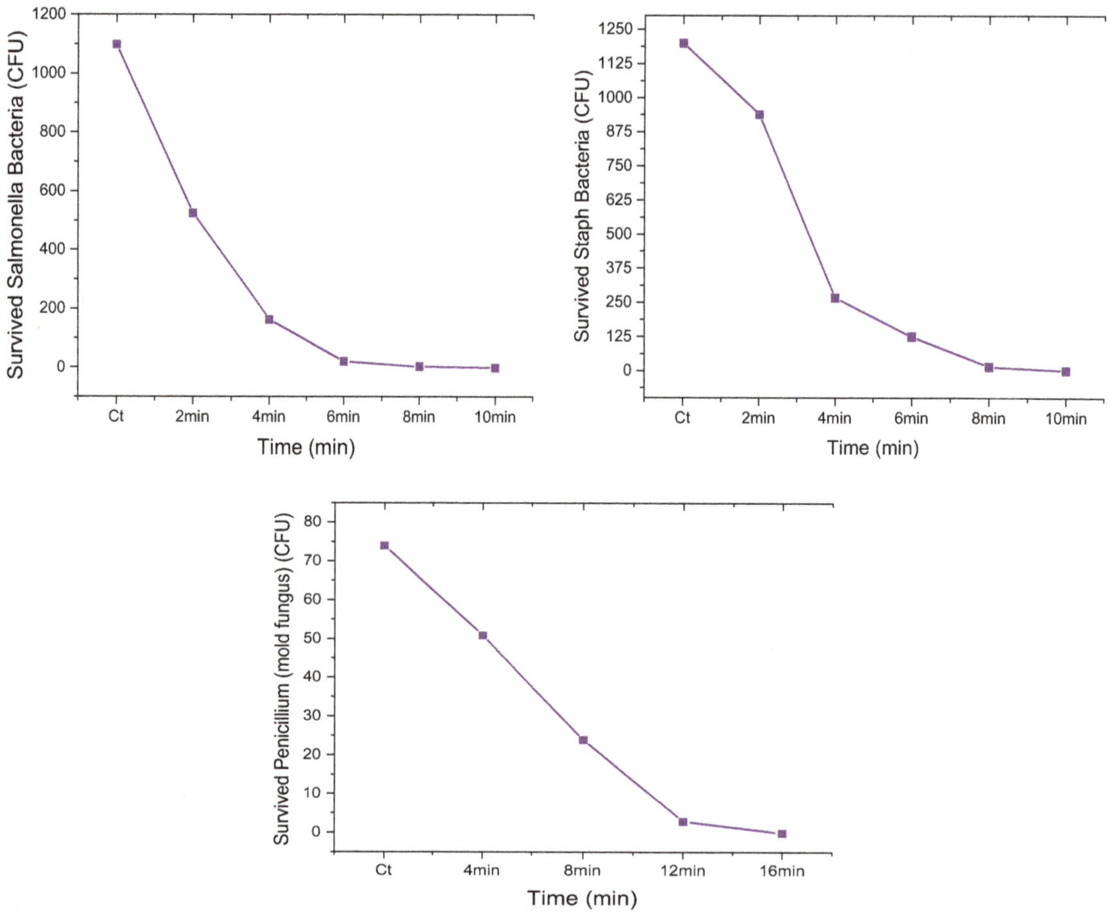

Fig. 9 Results of G Arc plasma on the *Penicillium, Staphylococcus,* and *Salmonella* bacteria

stage took 12 min while the second stage took only 3 min. It seems that the shield of this fungus is tighter against the UV radiation in comparison with two other agents. Thus, it took a longer time for destroying the shield of this fungus. This process occurred uniformly and in the first 12 min the slope of the survival curve is constant. Results show that after shielding destruction, this fungus dies down very fast and in only 3 min we do not have noticeable amount of them in the medium.

Results of scanning electron microscopy analysis (SEM)

Results were also checked using SEM images. Performing SEM images, at first, 1 cc of water sample before and after treatment was pipetted into a small container. Samples were centrifuged for 15 min till all the spores and bacteria were deposited. Then instead of water, glutaraldehyde 2% was added to the deposited part. After 24 h of fixing with glutaraldehyde, samples were centrifuged again until spores and bacteria sedimented. Then glutaraldehyde was drained. Again samples were centrifuged with 95, 70, 50, 30, and 100% ethanol for 15 min, respectively. At this step samples were dehydrated. Finally, dried samples on aluminum foil were coated with gold to perform SEM images.

In Fig. 10 the effect of 10-min G Arc plasma treatment on the Staphylococcus bacteria is shown. All bacteria were affected by plasma treatment. Before treatment (Fig. 10a) bacteria are spherical with about 800 nm diameter. The shell of bacterial is shinny in the scanning electron probe beam. After treatment, the shell of the bacteria was deformed. They were not spherical anymore. Some holes and cracks appeared on the body of bacteria. Their shells were not shiny and the edge of bacteria was deformed to a sawtooth-like structure. According to images in the case of Staphylococcus bacteria, the shell of bacteria was destroyed by plasma agents.

Effect of 6 and 10 min plasma treatment on the Salmonella bacteria is shown in Fig. 11. After 6-min treatment the shell of bacteria was destroyed. Some parts of its body were destructed. After 10-min treatment the shell of bacteria was removed completely. Their cores and internal parts of their bodies may be seen in the SEM image.

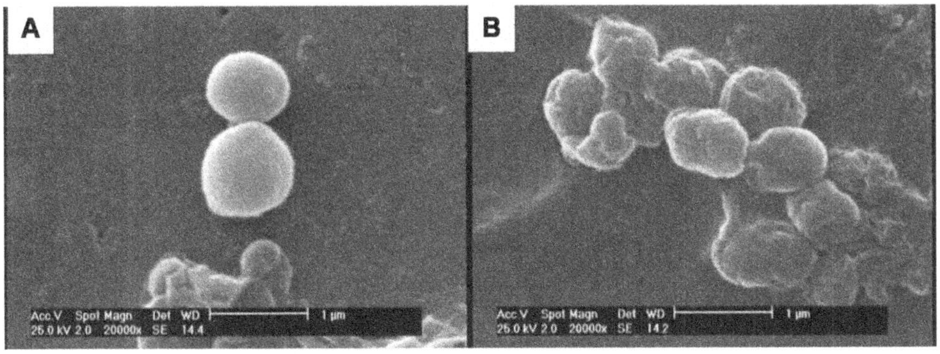

Fig. 10 Results of scanning electron microscopy on the staphylococcus aureus bacteria: control sample are completely spherical and symmetric (**a**) after 10-min treatment by G Arc plasma; the staph bacteria got deformed (**b**)

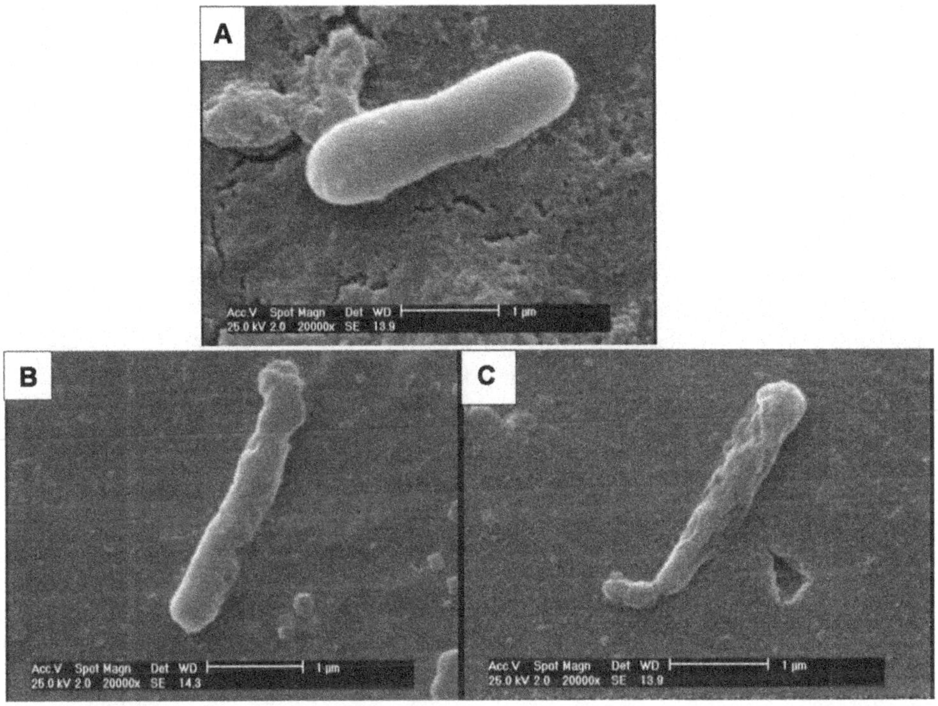

Fig. 11 Results of scanning electron microscopy on the *Salmonella* bacteria: control sample is of bacillus form (**a**) after 6-min treatment by plasma (**b**) after 10-min treatment by G Arc plasma (**c**)

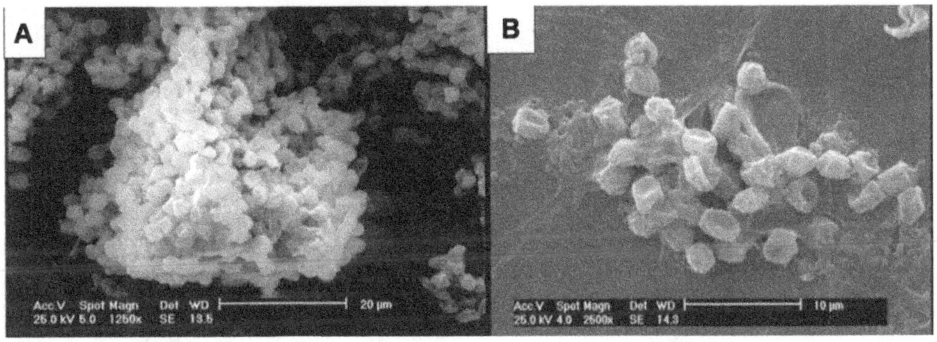

Fig. 12 Results of scanning electron microscopy on the Penicillium Green (mold fungi): spores in control sample are like chain or cluster (**a**) after treated by G Arc plasma, spores got deformed (**b**)

Figure 12a, b shows the evolution of Penicillium (mold fungus) after G Arc plasma treatment. Many spores of fungi were removed completely after plasma treatment. It seems that they have been squished and crushed because of plasma treatment. The numbers of remaining parts of spores are much smaller after the treatment and they got separated from each other, while they were like a chain or cluster before treatment. There are some holes in their cell shells and some parts of their internal organisms penetrated out. It is known that plasma, especially G Arc discharge, can produce shockwave in water which plays an important role in the detachment of fungal spores [18, 19].

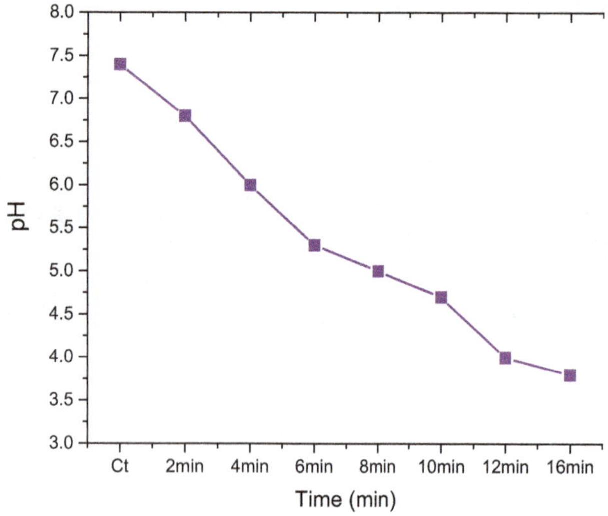

Fig. 13 Results of pH of water before and after using G Arc plasma

Results of measuring pH in water

pH of water before and after treatment was measured gradually during the plasma treatment process. Results are shown in Fig. 13. Within 16 min, the pH of water was decreased from 7.4 to 3.8. In other words the plasma treatment led to acidification of water.

Results of producing H_2O_2 (hydrogen peroxide) by G Arc plasma in water

Using the special kits (peroxide test) for measuring H_2O_2 (hydrogen peroxide) through the inserted directives on it and comparing with color change, the target numbers were obtained. Results are shown in Fig. 14. Hydrogen peroxide may be produced by combination of hydroxyl radicals and also can increase the ability of acidity of plasma, and the wavelength is approximately in the range of 379–430 nm [20, 21]. By increasing hydrogen peroxide as indicated in Figs. 13 and 14 the pH was increased.

Results of measuring NO_3 (nitrate) in water before and after plasma

The amount of nitrate in treated water during the treatment process is shown in Fig. 15. Results show that this amount was increased in the treated water in the first 6 min and was gradually decreased in the next 10 minutes. However, we still had about 10 mg/l NO_3 in the water after the treatment. Comparing the pH measurement with NO_3 results shows that in the plasma treatment process in the second part of the experiment because of H radical production, NO_3 molecules were changed to HNO_3 molecules.

Conclusion

The present study presents the effects of air plasma which was generated by G Arc discharge system for reducing microorganisms in water using different physical conditions. A power supply of 15 kV at 50 Hz frequency was employed to preform discharge with compressed air flow. Characteristics of plasma were measured using spectroscopic method. During this study important parameters of water including pH, H_2O_2, and NO_3 were measured. Effect of plasma discharge, especially on two different kinds of bacteria (staphylococcus aureus and Salmonella) and a type of fungus [Penicillium (mold fungus)] were studied. Results show that G Arc is one of the most considerable

Fig. 14 Results of producing H_2O_2 (hydrogen peroxide) in water by G Arc plasma

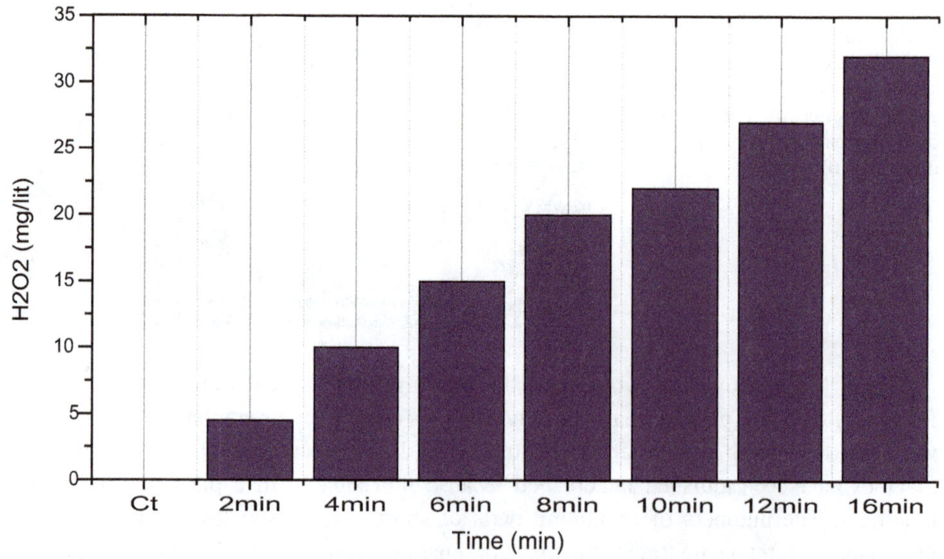

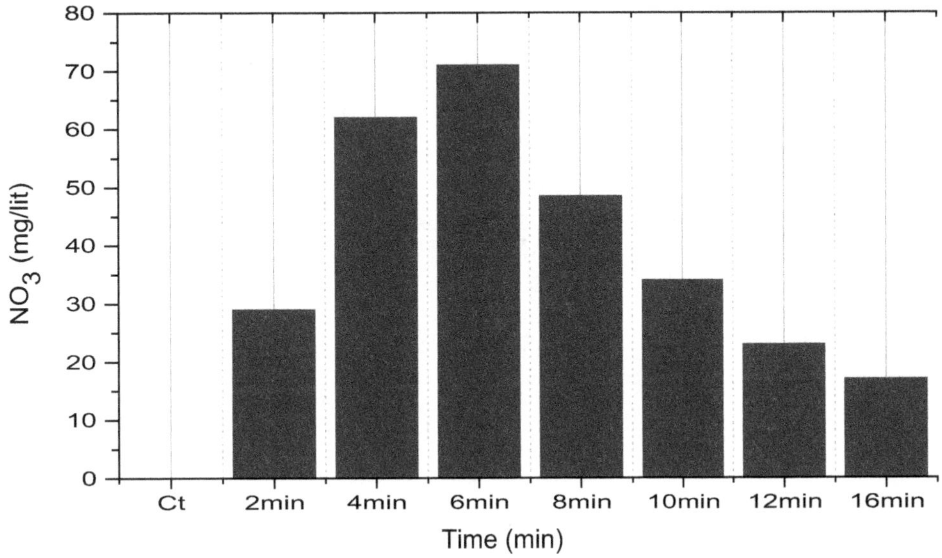

Fig. 15 Results of measuring NO_3 (nitrate) in water before and after using G Arc plasma

systems with more efficiency compared to the advanced oxidation techniques because of low equipment and energy costs [22, 23]. Comparing to other plasmas, G Arc plasma is more applicable and appropriate to the industry [22, 24]. Another feature of G Arc plasma comes from its plasma temperature. G Arc plasma in this working regime generates thermal plasma but it did not increase the temperature of water during the experiment. All useful plasma species, such as hydrogen peroxide, H radicals, ultraviolet radiations, nitrogen, and oxygen, may generate in this plasma, which makes it an efficient tool to treat polluted water.

References

1. Salarieh, S., Dorranian, D.: Sterilization of turmeric by atmospheric pressure dielectric barrier discharge plasma. Plasma Sci. Technol **15**, 1122–1126 (2013)
2. Kim, H.S., Cho, Y.I., Hwang, I.H., Lee, D.H., Cho, D.J., Rabinovic, A., Fridman, A.: Use of plasma gliding arc discharges on the inactivation of *E. Coli* in water. Separ. Purif. Technol. **120**, 423–428 (2013)
3. Gutsol, A., Vaze, N., Arjunan, K., Gallagher, M., Yang, Y., Zhu, J, et al.: Plasma for air and water sterilization. Plasma Assisted Decontamination of Biological and Chemical Agents, pp. 21–39 (2008)
4. Locke, B., Sato, M., Sunka, P., Hoffmann, M., Chang J.S.: Electrohydraulic discharge and nonthermal plasma for water treatment. Ind. Eng. Chem. Res. **45**, 882–905 (2006)
5. Fridman, A.: Plasma Chemistry. Cambridge Univ Press, Cambridge (2008)
6. Jiang, B., Zheng, J., Qiu, S., Wu, M., Zhang, Q., Yan, Z., Xue, Q.: Review on electrical discharge plasma technology for wastewater remediation. Chem. Eng. J. **236**, 348–368 (2014)
7. Korachia, M., Turan, Z., Senturk, K., Sahin, F., Aslan, N.: An investigation into the biocidal effect of high voltage AC/DC atmospheric corona discharges on bacteria, yeasts, fungi and algae. J. Electrostat. **67**, 678–685 (2009)
8. The Journal of Undergraduate Biological Studies. Pathogen Profile Dictionary: *Salmonella* typhimurium (2010)
9. Andersen, B., Frisvad, J.C., Søndergaard, I., Rasmussen, I.S., Larsen, L.S.: Associations between fungal species and water-damaged building materials. Appl. Environ. Microbiol. **77**(12), 4180–4188 (2011)
10. Siqueira, V.M., Oliveira, H.M.B., Santos, C., Paterson, R.R.M., Gusmão, N.B., Lima, N.: Filamentous fungi in drinking water, particularly in relation to biofilm formation. Int. J. Environ. Res. Public Health **8**(2), 456–469 (2011)
11. Simon, A., Anghel, S.D., Papiu, M., Dinu, O.: Physical and analytical characteristics of an atmospheric pressure argon-helium radiofrequency capacitively coupled plasma. Spectrochim. Acta Part B **65**, 272–278 (2010)
12. Jo, Y.K., Cho, J., Tsai, T.C., Staack, D., Kang, M.H., Roh, J.H., Shin, D.B., Cromwell, W., Gross, D.: A Non-thermal plasma seed treatment method for management of a seedborne fungal pathogen on rice seed. Crop Sci. **54**(2), 796–803 (2014)
13. Bingyan, C., Yulin, G., Yeqian, W., Changping, Z., Juntao, F., Feng, Z., Jingyi, W., Jianku, W.: Yield of hydrogen peroxide, ozone and nitrite nitrogen with DBD arrays in water mist spray. Plasma Sci. (ICOPS) (2015)
14. Geeorgescu, N., Lupu, A.R.: Tumoral and normal cells treatment with high voltage pulsed cold atmospheric plasma jets. IEEE Trans. Plasma Sci. **38**(8), 1949–1955 (2010)
15. Marco, A.G., Valentin, C.: New plasma diagnosis tables of hydrogen stark broadening including ion dynamic. J. Phys. B Atom. Mol. Opt. Phys. **29**(20), 4795 (1996)
16. Jiang, B., Zheng, J., Qiu, S., Wu, M., Zhang, Q., Yan, Z., Xue, Q.: Review on electrical discharge plasma technology for wastewater remediation. Chem. Eng. J. **236**, 348–368 (2014)
17. Du, C.M., Wang, J., Zhang, L., Li, H.X., Liu, H., Xiong, Y.: The application of a non-thermal plasma generated by gas–liquid gliding arc discharge in sterilization. New J. Phys. **14**, 013010 (2012)
18. Kang, M.H., Pengkit, A., Choi, K., Jeon, S.S., Choi, H.W., Shin, D.B., Choi, E.H., Uhm, H.S., Park, G.: Differential inactivation of

fungal spores in water and on seeds by ozone and arc discharge plasma. PLoS One **10**(9), e0139263 (2015)
19. Lee, H.Y., Kang, B.K., Uhm, H.S.: Underwater discharge and cell destruction by shockwave. J Korean Phys Soc. **42**, S880–S884 (2003)
20. Dey, S.K., Mukherjee, A.: Investigation of 3d-transition metal acetates in the oxidation of substituted dioxolene and phenols. J. Mol. Catal. A Chem. **407**, 93–101 (2015)
21. Huang, B.K., Sikes, H.D.: Quantifying intracellular hydrogen peroxide perturbations in terms of concentration. Redox Biol., 955–962 (2014)
22. El-Aragi, G.M.: Gliding arc discharge (GAD) experiment. Plasma Physics and Nuclear Fusion Dept., Nuclear Research Center, AEA, PO 13759 Cairo, Egypt
23. Benstaali, B., Moussa, D., Addou, A., Brisset, J.L.: Plasma treatment of aqueous solutes: some chemical properties of a gliding arc in humid air. Eur. Phys. J. Appl. Phys. **4**, 171–179 (1998)
24. Moussa, D., Brisset, J.-L.: Disposal of spent tributylphosphate by gliding arc plasma. J. Hazard. Mater. **B102**, 189–200 (2003)

14

Dispersion relation and growth rate in thermal plasma-loaded traveling wave tube with corrugated waveguide hollow electron beam

Samina Dehghanizadeh · Shahrooz Saviz

Abstract A theory of relativistic traveling wave tube with magnetized thermal plasma-filled corrugated waveguide with annular electron beam is given. The dispersion relation is obtained by linear fluid theory. The characteristic of the dispersion relation is obtained by numerical solutions. The effect of plasma density, corrugated period, waveguide radius and plasma thermal effect on the dispersion relation and growth rate are analyzed. Some useful results are given.

Keywords RTWT · Annular electron beam · Thermal plasma · Dispersion relation · Growth rate

Introduction

Relativistic traveling wave tube (RTWT) is an important high-power microwave (HPM) apparatus which has been developed in the past 20 years [1–3]. Most of the TWT mathematical analysis has been done by John Robinson Pierce and his colleagues at Bell Labs [4–6] and then has been developed by Chu et al. [7–9]. In TWT, sinusoidal corrugated slow wave structure (SWS) is used to reduce the phase velocity of the electromagnetic wave to synchronize it with the electron beam velocity, so that a strong interaction between the two can take place [10, 11]. TWT is extensively applied in satellite and airborne communications, radar, particle accelerators, cyclotron resonance and electronic warfare system. The plasma injection to TWT has been studied recently which can increase the growth rate and improve the quality of transmission of electron beam. We investigate the effect of thermal plasma and electron beam on the growth rate [12–17].

The use of plasmas for generating high-power microwaves is studied for more than 50 years. During the 1990s plasma-assisted slow-wave oscillators (SWO) were invented and actively developed at Hughes Research Lab (HRL) [22–27].

An analytical and numerical study is made on the dispersion properties of a cylindrical waveguide filled with plasma. An electron beam and static external magnetic field are considered as the mechanisms for controlling the field attenuation and possible stability of the waveguide [24–29].

In this paper, a RTWT with magnetized thermal plasma-filled corrugated waveguide with annular electron beam is studied. The dispersion relation of corrugated waveguide is derived from a solution of the field equations. By numerical computation, the dispersion characteristics of the RTWT are analyzed in detail in different cases of various geometric parameters of slow-wave structure.

In Sect. 2, the physical model of the RTWT filled with thermal plasma is established in an infinite longitudinal magnetic field. In Sect. 3, the dispersion relation of the RTWT is derived. In Sect. 4, the dispersion characteristics of the RTWT are analyzed by numerical computation.

S. Dehghanizadeh
Central Tehran Branch, Islamic Azad University, Tehran, Iran

S. Saviz (✉)
Science and Research Branch, Plasma Physics Research Center, Islamic Azad University, Tehran, Iran
e-mail: azarabadegan@gmail.com

Physical model

The analysis presented in this paper is based on the SWS shown in Fig. 1. The SWS of the system is the corrugated waveguide that reduce the speed of wave. Wave after collision

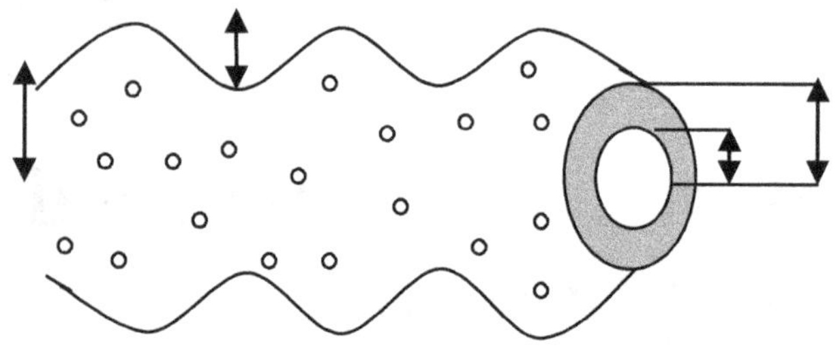

Fig. 1 Slow-wave structure and annular electron beam model of a plasma-filled relativistic traveling wave tube

with waveguide its speed is reduced and reaches to speed of the electron beam (synchronism), so the wave is amplified.

$$R(z) = R_0 + h\ \cos(\kappa_0 z), \tag{1}$$

$$K_n = k_z + n\kappa_0, \tag{2}$$

$$\kappa_0 = \frac{2\pi}{z_0}, \tag{3}$$

Cylindrical waveguide, whose wall radius $R(z)$, varies sinusoidal according to the relation (1), h is the corrugation amplitude, $\kappa_0 = 2\pi\ z_0$ is the corrugation wave number, and z_0 is the length of the corrugation period, R_0 is the waveguide radius and k_z is the axial wave number.

A finite annular relativistic electron beam with density n_b and radius R_b goes through the cylindrical waveguide, which is loaded completely with a thermal, uniform and collision less plasma of density n_p.The entire system is immersed in a strong, longitudinal magnetic field, which magnetizes both the beam and the plasma. Because dielectric constant is an anisotropic so it will be a tensor. In the beam–plasma case in a linearized scheme, the dielectric tensor, in cylindrical coordinates, may be given by:

$$\sigma = \begin{bmatrix} \frac{i\omega}{\omega^2-\omega_c^2} & \frac{-\omega_c}{\omega^2-\omega_c^2} & 0 \\ \frac{\omega_c}{\omega^2-\omega_c^2} & \frac{i\omega}{\omega^2-\omega_c^2} & 0 \\ 0 & 0 & \frac{1}{-i\omega + \frac{ik_n^2\gamma P_0}{mn\omega}} \end{bmatrix} \tag{4}$$

$$\varepsilon = \varepsilon\cdot\left(1 + \frac{i\sigma}{\varepsilon\cdot\omega}\right) \tag{5}$$

$$[\in] = \varepsilon_0 \begin{bmatrix} 1-\frac{\omega_p^2}{\omega^2-\omega_c^2} & -\frac{i\omega_p^2\omega_c}{\omega(\omega^2-\omega_c^2)} & 0 \\ \frac{i\omega_p^2\omega_c}{\omega(\omega^2-\omega_c^2)} & 1-\frac{\omega_p^2}{\omega^2-\omega_c^2} & 0 \\ 0 & 0 & 1-\frac{\omega_p^2}{\omega^2-\frac{k_n^2\gamma P_0}{mn}} \end{bmatrix} = 0 \tag{6}$$

$$[\in] = \varepsilon_0 \begin{bmatrix} \varepsilon_1 & \varepsilon_2 & 0 \\ -\varepsilon_2 & \varepsilon_1 & 0 \\ 0 & 0 & \varepsilon_3 \end{bmatrix} = 0 \tag{7}$$

We assume that B_0 is very strong that $\varepsilon_1 = 1$ and $|\varepsilon_2|$ is negligibly small.

$$[\in] = \varepsilon_0 \times \begin{bmatrix} 1 & 0 & 0 \\ 0 & 1 & 0 \\ 0 & 0 & 1-\frac{\omega_b^2}{\gamma^3(\omega-K_n v)^2-3K_n^2 v_{th_b}^2} - \frac{\omega_p^2}{\omega^2-3K_n^2 v_{th_p}^2} \end{bmatrix}, \tag{8}$$

$$\varepsilon_3 = 1 - \frac{\omega_b^2}{\gamma^3(\omega-K_n v)^2-3K_n^2 v_{th_b}^2} - \frac{\omega_p^2}{\omega^2-3K_n^2 v_{th_p}^2} \tag{9}$$

Here, $\omega_p = (\frac{n_p e^2}{m\varepsilon_0})^{1/2}$ is the plasma frequency, $\omega_b = (\frac{n_b e^2}{m\varepsilon_0})^{1/2}$ is the beam frequency, ω is the angular frequency of the electromagnetic wave, γ is the relativistic factor, $v_{th_p} = \sqrt{\frac{KT_p}{m}}$ is the thermal velocity of the plasma, $v_{th_b} = \sqrt{\frac{KT_b}{m}}$ is the thermal velocity of the beam, v is the velocity of the beam, T_b is the beam temperature, T_p is the plasma temperature and K is the Boltzmann constant.

Dispersion equation

In the above physical model, its Maxwell equations can be written as:

$$\nabla \times B = \frac{-i\omega}{c}\varepsilon\cdot E, \tag{10}$$

$$\nabla \times E = \frac{i\omega}{c}B, \tag{11}$$

$$\nabla\cdot B = 0, \tag{12}$$

$$\nabla\cdot D = 0. \tag{13}$$

Suppose that every variable can be regarded as:

$$A = Ae^{i(K_n z-\omega t)}. \tag{14}$$

From Eqs. (10) and (11) we can obtain:

$$\nabla\nabla\cdot E - \nabla^2 E = \frac{\omega^2}{c^2}\varepsilon\cdot E, \tag{15}$$

$$\nabla^2 E_z - [\nabla(\nabla\cdot E)]_z + \frac{\omega^2}{c^2}\varepsilon_3 E_z = 0, \tag{16}$$

Now Substituting Eq. (9) into Eq. (16), we have:

$$\nabla^2 E_z - [\nabla(\nabla\cdot E)]_z + \frac{\omega^2}{c^2} \times \left(1 - \frac{\omega_b^2}{\gamma^3(\omega-K_n v)^2-3K_n^2 v_{th_b}^2} - \frac{\omega_p^2}{\omega^2-3K_n^2 v_{th_p}^2}\right)E_z = 0. \tag{17}$$

From Eq. (13), we have:

$$\nabla \cdot \varepsilon_3 E_z = 0. \tag{18}$$

Substituting Eq. (9) into Eq. (18), we obtain:

$$\nabla \cdot E_z = \frac{\partial}{\partial z} E_z \left[\frac{\omega_b^2}{\gamma^3(\omega - K_n v)^2 - 3K_n^2 v_{th_b}^2} + \frac{\omega_p^2}{\omega^2 - 3K_n^2 v_{th_p}^2} \right]. \tag{19}$$

From Eqs. (16) and (19), the wave equation is obtained in the area of plasma-beam as follows:

$$\left[\nabla_\perp^2 + \left(\frac{\omega^2}{c^2} - K_n^2 \right) \left(1 - \frac{\omega_b^2}{\gamma^3(\omega - K_n)^2 - 3K_n^2 v_{th_b}^2} - \frac{\omega_p^2}{\omega^2 - 3K_n^2 v_{th_p}^2} \right) \right] E_z, \tag{20}$$

$$\nabla_\perp^2 = \nabla^2 - \frac{\partial^2}{\partial z^2} \tag{21}$$

$$E_z(r,z) = E_{zn}(r) e^{i(K_n z - \omega t)}, \tag{22}$$

$$E_r(r,z) = \frac{iK_n}{\frac{\omega^2}{c^2} - K_n^2} e^{i(K_n z - \omega t)} \frac{dE_{zn}}{dr}. \tag{23}$$

We investigated ground state ($n = 0$) in solving the equation, substituting Eq. (21) into Eq. (20), we have:

$$E_{zn} = \begin{cases} A_0 J_0(T_1 r) & 0 \le r \le R_b; \\ B_0 J_0(T_2 r) + C_0 N_0(T_2 r) & R_b \le r \le R_c; \\ D_0 J_0(T_1 r) + F_0 N_0(T_1 r) & R_c \le r \le R(z); \end{cases} \tag{24}$$

According to Fig. 1, in the area of $0 \le r \le R_b$, there is Plasma only, and in the area of $R_b \le r \le R_c$, there are plasma + beam, and in the area of $R_c \le r \le Rz$, there is plasma only.

$$T_1^2 = \left(\frac{\omega^2}{c^2} - K_n^2 \right) \left(1 - \frac{\omega_p^2}{\omega^2 - 3K_n^2 v_{th_p}^2} \right), \tag{25}$$

$$T_2^2 = \left(\frac{\omega^2}{c^2} - K_n^2 \right) \times \left(1 - \frac{\omega_b^2}{\gamma^3(\omega - K_n v)^2 - 3K_n^2 v_{th_b}^2} - \frac{\omega_p^2}{\omega^2 - 3K_n^2 v_{th_p}^2} \right). \tag{26}$$

The field components must satisfy the following continuity equations (first boundary condition):

$$E_z(r = R_b^-) = E_z(r = R_b^+), \tag{27}$$

$$E_z(r = R_c^-) = E_z(r = R_c^+) \tag{28}$$

$$E_r(r = R_b^-) = E_r(r = R_b^+) \tag{29}$$

$$E_r(r = R_c^-) = E_r(r = R_c^+) \tag{30}$$

As a result, the field components are obtained as follows:

$$B_0 = -\frac{\pi}{2} R_b s_0 A_0 \tag{31}$$

$$C_0 = -\frac{\pi}{2} R_b l_0 A_0 \tag{32}$$

$$F_0 = \frac{\pi^2}{4} R_b R_c U_0 A_0 \tag{33}$$

$$D_0 = \frac{\pi^2}{4} R_b R_c H_0 A_0 \tag{34}$$

where

$$s_0 = T_2 J_0(T_1 R_b) N_1(T_2 R_b) - T_1 J_1(T_1 R_b) N_0(T_2 R_b) \tag{35}$$

$$l_0 = T_1 J_1(T_1 R_b) J_0(T_2 R_b) - T_2 J_0(T_1 R_b) J_1(T_2 R_b) \tag{36}$$

$$H_0 = s_0(T_1 J_0(T_2 R_c) N_1(T_1 R_c) - T_1 J_1(T_2 R_c) N_0(T_1 R_c)) + l_0(T_1 N_0(T_2 R_c) N_1(T_1 R_c) - T_2 N_1(T_2 R_c) N_0(T_1 R_c)) \tag{37}$$

$$U_0 = s_0(T_2 J_1(T_2 R_c) J_0(T_1 R_c) - T_1 J_1(T_1 R_c) J_0(T_2 R_c)) + l_0(T_2 N_1(T_2 R_c) J_0(T_1 R_c) - T_1 N_0(T_2 R_c) J_1(T_1 R_c)) \tag{38}$$

At the perfectly conducting corrugated waveguide surface (second boundary condition), the tangential electric field must be zero,

$$E_z(r = R(z)) + E_r(r = R(z)) \frac{dR(z)}{dz} = 0. \tag{39}$$

Substituting Eqs. (22)–(24) into Eq. (39), we investigate second boundary condition in ground state ($n = 0$) to achieve the dispersion equation.

$$D_0 J_0(T_1 R(z)) + F_0 N_0(T_1 R(z)) e^{i(k_0 z - \omega t)} + \frac{ik_0}{\frac{\omega^2}{c^2} - k_0^2} e^{i(k_0 z - \omega t)} \times \frac{d}{dR(z)} \left[D_0 J_0(T_1 R(z)) + F_0 N_0(T_1 R(z)) e^{i(k_0 z - \omega t)} \right] \frac{dR(z)}{dz} = 0 \tag{40}$$

Using the factorization of Eq. (40) and substituting B_0 and C_0, we obtain:

$$-e^{i(k_0 z - \omega t)} \frac{\pi}{2} R_b A_0 \left(1 + \frac{ik_0}{\frac{\omega^2}{c^2} - k_0^2} \frac{d}{dz} \right) \times \left[H_0 J_0(T_1 R(z)) + U_0 N_0(T_1 R(z)) \right] = 0 \tag{41}$$

$$D \cdot A = D_n A_n = D_0 A_0 = 0, \tag{42}$$

"A" is a vector with element A_0 and "D" is a matrix with element D_0. With the help of derivative of Bessel functions and substituting Eq. (1), the dispersion relation can be obtained and written as [18–22],

$$D_0 = \left(1 + \frac{ik_0}{\frac{\omega^2}{c^2} - k_0^2}\right) \times [T_1 h\kappa_0 (H_0 J_1 \sin(\kappa_0 z) + U_0 N_1 \sin(\kappa_0 z))] \quad (43)$$

Numerical result and discussion

The analysis of the dispersion relation is obtained by Eq. 40. First, we consider the dispersion analysis in the absence of the electron beam.

In Fig. 2, the chosen parameters are as follows: $n_p = 3 \times 10^{11}\text{cm}^{-3}$, $R_0 = 1.60$ cm, $h = 0.7$ cm, $R_b = 0.7$ cm, $z = 0.33$ cm and $T_p = 30 \times 10^8$ °K, Fig. 2 shows the variation of normalized frequency Re($\frac{\omega}{c\kappa_0}$) versus wave number ($\frac{k_z}{\kappa_0}$) for several values of the corrugation periods (z_0). As seen in this figure the effect of z_0 increases the frequency.

The effect of the plasma temperature on the frequency of the system as a function of k_z is shown in Fig. 3. The chosen parameters are as follows: $n_p = 3 \times 10^{11}$ cm^{-3}, $R_0 = 1.60$ cm, $h = 0.7$ cm, $R_b = 0.7$ cm, $z = 0.33$ cm and $z_0 = 0.66$ cm. Figure 3 shows that the effect of plasma temperature increases the frequency. This effect is in good agreement with the thermal plasma dispersion relation $\omega^2 = \omega_p^2 + 3k_z^2 v_{th}^2$.

The effect of waveguide radius on the frequency of the wave as a function of k_z is shown in Fig. 4. The chosen parameters are as follows:$n_p = 3 \times 10^{11}\text{cm}^{-3}$, $h = 0.7$ cm, $R_b = 0.7$ cm, $z = 0.33$ cm, $T_p = 30 \times 10^8$ °K and $z_0 = 0.66$ cm. As illustrated in this figure, the frequency decreases with increase in the waveguide radius.

The variations of the frequency as a function of the wave number for different values of the plasma density are shown in Fig. 5. The chosen parameters are as follows: $R_0 = 1.60$ cm, $h = 0.7$ cm, $R_b = 0.7$ cm, $z = 0.33$ cm, $z_0 = 0.66$ cm and $T_p = 30 \times 10^8$ °K. It is clear that from the figure, the effect of plasma considerably increases the frequency. This effect is in good agreement with the simple relation of the plasma ($\omega^2 = \omega_p^2 + 3k_z^2 v_{th}^2$).

Now, we consider the analysis of the growth rate in the presence of the electron beam.

Figure 6, shows the effect of corrugation period on the normalized growth rate Im ($\frac{\omega}{c\kappa_0}$) as a function of the normalized wave number ($\frac{k_z}{\kappa_0}$). The chosen parameters are as follows: $\gamma = 1.001$, $h = 0.7$ cm, $R_0 = 1.6$ cm, $R_b = 0.7$ cm, $z = 0.33$ cm, $n_b = 10 \times 10^{11}$ cm^{-3}, $n_p = 3 \times 10^{11}$ cm^{-3}, $T_p = 30 \times 10^8$ °K and $T_b = 20 \times 10^7$ °K. It is clear that from Fig. 6, the growth rate increases by increasing the corrugation period.

The effect of electron beam internal radius on the growth rate on the frequency of the wave as a function of k_z is shown in Fig. 7. The chosen parameters are as follows $n_p = 3 \times 10^{11}$ cm^{-3}, $R_0 = 1.6$ cm, $z_0 = 0.66$ cm, $\gamma = 1.001$, $z = 0.33$ cm, $h = 0.7$ cm, $n_b = 10 \times 10^{11}$ cm^{-3}, $T_p = 30 \times 10^8$ °K and $T_b = 20 \times 10^7$ °K. As

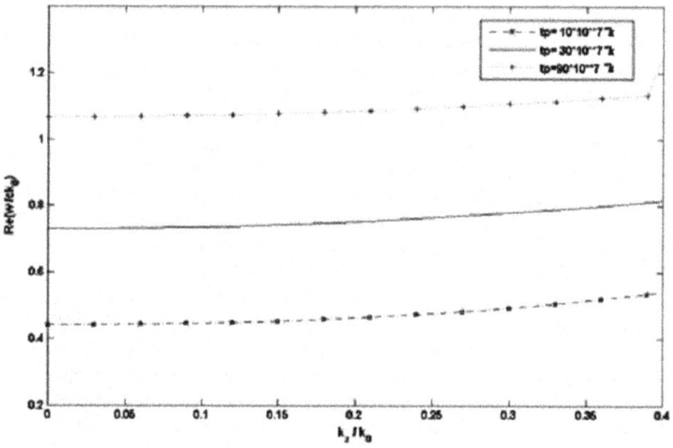

Fig. 3 Dispersion relation Re(ω)–k_z for various plasma temperatures

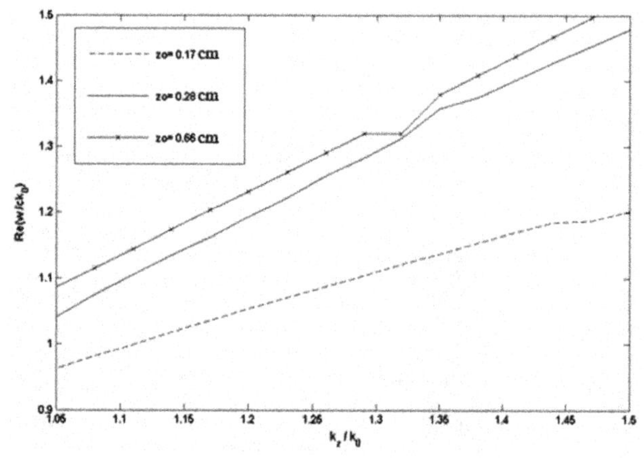

Fig. 2 Dispersion relation Re(ω)–k_z for various corrugation periods

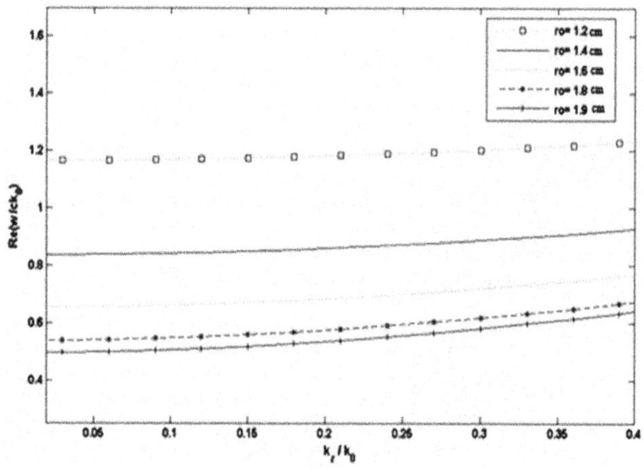

Fig. 4 Dispersion relation Re(ω)–k_z for various waveguideradiuses

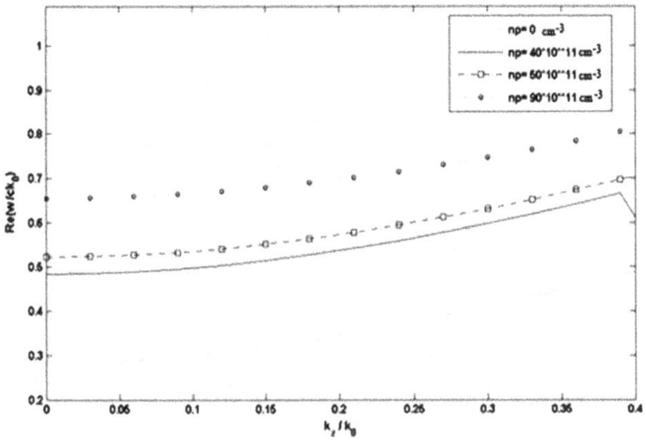

Fig. 5 Dispersion ration $\text{Re}(\omega)$–k_z for various plasma densities

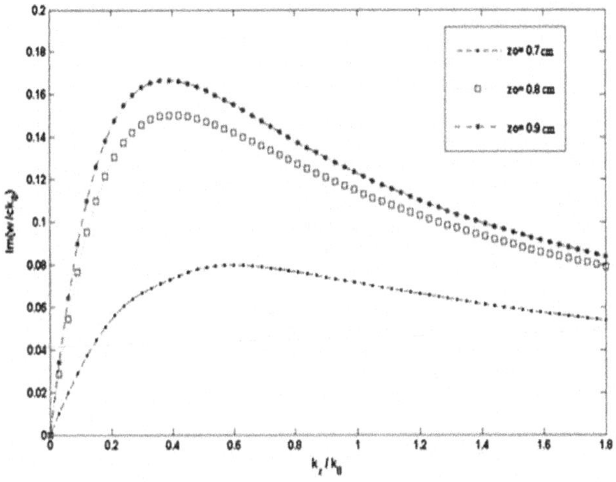

Fig. 6 Growth rate for various corrugation periods

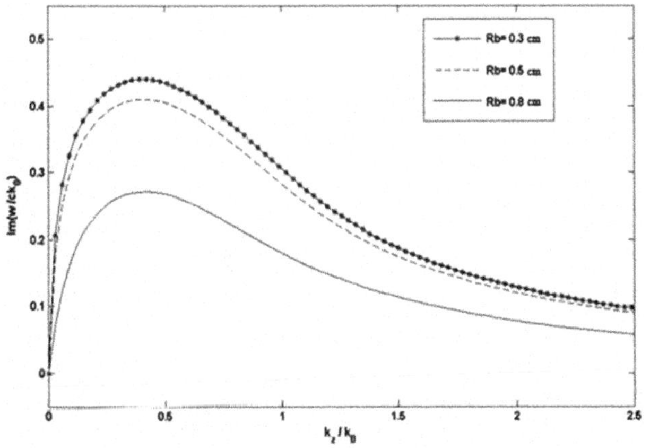

Fig. 7 Growth rate for various electron beam internal radius

illustrated in this figure, the frequency decreases by increasing the electron beam internal radius.

Figure 8, shows the effect of waveguide radius on the growth rate as a function of wave number for several values of waveguide radius. The chosen parameters are as

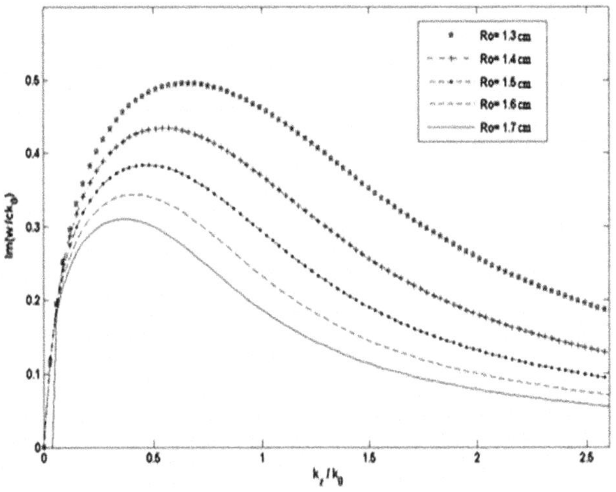

Fig. 8 Growth rate for various waveguide radiuses

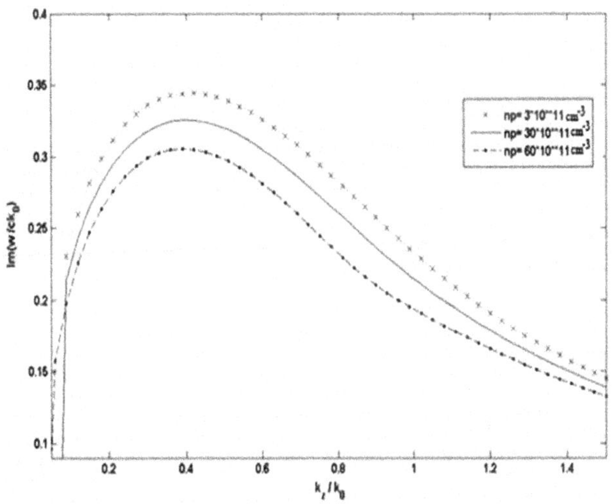

Fig. 9 Growth rate for various plasma densities

follows: $n_p = 3 \times 10^{11}$ cm^{-3}, $h = 0.7$ cm, $T_p = 30 \times 10^8$ °K, $R_b = 0.7$ cm, $z = 0.33$ cm, $\gamma = 1.001$, $z_0 = 0.66$ cm, $n_b = 10 \times 10^{11}$ cm^{-3} and $T_b = 20 \times 10$ °K. As seen in this figure, the growth rate decreases by increasing the R_0.

The effect of the plasma density on the growth rate of the system as a function of the wave number is shown in Fig. 9. It is clear that in this frequency range the effect of plasma density decreases the growth rate of the system. The chosen parameters are as follows: $R_0 = 1.6$ cm, $h = 0.7$ cm, $T_p = 30 \times 10^8$ °K, $R_b = 0.7$ cm, $z = 0.33$ cm, $\gamma = 1.001$, $z_0 = 0.66$ cm, $n_b = 10 \times 10^{11}$ cm^{-3} and $T_b = 20 \times 10^7$ °K. Figure 10 illustrates the variation of the growth rate for several values of the electron beam density. It is clear that from the figure, because of bunching effect, the increasing e-beam density increases the growth rate. The chosen parameters are as

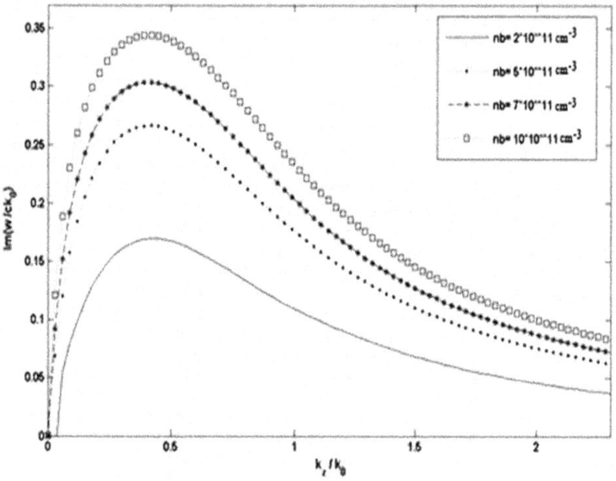

Fig. 10 Growth rate for various electron beam densities

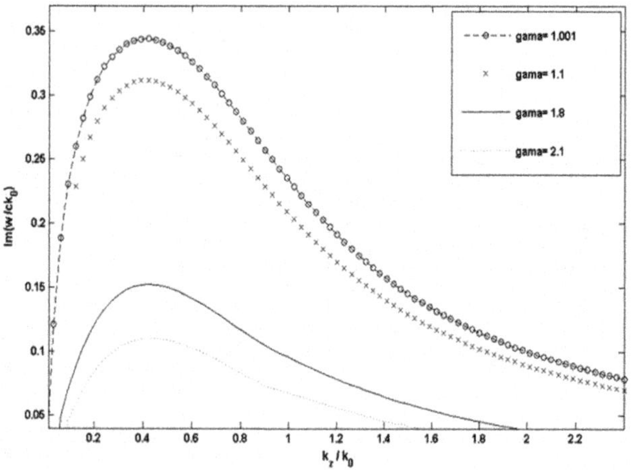

Fig. 11 Growth rate for various electron beam energies

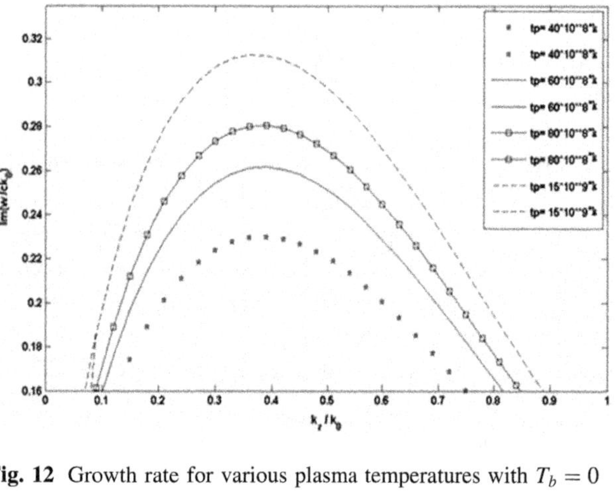

Fig. 12 Growth rate for various plasma temperatures with $T_b = 0$

follows: $R_0 = 1.6cm, h = 0.7cm, T_p = 30 \times 10^8$ °K, $R_b =$ 0.7cm , $z = 0.33$ cm, $\gamma = 1.001$, $z_0 = 0.66$ cm, $n_p = 3 \times 10^{11}$ cm^{-3}, and $T_b = 20 \times 10^7$ °K.

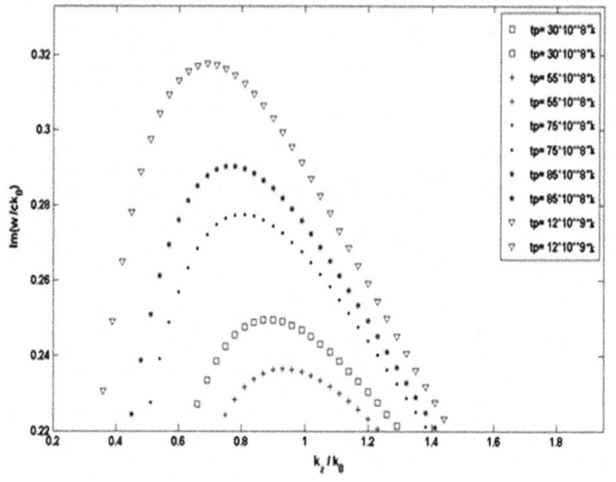

Fig. 13 Growth rate for various plasma temperatures with $T_b = 50 \times 10^8$ °K

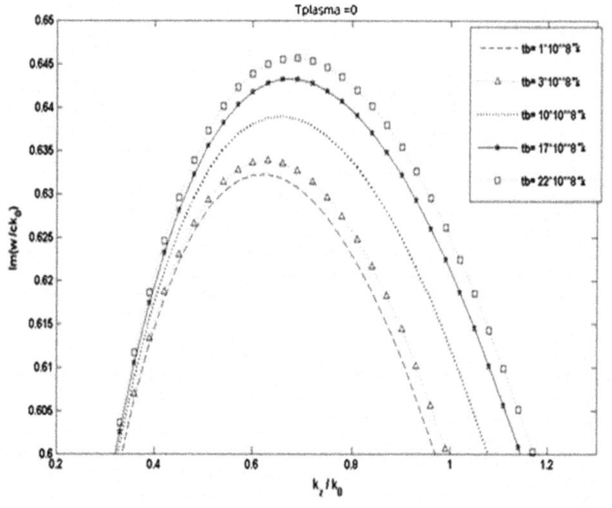

Fig. 14 Growth rate for various beam temperatures with $T_p = 0$

Figure 11 shows the variation of the growth rate as a function of the wave number for several values of the electron beam energy. The chosen parameters are as follows: $R_0 = 1.6$ cm, $h = 0.7$ cm, $z_0 = 0.66$ cm, $R_b = 0.7$ cm, $z = 0.33$ cm, $n_b = 10 \times 10^{11}$ cm^{-3}, $T_p = 30 \times 10^8$ °K, $n_p = 3 \times 10^{11}$ cm^{-3} and $T_b = 20 \times 10^7$ °K. As seen this figure because of the synchronism condition, the effect of γ decreases the growth rate.

The plasma temperature effect on the growth rate as a function of the normalized wave number is given in Fig. 12. The figure shows, the plasma temperature considerably increases the growth rate. The chosen parameters are as follows: $n_p = 3 \times 10^{11}$ cm^{-3}, $h = 0.7$ cm, $R_0 = 1.6$ cm, $R_b = 0.7$ cm, $z = 0.33$ cm, $\gamma = 1.001$, $z_0 = 0.66$ cm, $n_b = 0$

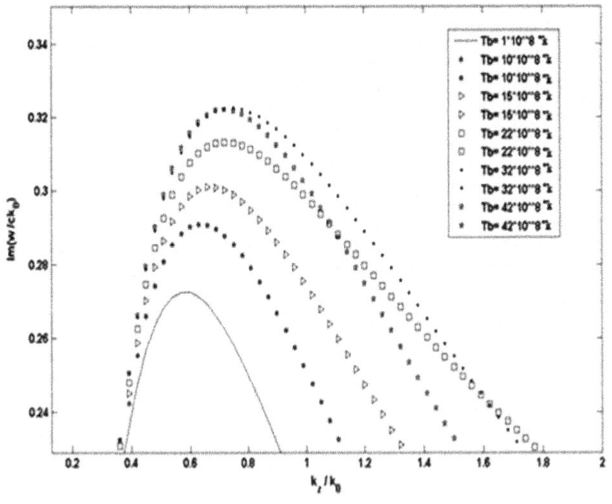

Fig. 15 Growth rate for various beam temperatures with $T_p = 12 \times 10^9$ °K

The plasma temperature effect on the growth rate with constant value $T_b = 50 \times 10^8$ °K, is given in Fig. 13. The figure shows, the plasma temperature at first decrease to the point $T_p = 55 \times 10^8$ °K, and then the graph increases. The chosen parameters are as follows: $n_p = 3 \times 10^{11}\text{cm}^{-3}, h = 0.7\text{cm}$, $R_0 = 1.6$ cm, $R_b = 0.7\text{cm}, z = 0.33$ cm, $\gamma = 1.001$, $z_0 = 0.66$ cm, $n_b = 10 \times 10^{11}\text{cm}^{-3}$ and $T_b = 50\times 10^8$ °K.

The beam temperature effect on e growth rate as a function of the normalized wave number is given in Fig. 14. It is clear that from the figure, the beam temperature considerably increases the growth rate. The chosen parameters are as follows: $h = 0.7$ cm, $R_0 = 1.6$ cm, $R_b = 0.7$ cm, $z = 0.33$ cm, $\gamma = 1.001$, $z_0 = 0.66$ cm, $n_b = 10\times10^{11}$ cm^{-3}, $T_p = 0$.

The beam temperature effect on the growth rate as a function of the normalized wave number is given in Fig. 15. It is clear from the figure that the beam temperature at first increase but from point $T_b = 32 \times 10^8$ °K the graph decreases. The chosen parameters are as follows: $h = 0.7$ cm, $R_0 = 1.6$ cm, $R_b = 0.7$ cm, $z = 0.33$ cm, $\gamma = 1.001$, $z_0 = 0.66$ cm, $n_b = 10\times10^{11}$ cm^{-3}, $T_p = 12\times 10^9$ °K, $n_p = 3 \times 10^{11}$ cm^{-3}.

Conclusions

In this paper, useful results are obtained as:

1. In the absence of the electron beam, the frequency increases by increasing the length of the corrugation period, plasma temperature and plasma density.
2. The frequency decreases by increasing the waveguide radius in the absence of the electron beam.
3. In the presence of the electron beam, the growth rate increases by increasing the corrugation period and e-beam density.
4. The growth rate decreases by increasing the waveguide radius, plasma density and e-beam energy in the presence of the electron beam.

References

1. Shiffler, D., Nation, J.A., Graham, S.K.: A high-power, traveling wave tube amplifier. IEEE Trans. Plasma Sci. **18**, 546 (1990)
2. Nusinovich, G.S., Carmel, Y., Antonsen, Jr., T.M.: Recent progress in the development of plasma-filled traveling-wave tubes and backward-wave oscillation. IEEE Trans. Plasma Sci. **26**, 628 (1998)
3. Kobayashi Jr, S., Antonsen, T.M., Nusinovich, G.S.: IEEE Trans. Plasma Sci. **26**, 669 (1998)
4. Pierce, J.R., Field, L.M.: Traveling-wave tubes. Proc. IRE **35**, 108 (1947)
5. Pierce, J.R.: Travelling-Wave Tubes. Van Nostrand Reinhold, New York (1950). (ch. 3)
6. Pierce, J.R.: Theory of the beam-type traveling wave tube. Proc. IRE **35**, 111 (1947)
7. Chu, L.J., Jackson, J.D.: Field theory of traveling-wave tubes. Proc. IRE **36**, 853 (1948)
8. Freund, H.P., Ganguly, A.K.: Three-dimensional theory of the free electron laser in the collective regime. Phys. Rev. A **28**(6), 3438–3449 (1983)
9. Freund, H.P.: Multimode nonlinear analysis of free-electron laser, amplifiers in three dimensions. Phys. Rev. A **37**(9), 3371–3380 (1988)
10. Brillouin, L.: Wave Propagation in Periodic Structures. Dover, New York (1953)
11. Collin, R.E.: Field Theory of Guided Waves. McGraw-Hill, New York (1960)
12. Saviz, S.: The effect of beam and plasma parameters on the four modes of plasma-loaded traveling-wave tube with tape helix. J. Theor. Appl. Phys. **8**, 135 (2014)
13. Saviz, S., Salehizadeh, F.: Plasma effect in tape helix traveling-wave tube. J. Theor. Appl. Phys. **8**, 1 (2014)
14. Wahley, D.R., et al.: Operation of a low voltage high-transconductance field emitter array TWT. In: Proceedings of IEEE Vacuum Electronics Conference, pp. 78–79 (2008)
15. Saviz, S., Shahi, F.: Analysis of axial electric field in thermal plasma-loaded helix traveling-wave tube with dielectric-loaded waveguide. IEEE Trans. Plasma Sci. **42**, 917 (2014)
16. Zavyalov, M.A., Mitin, L.A., Perevodchikov, V.I., Tskhai, V.N., Shapiro, A.L.: Powerful wideband amplifier based on hybrid plasma-cavity slow-wave structure. IEEE Trans. Plasma Sci. **22**, 600 (1994)
17. Saviz, S.: Plasma thermal effect on the growth rate of the helix traveling wave tube. IEEE Trans. Plasma Sci. **42**, 2023 (2014)
18. Miyamoto, K.: Parameter sensitivity of ITER type experimental tokamak reactor toward compactness. J. Plasma Fusion Res. **76**, 166 (2000)
19. Hong-Quan, X., Pu-Kun, L.: Theoretical analysis of a relativistic travelling wave tube filled with plasma. Chin. Phys. Soc. **16**(3), 766 (2007)
20. Chen, F.F.: Introduction to Plasma Physics and Controlled Fusion. Plenum Press, New York (1984)
21. Krall, N.A., Trivelpiece, A.W.: Principles of Plasma Physics. McGraw-Hill, New York (1973)

22. Goldston, R.J., Rutherford, P.H.: Introduction to Plasma Physics. Institute of Phyics Publishing, London (1995)
23. Kuzelev, M.V., Panin, V.A., Plotnikov, A.P., Rukhadze, A.A., M. V. Lomonosov State University, Moscow: The theory of transverse-unhomogeneous beam-plasma amplifiers. Zh. Eksp. Teor. Fiz. **101**, 460–478 (1992)
24. Novgorod, N., Abubakirov, R.E.B.: Peculiarities of backward-wave amplification by relativistic high-current electron beams. Inst. Appl. Phys. Russ. Acad. Sci. IEEE Trans. Plasma Sci. (Impact Factor: 0.95) (2010). doi:10.1109/TPS.2010.2043120
25. Chelpanov, V.I., Dubinov, A.E., Dubinov, E.E., Babkin, A.L: Pulsed power conference, digest of technical papers. In: 1997 IEEE International, Institute of Experimental Physics, Federation Nuclear Centre, Sarov, Russia (1997)
26. Selivanov, I.A., Shkvarunets, A.G.: Harmonic gyro-TWT amplifier for high power. In: High-Power Particle Beams, 1992 9th International Conference on IEEE, 25–29 May (1992)
27. Khalil, S.M., Mousa, N.M.: Dispersion characteristics of plasma-filled cylindrical waveguide. J Theor Appl Phys **8**, 111 (2014). doi:10.1007/s40094-014-0111-2
28. Babkin, A.L., Chelpanov, V.I., Dubinov, A.E., Dubinov, E.E., Hizhnyakov, A.A., Konovalov, I.V., Komilov, V.G., Selemir, V.D., Zhdanov, V.S.: Powerful electron accelerator "COV-CHEG": status, parameters and physical experiments. In: Pulsed Power Conference, 1997. Digest of Technical Papers. 11th IEEE International (1997)
29. Almazova, K.I., Borovkov, V.V., Komilov, V.G., Markevtsev, I.M., Tatsenko, O.M., Selemir, V.D.: Optical diagnostics methods of plasma current channels in plasma current open switches of the developed EMIR complex. In: Pulsed Power Plasma Science. IEEE Conference Record—Abstracts (2001)

15

Effect of obliqueness and external magnetic field on the characteristics of dust acoustic solitary waves in dusty plasma with two-temperature nonthermal ions

Akbar Sabetkar · Davoud Dorranian

Abstract In this paper, a theoretical investigation has been made of obliquely propagating dust acoustic solitary wave (DASW) structures in a cold magnetized dusty plasma consisting of a negatively charged dust fluid, electrons, and two different types of nonthermal ions. The Zakharov–Kuznetsov (ZK) and modified Zakharov–Kuznetsov (MZK) equations, describing the small but finite amplitude DASWs, are derived using a reductive perturbation method. The combined effects of the external magnetic field, obliqueness (i.e. the propagation angle), and the presence of second component of nonthermal ions, which are found to significantly modify the basic features (viz. amplitude, width, polarity) of DASWs, are explicitly examined. The results show that the external magnetic field, the propagation angle, and the second component of nonthermal ions have strong effects on the properties of dust acoustic solitary structures. The solitary waves may become associated with either positive potential or negative potential in this model. As the angle between the direction of external magnetic field and the propagation direction of solitary wave increases, the amplitude of the solitary wave (for both positive potential and negative potential) increases. With changing this angle, the width of solitary wave shows a maximum. The magnitude of the external magnetic field has no direct effect on the solitary wave amplitude. However, with decreasing the strength of magnetic field, the width of DASW increases.

A. Sabetkar · D. Dorranian (✉)
Laser Laboratory, Plasma Physics Research Center, Science and Research Branch, Islamic Azad University, Tehran, Iran
e-mail: d.dorranian@gmail.com

Keywords Magnetized dusty plasma · KP equation · ZK equation · Nonthermal ion · Soliton · Reductive perturbation method

Introduction

Dusty plasma is an ionized gas containing small particles of solid matter which acquire a large electric charge by collecting electrons and ions from the plasma. This state of plasma is ubiquitous in the universe, e.g., in interstellar clouds, in interplanetary space, in cometary tails, in ring systems of giant planets (like Saturn F-ring's), in mesospheric noctilucent clouds, as well as in many Earth bound plasma [1, 2].

Recently, nonlinear wave propagation in plasmas has become one of the most important subjects of plasma. One of the most common and most fascinating eigen-modes that exist in space and laboratory dusty plasmas is dust acoustic wave (DASW), which was first theoretically predicted by Rao et al. [3] and experimentally verified by Barkan et al. [4]. The properties of these nonlinear plasma waves can be described by different nonlinear differential equations, e.g., the Korteweg–de Vries (KdV) equation, Zakharov–Kuznetsov (ZK) equation, and so on [5–7].

Over the last few years, a great deal of attention has been paid to non-Maxwellian particle distributions. The Maxwellian distribution is applicable to systems in thermodynamic equilibrium. However, astrophysical systems and space plasmas are observed to have particle distributions that depart from the Maxwellian distribution due to non-equilibrium stationary state. This state may arise due to a number of physical mechanisms such as external force field present in natural space plasma environments, wave–particle interaction, and turbulence. Spatial observations revealed the existence of non-Maxwellian distribution functions with high-energy tails or pronounced shoulders. Such type of distributions is frequently called superthermal or nonthermal distributions and occurs in space/astrophysical environments. Many authors studied solitary wave propagation in such non-Maxwellian distributions.

For example, Lin and Duan [8] considering the nonthermal ion in a dusty plasma derived a Korteweg–de Vries (KdV) equation for DA wave and it was found that the nonthermal ions have very important effect on the propagation of DA solitary waves. In a recent report Dorranian and Sabetkar [9] reported that with increasing nonthermal ion population, the amplitude of solitary wave decreases, while the width of solitary waves increases. It is also reported that in nonthermal dusty plasmas, the DASW have larger amplitude, smaller width, and higher propagation velocity than those involving adiabatic ions. Zhang and Wang [10] in their report observed the possibility of co-existence of both compressive and rarefactive solitary wave depending on a critical value of nonthermal ion population.

In continuation of that a number of theoretical investigations have been made on DASWs. Recently, the effects of the dust fluid temperature on the DASWs have been investigated by a number of authors [11, 12]. Sayed and Mamun [13] assumed a dusty plasma containing the adiabatic dust fluid and non-adiabatic (isothermal) inertialess electron and ion fluid, and studied the effect of the dust fluid temperature on the DASWs.

Many authors derived the ZK equation to study the solitary wave structures in magnetized plasma in different environments [14–16]. Recently, Mahmood et al. [17] have derived the ZK equation for nonlinear acoustic waves in dense magnetized electron–positron (e–p) plasmas. They showed that an increase in the strength of the magnetic field may lead to a significant decrease of the width of the solitons. More recently Shahmoradi et al. [18] have investigated the effect of dust size, mass, and charge distributions on the characteristics of nonlinear dust acoustic solitary waves (DASWs) in magnetized dusty plasma. They found that at each strength of the external magnetic field, there is an optimum magnitude for its direction at which the width of DASW is maximum.

This paper is organized in the following fashion. Following the introduction in Sect. 1, the governing equations and model description of dusty plasma system are presented in Sect. 2. The Zakharov–Kuznetsov (ZK) equation and its modified form with their solitary wave solution are derived by employing the reductive perturbation method in Sects. 3 and 4, respectively. Results and discussion are presented in Sects. 5 and 6 is devoted to conclusion.

Model description

We consider a three-component dusty plasma system, which consists of negatively charged dust fluid, Boltzmann distributed electrons, and two-temperature nonthermal ions with fast particles in the presence of external static magnetic field $\mathbf{B} = B_0\hat{z}$, where $\hat{z}$ is the unit vector along z direction. Charge neutrality at equilibrium reads

$$n_{\mathrm{e}0} + n_{\mathrm{d}0}Z_{\mathrm{d}0} = n_{i\mathrm{l}0} + n_{i\mathrm{h}0} \tag{1}$$

where $n_{i\mathrm{lo}}$ ($n_{i\mathrm{h}0}$), $n_{\mathrm{e}0}$, and $n_{\mathrm{d}0}$ are the unperturbed lower (higher) temperature ion, electron, and dust number densities, respectively, and $Z_{\mathrm{d}0}$ is the unperturbed number of charges residing on the dust grain measured in the unit of electron charge. The normalized equations governing the dust acoustic wave dynamics are

$$\frac{\partial n_{\mathrm{d}}}{\partial t} + \nabla\cdot(n_{\mathrm{d}}\mathbf{u}_{\mathrm{d}}) = 0 \tag{2a}$$

$$\frac{\partial \mathbf{u}_{\mathrm{d}}}{\partial t} + (\mathbf{u}_{\mathrm{d}}\cdot\nabla)\mathbf{u}_{\mathrm{d}} = \nabla\phi - \omega_{\mathrm{cd}}(\mathbf{u}_{\mathrm{d}}\times\hat{z}) \tag{2b}$$

$$\nabla^2\phi = n_{\mathrm{d}} + n_{\mathrm{e}} - n_{i\mathrm{l}-n_{i\mathrm{h}}} \tag{2c}$$

where $\mathbf{u}_{\mathrm{d}} = (u_{\mathrm{d}x}, u_{\mathrm{d}y}, u_{\mathrm{d}z})$. The time and space variables are normalized by the dust plasma period $\omega_{\mathrm{pd}}^{-1} = \sqrt{m_{\mathrm{d}}/4\pi n_{\mathrm{d}0}Z_{\mathrm{d}0}^2e^2}$ and the Debye length $\lambda_{\mathrm{d}} = \sqrt{T_{\mathrm{eff}}/4\pi n_{\mathrm{d}0}Z_{\mathrm{d}0}e^2}$, respectively. n_{d} is the dust particle number density normalized to $n_{\mathrm{d}0}$; $\mathbf{u}_{\mathrm{d}}$ is the dust fluid velocity normalized to the dust acoustic speed $C_{\mathrm{d}} = \sqrt{Z_{\mathrm{d}0}T_{\mathrm{eff}}/m_{\mathrm{d}}}$ in which T_{eff} is the effective temperature which is defined below and m_{d} is the mass of the dust particles. ϕ is the electrostatic potential of plasma medium normalized by T_{eff}/e and n_i is the ion number density normalized to n_{i0}. For the case of dusty plasma with two kinds of ions at different temperatures, the effective temperature is

$$T_{\mathrm{eff}} = \left[\frac{1}{n_{\mathrm{d}0}Z_{\mathrm{d}0}}\left(\frac{n_{\mathrm{e}0}}{T_{\mathrm{e}}} + \frac{n_{i\mathrm{l}0}}{T_{i\mathrm{l}}} + \frac{n_{i\mathrm{h}0}}{T_{i\mathrm{h}}}\right)\right]^{-1} \tag{3}$$

in which T_{e}, $T_{i\mathrm{l}}$, and $T_{i\mathrm{h}}$ are the plasma electron temperature and the temperatures of plasma ions at lower and higher temperatures, respectively. $\omega_{\mathrm{cd}} = Z_{\mathrm{d}0}eB_0/m_{\mathrm{d}}\omega_{\mathrm{pd}}$ is the dust cyclotron frequency normalized to ω_{pd}. The distribution function that was chosen by Cairns et al. [19] is chosen to model an ion distribution with a population of fast particles. Therefore, the lower temperature ion density, $n_{i\mathrm{l}}$, and higher temperature ion density, $n_{i\mathrm{h}}$, are directly given by

$$n_{i\mathrm{l}} = \frac{\delta_1}{\delta_1+\delta_2-1}\left(1+\mu(s\phi+s^2\phi^2)\right)\exp(-s\phi), \tag{4a}$$

$$n_{ih} = \frac{\delta_2}{\delta_1+\delta_2-1}\left(1+\mu(\beta s\phi+\beta^2 s^2\phi^2)\right)\exp(-\beta s\phi), \tag{4b}$$

$$n_e = \frac{1}{\delta_1+\delta_2-1}\exp(\beta_1 s\phi), \tag{4c}$$

where $\mu = 4\alpha/(1+3\alpha)$, α is parameter determining the number of fast (nonthermal) ions.

It should be noted here that if we neglect the number of nonthermal ions in comparison with that of thermal ions, i.e. we put $\alpha = 0$, this dusty plasma model reduces to the model considered by Rao et al. [3]. From Eq. 1 one can write

$$\delta_1+\delta_2-1 \geq 0. \tag{5}$$

In above equations $\beta_1 = \frac{T_{il}}{T_e}$, $\beta_2 = \frac{T_{ih}}{T_e}$, $\frac{\beta_1}{\beta_2} = \frac{T_{il}}{T_{ih}}$, $\delta_1 = \frac{n_{il}}{n_e}$, $\delta_2 = \frac{n_{ih}}{n_e}$, and $S = \frac{T_{eff}}{T_{il}} = \frac{\delta_1+\delta_2-1}{\delta_1+\delta_2\beta+\beta_1}$.

Zakharov–Kuznetsov (ZK) equation

Derivation of the ZK equation

To investigate the nonlinear propagation of DASWs in magnetized plasma, we employ the standard RPM [7] to obtain the appropriate ZK equation. The independent variables are stretched as

$$x' = \varepsilon^{1/2}x \tag{6a}$$

$$y' = \varepsilon^{1/2}y \tag{6b}$$

$$z' = \varepsilon^{1/2}(z-\lambda t), \tag{6c}$$

$$t' = \varepsilon^{3/2}t, \tag{6d}$$

where ε is a small parameter measuring the strength of nonlinearity and λ the phase velocity of waves normalized by the dust acoustic speed C_d, which is determined later. The physical quantities are expanded about their equilibrium values in a power series of ε as

$$f = f^{(0)} + \sum_{i=1}^{\infty}\epsilon^m f^{(m)} \tag{7a}$$

$$u_{dx,y} = \sum_{i=1}^{\infty}\epsilon^{1+m/2}u_{dx,y}^{(m)}. \tag{7b}$$

The variables $f = (n_d, \phi, u_{dz})$ describe the state of the system with $f^{(0)} = (1,0,0)$. Substituting Eqs. 6 and 7 into the basic set of Eq. 2 and then equating the coefficient powers of ϵ in the lowest order, i.e. $O(\epsilon^{3/2})$, we obtain the followings:

$$n_d^{(1)} = \frac{u_{dz}^{(1)}}{\lambda} \tag{8a}$$

$$u_{dz}^{(1)} = \frac{-\phi^{(1)}}{\lambda} \tag{8b}$$

$$u_{dy}^{(1)} = \frac{1}{\omega_{cd}}\frac{\partial\phi^{(1)}}{\partial x'} \tag{8c}$$

$$u_{dx}^{(1)} = \frac{-1}{\omega_{cd}}\frac{\partial\phi^{(1)}}{\partial y'} \tag{8d}$$

Now, substituting Eq. 8 into the lowest order Poisson equation, we get the phase velocity of DASWs as

$$\lambda = \left[\frac{(\delta_1+\delta_2\beta+\beta_1)(1+3\alpha)}{(\delta_1+\delta_2\beta)(1-\alpha)+\beta_1(1+3\alpha)}\right]^{1/2} \tag{9}$$

which agrees exactly with the phase velocity of the perturbation mode derived by Dorranian et al. [9]. Similarly, to the next higher order of ϵ, i.e. $O(\epsilon^2)$ we obtain the second-order x and y components of the momentum and Poisson equation as

$$u_{dy}^{(2)} = \frac{-\lambda}{\omega_{cd}^2}\frac{\partial^2\phi^{(1)}}{\partial z'\partial y'} \tag{10a}$$

$$u_{dx}^{(2)} = \frac{-\lambda}{\omega_{cd}^2}\frac{\partial^2\phi^{(1)}}{\partial z'\partial x'} \tag{10b}$$

$$\left(\frac{\partial^2}{\partial x'^2}+\frac{\partial^2}{\partial y'^2}+\frac{\partial^2}{\partial z'^2}\right)\phi^{(1)} = n_d^{(2)} + \frac{1}{\lambda^2}\phi^{(2)} + \left(\frac{-1}{2}\frac{(\delta_1+\delta_2\beta^2-\beta_1^2)s^2}{\delta_1+\delta_2-1}\right) \times \left(\phi^{(1)}\right)^2 \tag{10c}$$

Also, following the same procedure, we can obtain the next higher-order continuity equation and z component of momentum equation as, i.e. $O(\epsilon^{5/2})$

$$\frac{\partial n_d^{(1)}}{\partial t'} - \lambda\frac{\partial n_d^{(2)}}{\partial z'} + \frac{\partial u_{dx}^{(2)}}{\partial x'} + \frac{\partial u_{dy}^{(2)}}{\partial y'} + \frac{\partial}{\partial z'}(u_{dz}^{(2)} + n_d^{(1)}u_{dz}^{(1)}) = 0 \tag{11a}$$

$$\frac{\partial u_{dz}^{(1)}}{\partial t'} - \lambda\frac{\partial u_{dz}^{(2)}}{\partial z'} + u_{dz}^{(1)}\frac{\partial u_{dz}^{(1)}}{\partial z'} - \frac{\partial\phi^{(2)}}{\partial z'} = 0 \tag{11b}$$

Solving the system of Eq. 11, with the aid of Eqs. 8, 9, and 10, we can readily obtain

$$\frac{\partial\phi^{(1)}}{\partial t'} + AB\phi^{(1)}\frac{\partial\phi^{(1)}}{\partial z'} + \frac{A}{2}\frac{\partial^3\phi^{(1)}}{\partial z'^3} + \frac{AD}{2}\left(\frac{\partial^3\phi^{(1)}}{\partial z' x'^2} + \frac{\partial^3\phi^{(1)}}{\partial z' y'^2}\right) = 0 \tag{12}$$

where

$$A = \lambda^3 \tag{13a}$$

$$D = \frac{\omega_{cd}^2 + 1}{\omega_{cd}^2} \tag{13b}$$

$$B = \frac{1}{2}\frac{(\delta_1 + \delta_2\beta^2 - \beta_1^2)(\delta_1 + \delta_2 - 1)}{(\delta_1 + \delta_2\beta + \beta_1)^2} - \frac{3}{2\lambda^4}. \tag{13c}$$

Equation 12 is known as the Zakharov–Kuznetsov (ZK) equation or the Korteweg–de Vries (KdV) equation in three dimensions.

Solitary wave solution of the ZK equation

To study the propagation of DASWs in a direction making an angle θ with the external magnetic field **B**, lying in the $x' - z'$ plane, we transform the coordinate system x', y', z' into the new coordinate system ζ, ξ, η by a rotation around the y' axis through an angle θ. The relations between the new and old coordinates become

$$\zeta = x' \cos\theta - z' \sin\theta \tag{14a}$$

$$\eta = y' \tag{14b}$$

$$\xi = x' \sin\theta + z' \cos\theta \tag{14c}$$

$$\tau = t' \tag{14d}$$

Under these changes of the independent variables, the ZK Eq. 12 becomes

$$\begin{aligned} &\frac{\partial\phi^{(1)}}{\partial\tau} + h_1\phi^{(1)}\frac{\partial\phi^{(1)}}{\partial\xi} + h_2\frac{\partial^3\phi^{(1)}}{\partial\xi^3} + h_3\phi^{(1)}\frac{\partial\phi^{(1)}}{\partial\zeta} \\ &\quad + h_4\frac{\partial^3\phi^{(1)}}{\partial\zeta^3} + h_5\frac{\partial^3\phi^{(1)}}{\partial\xi^2\partial\zeta} + h_6\frac{\partial^3\phi^{(1)}}{\partial\xi\partial\zeta^2} \\ &\quad + h_7\frac{\partial^3\phi^{(1)}}{\partial\xi\partial\eta^2} + h_8\frac{\partial^3\phi^{(1)}}{\partial\zeta\partial\eta^2} = 0. \end{aligned} \tag{15}$$

Coefficients of Eq.15 are

$$h_1 = AB\cos\theta, \quad h_2 = \frac{A}{2}\cos\theta(\cos^2\theta + D\sin^2\theta),$$

$$h_3 = -AB\sin\theta, \quad h_4 = -\frac{A}{2}\sin\theta(\sin\theta + D\cos^2\theta),$$

$$h_5 = A\sin\theta\left(D\left(\cos^2\theta - \frac{1}{2}\sin^2\theta\right) - \frac{3}{2}\cos^2\theta\right),$$

$$h_6 = A\cos\theta\left(D\left(\frac{1}{2}\cos^2\theta - \sin^2\theta\right) + \frac{3}{2}\sin^2\theta\right),$$

$$h_7 = \frac{AD}{2}\cos\theta, \qquad h_8 = -\frac{AD}{2}\sin\theta.$$

Defining the new variables $Z = \xi - u_0\tau$ and $\phi^{(1)} = \phi^{(0)}(Z)$ in Eq. 15, the ZK equation in a steady state can be written as

$$h_2\frac{d^3\phi^{(0)}}{dZ^3} - \left(u_0 - h_1\phi^{(0)}\right)\frac{d\phi^{(0)}}{dZ} = 0 \tag{16}$$

in which u_0 is the constant velocity normalized to the dust acoustic speed. Using the appropriate boundary conditions ($\phi^{(0)}$ and its two successive derivatives tend to zero when $Z \to \infty$) the solution of the Eq. 16 is given by

$$\phi^{(0)}(Z) = \phi_m^{(0)}\operatorname{sech}^2\left(\frac{Z}{W}\right) \tag{17}$$

where $\phi_m^{(0)}$ and W are the amplitude and width of the DAWs, respectively, given by

$$\phi_m^{(0)} = \frac{3u_0}{AB\cos\theta} \tag{18a}$$

$$W = \left(\frac{2A\cos\theta(\cos^2\theta + D\sin^2\theta)}{u_0}\right)^{1/2}. \tag{18b}$$

Modified Zakharov–Kuznetsov (MZK) equation

Derivation of the MZK equation

It is obvious that the propagation of compressive and rarefactive solitons depends on the sign of the nonlinear coefficient, AB, of the ZK equation. If we assume that the dispersion coefficient of the ZK equation $AB > 0$ ($AB < 0$), the dust acoustic solitary waves are compressive (rarefactive) waves. When the density of low (high) temperature ions $\delta_1(\delta_2)$ reaches the so-called $\delta_{1c}(\delta_{2c})$ critical density of low- (high-) temperature ions, the nonlinear coefficient of the ZK equation vanishes, i.e., $AB = 0$. Therefore, ZK equation breaks down and one has to seek for another equation suitable for describing the evolution of the system. At the critical density of low- (high-) temperature ions, the general method of the reductive perturbation method introduces the modified stretched variables defined by

$$x' = \epsilon x \tag{19a}$$

$$y' = \epsilon y \tag{19b}$$

$$z' = \epsilon(z - \lambda t) \tag{19c}$$

$$t' = \varepsilon^3 t. \tag{19d}$$

The dependent variables n_d, u_{dz}, and ϕ are expanded the same as Eq. 7 but u_{dy} and u_{dx} are expanded as follows:

$$u_{dx} = \epsilon^2 u_{dx}^{(1)} + \epsilon^3 u_{dx}^{(2)} + \epsilon^4 u_{dx}^{(3)} + \cdots \tag{20a}$$

$$u_{dy} = \epsilon^2 u_{dy}^{(1)} + \epsilon^3 u_{dy}^{(2)} + \epsilon^4 u_{dy}^{(3)} + \cdots. \tag{20b}$$

Using Eqs. 19, 20, and 7 in the main Eq. 2 and collecting terms with the same powers of expanding parameter ϵ we have Eqs. 8 and 9 again, for the lowest order, i.e. $O(\epsilon^2)$. Similarly, to the next higher order of ϵ, i.e. $O(\epsilon^3)$ we obtain the second-order x, y, and z components of the momentum and continuity equation and Poisson equation as

$$u_{\mathrm{dx}}^{(2)} = \frac{-\lambda}{\omega_{\mathrm{cd}}^2}\frac{\partial^2\phi^{(1)}}{\partial z'\partial x'} - \frac{1}{\omega_{\mathrm{cd}}^2}\frac{\partial\phi^{(2)}}{\partial y'} \tag{21a}$$

$$u_{\mathrm{dy}}^{(2)} = \frac{-\lambda}{\omega_{\mathrm{cd}}^2}\frac{\partial^2\phi^{(1)}}{\partial z'\partial y'} + \frac{1}{\omega_{\mathrm{cd}}^2}\frac{\partial\phi^{(2)}}{\partial x'} \tag{21b}$$

$$u_{\mathrm{dz}}^{(2)} = \frac{(\phi^{(1)})^2}{2\lambda^3} - \frac{\phi^{(2)}}{\lambda} \tag{21c}$$

$$n_{\mathrm{d}}^{(2)} = \frac{(\phi^{(1)})^2}{2\lambda^4} - \frac{\phi^{(2)}}{\lambda^2} \tag{21d}$$

$$\left(\frac{\partial^2}{\partial x'^2}+\frac{\partial^2}{\partial y'^2}+\frac{\partial^2}{\partial z'^2}\right)\phi^{(1)} = n_{\mathrm{d}}^3 + \frac{\phi^{(3)}}{\lambda^2} + \frac{(\delta_1+\delta^2-1)^2}{(\delta_1+\delta^2\beta+\beta_1)^3}\frac{(1+15\alpha)(\delta_1+\delta_2\beta^3)+(1+3\alpha)\beta_1^3}{6(1+3\alpha)} \times (\phi^{(1)})^3 + \frac{(\delta_1+\delta^2-1)}{(\delta_1+\delta_2\beta+\beta_1)^2}(\beta_1^2-\delta_1-\delta_2\beta^2)\phi^{(1)}\phi^{(2)}. \tag{21e}$$

Following the same procedure, we can obtain the next higher-order continuity equation and z component of momentum equation as, i.e. $O(\epsilon^4)$.

$$\frac{\partial n_{\mathrm{d}}^{(1)}}{\partial t'} - \lambda\frac{\partial n_{\mathrm{d}}^{(3)}}{\partial z'} + \frac{\partial}{\partial x'}(u_{\mathrm{dx}}^{(2)} + n_{\mathrm{d}}^{(1)}u_{\mathrm{dx}}^{(1)}) + \frac{\partial}{\partial y'}(u_{\mathrm{dy}}^{(2)} + n_{\mathrm{d}}^{(1)}u_{\mathrm{dy}}^{(1)}) + \frac{\partial}{\partial z'}(u_{\mathrm{dz}}^{(3)} + n_{\mathrm{d}}^{(1)}u_{\mathrm{dz}}^{(2)} + n_{\mathrm{d}}^{(2)}u_{\mathrm{dz}}^{(1)}) = 0 \tag{22a}$$

$$\frac{\partial u_{\mathrm{dz}}^{(1)}}{\partial t'} - \lambda\frac{\partial u_{\mathrm{dz}}^{(3)}}{\partial z'} + u_{\mathrm{dx}}^{(1)}\frac{\partial u_{\mathrm{dz}}^{(1)}}{\partial x'} + u_{\mathrm{dy}}^{(1)}\frac{\partial u_{\mathrm{dz}}^{(1)}}{\partial y'} + u_{\mathrm{dz}}^{(1)}\frac{\partial u_{\mathrm{dz}}^{(2)}}{\partial z'} + u_{\mathrm{dz}}^{(2)}\frac{\partial u_{\mathrm{dz}}^{(1)}}{\partial z'} - \frac{\partial\phi^{(3)}}{\partial z'} = 0. \tag{22b}$$

Solving the system of Eq. 22, using Eqs. 8, 9, and 21, we finally obtain the modified Zakharov–Kuznetsov (MZK) equation:

$$\frac{\partial\phi^{(1)}}{\partial t'} + AE\left(\phi^{(1)}\right)^2\frac{\partial\phi^{(1)}}{\partial z'} + \frac{A}{2}\frac{\partial}{\partial z'}\left(\frac{\partial^2}{\partial z'^2} + D\left(\frac{\partial^2}{\partial x'^2}+\frac{\partial^2}{\partial y'^2}\right)\right) \times \phi^{(1)} = 0 \tag{23}$$

A and D were introduced in the previous section and E is

$$E = \frac{15}{4\lambda^6} - \frac{1}{4}\frac{(\delta_1+\delta_2-1)^2((1+15\alpha)(\delta_1+\delta_2\beta^3)+(1+3\alpha)\beta_1^3)}{(\delta_1+\delta_2\beta+\beta_1)^3(1+3\alpha)} \tag{24}$$

Solitary wave solution of the MZK equation

To study the propagation of DASW in a direction making an angle θ with the external magnetic field **B**, lying in the $x'-z'$ plane, we transform the coordinate system x', y', z' into the new coordinate system ζ, ξ, η by a rotation around the y' axis through an angle θ. The relations between the new and old coordinates were introduced in Eq. 14. In this case, the new form of Eq. 23 will be

$$\frac{\partial\phi^{(1)}}{\partial\tau} + \Omega_1\left(\phi^{(1)}\right)^2\frac{\partial\phi^{(1)}}{\partial\xi} + \Omega_2\frac{\partial^3\phi^{(1)}}{\partial\xi^3} + \Omega_3\left(\phi^{(1)}\right)^2\frac{\partial\phi^{(1)}}{\partial\zeta} + \Omega_4\frac{\partial^3\phi^{(1)}}{\partial\zeta^3} + \Omega_5\frac{\partial^3\phi^{(1)}}{\partial\xi^2\partial\zeta} + \Omega_6\frac{\partial^3\phi^{(1)}}{\partial\xi\partial\zeta^2} + \Omega_7\frac{\partial^3\phi^{(1)}}{\partial\xi\partial\eta^2} + \Omega_8\frac{\partial^3\phi^{(1)}}{\partial\zeta\partial\eta^2} = 0. \tag{25}$$

where

$$\Omega_1 = AE\cos\theta, \quad \Omega_2 = \frac{1}{2}A\cos\theta(\cos^2\theta + D\sin^2\theta),$$

$$\Omega_3 = -AE\sin\theta \quad \Omega_4 = -\frac{1}{2}A\sin\theta(\sin^2\theta + D\cos^2\theta),$$

$$\Omega_5 = A\sin\theta\left(D\left(\cos^2\theta - \frac{1}{2}\sin^2\theta\right) - \frac{3}{2}\cos^2\theta\right)$$

$$\Omega_6 = A\cos\theta\left(D\left(\frac{1}{2}\cos^2\theta - \sin^2\theta\right) + \frac{3}{2}\sin^2\theta\right),$$

$$\Omega_7 = \frac{1}{2}AD\cos\theta, \qquad \Omega_8 = \frac{-1}{2}AD\sin\theta.$$

Using again the new variables $Z = \xi - u_0\tau$ and $\phi^{(1)} = \phi^{(0)}(Z)$ in Eq. 25, the MZK equation in a steady-state forms as

$$\Omega_2\frac{d^3\phi^{(0)}}{dZ^3} - \left(u_0 - \Omega_1\left(\phi^{(0)}\right)^2\right)\frac{d\phi^{(0)}}{dZ} = 0 \tag{26}$$

in which u_0 is constant velocity normalized to the dust acoustic speed. Now, using the appropriate boundary conditions ($\phi^{(0)}$ and its two successive derivatives tend to zero when $Z \to \infty$) the solution of the Eq. 26 is given by

$$\phi^{(0)}(Z) = \phi_m^{(0)}\operatorname{sech}\left(\frac{Z}{W}\right) \tag{27}$$

where $\phi_m^{(0)}$ and W are the amplitude and width of the DASWs, respectively, given by

$$\phi_m^{(0)} = \left(\frac{6u_0}{AE\cos\theta}\right)^{1/2} \tag{28a}$$

$$W = \left(\frac{A\cos\theta(\cos^2\theta + D\sin^2\theta)}{2u_0}\right)^{1/2} \tag{28b}$$

Solitons exist when $E > 0$.

Results and discussion

The nonlinear propagation of dust acoustic solitary waves (DASW) in a magnetized dusty plasma which consists of two different types of nonthermal ions, electrons, and mobile negatively charged dust particles is studied. The Zakharov–Kuznetsov (ZK) and modified Zakharov–Kuznetsov (MZK) equations, describing the small but finite amplitude DASW, are derived using a reductive perturbation method. The combined effects of the external magnetic field, obliqueness (i.e., the propagation angle), and the presence of second component of nonthermal ions, which are found to significantly modify the basic properties of DASWs, are explicitly examined. For $\alpha = 0$, and by assuming collisionless dusty plasma systems (such as outside ionopause of Halley's comet, Saturn's F-ring, Saturn's spokes, zodiacal dust disc (IAU) and supernovae shells), our results would coincide with the results obtained by Moslem [20]. It is obvious that if we neglect the contributions of external magnetic field, $\omega_{cd} = 0$, our present dusty plasma model corresponds to the dusty plasma system considered in a recent published work by Dorranian et al. [9]. As $u_0 > 0$, it is clear from Eqs. 13 and 18 that depending on whether AB is positive or negative, the solitary waves will be associated with either positive potential $\phi_m^{(0)} > 0$ or negative potential $\phi_m^{(0)} < 0$. Therefore, there exist solitary waves associated with positive (negative) potential when $AB > 0$ ($AB < 0$).

Figure 1 shows the variation of the nonlinear coefficient AB, with δ_1 and δ_2 for $u_0 = 1$, $\alpha = 0.4$, $\beta_1 = 0.1$, and $\beta_2 = 0.4$. This figure also indicates the critical value of δ_1 and δ_2 (which are $\delta_{2c} = 0.93$, and $\delta_{1c} = 0.57$). For the region of positive nonlinear coefficient AB, the nonlinear coefficient AB decreases as the values of both δ_1 and δ_2 decrease. AB is effectively changed with δ_1 and δ_2 when they are smaller than 5 and for larger magnitudes of δ_1 and δ_2, the nonlinear coefficient is almost constant. It is observed from Eqs. 13 and 18 that the amplitude of DASW $\phi_m^{(0)}$ is a nonlinear function of δ_1, δ_2, β_1, β, θ, and α. The variation of $\phi_m^{(0)}$ (for positive and negative potential) with

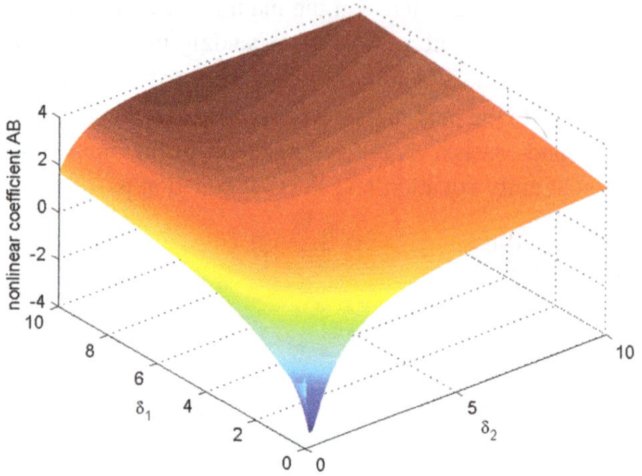

Fig. 1 The variation of the nonlinear coefficient AB with δ_1 and δ_2 for $u_0 = 1, \alpha = 0.4, \beta_1 = 0.1$ and $\beta_2 = 0.4$

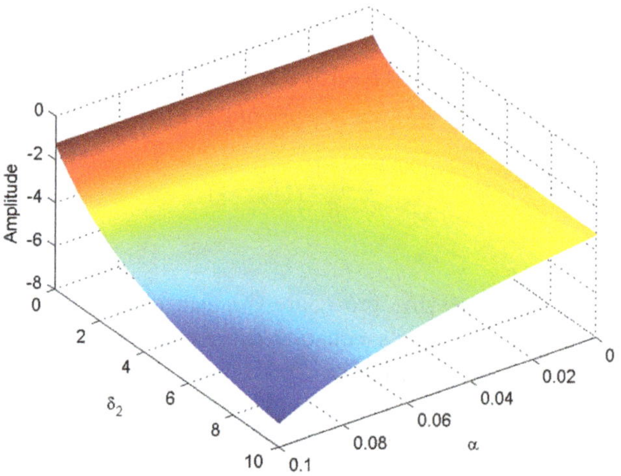

Fig. 2 The variation of the amplitude of the negative ZK solitary potential $\left(\phi_m^{(0)} < 0\right)$ with δ_2 and α for $u_0 = 1, \delta_1 = 0.1, \beta_1 = 0.1$, $\beta_2 = 0.4$, and $\theta = 30°$

δ_2, β_2, θ, and α are shown in Figs. 2–7. Figure 2(3) shows the variation of the amplitude of the negative solitary waves (positive solitary waves) with δ_2 and α for $u_0 = 1, \delta_1 = 0.1$, $\beta_1 = 0.1$, $\beta_2 = 0.4$, and $\theta = 30^o$ (for $u_0 = 1$, $\delta_1 = 1.9$, $\beta_1 = 0.1$, $\beta_2 = 0.4$, and $\theta = 30°$). Figure 2 shows that the amplitude increases as the values of both δ_2 and α increase. Figure 3 shows that the amplitude increases as the values of both δ_2 and α decrease. With decreasing the nonthermal coefficient α, the amplitude of soliton increases, while with decreasing δ_2 amplitude increases very slightly. In other words, the influence of nonthermal coefficient on the amplitude is more larger. This effect was observed in Ref. [9].

Figure 4(5) shows the variation of the amplitude of the negative solitary waves (positive solitary waves) with β_2 and θ for $u_0 = 1$, $\alpha = 0.1$, $\beta_1 = 0.1$, $\delta_1 = 0.1$, and $\delta_2 = 2.1$

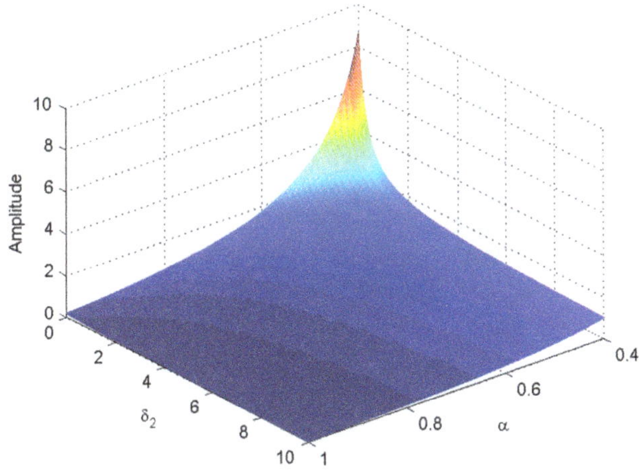

Fig. 3 The variation of the amplitude of the positive ZK solitary potential $\left(\phi_m^{(0)} > 0\right)$ with δ_2 and α for $u_0 = 1, \delta_1 = 1.9, \beta_1 = 0.1, \beta_2 = 0.4$, and $\theta = 30°$

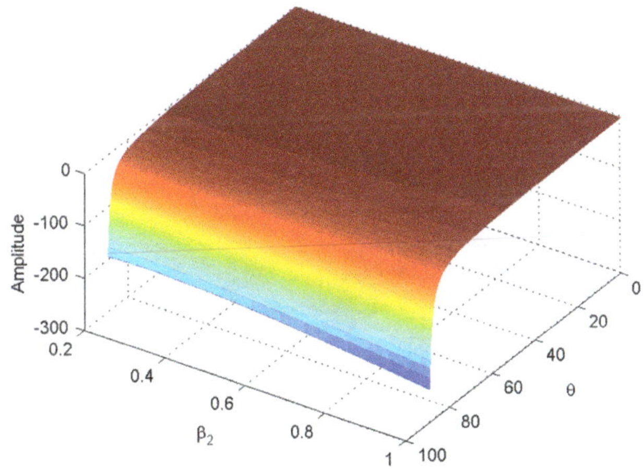

Fig. 4 The variation of the amplitude of the negative ZK solitary potential $\left(\phi_m^{(0)} < 0\right)$ with β_2 and θ for $u_0 = 1, \delta_1 = 0.1, \delta_2 = 2.1, \beta_1 = 0.1$, and $\alpha = 0.1$

(for $u_0 = 1$, $\alpha = 0.5, \beta_1 = 0.1, \delta_1 = 1.2$, and $\delta_2 = 2.1$). In Fig. 4 the rate of the increase is low (high) in the region about $48° < \theta < 70°$ $(70° < \theta < 89°)$ for all of β_2. In Fig. 5 the rate of the increase is low (high) in the region at about $50° < \theta < 77°$ $(77° < \theta < 89°)$ for all β_2. Amplitude of solitons are directly proportional with the angle of propagation, but in the region $0° < \theta < 50°$, the soliton amplitude changes very slightly both ZK and MZK solitons. For the propagation angle almost greater than 50°, variation of amplitude is noticeable in Figs. 4 and 5.

By considering $u_0 > 0$, it is clear from Eqs. 24 and 28 that MZK solitons exist when $E > 0$. Therefore, there always exist solitary waves associated with positive potential. In Fig. 6(7), the variation of the amplitude of the positive solitary waves with δ_2 and α is presented for $u_0 = 1$, $\delta_1 = 0.4$, $\beta_1 = 0.1$, $\beta_2 = 0.4$, and $\theta = 30°$ (with β_2

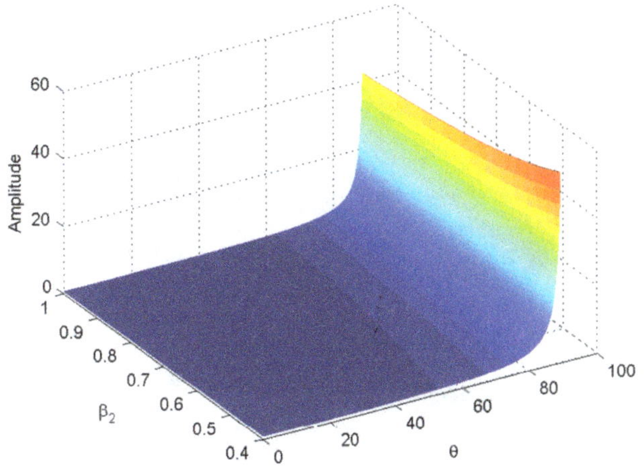

Fig. 5 The variation of the amplitude of the positive ZK solitary potential $\left(\phi_m^{(0)} > 0\right)$ with β_2 and θ for $u_0 = 1, \delta_1 = 1.2, \delta_2 = 2.1, \beta_1 = 0.1$, and $\alpha = 0.5$

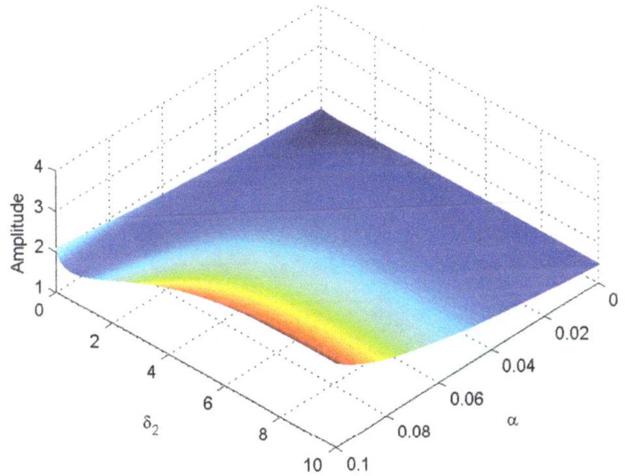

Fig. 6 The variation of the amplitude of the positive MZK solitary potential $\left(\phi_m^{(0)} > 0\right)$ with δ_2 and α for $u_0 = 1, \delta_1 = 0.4, \beta_1 = 0.1, \beta_2 = 0.4$, and $\theta = 30°$

and θ for $u_0 = 1$, $\delta_1 = 1.1$, $\beta_1 = 0.1$, $\alpha = 0.1$ and $\delta_2 = 2.1$). Figure 6 shows that the amplitude decreases for the lower regions of δ_2 (for about $\delta_2 < 0.76$), its magnitude increases rapidly with increasing the value of δ_2 (for the region $\delta_2 > 0.76$) to 3.05 and this figure also indicates that the amplitude increases with increasing α. Figure 7 shows that the amplitude increases slowly as the value of β_2 increases. The rate of increase is very high in the region at about $84° < \theta < 89°$.

The magnitude of the external magnetic field has a significant effect only on the width, but not on the amplitude of solitary waves. It is found from Eqs. 18 and 28 that the width is a nonlinear function of $\delta_1, \delta_2, \beta_2, \beta, \theta, \alpha$, and ω_{cd}. The variation of the width (for positive and negative

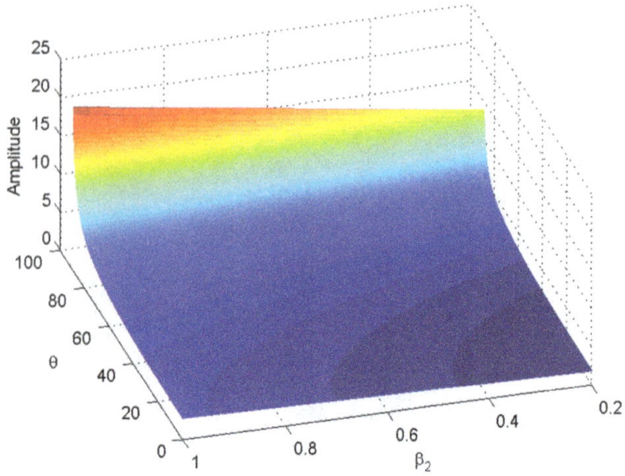

Fig. 7 The variation of the amplitude of the positive MZK solitary potential $\left(\phi_m^{(0)} > 0\right)$ with β_2 and θ for $u_0 = 1, \delta_1 = 1.1, \beta_1 = 0.1, \delta_2 = 2.1$, and $\alpha = 0.1$

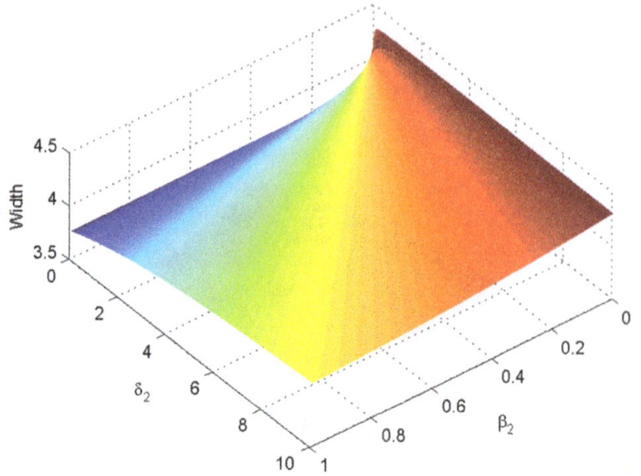

Fig. 9 The variation of the width of positive ZK solitary potential $\left(\phi_m^{(0)} > 0\right)$ with δ_2 and β_2 for $u_0 = 1, \delta_1 = 0.4, \beta_1 = 0.1, \alpha = 0.2, \theta = 30^o$, and $\omega_{cd} = 0.3$

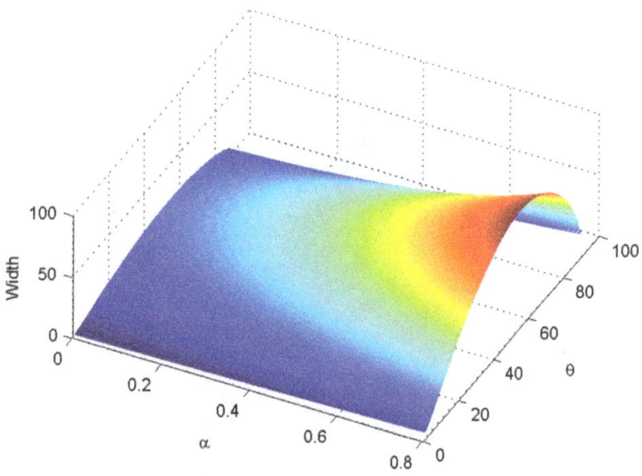

Fig. 8 The variation of the width of positive ZK solitary potential $\left(\phi_m^{(0)} > 0\right)$ with θ and α for $u_0 = 1, \delta_1 = 0.6, \delta_2 = 2.1, \beta_1 = 0.1, \beta_2 = 0.4$, and $\omega_{cd} = 0.6$

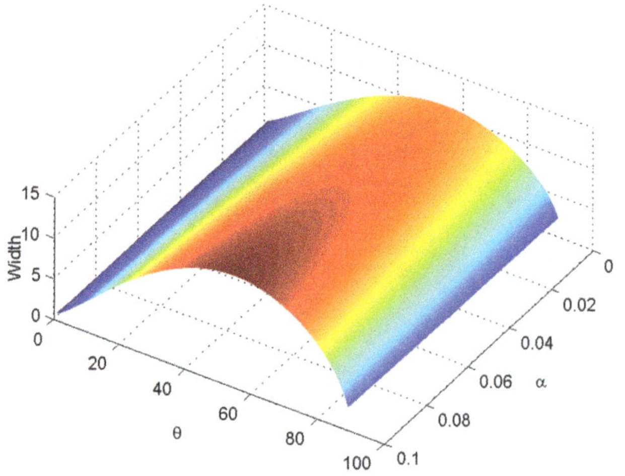

Fig. 10 The variation of the width of negative MZK solitary potential $\left(\phi_m^{(0)} < 0\right)$ with θ and α for $u_0 = 1, \delta_1 = 0.6, \delta_2 = 2.1, \beta_1 = 0.1, \beta_2 = 0.4$, and $\omega_{cd} = 0.6$

solitary waves) with δ_2, β_2, θ, and ω_{cd} is presented in Figs. 8, 9, 10, 11 and Fig. 12. Dependence of soliton width on θ and α (δ_2 and β_2) is shown in Fig. 8(9). In Fig. 8(9) we have $u_0 = 1, \delta_1 = 0.6, \beta_1 = 0.1, \beta_2 = 0.4, \delta_2 = 2.1$, and $\omega_{cd} = 0.6$ $(u_0 = 1, \delta_1 = 0.4, \beta_1 = 0.1, \alpha = 0.2, \theta = 30°$, and $\omega_{cd} = 0.3)$. Figure 8 shows that the width increases with θ for the lower range, i.e. $0 < \theta < 53.6°$, but decreases for its higher range, i.e. $53.6° < \theta < 89°$, and as $\theta \to 90°$, the width goes to 0. The maximum of width is at $\theta = 53.6°$. This also shows that the width increases with increasing α. The width increases slowly with increasing (decreasing) δ_2 (β_2) as shown in Fig. 9.

Figure 10(11) shows variation of soliton width with θ and α (δ_2 and β_2) for $u_0 = 1, \delta_1 = 0.6, \beta_1 = 0.1, \beta_2 = 0.4, \delta_2 = 2.1$, and $\omega_{cd} = 0.6$ (for $u_0 = 1, \delta_1 = 0.6, \beta_1 = 0.1, \alpha = 0.1, \theta = 30°$, and $\omega_{cd} = 0.4$). Figure 10 shows that the width increases with θ for the lower range, i.e., $0° < \theta < 55.11°$, but decreases for its higher range, i.e., $55.11° < \theta < 89°$ and as θ tends to 90°, the width goes to 0. This also shows that the width increases with increasing α. Figure 11 shows that the width remains almost unchanged for all value of β_2, but increases rapidly with increasing δ_2. Figure 12 shows the variation of width with θ and ω_{cd}. In this figure we have $u_0 = 1, \delta_1 = 0.1, \beta_1 = 0.1, \beta_2 = 0.4, \delta_2 = 2.1$, and $\alpha = 0.5$. The width of soliton for the case of MZK solution is half of the ZK solitons. Amplitude of soliton is independent of magnetic field intensity while is influenced by the direction of applied external magnetic

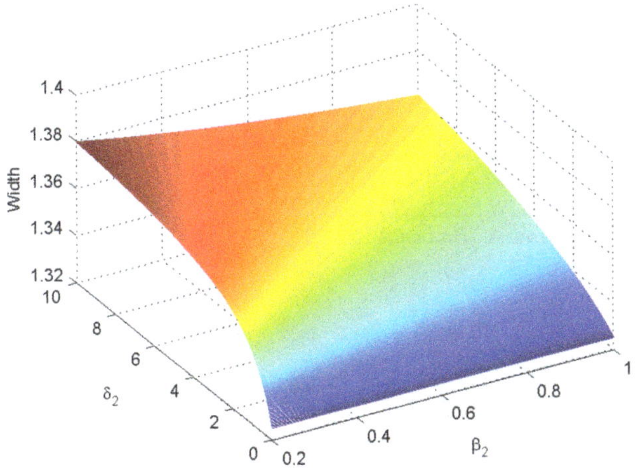

Fig. 11 The variation of the width of negative MZK solitary potential $\left(\phi_m^{(0)}<0\right)$ with δ_2 and β_2 for $u_0=1, \delta_1=0.6, \beta_1=0.1, \alpha=0.1$, $\theta=30°$, and $\omega_{cd}=0.4$

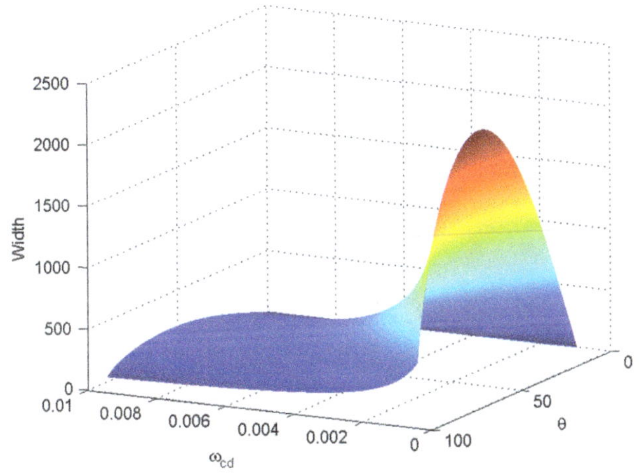

Fig. 12 The variation of the width of positive ZK solitary potential $\left(\phi_m^{(0)}>0\right)$ with ω_{cd} and θ for $u_0=1, \delta_1=0.1, \delta_2=2.1, \beta_1=0.1$, $\beta_2=0.4$, and $\alpha=0.5$

field. The effective variation of ω_{cd} on the width of soliton is when $\omega_{cd} \leq 0.01$. In the case of larger magnitude, width of solitons does not change with ω_{cd} noticeably, and the same results have been reported by Sabetkar et al. [21]. Figure 12 shows that the width increases with θ for the lower range, i.e., $0° < \theta < 56.5°$, but decreases for its higher range, i.e., $56.5°, < \theta < 89°$ and as θ tends to 90°, the width tends to a constant value. Also, with decreasing ω_{cd}, the width of soliton increases.

Conclusion

In present paper, nonlinear propagating dust acoustic solitary waves (DASWs) in a cold magnetized dusty plasma containing negatively charged dust particles, electrons, high- and low-temperature nonthermal ions in the presence of an external static magnetic field are investigated. For this purpose, a reasonable normalization of the hydrodynamic and Poisson equations is used to derive the Zakharov–Kuznetsov (ZK) and (MZK) equation for our dusty plasma system. We have found that depending on the values of $\delta_1, \delta_2, \beta_1, \beta_2$, and α, the solitary waves may become associated with either positive potential or negative potential. We have seen that as the values of δ_2 and α increase, the amplitude of the positive (negative) solitary waves decreases (increases). The width of soliton also increases with the increasing δ_2 and α. For higher-order approximation, i.e. MZK case, we have found that the amplitude of the positive solitary waves increases with increasing α. The amplitude of solitary waves decreases with δ_2 when $\delta_2 <$ 0.76 and increases rapidly for $\delta_2 > 0.76$.

We have found that as the angle between the direction of external magnetic field with the propagation direction of solitary wave, θ, increases, the amplitude of the solitary wave (for both positive potential and negative potential) increases. The width of solitary wave has a maximum magnitude when $\theta = 55.11°$. As θ tends to 90°, the width goes to 0, and the amplitude goes to infinity. It is found that (for both positive potential and negative potential) the amplitude increases slowly as the value of β_2 increases, but the width increases slowly. Taking into account the higher order of ϵ (i.e. MZK equation) the width remains almost unchanged for all values of β_2. The magnitude of the external magnetic field has no direct effect on the solitary wave amplitude. However, it has a direct effect on the width of the solitary waves, and also with decreasing ω_{cd}, the width of DASW increases. In fact, the external magnetic field makes the solitary structures more spiky.

References

1. Verheest, F.: Waves in dusty space plasmas. Kluwer Academic Publishers, (2000)
2. Shukla, P.K., Mamun, A.A.: Introduction to dusty plasma physics. Institute of Physics, Bristol (2002)
3. Rao, N.N., Shukla, P.K., Yu, M.Y.: Planet. Space. Sci. **38**, 543 (1990)
4. Barkan, A., Merlino, R.L., N.D., Angelo. Phys. Plasmas. **2**, 3563 (1995)
5. Dodd, R.K., Eilbeck, J.C., Gobbon, J.D., Morins, H.C.: Solitons and nonlinear waves equations. Academic, London (1982)
6. Zakharov, V.E., Kuznetsov, E.A.: Sov. Phys. JETP. **39**, 285 (1974)
7. Washimi, H., Taniuti, T.: Phys. Rev. Lett. **17**, 996 (1996)
8. Lin, M.-M., Duan, W.-S.: Chaos. Solitons fractals **33**, 1189 (2007)

9. Dorranian, D., Sabetkar, A.: Phys. Plasmas. **19**, 013702 (2012)
10. Zhang, K., Wang, H.: J. Korean. Phys. Soc. **55**, 1461 (2009)
11. Mendoza-Briceno, C.A., Russel, S.M., Mamun, A.A.: Planet. Space. Sci. **48**, 599 (2000)
12. Gill, T.S., Kaur, H., Saini, N.S.: J. Plasma. Phys. **70**, 481 (2004)
13. Sayed, F., Mamun, A.A.: Phys. Plasmas. **14**, 014502 (2007)
14. Mace, R.L., Hellberg, M.A.: Phys. Plasmas. **8**, 2649 (2001)
15. Xue, J.K., Lang, H.E.: Chin. Phys. **13**, 0060 (2004)
16. Mushtaq, A., Shah, H.A.: Phys. Plasmas. **12**, 072306 (2005)
17. Mahmood, S., Akhtar, N., Ur-Rehman, H.: Phys. Scr. **83**, 035505 (2011)
18. Shahmoradi, N., Dorranian, D.: Phys. Scr. **89**, 065602 (2014)
19. Cairns, R.A., Mamun, A.A., Bingham, R., Dendy, R.O., Bostrum, R., Nairn, C.M.C., Shukla, P.K.: Geophys. Res. Lett. **22**, 2709 (1995). doi:10.1029/95GL02781
20. Moslem, W.M.: Chaos. Solitons and fractals **23**, 939 (2005)
21. Sabetkar, A., Dorranian, D.: Phys. Scr. **90**, 035603 (2015)

16

Korteweg–deVries–Burgers (KdVB) equation in a five component cometary plasma with kappa described electrons and ions

Manesh Michael[1] · Neethu T. Willington[2] · Neethu Jayakumar[1] · Sijo Sebastian[1] · G. Sreekala[1] · Chandu Venugopal[1]

Abstract We investigate the existence of ion-acoustic shock waves in a five component cometary plasma consisting of positively and negatively charged oxygen ions, kappa described hydrogen ions, hot solar electrons, and slightly colder cometary electrons. The KdVB equation has been derived for the system, and its solution plotted for different kappa values, oxygen ion densities, as well as the temperature ratios for the ions. It is found that the amplitude of the shock wave decreases with increasing kappa values. The strength of the shock profile decreases with increasing temperatures of the positively charged oxygen ions and densities of negatively charged oxygen ions.

Keywords Cometary plasma · Ion-acoustic solitary wave · Shock wave · KdVB equation

Introduction

Propagation of nonlinear waves in astrophysical plasmas has generated a lot of interest in the plasma community: a large number of investigations are on-going on nonlinear waves, such as solitons, shocks, double layers, and so on, which are observed in space, astrophysical, and laboratory plasmas.

✉ Chandu Venugopal
cvgmgphys@yahoo.co.in

1 School of Pure and Applied Physics, Mahatma Gandhi University, Priyadarshini Hills, Kottayam 686 560, Kerala, India

2 Department of Physics, C. M. S. College, Kottayam 686001, Kerala, India

Most of these studies have focussed on the nonlinearity of the ion acoustic wave, which is an important wave in plasma. The first nonlinear analysis of the ion acoustic wave was by Sagdeev [1]; its first experimental observation was by Ikezi et al. [2]. Outside of the laboratory, the observations by Viking and Freja spacecrafts have identified solitary structures in the magnetosphere as density depressions.

A nonlinear and dispersive medium always supports a solitary wave. In a soliton structure, the nonlinearity is balanced by the dispersion. The Korteweg–deVries (KdV) equation, having negative nonlinear and positive dispersive effects, will give rise to rarefactive solitons. When both nonlinearity and dispersive effects are positive, the KdV equation may result in a compressive soliton. Such compressive solitons have been studied, for example, in a moving electron–positron plasma containing positively and negatively charged dust [3]. However, a medium having significant dissipative effect and dispersion supports the formation of shock waves instead of solitons which is best described by the Korteweg–deVries–Burgers (KdVB) equation. The dissipative Burger term in the nonlinear KdVB equation arises due to dissipative mechanisms, such as wave–particle interactions, turbulence, dust charge fluctuations in a dusty plasma, multi-ion streaming, Landau damping, anomalous viscosity, etc. [4–6]. When wave breaking due to nonlinearity is balanced by the combined effect of dispersion and dissipation, a monotonic or oscillatory dispersive shock wave is generated in a plasma [7]. If the dissipative effect is negligible the KdVB equation transforms into the KdV equation which permits a soliton solution. The KdVB equation has thus been extensively used to study the properties of solitons and shock waves in dusty plasmas [8–12].

The presence of high energy particles in the tail of a distribution causes plasmas observed in different space environments to deviate significantly from the well-known Maxwellian distribution. Using solar wind data, Vasyliunas first predicted a non-Maxwellian distribution [13]; this distribution, which later came to be known as the "kappa distribution," has been found in many magnetospheric and astrophysical environments. Ion-acoustic shock waves in plasmas consisting of superthermal electrons and positrons have been studied recently by many authors [14–20]; the effects of electrons and positrons on ion acoustic solitary waves have also been studied [21].

A cometary plasma is composed of hydrogen, and new born heavier ions and electrons with relative densities depending on their distances from the nucleus. Initially, positively charged oxygen ions were treated as the main heavier ion [22, 23]. However, the discovery of negatively charged oxygen ions [24] enables one to consider the plasma environment around a comet as a pair-ion plasma (O^+, O^-) with other ions (both lighter and heavier) constituting the other components of the plasma.

Thus, a cometary plasma is a true multi-ion plasma consisting of both lighter and heavier ions and electrons with different temperatures. We thus model our plasma as consisting of a pair-ion plasma of oxygen ions, lighter hydrogen ions, and two components of electrons with different temperatures. The lighter hydrogen ions and electrons are modeled by kappa distributions.

A very complex structure of multiple sub-shocks and interplanetary structures of comet Halley was reported by Giotto. In addition, unambiguous observations of bow shock crossings were provided by Vega-1 [25]. These shock structures are seriously affected by heavier ions due to mass loading of the solar wind and pickup driven ion instabilities [25]. And these heavy ions can be both positively and negatively charged as discussed above. In addition, in a recent study, Voelzke and Izaguirre, analyzing 886 images of comet Halley, identified 41 solitary structures [26].

The stability of ion acoustic waves in a four component dusty plasma has been investigated recently by several authors. Early studies on nonlinear wave propagation in four component dusty plasmas were by Sakanaka and Spassovska [27] and Verheest [28]. Various aspects of dust acoustic waves have been considered in the studies by several authors [29–36]. Other nonlinear electrostatic waves have been considered by Ghosh et al. [37, 38] and Dutta et al. [39].

For reasons given at the middle of this section, a reasonably accurate modeling of a cometary plasma requires at least five components. We, therefore, investigate the existence of ion-acoustic shock waves in a five component cometary plasma consisting of positively and negatively charged oxygen ions, kappa described hydrogen ions, hot solar electrons and slightly colder cometary electrons. Related studies of the effects of two types of electrons on ion-acoustic solitary waves are those by Khaled [40] and Shahmansouri et al. [41]. Similarly, the dust acoustic shock wave has also been studied in a dusty plasma with kappa distributed ions [11]. The KdVB equation is derived for this system. It is found that the amplitude of shock wave decreases with increasing kappa values. Its strength decreases with increasing temperatures of positively charged oxygen ions and densities of negatively charged oxygen ions.

Basic equations

We consider the existence of ion-acoustic shock waves in a five component plasma consisting of negatively and positively charged oxygen ions (represented, respectively, by subscripts '1' and '2'), kappa described hydrogen ions, hot electrons of solar origin and colder electrons of cometary origin. At equilibrium, charge neutrality requires that

$$n_{ce0} + n_{se0} + Z_1 n_{10} = n_{H0} + Z_2 n_{20}.$$

In the above relation, n_{ce0} and n_{he0} represent the equilibrium densities of cometary electrons and solar electrons, respectively. In addition, n_{10}, n_{20}, and n_{H0} are, respectively, the equilibrium densities of negatively charged oxygen (O^-) ions, positively charged oxygen (O^+) ions, and hydrogen ions. Z_1 and Z_2 denote the charge numbers of O^- and O^+ ions, respectively.

The kappa distribution of species 's' is given by:

$$n_s = n_{s0}\left[1 + \frac{e_s\varphi}{k_B T_s(\kappa_s - 3/2)}\right]^{-\kappa_s+1/2}. \tag{1}$$

In (1), s = H for hydrogen, = se for solar electrons, and = ce for cometary photo-electrons.n_s denotes the density (with the subscript '0' denoting the equilibrium value), e_s is the charge, T_s is the temperature, κ_s is the spectral index for species 's', k_B is Boltzmann's constant, and φ is the potential.

The dynamics of the heavier ions can be described by the following hydrodynamic equations:

$$\frac{\partial n_j}{\partial t} + \frac{\partial(n_j v_j)}{\partial x} = 0, \tag{2}$$

$$\left(\frac{\partial}{\partial t} + v_j\frac{\partial}{\partial x}\right)v_j = \mp\frac{Z_j e}{m_j}\frac{\partial\varphi}{\partial x} - \frac{1}{m_j n_j}\frac{\partial P_j}{\partial x} + \mu_j\frac{\partial^2 v_j}{\partial x^2}, \tag{3}$$

where '−' sign refers to positively charged oxygen ions (and vice versa) and v_j and m_j, respectively, denote the fluid velocity and mass of the j-species of ions ($j = O^-$, O^+). In (3), the adiabatic equation of state for ions is

$\frac{P_j}{P_{j0}} = \left(\frac{n_j}{n_{j0}}\right)^{\gamma} = N_j^{\gamma}$, where $P_{j0} = n_{j0}k_B T_j$, $\gamma = (N+2)/N$ for an N dimensional system, and μ_j is the ion kinematic viscosity. Here, we are considering a one-dimensional system, and hence, $\gamma = 3$.

Poisson's equation is given by

$$\frac{\partial^2 \varphi}{\partial x^2} = -4\pi e(n_H + Z_2 n_2 - Z_1 n_1 - n_{ce} - n_{se}). \quad (4)$$

We normalize (2)–(4) using the parameters of O^- ions according to $\varphi = \frac{e\phi}{k_B T_1}$, $V_j = \frac{v_j}{c_s}$, and $\tau = \frac{t}{\omega_{p1}^{-1}}$, where $c_s = \left(\frac{Z_1 k_B T_1}{m_1}\right)^{1/2}$ and $\omega_{p1} = \left(\frac{4\pi Z_1^2 e^2 n_{10}}{m_1}\right)^{1/2}$. The variable x is normalized using $\lambda_{D1} = \left(\frac{Z_1 k_B T_1}{4\pi Z_1^2 e^2 n_{10}}\right)^{1/2}$, while $N_j = \frac{n_j}{n_{j0}}$.

Thus, Eqs. (2)–(4) can be rewritten as:

$$\frac{\partial N_1}{\partial \tau} + \frac{\partial (N_1 V_1)}{\partial x} = 0, \quad (5)$$

$$\frac{\partial N_2}{\partial \tau} + \frac{\partial (N_2 V_2)}{\partial x} = 0, \quad (6)$$

$$\frac{\partial V_1}{\partial \tau} + V_1 \frac{\partial V_1}{\partial x} = \frac{\partial \phi}{\partial x} - \frac{3N_1}{Z_1}\frac{\partial N_1}{\partial x} + \rho_1 \frac{\partial^2 V_1}{\partial x^2}, \quad (7)$$

$$\frac{\partial V_2}{\partial \tau} + V_2 \frac{\partial V_2}{\partial x} = \frac{-Z_2 m}{Z_1}\frac{\partial \phi}{\partial x} - \frac{3m\beta N_2}{Z_1}\frac{\partial N_2}{\partial x} + \rho_2 \frac{\partial^2 V_2}{\partial x^2}, \quad (8)$$

where $m = \frac{m_1}{m_2}$, $\beta = \frac{T_2}{T_1}$, $\rho_1 = \frac{\mu_1}{\omega_{p1}\lambda_D^2}$, and $\rho_2 = \frac{\mu_2}{\omega_{p1}\lambda_D^2}$.

ρ_1 and ρ_2 now represent the normalized kinematic viscosities of the pair ions.

The normalized Poisson's equation after substitution of (1) is:

$$\begin{aligned}\frac{\partial^2 \phi}{\partial x^2} &= N_1 - N_2(1 + \mu_{ce} + \mu_{se} - \mu_H) \\ &+ \mu_{ce}\left(1 - \frac{\phi}{\sigma_{ce}(\kappa_{ce} - 3/2)}\right)^{-(\kappa_{ce}-1/2)} \\ &+ \mu_{se}\left(1 - \frac{\phi}{\sigma_{se}(\kappa_{se} - 3/2)}\right)^{-(\kappa_{se}-1/2)} \\ &- \mu_H\left(1 + \frac{\phi}{\sigma_H(\kappa_H - 3/2)}\right)^{-(\kappa_H-1/2)},\end{aligned} \quad (9)$$

where $\mu_{ce} = \frac{n_{ce0}}{Z_1 n_{10}}$, $\mu_{se} = \frac{n_{se0}}{Z_1 n_{10}}$, $\mu_H = \frac{n_{H0}}{Z_1 n_{10}}$, $\sigma_{ce} = \frac{T_{ce}}{T_1}$, $\sigma_{se} = \frac{T_{se}}{T_1}$, and $\sigma_H = \frac{T_H}{T_1}$.

Derivation of KdVB equation

We use the reductive perturbation method to derive the KdVB equation from (5) to (9) by introducing the transformations

$$\xi = \varepsilon^{1/2}(x - \lambda t), \quad \tau = \varepsilon^{3/2} t, \quad \rho_j = \varepsilon^{1/2}\rho_{j0}$$

where ε is a smallness parameter and λ is the wave phase speed.

To apply the reductive perturbation technique, the various parameters are expanded as:

$$N_{1,2} = 1 + \varepsilon N_{1,2}^{(1)} + \varepsilon^2 N_{1,2}^{(2)} + \cdots \quad (10)$$

$$V_{(1,2)} = \varepsilon V_{(1,2)}^{(1)} + \varepsilon^2 V_{(1,2)}^{(2)} + \cdots \quad (11)$$

$$\phi = \varepsilon\phi^{(1)} + \varepsilon^2\phi^{(2)} + \cdots \quad (12)$$

We substitute (10)–(12) in (5)–(9) and equate the coefficients of different powers of ε. From the coefficients of order $\varepsilon^{3/2}$, we get the first-order terms as:

$$N_1^1 = \frac{\phi^1}{\left(\frac{3}{Z_1} - \lambda^2\right)} \quad (13)$$

and

$$N_2^1 = \frac{(Z_2/Z_1)m\phi^1}{\left(\lambda^2 - \frac{3m\beta}{Z_1}\right)}. \quad (14)$$

Expressions for V_1^1 and V_2^1 can be obtained by multiplying (13) and (14) by λ.

In addition, the linear dispersion relation is:

$$\lambda^2 = \frac{S \pm \sqrt{S^2 - 12mZ_1^2 T[3m\beta T + mZ_2(1 + \mu_{ce} + \mu_{se} - \mu_H) + m\beta Z_1]}}{2TZ_1^2}, \quad (15)$$

where

$$S = Z_1^2 + mZ_1 Z_2(1 + \mu_{ce} + \mu_{se} - \mu_H) + 3Z_1 T(1 + m\beta), \quad (16)$$

and

$$T = \frac{\mu_{ce}(\kappa_{ce} - 1/2)}{(\kappa_{ce} - 3/2)\sigma_{ce}} + \frac{\mu_{se}(\kappa_{se} - 1/2)}{(\kappa_{se} - 3/2)\sigma_{se}} + \frac{\mu_H(\kappa_H - 1/2)}{(\kappa_H - 3/2)\sigma_H}. \quad (17)$$

Equating the coefficients of $\varepsilon^{5/2}$ in (5) and (6), we get

$$\frac{\partial N_i^1}{\partial \tau} - \lambda\frac{\partial N_i^2}{\partial \xi} + \frac{\partial V_i^2}{\partial \xi} + \frac{\partial (N_i^1 V_i^1)}{\partial \xi} = 0 \quad i = 1, 2. \quad (18)$$

And equating the coefficient of order $\varepsilon^{5/2}$ in (7) and (8) results in:

$$\frac{\partial V_1^1}{\partial \tau} - \lambda\frac{\partial V_1^2}{\partial \xi} + V_1^1\frac{\partial V_1^1}{\partial \xi} = \frac{\partial \phi^2}{\partial \xi} - \frac{3N_1^1}{Z_1}\frac{\partial N_1^1}{\partial \xi} - \frac{3}{Z_1}\frac{\partial N_1^2}{\partial \xi} + \rho_{10}\frac{\partial^2 V_1^1}{\partial \xi^2}, \quad (19)$$

$$\frac{\partial V_2^1}{\partial \tau} - \lambda\frac{\partial V_2^2}{\partial \xi} + V_2^1\frac{\partial V_2^1}{\partial \xi} = -\frac{Z_2 m}{Z_1}\frac{\partial \phi^2}{\partial \xi} - \frac{3m\beta N_2^1}{Z_1}\frac{\partial N_2^1}{\partial \xi} - \frac{3m\beta}{Z_1}\frac{\partial N_2^2}{\partial \xi} + \rho_{20}\frac{\partial^2 V_2^1}{\partial \xi^2}. \quad (20)$$

Finally, equating the coefficients of terms of order ε^2 from Poisson's equation (9) gives:

$$\frac{\partial^2\phi^1}{\partial\xi^2} = N_1^2 - N_2^2(1+\mu_{ce}+\mu_{se}-\mu_H) + T\phi^2 + \frac{\mu_{ce}}{2}\frac{(\kappa_{ce}^2-1/4)}{(\kappa_{ce}-3/2)^2\sigma_{ce}^2}(\phi^1)^2 + \frac{\mu_{se}}{2}\frac{(\kappa_{se}^2-1/4)}{(\kappa_{se}-3/2)^2\sigma_{se}^2}(\phi^1)^2 - \frac{\mu_H}{2}\frac{(\kappa_H^2-1/4)}{(\kappa_H-3/2)^2\sigma_H^2}(\phi^1)^2. \quad (21)$$

Substituting the values from (13) and (14) into (18) to (21) and eliminating the second-order terms, we obtain the KdVB equation as:

$$A\frac{\partial\phi^1}{\partial\tau} + \phi^1\frac{\partial\phi^1}{\partial\xi} + B\frac{\partial^3\phi^1}{\partial\xi^3} - C\frac{\partial^2\phi^1}{\partial\xi^2} = 0. \quad (22)$$

In (22), the coefficients A, B, and C are given by:

$$A = \frac{-2Z_1^2\lambda(Z_1\lambda^2-3)(Z_1\lambda^2-3m\beta)\left[(Z_1\lambda^2-3m\beta)^2 + m\left(\frac{Z_2}{Z_1}\right)(1+\mu_{ce}+\mu_{se}-\mu_H)(Z_1\lambda^2-3)^2\right]}{D}$$

$$B = \frac{-(Z_1\lambda^2-3)^3(Z_1\lambda^2-3m\beta)^3}{D}$$

$$C = \frac{-Z_1^2\lambda(Z_1\lambda^2-3)(Z_1\lambda^2-3m\beta)\left[\rho_{10}(Z_1\lambda^2-3m\beta)^2 - m\rho_{20}\left(\frac{Z_2}{Z_1}\right)(1+\mu_{ce}+\mu_{se}-\mu_H)(Z_1\lambda^2-3)^2\right]}{D}$$

where

$$L = \frac{\mu_{ce}(\kappa_{ce}^2-1/4)}{(\kappa_{ce}-3/2)^2\sigma_{ce}^2} + \frac{\mu_{se}(\kappa_{se}^2-1/4)}{(\kappa_{se}-3/2)^2\sigma_{se}^2} - \frac{\mu_H(\kappa_H^2-1/4)}{(\kappa_H-3/2)^2\sigma_H^2}$$

and

$$D = L(Z_1\lambda^2-3)^3(Z_1\lambda^2-3m\beta)^3 + 3Z_1^2(1+Z_1\lambda^2) \times (Z_1\lambda^2-3m\beta)^3 - 3m^2Z_2^2(1+\mu_{ce}+\mu_{se}-\mu_H) \times (Z_1\lambda^2-3)(Z_1\lambda^2+m\beta). \quad (23)$$

Solution of KdVB equation

To find the solution of (22), we use the transformed coordinate $\chi = (\xi - V\tau)$ of the comoving frame with speed V and use the boundary conditions: $\phi^1 \to 0$ and $\frac{\partial\phi^1}{\partial\chi}$, $\frac{\partial^2\phi^1}{\partial\chi^2}$, $\frac{\partial^3\phi^1}{\partial\chi^3} \to 0$ as $\chi \to \infty$ for a localized solution [42].

When the partial differential equation of a system is formed by the combined effect of dispersion and dissipation, a convenient method to solve it is "the tanh method" [43, 44]. Using the above transformation, (22) can be written as:

$$-AV\frac{\partial\phi^1}{\partial\chi} + \phi^1\frac{\partial\phi^1}{\partial\chi} + B\frac{\partial^3\phi^1}{\partial\chi^3} - C\frac{\partial^2\phi^1}{\partial\chi^2} = 0. \quad (24)$$

Again using the transformation $\alpha = \tanh\chi$ and assuming a series solution of the form $\phi^1(\alpha) = \sum_{i=0}^{n} a_i\alpha^i$, we arrive at the solution of (22) as:

$$\phi^1 = AV + 8Bk^2 + \frac{C^2}{25B} - 12k^2\tanh^2[k(\xi - V\tau)] - \frac{12}{5}kC[1-\tanh[k(\xi - V\tau)]]. \quad (25)$$

The speed of comoving frame is related to the coefficients A, B, and C as $V = \frac{(100B^2k^2 - C^2 + 60kBC)}{25AB}$ and $k = \frac{\pm C}{10B}$ which can be obtained using the boundary conditions.

Results

Solution (25) is applicable to any astrophysical plasma. However, in this study, we concentrate on parameters relevant to comet Halley: the observed value of the density of hydrogen ions was $n_H = 4.95\text{cm}^{-3}$; their temperature was

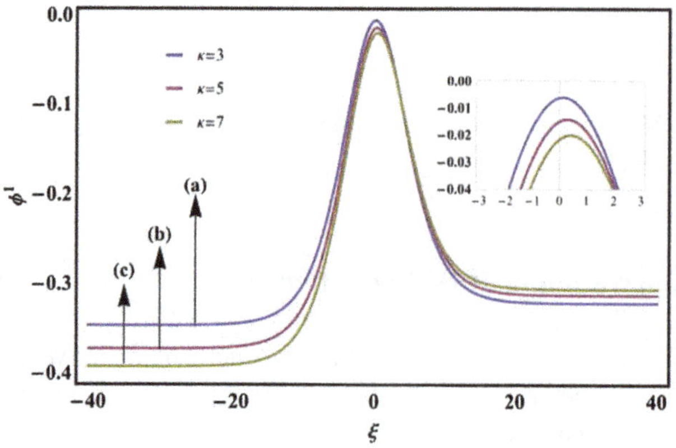

Fig. 1 ϕ^1 vs χ as a function of kappa indices

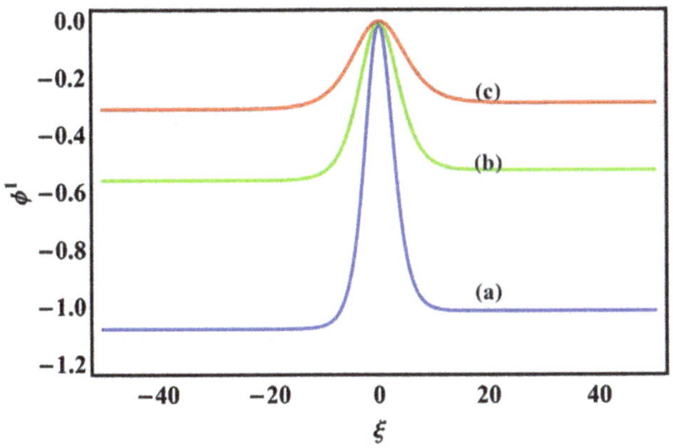

Fig. 2 ϕ^1 vs χ as a function of temperature ratio β

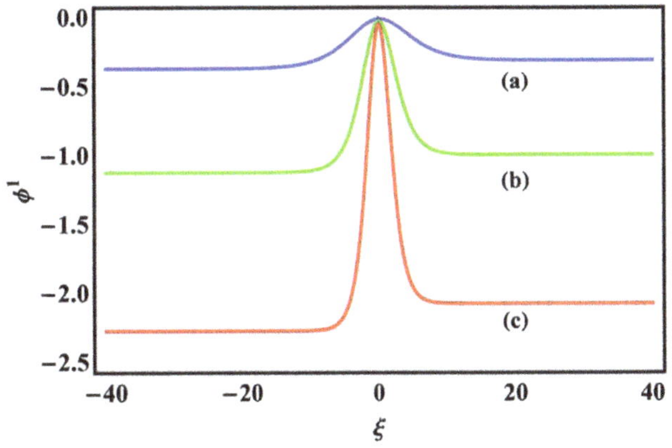

Fig. 4 ϕ^1 vs χ as a function of O^+ density

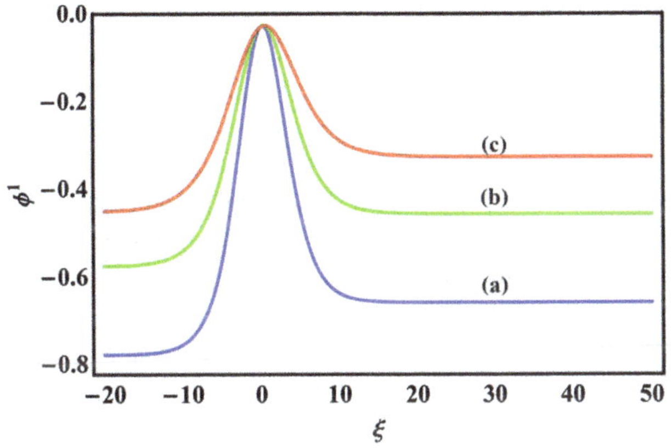

Fig. 3 ϕ^1 vs χ as a function of O^- density

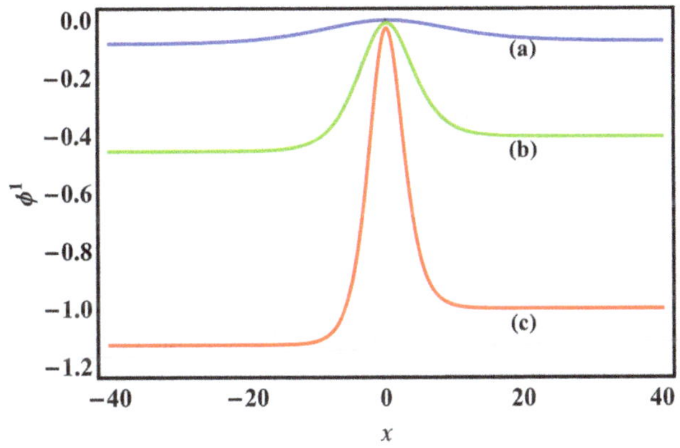

Fig. 5 ϕ^1 vs χ as a function of kinematic viscosity of O^+ ions

$T_H = 8 \times 10^4 K$. The temperature of the solar (or hot) electrons was $T_{se} = 2 \times 10^5$K [45]. The temperature of the second component of electrons, namely photoelectrons was set at $T_{ce} = 2 \times 10^4$K. Negatively charged oxygen ions with an energy $\sim$1 eV and densities $\leq$1 cm^{-3} was identified by Chaizy et al. [24]. We thus set the densities of positively charged oxygen ions at $n_{20} = 0.5$cm^{-3} and that of negatively charged oxygen ions at $n_{10} = 0.05$cm^{-3} [24, 45].

Figure 1 is a plot of the solution (25) of the KdVB Eq. (22), and shows the variation of the potential φ^1 versus χ as a function of the spectral indices; the parameters for the figure are $n_{10} = 0.05$cm^{-3}, $n_{20} = 0.5$cm^{-3}, $n_H = 4.95$cm^{-3}, $T_{ce} = 2 \times 10^4$K, $T_{se} = 2 \times 10^5$K, $T_1 = T_2 = 1.16\times 10^4 K$, and $Z_1 = Z_2 = 1$. Curve (a) is for the spectral index $\kappa = 3$, curve (b) is for $\kappa = 5$, and curve (c) is for $\kappa = 7$. It is clear from the figure that the amplitude of shock wave increases with a decrease of kappa indices.

Figure 2 is a plot of the potential φ^1 versus χ as a function of temperature ratio ($\beta = \frac{T_2}{T_1}$) of the pair ions. We keep the spectral indices at $\kappa_{ce} = \kappa_{se} = \kappa_H = 3$; the other parameters are the same as in Fig. 1. Curve (a) is for $\beta = 2$, curve (b) is for $\beta = 2.2$, and curve (c) is for $\beta = 2.4$. We find that the strength of shock profile decreases with an increase of the temperature of positively charged oxygen ions.

Figure 3 is again a plot of the potential φ^1 versus χ as a function of the negative oxygen ion densities n_{10}. The parameters used in this case are $n_{20} = 0.5$cm^{-3}, $n_H = 4.95$cm^{-3}, $T_{ce} = 2 \times 10^4$K, $T_{se} = 2 \times 10^5$K, $T_1 = T_2 = 1.16\times 10^4 K$, $Z_1 = Z_2 = 1$, and $\kappa_{ce} = \kappa_{se} = \kappa_H = 5$. Curve (a) is for $n_{10} = 0.06$ cm^{-3}, curve (b) is for $n_{10} = 0.07$ cm^{-3}, and curve (c) is for $n_{10} = 0.08$ cm^{-3}. We find that the strength of shock wave decreases with an increase of the negatively charged oxygen ion densities.

Figure 4 shows the variation of the amplitude of shock wave as a function of the positively charged oxygen ion densities. The parameters chosen are the same as in Fig. 3. Here, the negatively charged oxygen ion density is $n_{10} = 0.05$ cm^{-3}. Curve (a) is for $n_{20} = 0.3$ cm^{-3}, curve (b) is for $n_{20} = 0.5$ cm^{-3}, and curve (c) is for $n_{20} = 0.7$ cm^{-3}. We find that increasing positively charged

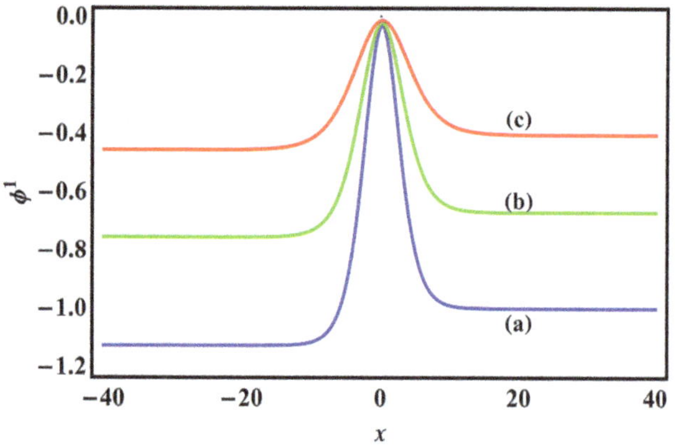

Fig. 6 ϕ^1 vs χ as a function of kinematic viscosity of O^- ions

oxygen ion density will increase the strength of the shock wave profile.

Figure 5 is a plot of the solution of the KdVB Eq. (22) and shows the variation of the potential φ^1 versus χ as a function of the normalized kinematic viscosity of positively charged oxygen ions (ρ_2); the parameters for the figure are $n_{10} = 0.05\text{cm}^{-3}$, $n_{20} = 0.5\text{cm}^{-3}$, $n_H = 4.95\text{cm}^{-3}$, $T_{ce} = 2\times\ 10^4\text{K}$, $T_{se} = 2 \times 10^5\text{K}$, $T_1 = T_2 = 1.16 \times 10^4 K$, and $Z_1 = Z_2 = 1$. Curve (a) is for the kinematic viscosity $\rho_2 = 0.25$, curve (b) is for $\rho_2 = 0.5$, and curve (c) is for $\rho_2 = 0.75$. It is clear from the figure that the strength of shock wave increases with an increase in kinematic viscosity of positively charged oxygen ions.

Figure 6 shows the variation of the potential φ^1 versus χ as a function of the normalized kinematic viscosity of negatively charged oxygen ions (ρ_1). The parameters used in this case are the same as in Fig. 5. Curve (a) is for the kinematic viscosity $\rho_1 = 0.1$, curve (b) is for $\rho_1 = 0.3$, and curve (c) is for $\rho_1 = 0.5$. We find that the strength of shock wave decreases with an increase in the kinematic viscosity of negatively charged oxygen ions.

In a study of electron acoustic solitary and shock waves in dissipative space plasmas with superthermal, hot electrons, Han et al. [46] found that the amplitude of the shock wave decreased with increasing spectral index κ. They also found that the strength of the shock wave increased with increasing ratios of the cold to hot electron densities and electron kinematic viscosities. In another study on drift ion-acoustic shocks in an inhomogeneous 2-D quantum plasma, Masood et al. [47] found that the strength of the shock waves increased with increasing densities of ions.

The results of Fig. 1, which are in agreement with that of Han et al. [46], can be considered as extending their result to a multi-component plasma as we now have three components being described by kappa distributions. Similarly, the conclusions of Fig. 4 where the strength of the shock wave increases with increasing positively charged oxygen ion density is in agreement with the result of Masood et al. [47].

Finally, the other observation of Han et al. [46] that the strength of shock wave increased with increasing electron kinematic viscosity was interpreted as being due to increasing dissipation among the constituents. Increasing the density of positively charged oxygen ions would lead to an increase in the electron density; conversely, an increase in the negatively charged oxygen ion densities would lead to a decrease in the electron densities. The results of Figs. 5 and 6 are consistent with this observation of Han et al. [46].

More recently, Sabetkar and Dorranian [48] investigated the characteristics of dust acoustic solitary waves (DASWs) in a plasma composed of two populations of ions, electrons and negatively charged dust. The ions and electrons were modeled by three-dimensional non-extensive and kappa distributions, respectively. They derived the Zakharov–Kuznetsov (ZK) and modified Zakharov–Kuznetsov (mZK) equations to describe dust acoustic waves in the system. Both positive and negative polarity solutions were found to exist. They found that the amplitude of the ZK solitons decreased with increasing κ values, which is the conclusion from Fig. 1. In addition, the amplitude of these solitons increased with an increase in the temperature ratio of the cold to hot ions, similar to the conclusion from Fig. 2.

In yet another study, the same authors [49] studied the "fast" and "slow" dust acoustic waves in a plasma composed of Maxwellian electrons, kappa distributed positive ions, kappa-Schamel distributed negative ions and negatively charged dust particles. The 1 and 3D Schamel–Korteweg–deVries (S–KdV) equation that they derived admitted only compressive solitons. A point of convergence with their conclusion is the decrease in the amplitude of the soliton with the temperature ratio β_p $(=\frac{T_p}{T_n})$. Further comparisons are not possible because of the different distributions used and also the different equations studied.

Conclusion

We have, in this paper, studied ion-acoustic shock waves in a five component plasma of positively and negatively charged oxygen ions, lighter hydrogen ions, and hot and cold electrons by deriving the KdVB equation. The impact of spectral indices kappa, temperature of positively charged oxygen ions, and density of oxygen ions (O^+, O^-) on the shock wave amplitude is studied. We find that in a cometary plasma with the above components, the nonlinear wave is in a transition state from shock to soliton [20]. A reduction in the shock wave amplitude is seen with increasing spectral indices and positively charged oxygen ion density. The strength of the shock profile also decreases with increasing positively charged oxygen ion temperatures and negatively charged oxygen ion densities. These results can be expected to contribute to an understanding of shocks in comets as we have two heavy ion components and heavy ions were surmised to affect shocks in cometary plasmas [25].

Acknowledgments The authors thank the referees for their valuable comments. Financial assistance from Kerala State Council for Science, Technology and Environment, Thiruvananthapuram, Kerala, India (JRFs for MM and SG) and the University Grants Commission (EF) is gratefully acknowledged.

References

1. Sagdeev, R.Z., Leontovich, M.A.: Cooperative phenomena and shock waves in collisionless plasmas. Rev. Plasma Phys. **4**, 23 (1966)
2. Ikezi, H., Taylor, R., Baker, D.: Formation and interaction of ion-acoustic solitions. Phys. Rev. Lett. **25**(1), 11 (1970)
3. Malik, R., Malik, H.K.: Compressive solitons in a moving e-p plasma under the effect of dust grains and an external magnetic field. J. Theor. Appl. Phys. **7**, 65 (2013)
4. Mamun, A.A., Shukla, P.K.: Cylindrical and spherical dust ion–acoustic solitary waves. Phys. Plasmas **9**(4), 1468 (2002)
5. Xue, J.K.: Cylindrical and spherical dust–ion acoustic shock waves. Phys. Plasmas **10**, 4893 (2003)
6. Sahu, B., Roychoudhury, R.: Quantum ion acoustic shock waves in planar and nonplanar geometry. Phys. Plasmas **14** (7), 2310 (2007)
7. Shukla, P.K., Mamun, A.A.: Solitons, shocks and vortices in dusty plasmas. New J. Phys. **5**(1), 17 (2003)
8. Shukla, P.K., Mamun, A.A.: Dust-acoustic shocks in a strongly coupled dusty plasma. IEEE Tr. PS. **29**, 221 (2001)
9. Mamun, A.A., Eliasson, B., Shukla, P.K.: Dust-acoustic solitary and shock waves in a strongly coupled liquid state dusty plasma with a vortex-like ion distribution. Phys. Lett. A **332**, 412 (2004)
10. Ghosh, S., Gupta, M.R.: Charging-delay effect on longitudinal dust acoustic shock wave in strongly coupled dusty plasma. Phys. Plasmas **12**, 092306 (2005)
11. Pakzad, H.R.: Dust acoustic shock waves in strongly coupled dusty plasmas with kappa-distributed ions. Ind. J. Phys. **86**, 743 (2012)
12. El-Hanbaly, A.M., El-Shewy, E.K., Sallah, M., Darweesh, H.F.: Linear and nonlinear analysis of dust acoustic waves in dissipative space dusty plasmas with trapped ions. J. Theor. Appl. Phys. **9**, 167 (2015)
13. Vasyliunas, V.M.: Low-energy electrons on the day side of the magnetosphere. J. Geophys. Res. **73**(23), 7519 (1968)
14. Masood, W., Mahmood, S., Imtiaz, N.: Electrostatic shocks and solitons in pair-ion plasmas in a two-dimensional geometry. Phys. Plasmas **16**, 2306 (2009)
15. Masood, W., Rizvi, H.: Two dimensional nonplanar evolution of electrostatic shock waves in pair-ion plasmas. Phys. Plasmas **19**(12), 119 (2012)
16. Samanta, U.K., Chatterjee, P., Mej, M.: Soliton and shocks in pair ion plasma in presence of superthermal electrons. Astrophys. Space Sci. **345**(2), 291 (2013)
17. Sultana, S., Kourakis, I., Saini, N.S., Hellberg, M.A.: Oblique electrostatic excitations in a magnetized plasma in the presence of excess superthermal electrons. Phys. Plasmas **17**, 032310 (2010)
18. Kourakis, I., Sultana, S., Hellberg, M.A.: Dynamical characteristics of solitary waves, shocks and envelope modes in kappa-distributed non-thermal plasmas: an overview. Plasma Phys. Control. Fusion **54**(12), 124001 (2012)
19. Sultana, S., Kourakis, I.: Electron-scale electrostatic solitary waves and shocks: the role of superthermal electrons. Eur. Phys. J. **D.66**, 1 (2012)
20. Saeed, R., Shah, A.: Nonlinear Korteweg–de Vries–Burger equation for ion acoustic shock waves in a weakly relativistic electron–positron–ion plasma with thermal ions. Phys. Plasmas **17**, 032308 (2010)
21. Ghosh, B., Banerjee, S.: Effect of nonthermal electrons and positrons on ion-acoustic solitary waves in a plasma with warm drifting ions. Ind. J. Phys. **89**, 1307 (2015)
22. Ipavich, F.M., Galvin, A.B., Gloeckler, G., Hovestadt, D., Klecker, B., Scholer, M.: Comet Giacobini-Zinner: in situ observations of energetic heavy ions. Science **232**, 366 (1986)
23. Coplan, M.A., Ogilvie, K.W., A'Hearn, M.F., Bochsler, P., Geiss, J.: Ion composition and upstream solar wind observations at comet Giacobini-Zinner. J. Geophys. Res. **92**, 39 (1987)
24. Chaizy, P., Reme, H., Sauvaud, J.A., d'Uston, C., Lin, RP., Larson, D.E., Mitchell, D.L., Zwickl, R.D., Baker, D.N., Bame, S.J., Feldman, W.C., Fuselier, S.A., Huebner, W.F., McComas, D.J., Young, D.T.: Negative ions in the coma of comet Halley. Nature. **349**, 393 (1991)
25. Coates, A.J.: Heavy ion effects on cometary shocks. Adv. Space Res. **15**, 403 (1995)
26. Voelzke, M.R., Izaguirre, L.S.: Morphological analysis of the tail structures of comet P/Halley 1910 II. Planet. Space Sci. **65**(1), 104 (2012)
27. Sakanaka, P.H., Spassovska, I.: Study of nonlinear phenomena in four-component dusty plasma with charge fluctuation. Phys. Scr. **T131**, 014040 (2008)
28. Verheest, F., Hellberg, M.A., Kourakis, I.: Acoustic solitary waves in dusty and/or multi-ion plasmas with cold, adiabatic, and hot constituents. Phys. Plasmas **15**, 112309 (2008)
29. Dorranian, D., Sabetkar, A.: Dust acoustic solitary waves in a dusty plasma with two kinds of nonthermal ions at different temperatures. Phys. Plasmas **19**, 013702 (2012)
30. Shahmohammadi, N., Dorranian, D.: Effect of dust charge fluctuation on multidimensional instability of dust-acoustic solitary waves in a magnetized dusty plasma with nonthermal ions. Phys. Plasmas **22**, 103707 (2015)
31. Sharif Moghadam., S, Dorranian., D.: Effect of size distribution on the dust acoustic solitary waves in dusty plasma with two kinds of nonthermal ions. Adv. Mater. Sci. Eng (2013)
32. Shahmoradi, N., Dorranian, D.: Effects of variable dust size, charge and mass on the characteristics of dust acoustic solitary waves in a magnetized dusty plasma. Phys. Scr. **89**, 065602 (2014)
33. El-Taibany, W.F.: Nonlinear dust acoustic waves in inhomogeneous four-component dusty plasma with opposite charge polarity dust grains. Phys. Plasmas **20**, 093701 (2013)
34. Tomar, R., Malik, H.K., Dahiya, R.P.: Reflection of ion acoustic solitary waves in a dusty plasma with variable charge dust. J. Theor. Appl. Phys. **8**, 126 (2014)
35. Tomar, R., Bhatnagar, A., Malik, H.K., Dahiya, R.P.: Evolution of solitons and their reflection and transmission in a plasma having negatively charged dust grains. J. Theor. Appl. Phys. **8**, 138 (2014)
36. Sabetkar, A., Dorranian, D.: Effect of obliqueness and external magnetic field on the characteristics of dust acoustic solitary

waves in dusty plasma with two-temperature nonthermal ions. J. Theor. Appl. Phys. **9**, 141 (2015)
37. Ghosh, S., Sarkar, S., Khan, M., Gupta, M.R.: Nonlinear properties of small amplitude dust ion acoustic solitary waves. Phys. Plasmas **7.9**, 3594 (2000)
38. Ghosh, S., Bharuthram, R.: Ion acoustic solitons and double layers in electron–positron–ion plasmas with dust particulates. Astrophys. Space Sci. **314**, 121 (2008)
39. Dutta, D., Singha, P., Sahu, B.: Interlaced linear-nonlinear wave propagation in a warm multicomponent plasma. Phys. Plasmas **21**, 122308 (2014)
40. Khaled, M.A.H.: Dust ion acoustic solitary waves and their multidimensional instability in a weakly relativistic adiabatic magnetized dusty plasma with two different types of adiabatic electrons. Ind. J. Phys. **88**, 647 (2014)
41. Shahmansouri, M., Shahmansouri, B., Darabi, D.: Ion acoustic solitary waves in nonplanar plasma with two-temperature kappa distributed electrons. Ind. J. Phys. **87**, 711 (2013)
42. Hussain, S., Akhtar, N.: Korteweg de Vries Burgers equation in multi-ion and pair-ion plasmas with Lorentzian electrons. Phys. Plasmas **20**, 012305 (2013)
43. Malfliet, W.: Solitary wave solutions of nonlinear wave equations. Am. J. Phys. **60**(7), 650 (1992)
44. Malfliet, W.: The tanh method: a tool for solving certain classes of nonlinear evolution and wave equations. J. Comput. Appl. Math. **164**, 529 (2004)
45. Brinca, A.L., Tsurutani, B.T.: Unusual characteristics of the electromagnetic waves excited by cometary new born ions with large perpendicular energies. Astron. Astrophys. **187**, 311 (1987)
46. Han, J.N., Duan, W.S., Li, J.X., He, Y.L., Luo, J.H., Nan, Y.G., Han, Z.H., Dong, G.X.: Study of nonlinear electron-acoustic solitary and shock waves in a dissipative, nonplanar space plasma with superthermal hot electrons. Phys. Plasmas **21**, 012102 (2014)
47. Masood, W., Karim., Shah, H.A., Siddiq, M.: Drift ion acoustic shock waves in an inhomogeneous two-dimensional quantum magnetoplasma. Phys. Plasmas. **16**, 042108 (2009)
48. Sabetkar, A., Dorranian, D.: Role of superthermality on dust acoustic structures in the frame of a modified Zakharov–Kuznetsov equation in magnetized dusty plasma. Phys. Scr. **90**, 035603 (2015)
49. Sabetkar, A., Dorranian, D.: Parametric study of a Schamel equation for low-frequency dust acoustic waves in dusty electronegative plasmas. Phys. Plasmas **22**, 083705 (2015)

17

Amplitude modulation of three-dimensional low-frequency solitary waves in a magnetized dusty superthermal plasma

Shalini[1] · **A. P. Misra**[2] · **N. S. Saini**[1]

Abstract The amplitude modulation of three-dimensional (3D) dust ion-acoustic wave (DIAW) packets is studied in a collisionless magnetized plasma with inertial positive ions, superthermal electrons and negatively charged immobile dust grains. By using the reductive perturbation technique, a 3D-nonlinear Schrödinger equation is derived, which governs the slow modulation of DIAW packets. The latter are found to be stable in the low-frequency ($\omega < \omega_c$) regime, whereas they are unstable for $\omega > \omega_c$, and the modulational instability is related to the modulational obliqueness (θ). Here, $\omega(\omega_c)$ is the nondimensional wave (ion-cyclotron) frequency. It is shown that the superthermal parameter κ, the frequency ω_c as well as the charged dust impurity ($0 < \mu < 1$) shift the MI domains around the ω–θ plane, where μ is the ratio of electron-to-ion number densities. Furthermore, it is found that the decay rate of instability is quenched by the superthermal parameter κ with cutoffs at lower wave number of modulation (*K*); however, it can be higher (lower) with increasing values of μ (ω_c) having cutoffs at higher values of *K*.

Modulational instability · NLS equation · Superthermal plasma · Dusty plasma · Solitary wave

✉ N. S. Saini
nssaini@yahoo.com

Shalini
shal.phy29@gmail.com

A. P. Misra
apmisra@gmail.com

1 Department of Physics, Guru Nanak Dev University, Amritsar, India

2 Department of Mathematics, Siksha Bhavana, Visva-Bharati University, Santiniketan, India

Introduction

The nonlinear features of solitary waves in dusty plasmas have been of great importance over the last many years due to their wide range of applications in space, astrophysical and laboratory environments [1–3]. Dust grains are typically micron or sub-micron sized particles and are ubiquitous ingredients in our universe. The presence of charged dust grains in an electron–ion plasmas not only alters the characteristics of ion-acoustic solitary waves (IASWs), but also modifies the ion-acoustic wave as well as generates a new kind of mode, namely, the dust-acoustic (DA) wave. More than three decades ago [4], reported theoretically the existence of DIA waves in a dusty plasma. Later, in laboratory experiments [5], confirmed the existence of these waves. A large number of investigations on DIA waves in multicomponent plasmas have been reported in the framework of Sagdeev's approach as well as reductive perturbation technique.

Furthermore, in many observations it has been confirmed that superthermal particles exist in space plasmas [6–9] and laboratory environments [10]. These superthermal particles are described by Lorentzian (kappa) distribution which is more appropriate for analysis of data rather than a Maxwellian distribution [9]. Furthermore, such distribution has been widely used to investigate various collective modes as well as nonlinear coherent structures like solitons, shocks, envelope solitons through the description of Korteweg–de Vries (KdV), Korteweg–de Vries Burgers (KdVB) and nonlinear Schrödinger (NLS) equations [11–18].

On the other hand, there has also been a growing interest in investigating the nonlinear modulation of electrostatic waves in plasmas owing to their importance not only in space and astrophysical environments, but also in laboratory plasmas. The modulational instability (MI) of nonlinear waves in plasmas has been a well-known mechanism for the localization of wave energy, which leads to the formation of bright envelope solitons. However, in the absence of instability, the evolution of the system can be in the form of dark envelope solitons. Furthermore, due to a small plane wave perturbation, MI can have exponential growth which leads to the amplification of the sidebands, and thus break up the uniform waves into a train of oscillations. A large number of investigations on MI of electrostatic or electromagnetic waves can be found in the literature [14, 18–26]. To mention few, the MI of obliquely propagating DIA waves in an unmagnetized plasma containing positive ions, electrons and immobile dust grains was reported by [19]. It was observed from the stability analysis that the obliqueness in the modulation direction has a profound effect on the condition of MI. They also observed the influence of ion temperature on the amplitude modulation of wave and noticed that wave stability profile may be strongly modified by ion temperature [21]. Furthermore, the nonlinear propagation of wave envelopes in an unmagnetized superthermal dusty plasma was investigated by El-Labany [27] et al. They showed that the electron superthermality and the dust grain charge significantly modify the profiles of the wave envelope and the associated regions of instability. Ahmadihojatabad et al. [28] studied the influence of superthermal and trapped electrons on the obliquely propagating ion-acoustic waves (IAWs) in magnetized plasmas. Bains et al. [25] addressed the MI of ion-acoustic wave envelopes in a multicomponent magnetized plasma using a quantum fluid model. They observed that the ion number density, the constant magnetic field and the quantum coupling parameter have strong effects on the growth rate of MI. Also, the nonlinear propagation of DIA wave envelopes in a three-dimensional magnetized plasma containing nonthermal electrons featuring Tsallis distribution, both positive and negative ions, and immobile charged dust was investigated by [29].

To the best of our knowledge the investigation of MI of DIAWs in a magnetized dusty plasma containing superthermal electrons has not yet been reported. Our purpose in this investigation is to consider the propagation of DIA wave envelopes in a magnetized dusty plasma containing cold positive ions and superthermal electrons. We have employed the standard multiple-scale perturbation technique to derive the NLS equation. It was shown that in earlier investigations [30] that MI of ion acoustic waves is significantly influenced by the presence of superthermal electrons and growth rate is larger in the presence of more superthermal electrons. We have, however, investigated the combined effects of the external magnetic field, dust concentration and the superthermality of electrons on the MI of DIA wave packets. It is shown that the superthermality of electrons (via κ), the charged dust impurity and the external magnetic field shift the MI domains around the $\omega-\theta$ plane, where ω is the wave frequency and θ stands for modulational obliqueness. Further, we have also studied the decay rate of MI by different plasma parameters.

The paper is organized as follows: in Sect. 2, the basic equations governing the nonlinear dynamics of DIA wave envelopes in magnetized superthermal plasmas are presented and the three-dimensional NLS equation is derived. The effects of various physical parameters on the existence of stable/unstable regions for the modulation of DIA waves are investigated in Sect. 3. Finally, Sect. 4 contains the summary and conclusions of our results.

The model equations and derivation of the 3D-NLSE

We consider the nonlinear propagation of DIA waves in a magnetized plasma consisting of superthermal electrons, cold positive ions and negatively charged immobile dust grains. The plasma is immersed in the constant magnetic field $\mathbf{B}_0 = B_0\hat{z}$. We adopt a fluid model for the dynamics of DIA waves in a magnetized plasma which consists of the continuity, momentum and the Poisson's equations. Thus, we have

$$\frac{\partial n}{\partial t} + \nabla \cdot (n\mathbf{U}) = 0, \tag{1}$$

$$\frac{\partial \mathbf{U}}{\partial t} + (\mathbf{U}\cdot\nabla)\mathbf{U} = -\frac{e}{M}\nabla\phi + \frac{eB_0}{M}(\mathbf{U}\times\hat{z}), \tag{2}$$

$$\nabla^2\phi = 4\pi e(n_e - n + Z_d n_d 0), \tag{3}$$

where the superthermal electrons are given by the kappa distribution [31]

$$n_e = n_{e0}\left[1 - \frac{e\phi}{(\kappa - 3/2)k_B T_e}\right]^{-\kappa+1/2}. \tag{4}$$

The set of fluid Eqs. (1)–(3) in nondimensional forms are written as

$$\frac{\partial n}{\partial t} + \nabla \cdot (n\mathbf{U}) = 0, \tag{5}$$

$$\frac{\partial \mathbf{U}}{\partial t} + (\mathbf{U}\cdot\nabla)\mathbf{U} = \nabla\phi + \omega_c(\mathbf{U}\times\hat{z}), \tag{6}$$

$$\nabla^2\phi = n_e - n + (1-\mu), \tag{7}$$

where n_e and n are the number densities of electrons and ions normalized by the equilibrium number density of ions n_0, ϕ is the electric potential normalized by $k_B T_e/e$, $\mathbf{U} \equiv (u, v, w)$ is the ion fluid velocity normalized by the DIA speed $C_s (= \sqrt{k_B T_e/M})$. The space and time coordinates are normalized by the Debye length $\lambda_D[= (k_B T_e/4\pi n_0 e^2)^{1/2}]$ and the inverse of ion plasma frequency $\omega_{pi}[= (4\pi e^2 n_0/M)^{1/2}]$, respectively. Furthermore, $\omega_c = eB_0/cM$ is the ion gyrofrequency normalized by ω_{pi}. The charge neutrality condition yields $1 - \mu = Z_d n_{d0}/n_0$, where $\mu = n_{e0}/n_0$ is the ratio of equilibrium number densities of electrons and ions. Next, in the small-amplitude perturbations, i.e., $|\phi| \ll 1$, and, in particular, $|\phi/(\kappa - 3/2)| \ll 1$, Eq. (4) reduces to

$$n_e \approx \mu + q_1\phi + q_2\phi^2 + q_3\phi^3, \tag{8}$$

where the coefficients are given by

$$q_1 = \mu\frac{(\kappa - 1/2)}{(\kappa - 3/2)}, q_2 = \mu\frac{(\kappa^2 - 1/4)}{2(\kappa - 3/2)^2}, \\ q_3 = \mu\frac{(\kappa^2 - 1/4)(\kappa + 3/2)}{6(\kappa - 3/2)^3}. \tag{9}$$

In order to derive the evolution equation for weakly nonlinear DIA wave envelopes, we employ the standard multiple-scale technique [32, 33] in which the coordinates are stretched as

$$\xi = \epsilon x, \eta = \epsilon y, \zeta = \epsilon(z - V_g t), \tau = \epsilon^2 t. \tag{10}$$

Consider $A \equiv (n, w, \phi)$ and $B \equiv (u, v)$ as the state vectors which describe the state at a position z and time t. The perturbations from the equilibrium state $A^{(0)} = (1, 0, 0)^T$ and $B^{(0)} = (0, 0)^{T'}$ are considered by assuming $A = A^{(0)} + \sum_{m=1}^{\infty} \epsilon^m A^{(m)}$ and $B = B^{(0)} + \sum_{m=1}^{\infty} \epsilon^{m+1} B^{(m)}$. The slow-scale dependence of all perturbed state enter via the l-th harmonic amplitude $A_l^{(m)}$ and $B_l^{(m)}$ given as $A^{(m)} = \sum_{l=-m}^{m} A_l^{(m)}(\xi, \eta, \zeta, \tau)e^{il(kz-\omega t)}$ and $B^{(m)} = \sum_{l=-m}^{m} B_l^{(m)}(\xi, \eta, \zeta, \tau)e^{il(kz-\omega t)}$, where ω and k, respectively, represent the carrier wave frequency and the wavenumber. In order that n, $\mathbf{U}$, ϕ, etc. are all real, the state variables must satisfy the reality condition with respect to its complex conjugate parts. One should note that the transverse (to the magnetic field) velocity components u and v appear at higher order in ϵ than the parallel component w. The anisotropy and higher order effects are introduced via strong magnetic field and gyro-motion of fluid, respectively, in the presence of weak perturbations [34].

We substitute the stretched coordinates (10) and the expansions given above into Eqs. (5)–(8), and collect terms in different powers of ϵ to obtain a set of reduced equations. Thus, equating the coefficients for $m = 1$ and $l = 1$, we obtain the following first-order quantities in terms of $\phi_1^{(1)}$

$$n_1^{(1)} = \frac{k^2}{\omega^2}\phi_1^{(1)}, \tag{11}$$

$$w_1^{(1)} = \frac{k}{\omega}\phi_1^{(1)}, \tag{12}$$

together with the linear dispersion relation

$$\omega^2 = \frac{k^2}{k^2 + q_1}. \tag{13}$$

From the second-order reduced equations ($m = 2$, $l = 1$), the following compatibility condition in terms of the group velocity of waves is obtained as

$$V_g \equiv \frac{\partial\omega}{\partial k} = q_1\frac{\omega^3}{k^3}. \tag{14}$$

We have depicted the variation of the carrier wave frequency (Eq. 13) and the group velocity (Eq. 14) against the carrier wave number k in Fig. 1. From the upper panel, we find that as k increases, the frequency ω increases and it approaches a constant value at higher $k(>1)$. Furthermore, ω (normalized by the ion plasma frequency ω_{pi}) increases and approaches a constant value (close to 1) as the wave number k increases. Also, the value of ω increases with increasing values of the spectral index κ (i.e., when the superthermality of electrons is somewhat relaxed); however, it remains almost unaltered for $\kappa > 8$. Nevertheless, a reduction of the wave frequency is noticed with increasing values of the electron-to-ion number density ratio μ. This implies that as the number density of electrons increases, more electrons will flow out of the dust grains, i.e., dust charge number decreases in order to maintain the quasineutrality. Further increase of μ may eventually lead to the case similar to the dust-free electron–ion plasma. Thus, negatively charged dust impurity in the plasma with $n_0 > n_{e0}$ effectively increases the wave frequency. Such dust impurity has also a significant effect on the group velocity of waves as shown in the lower panel of Fig. 1 (see the solid and dashed lines). The group velocity of waves (V_g) decreases with an increase in the wave number (k) for different values of κ and μ. It is very interesting to see that for smaller $k<0.5$ (i.e., for larger wavelength), the group velocity reduces with an increase in μ; however, it increases with larger k. Furthermore, an increase in the parameter κ (e.g., from $\kappa = 4$ to $\kappa = 8$) leads to an enhancement of V_g as $k \to 0$. Here, note that further increase of $\kappa(>8)$ does not give any significant change in V_g.

For $l = 0$, 1 and 2, we can determine the second-order harmonic modes in terms of $\phi_1^{(1)}$. So, for $m = 2$, $l = 1$, we have the reduced equations

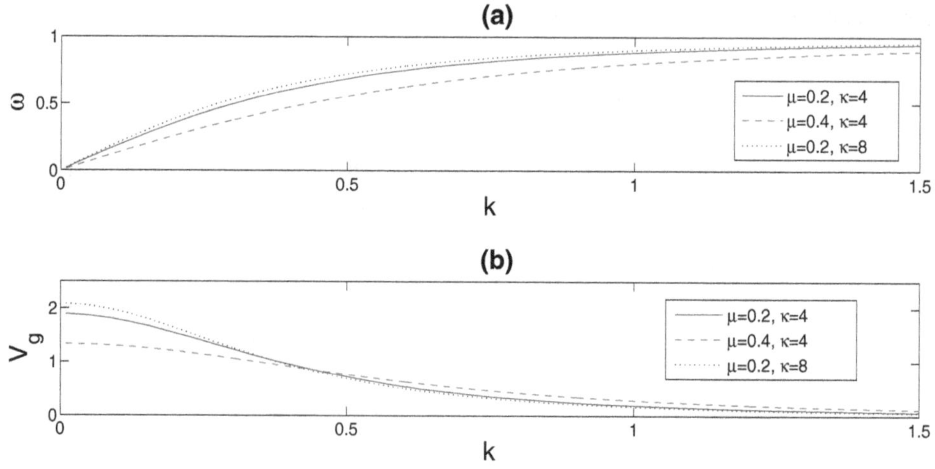

Fig. 1 Carrier wave frequency ω (*upper panel*) and the group velocity of the wave packet V_g (*lower panel*) are plotted against the wave number k for different values of κ and μ as shown in the legends

$$i\omega n_1^{(2)} + ikw_1^{(2)} = V_g \frac{\partial n_1^{(1)}}{\partial \zeta}, \tag{15}$$

$$i\omega w_1^{(2)} + ik\phi_1^{(2)} = V_g \frac{\partial w_1^{(1)}}{\partial \zeta} - \frac{\partial \phi_1^{(1)}}{\partial \zeta}, \tag{16}$$

with

$$u_1^{(1)} = \frac{\omega_c \frac{\partial \phi_1^{(1)}}{\partial \eta} - i\omega \frac{\partial \phi_1^{(1)}}{\partial \xi}}{\omega^2 - \omega_c^2}, \tag{17}$$

$$v_1^{(1)} = -\left[\frac{i\omega \frac{\partial \phi_1^{(1)}}{\partial \eta} + \omega_c \frac{\partial \phi_1^{(1)}}{\partial \xi}}{\omega^2 - \omega_c^2}\right]. \tag{18}$$

The second-order harmonic modes with $m = 2$ and $l = 2$ are given by

$$-\omega n_2^{(2)} + kw_2^{(2)} + kn_1^{(1)} w_1^{(1)} = 0, \tag{19}$$

$$-2\omega w_2^{(2)} + k(w_1^{(1)})^2 + k\phi_2^{(2)} = 0. \tag{20}$$

Thus, we obtain

$$n_2^{(2)} = C_1^{(22)} (\phi_1^{(1)})^2, u_2^{(2)} = C_2^{(22)} (\phi_1^{(1)})^2, \tag{21}$$

$$\phi_2^{(2)} = C_3^{(22)} (\phi_1^{(1)})^2, \tag{22}$$

where the coefficients are

$$C_1^{(22)} = (4k^2 + q_1) C_3^{(22)} + q_2, \tag{23}$$

$$C_2^{(22)} = \frac{\omega}{k}\left[C_1^{(22)} - (k^2 + q_1)^2\right], \tag{24}$$

$$C_3^{(22)} = \frac{q_2}{3k^2} + \frac{k^2}{2\omega^4}. \tag{25}$$

We note that the first-order zeroth harmonics ($n_0^{(1)}$, $w_0^{(1)}$, $\phi_0^{(1)}$) vanish [35], which gives $u_0^{(1)} = v_0^{(1)} = 0$. For $m = 2$, $l = 0$, we obtain the second-order and zeroth-order harmonic modes in the following forms:

$$n_0^{(2)} = C_1^{(20)} \left(\phi_1^{(1)}\right)^2, u_0^{(2)} = C_2^{(20)} \left(\phi_1^{(1)}\right)^2, \tag{26}$$

$$\phi_0^{(2)} = C_3^{(20)} \left(\phi_1^{(1)}\right)^2, \tag{27}$$

where the coefficients are

$$C_1^{(20)} = q_1 C_3^{(20)} + 2q_2, \tag{28}$$

$$C_2^{(20)} = V_g C_1^{(20)} - 2\frac{\omega}{k}(k^2 + q_1)^2, \tag{29}$$

$$C_3^{(20)} = \frac{2q_2 V_g^2 - (k^2 + 3q_1)}{1 - q_1 V_g^2}. \tag{30}$$

Proceeding to the next order ($m = 3$) and solving for the first harmonic equations ($l = 1$), an explicit compatibility condition is determined, from which we obtain the following NLS equation for $\Phi \equiv \phi_1^{(1)}$:

$$i\frac{\partial \Phi}{\partial \tau} + P\frac{\partial^2 \Phi}{\partial \zeta^2} + Q|\Phi|^2\Phi - S\left(\frac{\partial^2 \Phi}{\partial \xi^2} + \frac{\partial^2 \Phi}{\partial \eta^2}\right) = 0. \tag{31}$$

The coefficient of dispersion P and the nonlinearity Q are given by

$$P \equiv \omega''(k) = -\frac{3}{2} q_1 \frac{\omega^5}{k^4}, \tag{32}$$

$$Q = \frac{\omega^3}{k^2}\left[\frac{3}{2} q_3 + q_2 \{C_3^{(20)} + C_3^{(22)}\}\right] - \frac{\omega}{2}\{C_1^{(20)} + C_1^{(22)}\} - k\{C_2^{(22)} + C_2^{(20)}\}. \tag{33}$$

The coefficient S which accounts for the combined effects of transverse perturbations and the external magnetic field is given by

$$S = \frac{\omega^3}{2k^2(\omega_c^2 - \omega^2)}. \tag{34}$$

Stability analysis

We note that the amplitude modulation of DIA wave envelopes typically depend on the coefficients of the NLS equation (31), which parametrically depend on the density ratio μ, the superthermality of electrons (via κ) as well as the intensity of the magnetic field (via ω_c). Inspecting the coefficients P, Q and S, we find that $P=-(3/2)q_1\frac{\omega^5}{k^4}\equiv-(3/2)\frac{\omega^3}{k^2}(1-\omega^2)$, i.e., P is always negative for $\kappa>3/2$ (for which $q_1>0$) and $\omega<1$. However, Q can be positive or negative depending on the values of k, μ and κ. Also, $S>0 (S<0)$ according to when $\omega<\omega_c (\omega>\omega_c)$. We will find that the key elements responsible for the MI are the ratios P/Q and S/P together with their signs and magnitudes. Considering a harmonic wave solution of Eq. (31) of the form $\Phi=\Phi_0\exp(iQ|\Phi_0|^2\tau)$ with Φ_0 denoting the constant amplitude, one can obtain the following dispersion relation for the modulated DIA wave packets [25]:

$$\Omega^2 = K^4\left(\frac{P\alpha^2-S}{1+\alpha^2}\right)^2 \times\left(1-\frac{2(1+\alpha^2)|\Phi_0|^2}{K^2}\frac{Q/P}{\alpha^2-S/P}\right), \quad (35)$$

where Ω and $K\equiv\sqrt{K_\xi^2+K_\eta^2+K_\zeta^2}$, respectively, denote the wave frequency and the wave number of modulation. The parameter $\alpha\equiv K_\zeta/\sqrt{K_\xi^2+K_\eta^2}$ is related to the modulational obliqueness θ which the wave vector $\mathbf{K}$ makes with the resultant of $K_\xi\hat{x}$ and $K_\eta\hat{y}$, i.e., $\theta=\arctan(\alpha)$. From Eq. (35), we find that there exists a critical wave number K_c such that $K^2<K_c^2\equiv2|\Phi_0|^2(1+\alpha^2)(Q/P)/(\alpha^2-S/P)$, the MI sets in either for $PQ>0$, $\alpha_1^2-S/P>0$ or for $PQ<0$, $\alpha_1^2-S/P<0$ [25]. It is further found that a critical value of θ, i.e., $\theta_c\equiv\arctan(\sqrt{S/P})$ also exists for the occurrence of *MI*. Thus, the MI may occur either for $PQ>0$, $\theta>\theta_c$ or $PQ<0$, $\theta<\theta_c$, i.e., we have two possible cases:

- Case I: when $\omega<\omega_c$, the MI sets in for $Q<0$ and for any value of θ in $0\leq\theta\lesssim\pi/2$.
- Case II: when $\omega>\omega_c$, the MI sets in either for $Q<0$ and $\theta>\theta_c$ or $Q>0$ and $\theta<\theta_c$.

From the subsequent analysis and Fig. 2, it will be clear that the Case I is not admissible to the present study as there is no common region for which $\omega<\omega_c$ and $Q<0$ are satisfied. So, we will focus only on Case II. It turns out that when the DIA wave frequency is larger than the ion-cyclotron frequency ω_c, the MI is related to the obliqueness parameter θ; however, the instability disappears for $\omega<\omega_c$. We numerically investigate different stable and unstable regions in the $\omega-\theta$ plane as shown in Fig. 2. We find that the charged dust impurity (represented by the parameter μ with $0<\mu<1$), the superthermal parameter κ and the gyrofrequency ω_c shift the stable/unstable regions around the $\omega-\theta$ plane. From panels (a, b) it is clear that as μ increases, i.e., as the number of charged dust grains decreases, a part of the instability region (with $\theta<\theta_c$) shifts to a stable one and the region of stability in the $\omega-\theta$ plane increases. However, the instability region with $\theta>\theta_c$ increases slightly with increasing values of μ. This implies that when the obliqueness parameter θ is below its critical value θ_c, the presence of charged dust impurity in the plasma favors the instability of modulated wave packets. Comparing panel (c) with panel (a) we find that the superthermality of electrons (with lower values of κ) also favors the instability in the region with $\theta<\theta_c$. The instability region with $\theta>\theta_c$ remains almost unchanged. From panels (a) and (d) it is also evident that the external magnetic field significantly reduces the regions of instability both in the cases of $\theta<\theta_c$ and $\theta>\theta_c$.

The maximum growth/decay rate $\Gamma_{\max}=\mathrm{Im}\,(\Omega)_{max}$ can be obtained from Eq. (35) as $\Gamma_{max}=Q|\phi_0|^2$ provided $PK_\zeta^2-S(K_\xi^2+K_\eta^2)=Q|\phi_0|^2$ is satisfied. The decay rate of MI is depicted in Fig. 3 for different values of κ, μ and ω_c. Clearly, the effects of higher values of κ (less superthermality) suppresses the instability decay rate with cutoffs at significantly lower wave numbers of modulation (see the solid and dotted lines). However, the decay rate becomes higher with increasing values of the electron concentration (or decreasing the dust concentration) with cutoffs at higher K (see the solid and dashed lines). We find that in contrast to the unmagnetized plasmas, the effect of ω_c is to increase the cutoffs at higher wave numbers of modulation; however, the decay rate is slightly reduced (see the solid and dash-dotted lines).

Summary and conclusion

We have investigated the amplitude modulation of DIA wave packets in a magnetized multicomponent plasma consisting of singly charged positive ions, superthermal electrons featuring kappa distribution and negatively charged immobile dust grains. Using the multiple-scale technique, an NLS equation is derived which governs the evolution of DIA wave envelopes. It is shown that both the dispersive and the nonlinear coefficients of the NLS equation are significantly modified by the effects of charged dust impurity, the external magnetic field as well as the superthermality of electrons. Different stable and unstable regions under modulation are obtained in the plane of the carrier wave frequency (ω) and the obliqueness (θ) of modulation. It is found that the parameters κ, μ,

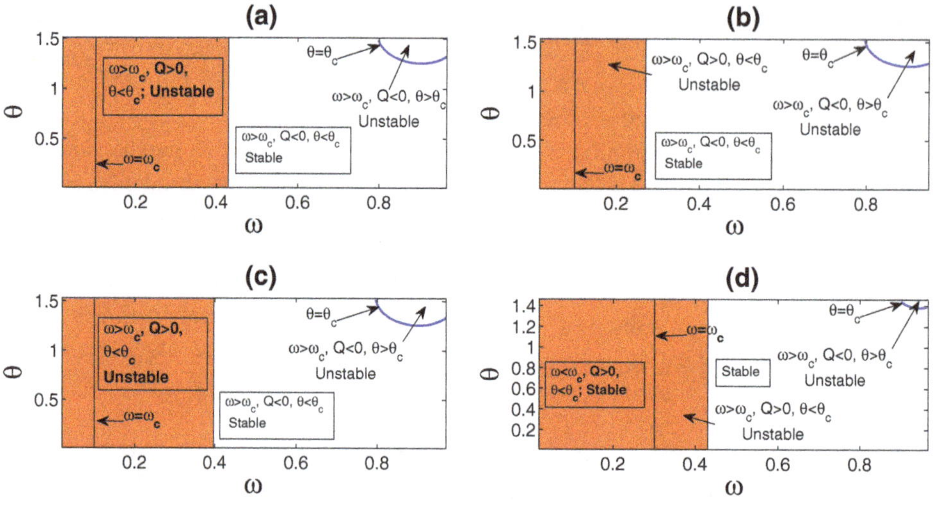

Fig. 2 The stable and unstable regions of wave modulation are shown by the contour plots of $Q=0, \omega=\omega_c$ and $\theta=\theta_c$ in the $\omega-\theta$ plane for different values of the parameters: **a** $\kappa=4, \omega_c=0.1$ and $\mu=0.2$, **b** $\kappa=4, \omega_c=0.1$ and $\mu=0.3$, **c** $\kappa=6, \omega_c=0.1$ and $\mu=0.2$, and **d** $\kappa=4, \omega_c=0.3$ and $\mu=0.2$. The *shaded* or *gray* (*blank* or *white*) region stands for $Q>0(Q<0)$. When $\omega>\omega_c$, the *MI* occurs either for $Q>0, \theta<\theta_c$, or for $Q<0, \theta>\theta_c$. No instability occurs in the regime $\omega<\omega_c$. In some other regions the modulated wave becomes stable

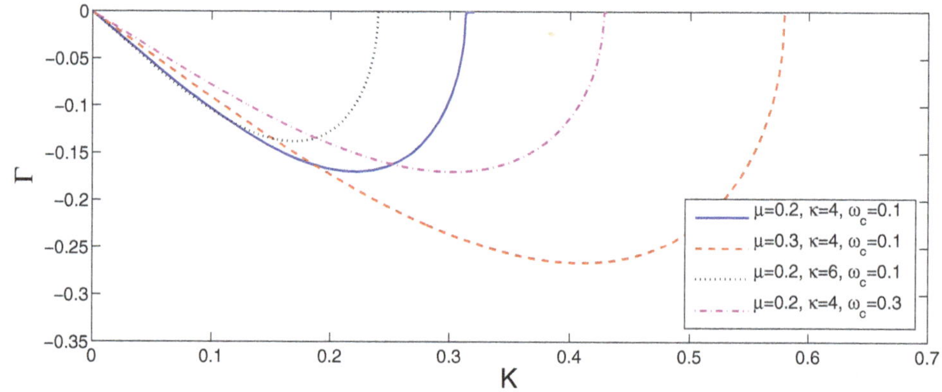

Fig. 3 The decay rate of modulational instability Γ is shown against the wave number of modulation K with the variation of the parameters as shown in the legend. The other parameter values are $k=0.5$ and $\phi_0=0.5$

ω_c and θ remarkably shift the stable/unstable regions around the ω–θ plane. The growth/decay rate of instability is also examined numerically with these plasma parameters. The main results are summarized as follows:

- Starting from a set of fluid equations, the dynamics of weakly nonlinear, slowly varying DIA wave packets is shown to be governed by a three-dimensional NLS equation in which the additional dispersive terms (leading to two more space dimensions in the equation) appear due to the combined effects of the transverse perturbations and the external magnetic field. The fluid model with κ-distributed electrons and stationary charged dust particles is valid for the plasma parameters satisfying $0<\mu<1$ and $\kappa>3/2$.
- The carrier wave frequency is seen to assume a constant value at large $k>1$, and approaches the ion plasma frequency with increasing values of κ. The wave frequency ω and hence the group velocity V_g get significantly reduced with higher values of μ.
- The group velocity dispersion P of the NLS equation is always negative irrespective of the values of k and the plasma parameters. The nonlinear coefficient Q is always negative for $k \gtrsim 1$; however, it can be either positive or negative in the range $0<k<1$ depending on the values of κ and μ. For propagation below the ion cyclotron frequency, the DIA wave packet is always stable. However, for $\omega>\omega_c$, it is unstable and the MI is related to the obliqueness parameter θ. The parameters κ, μ and ω_c are found to shift the instability regions around the $\omega-\theta$ plane significantly.
- The decay rate of MI is found to be significantly suppressed by the effects of κ, i.e., when κ increases with cutoffs at lower wave numbers of modulation. However, it can be higher with increasing values of the density ratio μ. The effect of the external magnetic field

is to decrease the decay rate with cutoffs at higher values of the wave number of modulation.

The findings of the present investigation may be useful for the modulation of dust–ion acoustic wave envelopes in dusty superthermal plasmas such as those in laboratory [36], space [37–39] and astrophysical [40] environments.

Acknowledgements This work was supported by DRS-II(SAP) no. F 530/17/DRS-II/2015(SAP-I) UGC, New Delhi. A. P. M. acknowledges support from UGC-SAP (DRS, Phase III) with Sanction order no. F.510/3/DRS-III/2015(SAPI) dated 25/03/2015, and UGC-MRP with F. no. 43-539/2014 (SR) and FD Diary no. 3668 dated 17.09.2015.

References

1. Shukla, P.K., Mendis, D.A., Desai, T.: Advances in dusty plasmas. World Scientific, Singapore (1999)
2. Shukla, P.K., Mamun, A.A.: Introduction to dusty plasma physics. Institute of Physics, Bristol (2002)
3. Boufendi, L., Mikikian, M., Shukla, P.K.: New vistas in dusty plasmas. In: AIP Proceeding, AIP, New York (2005)
4. Shukla, P.K., Silin, V.P.: Dust ion-acoustic wave. Phys. Scrip. **45**, 508 (1992)
5. Barkan, A., D'Angelo, N., Merlino, R.: Experiments on ion-acoustic waves in dusty plasmas. Planet. Space. Sci. **44**, 239 (1996)
6. Summers, D., Thorne, R.M.: The modified plasma dispersion function. Phys. Fluids B **3**, 1835 (1991)
7. Sittler Jr., E.C., Ogilvie, K.W., Scudder, J.D.: Survey of low-energy plasma electrons in Saturn's magnetosphere: Voyagers 1 and 2. J. Geophys. Res. **88**, 8847 (1983)
8. Mace, R.L., Hellberg, M.A.: A dispersion function for plasmas containing superthermal particles. Phys. Plasmas **2**, 2098 (1995)
9. Vasyliunas, V.M.: A survey of low-energy electrons in the evening sector of the magnetosphere with OGO 1 and OGO 3. J. Geophys. Res. **73**, 2839 (1968)
10. Hellberg, M.A., Mace, R.L., Armstrong, R.J., Karlstad, G.: Electron-acoustic waves in the laboratory: an experiment revisited. J. Plasma Phys. **64**, 433 (2000)
11. Saini, N.S., Kourakis, I., Hellberg, M.A.: Arbitrary amplitude ion-acoustic solitary excitations in the presence of excess superthermal electrons. Phys. Plasmas **16**, 062903 (2009)
12. Shah, A., Mahmood, S., Haque, Q.: Propagation of solitary waves in relativistic electron-positron-ion plasmas with kappa distributed electrons and positrons. Phys. Plasmas **18**, 114501 (2011)
13. El-Tantawy, S.A., El-Bedwehy, N.A., Moslem, W.M.: Nonlinear ion-acoustic structures in dusty plasma with superthermal electrons and positrons. Phys. Plasmas **18**, 052113 (2011)
14. Sultana, S., Kourakis, I.: Electrostatic solitary waves in the presence of excess superthermal electrons: modulational instability and envelope soliton modes. Plasma Phys. Control. Fusion **53**, 045003 (2011)
15. Shahmansouri, M., Tribeche, M.: Propagation properties of ion acoustic waves in a magnetized superthermal bi-ion plasma. Astrophys. Space Sci. **350**, 623 (2014)
16. Adnan, M., Mahmood, S., Qamar, A.: Small amplitude ion acoustic solitons in a weakly magnetized plasma with anisotropic ion pressure and kappa distributed electrons. Adv. Sp. Res. **53**, 845 (2014)
17. Shahmansouri, M., Astaraki, E.: Transverse perturbation on three-dimensional ion acoustic waves in electron–positron–ion plasma with high-energy tail electron and positron distribution. J. Theor. Appl. Phys. **8**, 189 (2014)
18. Shalini K, Saini N.S., Misra A.P.: Modulation of ion-acoustic waves in a nonextensive plasma with two-temperature electrons. Phys. Plasmas. **22**, 092124 (2015)
19. Kourakis, I., Shukla, P.K.: Ion-acoustic waves in a two-electron-temperature plasma: oblique modulation and envelope excitations. J. Phys. A Math. Gen. **36**, 11901 (2003)
20. Kourakis, I., Shukla, P.K.: Electron-acoustic plasma waves: oblique modulation and envelope solitons. Phys. Rev. E. **69**, 036411 (2004)
21. Kourakis, I., Shukla, P.K.: Finite ion temperature effects on oblique modulational stability and envelope excitations of dust-ion acoustic waves. Euro. Phys. J. D. **28**, 109 (2004)
22. Kourakis, I., Shukla, P.K.: Oblique amplitude modulation of dust-acoustic plasma waves. Phys. Scr. **69**, 316 (2004)
23. Misra, A.P., Bhowmik, C.: Nonlinear wave modulation in a quantum magnetoplasma. Phys. Plasmas **14**, 012309 (2007)
24. Saini, N.S., Kourakis, I.: Dust-acoustic wave modulation in the presence of superthermal ions. Phys. Plasmas **15**, 123701 (2008)
25. Bains, A.S., Misra, A.P., Saini, N.S., Gill, T.S.: Modulational instability of ion-acoustic wave envelopes in magnetized quantum electron-positron-ion plasmas. Phys. Plasmas **17**, 012103 (2010)
26. El-Taibany, W.F., Kourakis, I.: Modulational instability of dust acoustic waves in dusty plasmas: modulation obliqueness, background ion nonthermality, and dust charging effects. Phys. Plasmas **13**, 062302 (2006)
27. El-Labany, S.K., El-Shewy, E.K., Abd El-Razek, H.N., El-Rahman, A.A.: Wave propagation in strongly dispersive superthermal dusty plasma. Adv. Sp. Res. **59**, 1962 (2017)
28. Ahmadihojatabad, N., Abbasi, H., Pajouh, H.H.: Influence of superthermal and trapped electrons on oblique propagation of ion-acoustic waves in magnetized plasma. Phys. Plasmas **17**, 112305 (2010)
29. Guo, S., Mei, L.: Three-dimensional dust-ion-acoustic rogue waves in a magnetized dusty pair-ion plasma with nonthermal nonextensive electrons and opposite polarity dust grains. Phys. Plasmas **21**, 82303 (2014)
30. Gharaee, H., Afghah, S., Abbasi, H.: Modulational instability of ion-acoustic waves in plasmas with superthermal electrons. Phys. Plasmas **18**, 032116 (2011)
31. Hellberg, M.A., Mace, R.L., Baluku, T.K., Kourakis, I., Saini, N.S.: Comment on "Mathematical and physical aspects of Kappa velocity distribution". Phys. Plasmas **16**, 094701 (2009)
32. Taniuti, T., Yajima, N.: Perturbation method for a nonlinear wave modulation. J. Math. Phys. **10**, 1369 (1969)
33. Asano, N., Taniuti, T., Yajima, N.: Perturbation method for a nonlinear wave modulation. J. Math. Phys. **10**, 2020 (1969)
34. Xue, J-K: Modulation of magnetized multidimensional waves in dusty plasma. Phys. Plasmas **12**, 062313 (2005)
35. Taniuti, T.: Reductive perturbation method and far fields of wave equations. Progress Theoret. Phys. Suppl. **55**, 1 (1974)
36. Liu, J.M., DeGroot, J.S., Matte, J.P., Johnston, T.W., Drake, R.P.: Measurements of inverse bremsstrahlung absorption and non-maxwellian electron velocity distributions. Phys. Rev. Lett. **72**, 2717 (1994)
37. Montgomery, M.D., Bame, S.J., Hundhausen, A.J.: Solar wind electrons: vela 4 measurements. J. Geophys. Res. **73**, 4999 (1968)
38. Maksimovic, M., Pierrard, V., Riley, P.: A kinetic model of the solar wind with Kappa distribution functions in the corona. Geophys. Res. Lett. **24**, 1151 (1997)
39. Zouganelis, I.: Measuring suprathermal electron parameters in space plasmas: implementation of the quasi-thermal noise spectroscopy with kappa distributions using in situ Ulysses/URAP radio measurements in the solar wind. J. Geophys. Res. **113**, A08111 (2008)

18

Low and high frequency instabilities in an explosion-generated-plasma and possibility of wave triplet

O. P. Malik · Sukhmander Singh · Hitendra K. Malik · A. Kumar

Abstract An explosion-generated-plasma is explored for low and high frequency instabilities by taking into account the drift of all the plasma species together with the dust particles which are charged. The possibility of wave triplet is also discussed based on the solution of dispersion equation and synchronism conditions. High frequency instability (HFI) and low frequency instability (LFI) are found to occur in this system. LFI grows faster with the higher concentration of dust particles, whereas its growth rate goes down if the mass of the dust is higher. The ion and electron temperatures affect its growth in opposite manner and the electron temperature causes this instability to grow. In addition to the instabilities, a simple wave is also observed to propagate, whose velocity is larger for larger wave number, smaller mass of the dust and higher ion temperature.

Keywords Dust particles · Explosion-generated-plasma · Dispersion equation · Low frequency instability · High frequency instability

O. P. Malik
Department of ECE, Al-Falah University,
Dhauj, Faridabad, Haryana, India

S. Singh
Motilal Nehru College, South Campus, Delhi University,
New Delhi 110 021, India

H. K. Malik (✉)
PWAPA Laboratory, Department of Physics, Indian Institute of Technology Delhi, New Delhi 110 016, India
e-mail: hkmalik@hotmail.com

A. Kumar
Department of Applied Sciences, Al-Falah University,
Dhauj, Faridabad, Haryana, India

Introduction

An electromagnetic interference on electronic systems due to high-power microwaves (HPMs) introduces noise or signals into the electronic systems. This could cause a temporary system malfunction and component degradation; even a permanent physical damage is also possible at high levels of irradiation. The HPM sources have been under investigation for several years as potential weapons for a variety of combat, sabotage, and terrorist applications [1–3]. The pulse released by an electromagnetic weapon lasts for an extremely short time, i.e. around 100 picoseconds. Hence, the absorption of this blast of high energy by anything capable of conducting electricity (including nerves and neurons) overwhelms the recipient. As a result, the computers used in data processing systems, communications systems, satellites, industrial controls, displays, military systems, radar, HF, VHF, UHF, and television equipment are all susceptible to the electromagnetic pulses (EMPs).

If we think about the explosion-generated-plasma (EGP), we find that the beams of ions and electrons are a source of free energy which can be transferred to waves. So instabilities can evolve in the nonlinear systems. If conditions are favourable, the resonant interaction of the waves in plasma can lead to nonlinear instabilities, in which all the waves grow faster than exponentially and attain enormously large amplitudes in a finite time or after a finite distance, depending on whether temporal or spatial growth is considered. These instabilities are referred to as explosive instabilities. Such instabilities could be of considerable practical interest, as these seem to offer a mechanism for rapid dissipation of coherent wave energy into thermal motion, and hence may be effective for plasma heating [4–7]. A consistent theory of explosive instability shows that in the three-wave approximation amplitudes of all the waves tend to infinity over a finite time called explosion time. The simplest wave coupling process that can exhibit explosive character is the coupling of three-waves with fixed phases [5, 8, 9]. For a wave triplet, all the three waves grow simultaneously. This phenomenon was first described by Cairns [10] using the kinetic equation which takes into account the interactions of waves with random phases and different signs of energy. Fainshtein and Chernova [11] have investigated the high power electromagnetic radiation from the development of explosive and high-frequency instabilities in a system consisting of a relativistic ion beam and a nonisothermal plasma. Based on an asymptotic method they derived and analyzed truncated equations for the complex mode amplitudes, and showed that the explosion is stabilized by a nonlinear frequency shift, while the high-frequency instability is analogous to the decay of low-frequency modes.

This can be seen that in most of the investigations the researchers have used the Boltzmann distribution of the electrons keeping in mind their mass to be negligible. Moreover, the electrons are employed to make background only. However, we consider the finite mass of the electrons and also take into account the dust particles which are always present in most of such plasma and whose charge may fluctuate due to currents flowing into the dust [12, 13]. In view of a strong nonisothermal plasma, we consider the temperature of electrons (T_e) to be much higher than that of the ions (T_i), i.e. $T_e \gg T_i$ [14–17]. For the sake of generality, we consider initial drift of all the plasma species and derive the dispersion equation for this system and solve it numerically for investigating the instabilities.

Basic model

The plasma is taken to compose electrons, singly charged ions and negatively charged dust grains of uniform mass and charge. Hence, the quasineutrality condition reads $n_{i0} = n_{e0} - \alpha Z_d n_{d0}$, where n_{i0}, n_{e0} and n_{d0} are the unperturbed number density of the ions, electrons and dust grains, respectively, and Z_d is the magnitude of the charge on the dust. The parameter α represents the nature of charge on the dust, and it is positive (negative) for the positively (negatively) charged dust grains.

Basic equations and dispersion relation

If $n_i(n_e)$ is the density of ions (electrons), $M(m)$ is the mass of ion (electron) and $\vec{v}_i(\vec{u}_e)$ is the ion (electron) fluid velocity along with their unperturbed values as v_0 and u_0 in the x-direction, then one-dimensional continuity equation and momentum equation for the ion, electron and dust fluids can be written as

$$(n_i)_t + (n_i v_i)_x = 0 \quad (1)$$

$$(n_e)_t + (n_e u_e)_x = 0 \quad (2)$$

$$(n_d)_t + (n_d v_d)_x = 0 \quad (3)$$

$$(u_e)_t + u_e (u_e)_x + \frac{M}{m n_e}(n_e)_x - \frac{M}{m}\phi_x = 0 \quad (4)$$

$$(v_i)_t + v_i (v_i)_x + \frac{\sigma}{n_i}(n_i)_x + \phi_x = 0 \quad (5)$$

$$(v_d)_t + v_d (v_d)_x + \frac{\alpha M}{m_d}\phi_x = 0 \quad (6)$$

The system of equations can be closed with the following Poisson's equation.

$$\phi_{xx} - n_e + n_i + n_d Z_d \alpha = 0. \quad (7)$$

In the above equations, the subscripts x and t denote the respective differentiation. The densities are normalized by a background density n_0, potential ϕ by T_e/e, time t by the inverse of frequency $\omega_{pi} = \sqrt{e^2 n_0/\varepsilon_0 M}$, velocities v_i, u_i, v_d by the ion acoustic speed $C_s = \sqrt{T_e/M}$ and length x by the Debye length $\lambda_{De} = \sqrt{\varepsilon_0 T_e/e^2 n_0}$.

The solution of the above equations is obtained for the variation of perturbed quantities as $\psi_1 \sim \exp(i\omega t - ikx)$ together with $\psi_1 \equiv n_{i1}, n_{e1}, n_{d1}, \vec{v}_{i1}, \vec{u}_{e1}, \phi_1, \vec{v}_{d1}$. Here ω is the frequency of oscillations and k is the wave number. Hence, the following expressions for the perturbed densities are obtained from the basic fluid equations.

$$n_{i1} = \frac{k^2 n_{i0}\phi_1}{(\omega - k v_0)^2 - \sigma k^2} \quad (8)$$

$$n_{e1} = \frac{k^2 \phi_1}{\frac{m}{M}(\omega - k u_0)^2 - k^2} \quad (9)$$

$$n_{d1} = \frac{M}{m_d}\frac{k^2 \alpha Z_d n_{d0}\phi_1}{(\omega - k v_{d0})^2} \quad (10)$$

The use of these expressions in the Poisson's Eq. (7) yields

$$\frac{k^2\phi_1}{\frac{m}{M}(\omega - k u_0)^2 - k^2} - \frac{k^2 n_{i0}\phi_1}{(\omega - k v_0)^2 - \sigma k^2} - \frac{k^2 \alpha^2 Z_d^2 n_{d0} M \phi_1}{m_d(\omega - k v_{d0})^2} + k^2\phi_1 = 0 \quad (11)$$

After simplifying the above equation, we get the following dispersion equation

$$\frac{m m_d}{M^2}\omega^6 - b_0\omega^5 + \omega^4 b_1 + \omega^3 b_2 + \omega^2 b_3 + \omega b_4 + a_1 a_2 a_3 + a_2 a_4 + a_1 a_4 (n_{i0} - a_2) = 0 \quad (12)$$

Here

$$b_0 = \frac{m_d}{M}\left(a_{11} + \frac{2m}{M}kv_{d0}\right), b_1 = a_{15} + \frac{m_d}{M}\left(1 - \frac{n_{i0}m}{M}\right) - a_3\frac{m}{M}, b_2 = a_3a_{11} - a_{16} + n_{i0}a_8 - a_5,$$

$$b_3 = a_3a_{12} - a_{17} + a_6 + n_{i0}a_9, b_4 = a_{18} - a_7 + a_3a_{13} + n_{i0}a_{10}, a_1 = k^2\left(1 - \frac{m}{M}u_0^2\right),$$

$$a_2 = k^2(v_0^2 - \sigma), a_3 = \alpha^2 Z_d^2 n_{d0}, a_4 = \frac{m_d}{M}k^2 v_{d0}^2$$

together with

$$a_5 = \frac{m_d}{M}k(v_0 + v_{d0}), a_6 = \frac{m_d}{M}k^2\left[v_0(4v_{d0} + v_0) + v_{d0}^2\right],$$

$$a_7 = 2\frac{m_d}{M}k^3 v_0 v_{d0}(v_{d0} + v_0),$$

$$a_8 = \frac{2mm_d}{M^2}k(u_0 + v_{d0}), a_9 = \frac{m_d}{M}\left[a_1 - \frac{m}{M}v_{d0}k^2(4u_0 + v_{d0})\right],$$

$$a_{10} = \frac{2m_d}{M}kv_{d0}\left[\frac{k^2u_0m}{M}v_{d0} - a_1\right],$$

$$a_{11} = \frac{2m}{M^2}k(u_0 + v_0), a_{12} = a_1 - \frac{m}{M}v_0k^2(4u_0 + v_0),$$

$$a_{13} = 2kv_0\left(\frac{m}{M}u_0k^2v_0 - a_1\right),$$

$$a_{15} = \frac{m_d}{M}\left[kv_{d0}\left(\frac{m}{M}kv_{d0} + 2a_{11}\right) - a_{12}\right],$$

$$a_{16} = \frac{m_d}{M}\left[a_{13} + kv_{d0}(kv_{d0}a_{11} - 2a_{12})\right],$$

$$a_{17} = \frac{m_d}{M}k\left[a_1kv_0^2 - 2a_{13}v_{d0} + kv_{d0}^2a_{12}\right],$$

$$a_{18} = \frac{m_d}{M}v_{d0}k^2\left[2a_1kv_0^2 - v_{d0}a_{13}\right].$$

Results and discussion

The dispersion Eq. (12) carries a term α^2, it means the same results are expected for the cases of positively charged dust grains and negatively charged dust grains. We numerically solve Eq. (12) by giving typical values to various parameters in view of the plasma generated in an explosion. Hence, $k = 0.5$, $v_{d0} = 0.002$, $n_{i0} = 1.1$, $M = 23.38 \times 10^{-27}$ kg, $m_d = 10^{-20}$ kg, $u_0 = 0.4$, $v_0 = 0.04$, $n_{nd0} = 0.001$, $Z_d = 100$, $T_i = 1$ eV and $T_e = 10$ eV [18–22]. Our numerical calculations show that two types of instabilities occur in the plasma if we consider the role of dust mass, which is present in the said plasma. The growth rates of these instabilities are found to be influenced by density and mass of the dust grains. For the parameters within this range, we observe that there are two roots that satisfy the condition for unstable growth of the disturbances. However, in some cases more roots are found, but their growth is too low, showing that these roots are very small perturbations in the system. We solve Eq. (12) for the complex root of $\omega(\equiv \omega_R - i\gamma)$ and plot the normalized growth rate γ in Figs. 1, 2, 3, 4. Since the normalization is

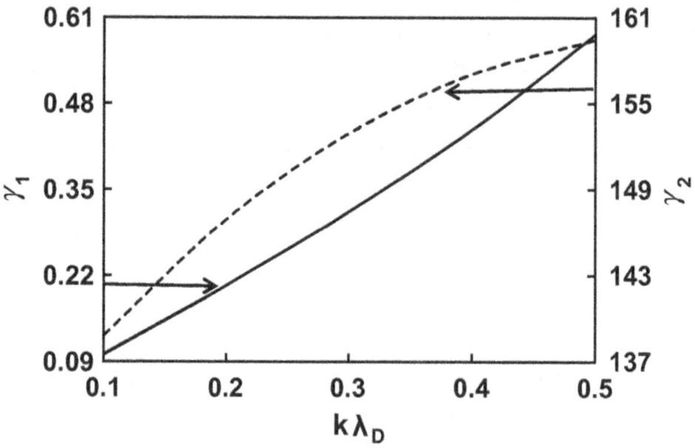

Fig. 1 Variation of growth rates with normalized wave number, when $v_{d0} = 0.002$, $n_{i0} = 1.1$, $M = 23.38 \times 10^{-27}$ kg, $m_d = 10^{-20}$ kg, $u_0 = 0.4$, $v_0 = 0.04$, and $n_{nd0} = 0.001$, $Z_d = 100$, $T_i = 1$ eV and $T_e = 10$ eV

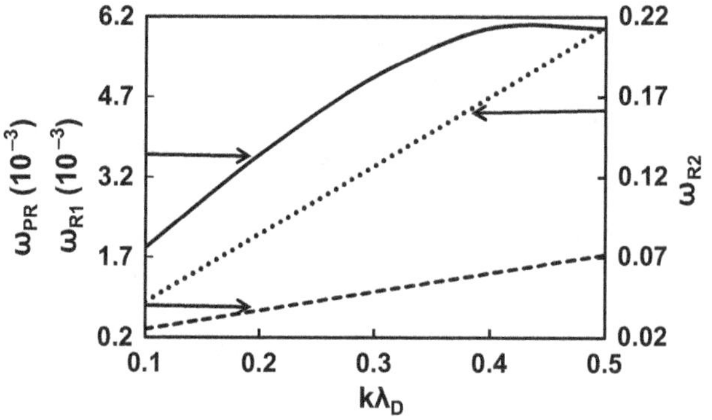

Fig. 2 Variation of propagating and real frequencies with normalized wave number (*dashed line*) when the other parameters are the same as in Fig. 1. The *solid line* (*dotted line*) corresponds to $\omega_{R1}(\omega_{R2})$

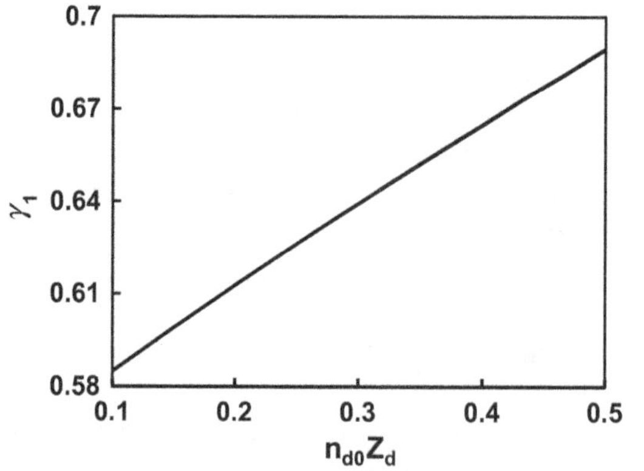

Fig. 3 Variation of growth rate γ_1 with dust density, when $k = 0.5$ and the other parameters are the same as in Fig. 1

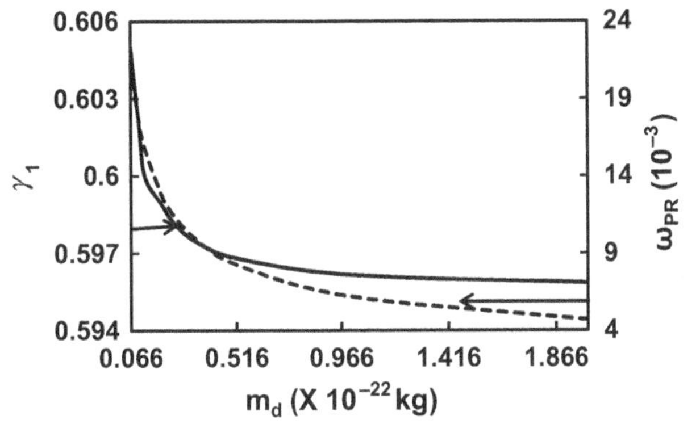

Fig. 4 Variation of growth rate γ_1 and propagating frequency with dust mass, when $k = 0.5$ and the other parameters are the same as in Fig. 1

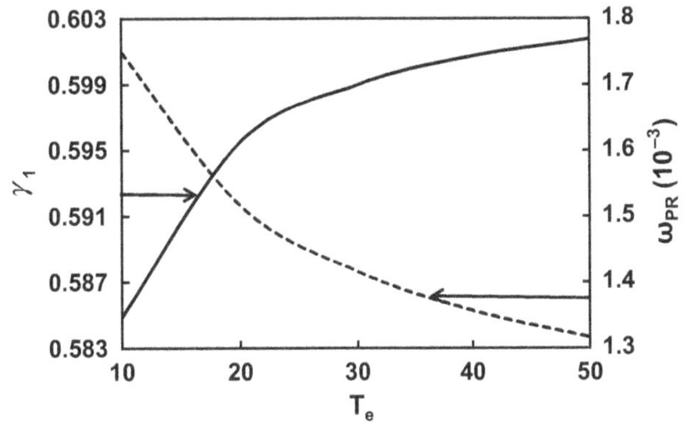

Fig. 6 Weak dependence of growth rate γ_1 and frequency of propagating mode on electron temperature, when $k = 0.5$ and the other parameters are the same as in Fig. 1

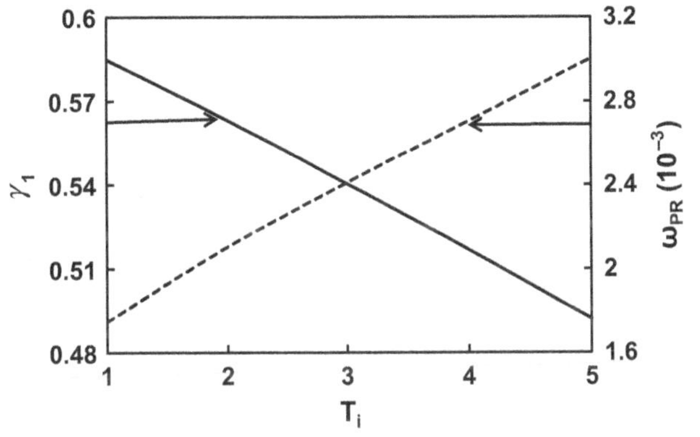

Fig. 5 Dependence of growth rate γ_1 on ion temperature, when $k = 0.5$ and the other parameters are the same as in Fig. 1

done with respect to ion plasma frequency, the growth rate $\gamma < 1$ is called to correspond to lower growth rate instability (LFI) whereas the growth rate $\gamma > 1$ to higher growth rate instability (HFI).

Figure 1 shows the variation of growth rates of the instabilities with wave number. Here γ_2 corresponds to the higher growth rate instability (called HFI) and γ_1 corresponds to the lower growth rate instability (LFI). Both the growth rates γ_1 (solid line) and γ_2 (dashed line) behave similarly and these go higher for the longer values of k.

Figure 2 shows that the higher rate instability is a constant velocity wave. However, the lower growth rate instability (the graph marked with ω_{R2}) is found to show wave number dependence behaviour. This becomes a constant frequency wave at the larger value of wave number. In addition to the two types of instabilities, this Figure shows that there exists a propagating mode also, whose velocity remains a constant.

Lower growth rate instability (LFI) is found to be influenced by the presence of dust grains. For example, Fig. 3 show that the growth rate γ_1 is increased when the dust grains in larger number are present. This is due to the negative charge on the dust, which provides larger restoring force to the oscillations of the species because of which the growth is enhanced. However, the mass of dust particles shows opposite effect on the growth rate γ_1, as expected. Similar effect of dust particles mass is observed on the frequency and hence, on the speed of propagating mode (please see the graph marked with ω_{PR} in Fig. 4).

The effect of ion and electron temperatures on the growth rate γ_1 is shown in Figs. 5 and 6, respectively. The ion temperature is found to reduce the growth of lower growth rate instability (LFI), whereas the electron temperature enhances the growth rate γ_1. However, the growth tries to saturate at larger value of electron temperature. Similar effect of electron temperature on the growth rate of dissipative instability has been observed by Keidar and Beilis [23]. The higher growth rate for the case of higher electron temperature can be explained based on the probability of the collisions. On the other hand, the propagating mode (marked with ω_{PR}) behaves oppositely with the ion and electron temperatures. The frequency ω_{PR} and hence, the speed of this wave is found to reduce with the electron temperature and to enhance with the ion temperature (Fig. 6).

Wave triplet analysis

Now we discuss the possibility of wave triplet in EGP having dust grains. For this, we reproduce Eq. (12) in the following form

$$\omega^3\left[\frac{mm_d}{M^2}\omega^3 - a_{14}\omega^2 + \omega b_1 + b_2\right] + \omega^2 b_3 + \omega b_4 + \{a_1a_2a_3 + a_2a_4 + a_1a_4(n_{i0} - a_2)\} = 0 \quad (13)$$

This equation shall yield three waves (triplet) if the coefficients b_3 and b_4 together with the term in curly bracket vanishes. It means

$$a_{12}a_3 - a_{17} + a_6 + n_{i0}a_9 = 0 \tag{14}$$

$$a_{18} - a_7 + a_3a_{13} + n_{i0}a_{10} = 0 \tag{15}$$

$$a_1a_2a_3 + a_2a_4 + a_1a_4(n_{i0} - a_2) = 0. \tag{16}$$

By solving these equations, we can get the following expression for the drift velocity of the dust grains.

$$\upsilon_{d0} = \frac{\frac{M}{m_d}a_3a_{13}\left[1 - \frac{m}{M}n_{i0} - \frac{a_{12}}{a_1a_2}(a_2 + n_{i0}a_1)\right] - (a_2 - a_1a_2 + n_{i0}a_1)\left(2\frac{m}{M}n_{i0}ku_0 - a_{13} - 2k\upsilon_0\right)}{2k(a_2 - a_1a_2 + n_{i0}a_1) - 2k\left(2\frac{m}{M}n_{i0}ku_0 - a_{13} - 2k\upsilon_0\right)^2} \tag{17}$$

This expression shows the dependence of the drift velocity on the densities, masses, temperatures and initial velocities of ions and electrons in addition to the mass of dust grains and wave number k.

Along with the use of this expression, Eq. (13) yields $\frac{mm_d}{M^2}\omega^3 - a_{14}\omega^2 + \omega b_1 + b_2 = 0$, which can be reformulated as

$$\omega^3 + a\omega^2 + b\omega + d = 0 \tag{18}$$

Here

$$a = -\frac{M^2a_0}{mm_d}, b = \frac{M^2b_1}{mm_d} \text{and} \quad d = \frac{M^2b_2}{mm_d}.$$

Every cubic equation with real coefficients has at least one solution ω among the real numbers. This is a consequence of the intermediate value theorem. We can distinguish several possible cases using the discriminant

$$\Delta = 18abd - 4a^3d + a^2b^2 - 4b^3 - 27d^2 \tag{19}$$

Since we are interested in propagating waves, three distinct real roots of Eq. (14), Δ should be positive.

Further, we assume these roots as α, β and γ. Hence

$$\omega^3 + a\omega^2 + b\omega + d = (\omega - \alpha)(\omega - \beta)(\omega - \gamma) \tag{20}$$

The roots of the above equation would satisfy

$$\alpha + \beta + \gamma = -a \tag{21}$$

$$\alpha\beta + \beta\gamma + \alpha\gamma = b \tag{22}$$

$$\alpha\beta\gamma = -d \tag{23}$$

For the occurrence of triplets, the roots α, β and γ should additionally meet the following condition

$$\alpha + \beta = \gamma. \tag{24}$$

Finally, we obtain

$$\alpha = \omega_1 \equiv -\frac{a^2 - \sqrt{a^4 - 32da}}{4a},$$

$\beta = \omega_2 \equiv -\frac{a^2 + \sqrt{a^4 - 32da}}{4a}$. The following condition is also required to be satisfied

$$a^3 + 8d - 4ab = 0.$$

The coefficients a, b and d show their dependence on the wave number k through the drift velocity υ_{d0} and other coefficients such as a_0, b_1 and b_2. Hence, it would be possible to achieve conditions for synchronism, i.e. $k_3 = k_1 + k_2$ and $\omega_1 = \omega_2 + \omega_3$ (which we have written as $\gamma = \alpha + \beta$) in the present explosion-generated-plasma. It means this plasma would support the excitation of wave triplet. This is also supported by Fig. 7, which is plotted based on the dispersion Eq. (12) and shows that a wave with negative energy is possible and the conditions for synchronism can be fulfilled [4, 15, 16, 23].

Conclusions

In an EGP, dispersion relation was derived and solved numerically to investigate the evolved instabilities under the effect of initial drifts of ions, electrons and charged dust particles. Two types of instabilities, naming LFI and HFI, were found in the said plasma in addition to a constant velocity propagating mode. The phase velocity of the mode and the growth rate of the LFI were observed to be reduced

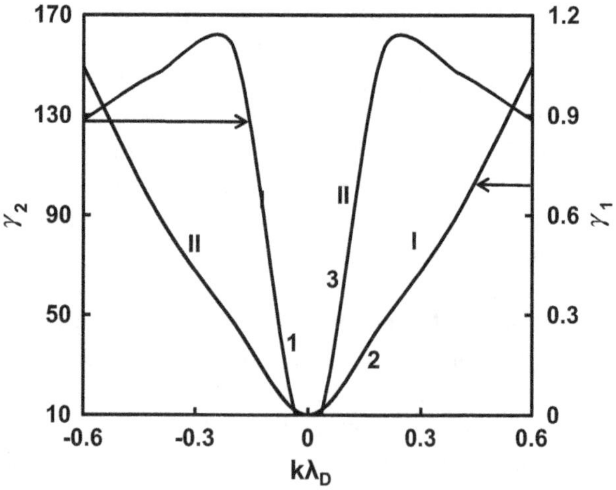

Fig. 7 Dispersion curves corresponding to Eq. (12)

for higher mass of the dust particles. Only the LFI showed the dependence on the ion and electron temperatures. These results are based on the fixed charge on the dust grains. However, we can estimate the consequences of dust charge fluctuations based on the nonlinear property of the plasma. This has been seen that the system becomes more nonlinear when the charge on the dust grains fluctuates. Since nonlinearity generally enhances the wave amplitude [12], higher amplitude instabilities are expected in the EGP if the charge on the dust grain is observed to fluctuate.

Acknowledgement The author HKM would like to thank CSIR, Government of India, for the financial support.

References

1. Giri, D.V., Tesche, F.M.: Classification of intentional electromagnetic environments. IEEE Trans. Electromagn. Compat. **46**, 323 (2004)
2. Efanov, V.: Gigawatt all solid state nano- and pico-second pulse generators for radar applications. In: Proceedings of 14th IEEE International Pulsed Power Conf., Dallas, TX (2003)
3. Staines, G.: Compact sources for tactical RF weapon applications (Diehl). In: Proceedings AMEREM, Annapolis, MD (2002)
4. Aamodt, R.E., Sloan, M.L.: Nonlinear interactions of positive and negative energy waves. Phys. Fluids **11**, 2218 (1968)
5. Wilhelmson, H.: On the explosive instabilities of waves in plasmas with special regard to dissipation and phase effects. Phys. Scr. **7**, 209 (1973)
6. Malik, H.K., Singh, S.: Resistive instability in a Hall plasma discharge under ionization effect. Phys. Plasmas **20**, 052115 (2013)
7. Singh, S., Malik, H.K.: Nishida.Y.: High frequency electromagnetic resistive instability in a Hall thruster under the effect of ionization. Phys. Plasmas **20**, 102109 (2013)
8. Wilhelmson, H., Weiland, J.: Coherent non-linear interaction of waves in plasmas. Pergamon Press, Oxford and New York (1977)
9. Wilhelmson, H., Stenflo, I., Engelmann, F.: Explosive instabilities in the well defined phase description. J. Math. Phys. **11**, 1738 (1970)
10. Cairns, R.A.: The role of negative energy waves in some instabilities of parallel flows. J. Fluid Meek **92**, 1 (1979)
11. Fainshtein, S.M., Chernova, E.A.: Generation of high-power electromagnetic radiation from the development of explosive and high-frequency instabilities in a system consisting of a relativistic ion beam and a nonisothermal plasma. JETP **84**, 442 (1996)
12. Tomar, R., Malik, H.K., Dahiya, R.P.: Reflection of ion acoustic solitary waves in a dusty plasma with variable charge dust. J. Theor. Appl. Phys. **8**, 126 (2014)
13. Malik, R., Malik, H.K.: Compressive solitons in a moving e-p plasma under the effect of dust grains and an external magnetic field. J. Theor. Appl. Phys. **7**, 65 (2013)
14. Landau, L.D., Lifshits, E.M.: Fluid Mechanics. Addison-Wesley, New York (1959)
15. Coppi, B., Rosenbluth, M.N., Sudan, R.N.: Nonlinear interactions of positive and negative energy modes in rarefied plasmas(I). Ann. Phys. **55**, 207 (1969)
16. Moiseev, S.S., Oraevsky, V.N., Pungin, V.G.: Nonlinear instabilities in plasmas and hydrodynamics. Taylor & Francis (1999)
17. Fainshtein, S.M.: On the possibility of generation of high-power low-frequency radiation as a result of evolution of explosive instability in the flow–nonisothermal plasma system. Radiophys. Quantum Electron J. **54**, 193 (2011)
18. Luo, Q.-Z., D'Angelo, N., Merlino, R.L.: Experimental study of shock formation in a dusty plasma. Phys. Plasmas **6**, 3455 (1999)
19. Merlino, R.L., Barkan, A., Thompson, C., D'Angelo, N.: Laboratory studies of waves and instabilities in dusty plasmas. Phys. Plasmas **5**, 1607 (1998)
20. Shukla, P.K., Mamun, A.A.: Dust-acoustic shocks in a strongly coupled dusty plasma. IEEE Trans. Plasma Sci. **29**, 221 (2001)
21. Nejoh, Y.N.: Double layers, spiky solitary waves, and explosive modes of relativistic ion acoustic waves propagating in a plasma. Phys. Fluids B **4**, 2830 (1992)
22. Esipchuk, Y.V., Tilinin, G.N.: Drift instability in a Hall-current plasma accelerator. Sov. Phys. Tech. Phys. **21**, 417 (1976)
23. Keidar, M., Beilis, I.I.: Electron transport phenomena in plasma devices with $E \times B$ Drift. IEEE Trans. Plasma Sci. **34**, 804 (2006)

19

Dynamic structures of nonlinear ion acoustic waves in a nonextensive electron–positron–ion plasma

Uday Narayan Ghosh[1] · Asit Saha[1,2] · Nikhil Pal[1] · Prasanta Chatterjee[1]

Abstract The dynamic structures of ion acoustic waves in an unmagnetized plasma with q-nonextensive electrons and positrons are investigated applying the bifurcation theory of planar dynamical systems through direct approach. Model equations are transformed to a planar dynamical system using a traveling wave transformation. Using the bifurcations of planar dynamical system, the existence of solitary and periodic waves is shown. We have obtained new analytical forms for solitary and periodic waves depending on parameters p, q, σ and v. Considering an external periodic perturbation, the chaotic behavior of nonlinear ion acoustic waves is presented. Depending upon different regimes of the nonextensive parameter q, the effect of q is shown on chaotic motions of ion acoustic waves with fixed values of other parameters p, σ and v. It is seen that the unperturbed system has the solitary and periodic wave solutions, but the perturbed dynamical system has chaotic motions for same values of parameters p, q, σ and v.

Keywords Solitary wave · Periodic wave · Chaotic behavior · Bifurcation theory

✉ Asit Saha
asit_saha123@rediffmail.com

1 Department of Mathematics, Siksha Bhavana, Visva Bharati University, Santiniketan 731235, India

2 Department of Mathematics, Sikkim Manipal Institute of Technology, Majitar, Rangpo, East-Sikkim 737136, India

Introduction

The nineteenth century and half of twentieth century can be viewed as the triumph of linear physics, which started with Maxwell's equations, based on a linear formalism emphasizing a superposition principle. But the physicists had noticed the importance of nonlinear phenomena which appeared in the momentum balance equation of electro-hydrodynamics, gravitational theory, etc. The importance of an intrinsic analysis of nonlinear phenomena has been gradually understood, and led to two concepts, the strange attractor and the soliton. Both are related to astonishing properties of nonlinear systems, the strange attractor is linked to the idea of chaos [1] in a system with small number of degree of freedom, while the solitons appear in the systems with the large number of degree of freedom. The study of interesting solitonic structures, periodic solution, and chaotic structures [2–5] in plasma dynamics is very important and curious. Therefore, the investigation of various structures like solitonic, periodic, quasi-periodic, and chaotic in nonlinear plasma dynamics is a growing research field of plasma physics. Some of the nonlinear evolution equations like Kortewg-de Vries (KdV), Kortewg-de Vries Burgers (KdVB), etc., arisen from many physical fields are completely integrable [6, 7]. It is known that a completely integrable nonlinear system possesses some nice properties like the Lax pair, N-soliton solutions, infinite conservation laws, Painlev property and bi-Hamiltonian structure. However, there often exist various perturbations in many real physical processes [8–10]. The addition of a perturbation or forcing term to an integrable equation can lead to chaotic dynamics [1], while deterministic chaos is one of the most interesting nonlinear phenomena. In the present paper, we want to study dramatic changes of structures from periodic to chaotic or solitonic to chaotic of ion acoustic waves in electron–positron–ion plasmas through direct approach. Indeed electrons are often accelerated to energies of tens of MeV by the electric field induced during the disruptive instability in tokamaks [11]. The resulting beam of runway electrons can carry up to about half of the original plasma current. At these high energies, electron–positron pairs can be created in collisions between the runaway electrons and background plasma ions and electrons. Helander and Ward [12] estimated the number of such pairs and discussed the fate of the positrons created in this way. The experiments [13–16] have established the possibility of creating a nonrelativistic electron–positron plasma in the laboratory. There are at least two schemes in which the nonrelativistic electron–positron plasma can be produced in the laboratory. In one scheme, a relativistic electron beam impinges on a high-Z target, where positrons are produced copiously. The relativistic pair plasma is then trapped in a magnetic mirror and is expected to cool rapidly by radiation [17]. In another scheme, positrons are accumulated from a radioactive source [15]. The production of pure positron plasmas [13, 15, 18] now makes it possible to perform laboratory experiments on electron–positron plasmas. A natural extension of this research is to learn how to accumulate and store sufficient numbers of positrons so that they behave as a collective, many-body system. Surko et al. [15] have developed a method to accumulate and store positrons in an electrostatic trap using a tungsten moderator and inelastic collisions with nitrogen gas. The resulting positron gas fulfills the requirements on density n and temperature T for it to act collectively as a classical, single-component positron plasma. The electron–positron plasmas occur in many astrophysical environments such as the inner regions of the accretion disks surrounding black holes [19], the center of our galaxy [20], the early universe [21], the polar regions of neutron stars [22], active galactic nuclei [23], or pulsar magnetosphere [24], and in solar atmosphere [25] together with small number of ions. These types of three-component e–p–i plasmas can also be found in the laboratory plasma, for example, during the propagation of a short relativistic strong laser pulse in matter, and photo production of pairs due to the photon scattering by nuclei can lead to the formation of e–p–i plasmas [26, 27]. Indeed, electron–positron plasmas represent a large class of equal-mass plasmas, a class of plasmas that may offer plasma physical properties quite different from those of conventional ion–electron plasmas. Clearly, the properties of wave motions in an electron–positron–ion plasma should be different from those in two-component electron–positron plasmas. A great deal of attention has been paid to study the electron–positron–ion plasmas during the last three decades [28–34].

Out of the existence of electron–positron–ion plasmas in various physical plasma situations, nonextensive electron–positron–ion plasmas is the most studied research field due to the limitation of proper implementation of Maxwell distribution in long-range interactions in unmagnetized collision less plasma where the nonequilibrium stationary state exists. Space plasma observations clearly indicate the presence of ion and electron populations that are far away from their thermodynamic equilibrium [35–39]. A new statistical approach, [40] namely nonextensive statistics or Tsallis statistics based on the derivation of Boltzmann–Gibbs–Shannon (BGS) entropic measure, [41] is proposed to the study the cases where Maxwell distribution is considered inappropriate. This was first acknowledged by Reni [40] and afterward proposed by Tsallis [41], where the entropic index q characterized the degree of non extensivity of the considered system. The parameter q that underpins the generalized entropy of Tsallis is linked to the underlying dynamics of the system and measures the amount of its nonextensivity. In statistical mechanics and thermodynamics, systems characterized by the property of nonextensivity are systems for which the entropy of the whole is different from the sum of the entropies of the respective parts. In other words, the generalized entropy of the whole is greater than the sum of the entropies of the parts if $q < 1$ (superextensivity), whereas the generalized entropy of the system is smaller than the sum of the entropies of the parts if $q > 1$ (subextensivity). In accordance with the evidences found earlier [40–52], the q-entropy may provide a convenient frame for the analysis of many astrophysical scenarios, such as stellar poly tropes, solar neutrino problem, and peculiar velocity distribution of galaxy cluster. To study all possible astrophysical scenarios, it is wise to follow the nonextensive distribution. As electrons and positrons have the same mass but opposite charge, it is expected that they will be described by a similar distribution. Shahmansouri and Alinejad [53] studied the effect of electron nonextensivity on oblique propagation of arbitrary ion acoustic waves in a magnetized plasma. Shahmansouri and Astaraki [54] investigated the transverse perturbation on three-dimensional ion acoustic waves in electron–positron–ion plasma with high-energy tail electron and positron distribution. Shahmansouri and Alinejad [55] also investigated arbitrary amplitude electron acoustic (EA) solitary waves in a magnetized nonextensive plasma comprising cool fluid electrons, hot nonextensive electrons, and immobile ions. Sabetkar and Dorranian [56] studied the nonextensive effects on the characteristics of dust-acoustic solitary waves in magnetized dusty plasma with two-temperature isothermal ions.

Recently, Samanta et al. [57] studied bifurcations of dust-ion acoustic traveling waves in a magnetized dusty plasma with a q-nonextensive electron velocity distribution using bifurcation theory of planar dynamical systems for the first time in the literature. A number works [58–66] on bifurcations of nonlinear waves in plasmas have been reported through perturbative and nonperturbative approaches. Saha and chatterjee [67] studied propagation and interaction of dust-acoustic multi-soliton in dusty plasmas with q-nonextensive electrons and ions. Very recently, Saha et al. [2] investigated the dynamic behavior of ion acoustic waves in electron–positron–ion magnetoplasmas with superthermal electrons and positrons. Sahu et al. [3] studied the quasi-periodic behavior in quantum plasmas due to the presence of bohm potential. Zhen et al. [4] studied dynamic behavior of the quantum ZK equation in dense quantum magnetoplasma. Zhen et al. [5] also studied soliton solution and chaotic motion of the extended ZK equations in a magnetized dusty plasmas with Maxwellian hot and cold ions.

The remaining part of the paper is organized as follows: In "Basic equations" section, we consider basic equations. In "Planar dynamical system and phase portraits" section, we obtain a planar dynamical system and corresponding phase portraits. New solitary and periodic wave solutions are derived in "New solitary and periodic wave solutions" section. We present the chaotic behavior of the perturbed system in "Chaos in the perturbed system" and "Conclusions" sections are kept for conclusions.

Basic equations

In this work, we consider a three-component collisionless unmagnetized plasma containing inertial ions, and q-nonextensive velocity distributed electrons and positrons. In equilibrium, the charge neutrality condition is $n_{e0} = n_{p0} + n_0$, where n_{e0}, n_{p0} and n_0 are the unperturbed number densities of electron, positron and ion, respectively. The dynamics of nonlinear ion acoustic waves in such plasma is described by the following normalized equations:

$$\frac{\partial n}{\partial t} + \frac{\partial (nu)}{\partial x} = 0, \tag{1}$$

$$\frac{\partial u}{\partial t} + u\frac{\partial u}{\partial x} = -\frac{\partial \phi}{\partial x}, \tag{2}$$

$$\frac{\partial^2 \phi}{\partial x^2} = n_e - n_p - n. \tag{3}$$

The density of the q-nonextensive electrons and positrons are given by

$$n_e = \frac{1}{1-p}\{1 + (q-1)\phi\}^{\frac{1}{q-1}+\frac{1}{2}}, \tag{4}$$

$$n_p = \frac{p}{1-p}\{1 - (q-1)\sigma\phi\}^{\frac{1}{q-1}+\frac{1}{2}}, \tag{5}$$

where n_e, n_p, and n are the number densities of electrons, positrons and ions, respectively, normalized by their unperturbed densities. In this case, u and ϕ are the ion fluid velocity and electrostatic potential, respectively, normalized by the ion acoustic speed $c = (T_e/m)^{1/2}$, and T_e/e, where e is the electron charge and m is the mass of ions. The time variable is normalized by inverse of ion plasma frequency $\omega^{-1} = (m/4\pi n_0 e^2)^{1/2}$ and the space variable is normalized by the Debye length $= (T_e/4\pi n_0 e^2)^{1/2}$, respectively. Here $p = n_{p0}/n_{e0}$, and $\sigma = T_e/T_p$.

The state of a plasma is kinetically characterized by the one-particle distribution function $f(\vec{x}, \vec{v}, t)$. The quantity $f(\vec{x}, \vec{v}, t)d^3xd^3v$ gives, at each time t, the number of particles in the volume element d^3xd^3v around the particle position $\vec{x}$ and velocity $\vec{v}$. In principle [46], this distribution function verifies the q-nonextensive Boltzmann transport equation or Vlasov equation

$$\frac{\partial f}{\partial t} + v\frac{\partial f}{\partial x} + \frac{k}{m_e}\frac{\partial f}{\partial v} = C_q(f),$$

where C_q denotes the q-collisional term. Here, nonextensivity effects can be incorporated only through the collisional term under the consideration that the C_q is consistent with the energy, momentum, and particle number conservation laws. To generalize the usual Boltzmann–Gibbs thermostatics according to the demand of thermodynamical or statistical description of nonextensive systems, the standard Boltzmann–Gibbs approach based on the extensive entropy measure $S = -k\sum_i p_i ln p_i$, where k is the Boltzmann constant and p_i denotes the probabilities of microscopic configurations modified by Tsallis [41, 42] in the following nonextensive form of entropy $S_q = k\frac{1-\sum_i p_i^q}{q-1}$, where q is a parameter quantifying the degree of nonextensivity. Also Tsallis [41, 42] measure verifies $S_q(A+B) = S_q(A) + S_q(B) + (1-q)S_q(A)S_q(B)$. In the limit $q \to 1$, S_q reduces to the standard logarithmic measure and the usual additivity of the entropy is recovered. Advancing in this manner [45], one can get the following q-distribution function

$$f_e(v) = C_q\left\{1 + (q-1)\left[\frac{m_e v^2}{2T_e} - \frac{e\phi}{T_e}\right]\right\}^{\frac{1}{q-1}}.$$

The variables or parameters have their usual meaning. It may be noted that $f_e(v)$ is the particular distribution that maximizes the Tsallis entropy and therefore conforms to the laws of thermodynamics. The normalization constant C_q is given by

$$C_q = n_{e0}\frac{\Gamma\left(\frac{1}{1-q}\right)}{\Gamma\left(\frac{1}{1-q} - \frac{1}{2}\right)}\sqrt{\frac{m_e(1-q)}{2\pi T_e}} \quad \text{for} \quad -1<q<1;$$

$$C_q = n_{e0}\left(\frac{1+q}{2}\right)\frac{\Gamma\left(\frac{1}{q-1}+\frac{1}{2}\right)}{\Gamma\left(\frac{1}{q-1}\right)}\sqrt{\frac{m_e(q-1)}{2\pi T_e}} \quad \text{for} \quad q>1;$$

where the parameter q stands for the strength of nonextensivity. It may be useful to note that $q< -1$, the q-distribution is unnormalizable. It should be noted that for $q>1$, the q-distribution function exhibits a thermal cutoff on the maximum value allowed for the velocity of the particles, which is given by

$$v_{\max} = \sqrt{\frac{2T_e}{m_e}\left(\frac{e\phi}{T_e}+\frac{1}{q-1}\right)};$$

we get

$$n_e(\phi) = \int_{-\infty}^{\infty} f_e(v)\mathrm{d}v, \quad \text{for} \quad -1<q<1;$$

$$n_e(\phi) = \int_{-v\max}^{+v_{\max}} f_e(v)\mathrm{d}v, \quad \text{for} \quad q>1.$$

The derivation of nonextensive distribution from the density function gives

$$n_e = [1+(q-1)\phi]^{\frac{1+q}{2(q-1)}}.$$

In stead of gaussian profile one, q-nonextensive electrons satisfy a power law distribution which reduces to the Maxwellian distribution as $q \to 1$. It should be emphasized that the physical state described by the q-distribution is not the thermodynamic equilibrium. The nonextensive parameter q was proved to relate to the temperature gradient and the potential energy of the system in terms of the formula $k_B\nabla T_e + (1-q)Q\nabla\phi = 0$. Thus, the deviation of q from unity qualifies the degree of the inhomogeneity of temperature or the deviation from the equilibrium [69]. It is shown clearly from the above formula that the nonextensive parameter is $q \neq 1$ if and only if the temperature gradient is $\nabla T \neq 0$, which gives a clear physics of $q \neq 1$ with regard to the nature of nonisothermal configurations of plasma systems with the Coulombian long-range interactions. The above formula is a mathematical expression of the nonextensive parameter q, and it gives a clearly physical meaning of $q \neq 1$ about temperature gradient and the Coulombian force on an electron in the nonisothermal plasma. If $\nabla T = 0$, the system becomes isothermal, and we have $q = 1$, which corresponds to the thermal equilibrium state for which B–G statistics has presented well description. While if $\nabla T \neq 0$, then $q \neq 1$, which corresponds to the case of Tsallis statistics. We therefore conclude that Tsallis statistics can deal with the nonisothermal nature in plasma systems with the Coulombian long-range interactions [68, 69]. The physical meaning of nonextensive parameter of electron (q) different from 1 can be explained [69], respectively, by the relations, $(1-q)e\nabla\phi = k_B\nabla T_e$.

Planar dynamical system and phase portraits

In this section, we transform our model equations into a planar dynamical system. To do so, we introduce a new variable $\xi = x - vt$, where v is the velocity of the ion acoustic traveling wave. Substituting the new variable ξ into Eqs. (1) and (2) and using the initial condition $u = 0, n = 1$, and $\phi = 0$, we can express the ion number density as

$$n = \frac{v}{\sqrt{v^2 - 2\phi}}. \tag{6}$$

Substituting Eqs. (4), (5), and (6) into Eq. (3) and considering the terms involving ϕ up to third degree, we have

$$\frac{\mathrm{d}^2\phi}{\mathrm{d}\xi^2} = a\phi + b\phi^2 + c\phi^3, \tag{7}$$

where $a = \frac{(q+1)(1+p\sigma)}{2(1-p)} - \frac{1}{v^2}$, $b = \frac{(q+1)(3-q)(1-p\sigma^2)}{8(1-p)} - \frac{3}{2v^4}$, and $c = \frac{(q+1)(3-q)(5-3q)(1+p\sigma^3)}{48(1-p)} - \frac{5}{2v^6}$.

Then, Eq. (7) is equivalent to the following planar dynamical system:

$$\begin{cases} \dfrac{\mathrm{d}\phi}{\mathrm{d}\xi} = z, \\ \dfrac{\mathrm{d}z}{\mathrm{d}\xi} = a\phi + b\phi^2 + c\phi^3. \end{cases} \tag{8}$$

It is important to note that a system of planar equations $\frac{\mathrm{d}\phi}{\mathrm{d}\xi} = f_1(\phi, z)$, $\frac{\mathrm{d}z}{\mathrm{d}\xi} = f_2(\phi, z)$ is called a Hamiltonian system (in classical mechanics) if there exists a function $H(\phi, z)$ such that $f_1 = \frac{\partial H}{\partial z}$ and $f_2 = -\frac{\partial H}{\partial \phi}$. A necessary and sufficient condition for a planar system $\frac{\mathrm{d}\phi}{\mathrm{d}\xi} = f_1(\phi, z)$, $\frac{\mathrm{d}z}{\mathrm{d}\xi} = f_2(\phi, z)$ to be Hamiltonian is that $\frac{\partial f_1}{\partial \phi} + \frac{\partial f_2}{\partial z} = 0$.

The system (8) is a planar Hamiltonian system with Hamiltonian function:

$$H(\phi, z) = \frac{z^2}{2} - a\frac{\phi^2}{2} - b\frac{\phi^3}{3} - c\frac{\phi^4}{4} = h, \text{say}. \tag{9}$$

The system Eq. (8) is a planar dynamical system with parameters q, p, σ and v. It is clear that the phase orbits defined by the vector fields of Eq. (8) will determine all traveling wave solutions of Eq. (7). We will study the bifurcations of phase portraits of Eq. (8) in the (ϕ, z) phase plane depending on the parameters. A homoclinic orbit of Eq. (8) gives a solitary wave solution of Eq. (7). Similarly, a periodic orbit of Eq. (8) gives a periodic traveling wave solution of Eq. (7).

We study the bifurcation set and phase portraits of the planar dynamical system (8). Clearly, on the (ϕ, z) phase plane, the abscissas of equilibrium points of system (8) are the zeros of $f(\phi) = \phi(\phi^2 + \frac{b}{c}\phi + \frac{a}{c})$. Let $E_i(\phi_i, 0)$ be an equilibrium point of the dynamical system (8) where $f(\phi_i) = 0$. When $b^2 - 4ac > 0$, there exist three equilibrium points at $E_0(\phi_0, 0)$, $E_1(\phi_1, 0)$, and $E_2(\phi_2, 0)$, where $\phi_0 = 0$, $\phi_1 = \frac{-b+\sqrt{b^2-4ac}}{2c}$, and $\phi_2 = \frac{-b-\sqrt{b^2-4ac}}{2c}$. If $M(\phi_i, 0)$ is the coefficient matrix of the linearized system of the dynamical system (8) at an equilibrium point $E_i(\phi_i, 0)$, then we get

$$J = \det M(\phi_i, 0) = -cf'(\phi_i). \tag{10}$$

By the theory of planar dynamical systems [70, 71], it is clear that the equilibrium point $E_i(\phi_i, 0)$ of the planar dynamical system 8 is a saddle point when $J < 0$ and the equilibrium point $E_i(\psi_i, 0)$ of the planar dynamical system (8) is a center when $J > 0$.

Applying the systematic analysis of parameters q, p, σ, and v, we have presented the phase portrait of the system (8) in Figs. 1 and 2. In Fig. 1, we have presented the phase portrait of the system (8) for $q = -0.8, p = 0.5, \sigma = 0.6$, and $v = 1.6$. Thus, the velocity of the ion acoustic traveling wave is sonic. There are three equilibrium points of the system (8) at $E_0(\phi_0, 0)$, $E_1(\phi_1, 0)$, and $E_2(\phi_2, 0)$ with $\phi_2 < 0 < \phi_1$. The equilibrium points $E_1(\phi_1, 0)$, $E_2(\phi_2, 0)$ are saddle points and $E_0(\phi_0, 0)$ is a center. There is a homoclinic orbit at the equilibrium point $E_2(\phi_1, 0)$ enclosing the center at $E_0(\phi_0, 0)$ which is surrounded by a family of periodic orbits. In Fig. 2, we have shown the phase portrait of the system (8) for $q = 0.1, p = 0.5, \sigma = 0.6$ and $v = 1$. In this case, there are three equilibrium points of the system (8) at $E_0(\phi_0, 0)$, $E_1(\phi_1, 0)$, and $E_2(\phi_2, 0)$ with $\phi_1 < 0 < \phi_2$. The equilibrium points $E_1(\phi_1, 0)$, $E_2(\phi_2, 0)$ are centers and $E_0(\phi_0, 0)$ is a saddle point. There is a pair of homoclinic orbits at the equilibrium point $E_0(\phi_1, 0)$ enclosing the centers at $E_1(\phi_1, 0)$ and $E_2(\phi_2, 0)$ which are surrounded by a family of periodic orbits.

It is to be noted that for $q > 1$ with fixed values of other parameters ($p = 0.5, \sigma = 0.6$, and $v = 1$), the type of the

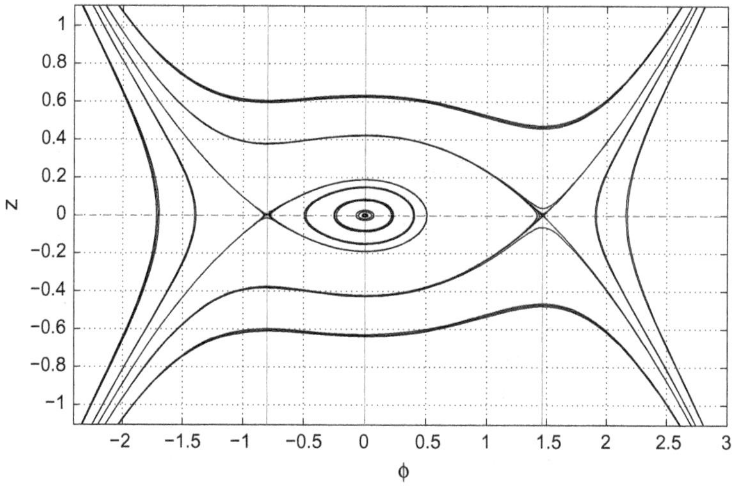

Fig. 1 Phase portrait of Eq. (8) for $q = -0.8, p = 0.5, \sigma = 0.6$ and $v = 1.6$

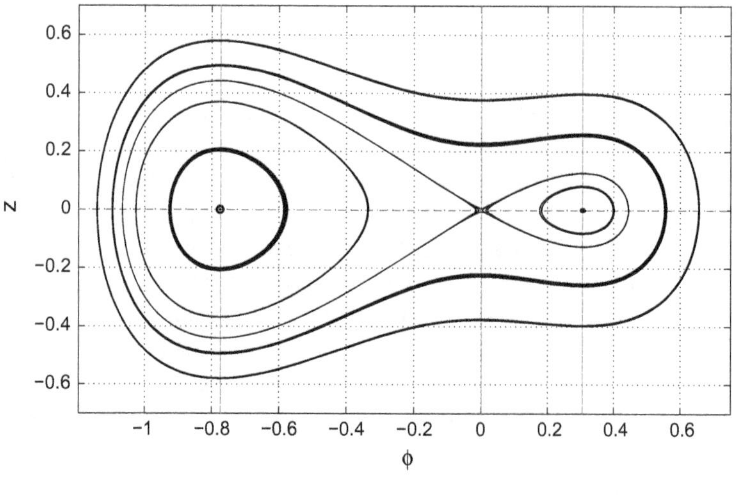

Fig. 2 Phase portrait of Eq. (8) for $q = 0.1, p = 0.5, \sigma = 0.6$, and $v = 1$

phase portrait is same as Fig. 2. So the phase portrait for $q > 1$ is not presented.

New solitary and periodic wave solutions

In this section, we present solitary wave solutions and periodic wave solutions with the help of the dynamical system (8) and the Hamiltonian function (9). It is important to note that if a phase portrait of a dynamical system has a homoclinic orbit at an equilibrium point of the system, then the system has a solitary wave solution corresponding to the homoclinic orbit at that point. If a phase portrait of a dynamical system has a family of periodic orbits about an equilibrium point of the system, then the system has a family of periodic wave solutions corresponding to the family of periodic orbits about that point. It should be noted that $\mathrm{sn}(\Omega\xi, k)$ is the Jacobian elliptic function [72] with the modulo k.

(1) The dynamical system (8) has a family of periodic orbits about the equilibrium point $E_0(\phi_0, 0)$ in Fig. 1 described by $H(\phi, z) = h$, $h \in (h_2, 0)$, where $h_2 = H(\phi_2, 0)$. Corresponding to this family of periodic orbits about $E_0(\phi_0, 0)$, our system has a family of periodic wave solutions:

$$\phi(\xi) = \frac{(\beta_1 - \gamma_1)\delta_1 \mathrm{sn}^2(\Omega\xi, k) - \gamma_1(\beta_1 - \delta_1)}{(\beta_1 - \gamma_1)\mathrm{sn}^2(\Omega\xi, k) - (\beta_1 - \delta_1)}, \tag{11}$$

with $\Omega = \sqrt{-\frac{c}{8}(\beta_1 - \delta_1)(\gamma_1 - \alpha_1)}$, $k = \sqrt{\frac{(\alpha_1-\delta_1)(\beta_1-\gamma_1)}{(\alpha_1-\gamma_1)(\beta_1-\delta_1)}}$, where $\alpha_1, \beta_1, \gamma_1$, and δ_1 are roots of the equation $h + \frac{c}{4}\phi^4 + \frac{b}{3}\phi^3 + \frac{a}{2}\phi^2 = 0$, with $\alpha_1 > \beta_1 > \gamma_1 > \delta_1$, $h \in (h_2, 0)$.

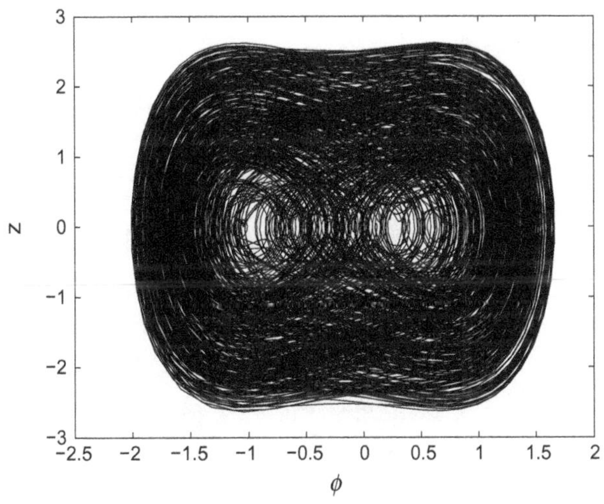

Fig. 3 Phase portrait of the perturbed system (13) for $q = -0.01$, $p = 0.5$, $\sigma = 0.6$, $v = 1$, $f_0 = 1$ and $\omega = 1$ (initial condition $\phi = 0.23, z = 0.1$)

(2) The dynamical system (8) has a pair of homoclinic orbits about the equilibrium point $E_0(\phi_0, 0)$ in Fig. 2 described by $H(\phi, z) = 0$. Corresponding to this pair of homoclinic orbits at $E_0(\phi_0, 0)$, our system has both compressive and rarefactive solitary wave solutions:

$$\phi(\xi) = \pm \frac{1}{\sqrt{2\left(1 - \frac{b^2}{9ac}\right)\sin\left(2\sqrt{\frac{a}{c}}\xi\right) + \frac{b}{6a}}}. \tag{12}$$

It is important to note that one can obtain solitary wave solution corresponding to the homoclinic orbit at $E_2(\phi_2, 0)$ in Fig. 1. Similarly, one can obtain two families of periodic wave solutions corresponding to two families of periodic

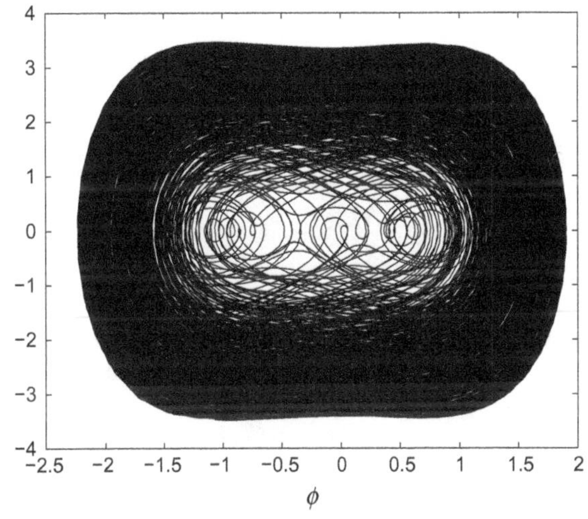

Fig. 4 Phase portrait of the perturbed system (13) for $q = 0.1$ with same values of other parameters as Fig. 3 (initial condition $\phi = 0.3, z = 0.1$)

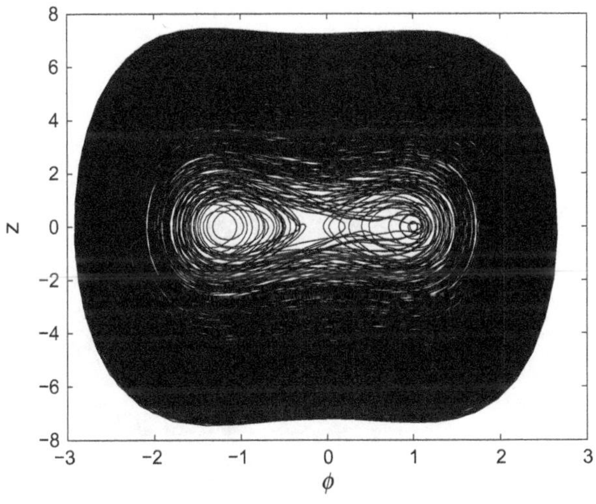

Fig. 5 Phase portrait of the perturbed system (13) for $q = 2$ with same values of other parameters as Fig. 3 (initial condition $\phi = 1, z = 0.8$)

orbits about $E_1(\phi_1, 0)$ and $E_2(\phi_2, 0)$ in Fig. 2. In the work [61], the authors derived compressive solitary wave solution involving $\mathrm{sech}^2\xi$ corresponding to the homoclinic orbit at the saddle point (see Fig. 4 in [61]) and periodic wave solutions involving $\sec^2\xi$ corresponding to the periodic orbits about the center (see Fig. 2 in [61]) of the dynamical system. But, in the present work, we obtain a family of periodic wave solutions (11) involving Jacobian elliptic function $\mathrm{sn}^2(\Omega\xi, k)$ corresponding to the family of periodic orbits about the center $E_0(\phi_0, 0)$ in Fig. 1. We also obtain both compressive and rarefactive solitary wave solutions (12) corresponding to the pair of homoclinic orbits at the saddle point $E_0(\phi_0, 0)$ in Fig. 2.

Chaos in the perturbed system

In this section, we will discuss the chaotic behavior of the following perturbed system:

$$\begin{cases} \dfrac{d\phi}{d\xi} = z, \\ \dfrac{dz}{d\xi} = a\phi + b\phi^2 + c\phi^3 + f_0\cos(\omega\xi), \end{cases} \tag{13}$$

where $f_0\cos(\omega\xi)$ is the external periodic perturbation, f_0 is strength of the external perturbation, and ω is the frequency. The difference between the system (8) and the system (13) is that only external periodic perturbation is added with the system (8). The system (13) depends on six independent parameters q, p, σ, v, f_0, and ω. An investigation of such a system for complete range of parametric space or the influence of each parameter is complicated and difficult. To simplify the analysis, all parameters are kept as constants except q to be changed. In order to explore the possible chaotic structure of the perturbed system (13), we consider special values of the parameter q with fixed values of p, σ, v, f_0, and ω in three possible regimes $-1<q<0, 0<q<1$ and $q>1$. We could in fact vary any of the other parameters, but this does not give us any significant different qualitative results.

In Figs. 3, 4, and 5, we have presented phase portraits of the perturbed dynamical system (13) for different values of q (-0.01 (see Fig. 3), 0.1 (see Fig. 4), 2 (see Fig. 5)) with fixed values of other parameters $p = 0.5, \sigma = 0.6, f_0 = 1, \omega = 1$, and $v = 1$. In this case, the velocity of the perturbed traveling wave is sonic. It is clear that the perturbed system (13) shows chaotic oscillations. Any periodic or quasi-periodic behaviors are not observed in Figs. 3, 4, and 5 even if the external periodic perturbation is considered. Furthermore, the developed chaotic motions occur (see Figs. 3, 4, and 5) and the solutions ignore the periodic motions and represent random sequences of uncorrelated

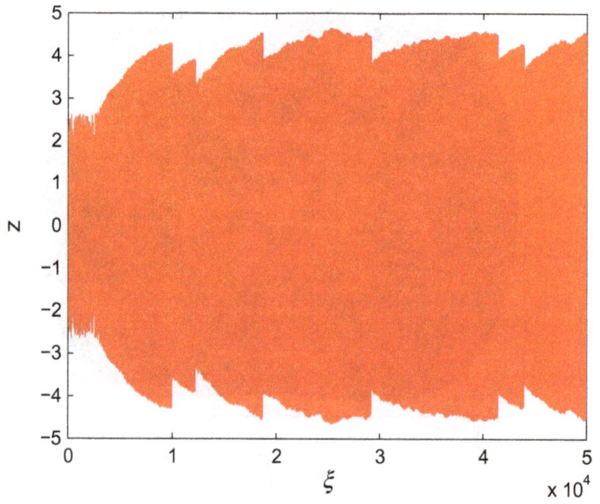

Fig. 6 Plot of z versus ξ of the perturbed system (13) for same values of parameters as Fig. 3

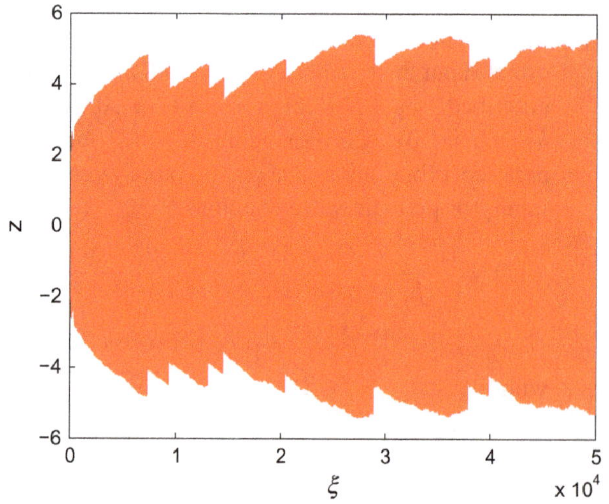

Fig. 7 Plot of z versus ξ of the perturbed system (13) for same values of parameters as Fig. 4

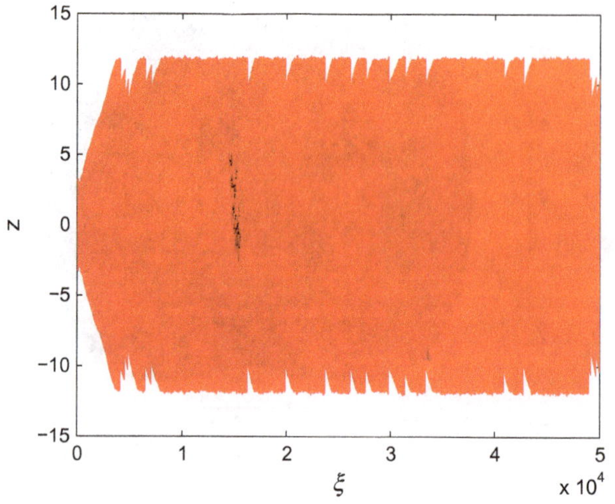

Fig. 8 Plot of z versus ξ of the perturbed system (13) for same values of parameters as Fig. 5

oscillations. For different ranges of the nonextensive parameter q, different developed chaotic motions(see Figs. 3, 4, and 5) are presented with suitable initial conditions. In Figs. 6, 7, and 8, we have plotted z vs. ξ for the perturbed system (13) for different values of q (-0.01 (see Fig. 6), 0.1 (see Fig. 7), 2 (see Fig. 8)) with same values of other parameters as Fig. 3. In other words, the perturbed system (13) shows chaotic behavior when electrons or positrons evolve away from their Maxwell–Boltzmann equilibrium. It is easily seen that chaotic behavior is visible in the system (13) for different values of q.

Conclusions

We have addressed the dynamic structures of ion acoustic waves in an unmagnetized plasma with q-nonextensive electrons and positrons using the bifurcation theory of planar dynamical systems through direct approach. We have transformed the model equations into a planar dynamical system using a traveling wave transformation. Using the bifurcations of planar dynamical system, we have presented the existence of solitary and periodic waves through phase portrait analysis. We have derived new analytical forms for solitary and periodic waves depending on parameters q, p, σ, and v. Considering an external periodic perturbation, chaotic structure of ion acoustic waves has been presented. Depending upon different regimes of the nonextensive parameter q, we have shown the effect of q on chaotic structures of ion acoustic waves with fixed values of other parameters p, σ and v. It has been observed that the unperturbed system has the solitary and periodic wave solutions, but the perturbed dynamical system has chaotic structures for same values of parameters q, p, σ, and v. Our present study could be helpful in understanding the solitary, periodic, and chaotic structures of ion acoustic nonlinear waves in space plasmas [19–25] as well as in laboratory plasmas [26, 27], where q-nonextensive electrons and positrons are present.

References

1. Lakshmanan, M., Rajasekar, S.: Nonlinear Dynamics Integrability, Chaos and Patterns. Springer, Heidelberg (2003)
2. Saha, A., Pal, N., Chatterjee, P.: Dynamic behavior of ion acoustic waves in electron-positron-ion magnetoplasmas with superthermal electrons and positrons. Phys. Plasma **21**, 102101 (2014)
3. Sahu, B., Poria, S., Roychoudhury, R.: Solitonic, quasi-periodic and periodic pattern of electron acoustic waves in quantum plasma. Astrophys. Space Sci. **341**, 567 (2012)
4. Zhen, H., Tian, B., Wang, Y., Zhong, H., Sun, W.: Dynamic behavior of the quantum Zakharov-Kuznetsov equations in dense quantum magnetoplasmas. Phys. Plasma **21**, 012304 (2014)
5. Zhen, H., Tian, B., Wang, Y., Sun, W., Liu, L.: Soliton solutions and chaotic motion of the extended Zakharov-Kuznetsov equations in a magnetized two-ion-temperature dusty plasma. Phys Plasma **21**, 073709 (2014)
6. Hong, W.P.: Comment on: "Spherical Kadomtsev–Petviashvili equation and nebulons for dust ion-acoustic waves with symbolic computation". Phys. Lett. A **340**, 243 (2005)
7. Tian, B., Gao, Y.T.: Cylindrical nebulons, symbolic computation and Bäcklund transformation for the cosmic dust acoustic waves. Phys. Plasma **12**, 070703 (2005)
8. Nozaki, K., Bekki, N.: Chaos in a perturbed nonlinear Schrödinger equation. Phys. Rev. Lett. **50**, 1226 (1983)
9. Williams, G.P.: Chaos Theory Tamed. Joseph Henry, Washington (1997)
10. Beiglbock, W., Eckmann, J.P., Grosse, H., Loss, M., Smirnov, S., Takhtajan, L., Yngvason, J.: Concepts and Results in Chaotic Dynamics. Springer, Berlin (2000)
11. Wesson, J.A., et al.: Disruptions in JET. Nucl. Fusion **29**, 641 (1989)
12. Helander, P., Ward, D.J.: Positron creation and annihilation in tokamak plasmas with runaway electrons. Phys. Rev. Lett. **90**, 135004 (2003)
13. Greaves, R.G., Tinkle, M.D., Surko, C.M.: Creation and uses of positron plasmas. Phys. Plasma **1**, 1439 (1994)
14. Greaves, R.G., Surko, C.M.: An electron-positron beam-plasma experiment. Phys. Rev. Lett. **75**, 3846 (1995)
15. Surko, C.M., Leventhal, M., Passner, A.: Positron plasma in the laboratory. Phys. Rev. Lett. **62**, 901 (1989)
16. Tsytovich, V., Wharton, C.B.: Laboratory electron-positron plasma - a new research object. Comments Plasma Physics Controlled Fusion **4**, 91 (1978)
17. Trivelpiece, A.W.: Nonneutral plasmas. Comments Plasma Physics Controlled Fusion **1**, 57 (1972)
18. Surko, C.M., Murphy, T.J.: Use of the positron as a plasma particle. Phys. Fluids B **2**, 1372 (1990)
19. Rees, M.J.: New Interpretation of Extragalactic Radio Sources. Nature **229**, 312 (1971)
20. Burns, M.L.: In Positron-Electron Pairs in Astrophysics. American Institute of Physics, New York (1983)
21. Rees, M.J.: In: The Very Early Universe. Gibbons, G.W., Hawking, S.W., Siklas, S. (eds), Cambridge University Press, Cambridge (1983)
22. Michel, F.C.: Theory of Neutron Star Magnetosphere. Chicago University Press, Chicago (1991)
23. Miller, H.R., Witta, P.J.: Active Galactic Nuclei, p. 202. Springer, Berlin (1987)
24. Michel, F.C.: Theory of pulsar magnetospheres. Rev. Mod. Phys. **54**, 1 (1982)
25. Hansen, E.T., Emshie, A.G.: The Physics of Solar Flares, p. 124. Cambridge University Press, Cambridge (1988)
26. Berezhiani, V.I., Tskhakaya, D.D., Shukla, P.K.: Pair production in a strong wake field driven by an intense short laser pulse. Phys. Rev. A **46**, 6608 (1992)
27. Liang, E.P., Wilks, S.C., Tabak, M.: Pair production by ultraintense lasers. Phys. Rev. Lett. **81**, 4887 (1998)
28. Mahmood, S., Ur-Rehman, H.: Electrostatic solitons in unmagnetized hot electron–positron–ion plasmas. Phys. Lett. A **373**, 2255 (2009)
29. Alinejad, H.: Effect of excavated trapped electron distributions on ion-acoustic solitary structures in an electron–positron–ion plasma. Phys. Lett. A **373**, 3663 (2009)
30. Popel, S.I., Vladimirov, S.V., Shukla, P.K.: Ion-acoustic solitons in electron–positron–ion plasmas. Phys. Plasma. **2**, 716 (1995)
31. Gill, T.S., Bains, A.S., Sainia, N.S., Bedi, C.: Ion-acoustic envelope excitations in electron–positron–ion plasma with non-thermal electrons. Phys. Lett. A **374**, 3210 (2001)
32. El-Awady, E.I., El-Tantawy, S.A., Moslema, W.M., Shukla, P.K.: Electron–positron–ion plasma with kappa distribution: Ion

acoustic soliton propagation. Phys. Lett. A **374**, 3216 (2010)

33. El-Shamy, E.F., El-Bedwehy, N.A.: On the linear and nonlinear characteristics of electrostatic solitary waves propagating in magnetized electron–positron–ion plasmas. Phys. Lett. A **374**, 4425 (2010)
34. Iqbal, M., Shukla, P.K.: Relaxation of a magnetized electron–positron–ion plasma with flows. Phys. Lett. A **375**, 2725 (2011)
35. Shukla, P.K., Rao, N.N., Yu, M.Y., Tsintsa, N.L.: Relativistic nonlinear effects in plasmas. Phys. Rep. **138**, 1 (1986)
36. Ghosh, S., Bharuthram, R.: Ion acoustic solitons and double layers in electron-positron-ion plasmas with dust particulates. Astrophys. Space Sci. **314**, 121 (2008)
37. Pakzad, H.R.: Ion acoustic solitary waves in plasma with non-thermal electron and positron. Phys. Lett. A **373**, 847 (2009)
38. Hamity, V.H., Barraco, D.E.: Generalized nonextensive thermodynamics applied to the cosmic background radiation in a Robertson-Walker universe. Phys. Rev. Lett. **76**, 4664 (1996)
39. Torres, D.F., Vucetich, H., Plastino, A.: Early universe test of nonextensive statistics. Phys. Rev. Lett. **79**, 1588 (1997)
40. Renyi, A.: On a new axiomatic theory of probability. Acta Math. Hung. **6**, 285 (1955)
41. Tsallis, C.: Possible generalization of Boltzmann-Gibbs statistics. J. Stat. Phys. **52**, 479 (1988)
42. Curado, E.M.F., Tsallis, C.: Generalized statistical mechanics: connection with thermodynamics. J. Phys. A: Math. Gen. **24**, L69 (1991)
43. Lima, J.A.S., Silva Jr, R., Santos, J.: Plasma oscillations and nonextensive statistics. Phys. Rev. E **61**, 3260 (2000)
44. Tsallis, C., Levy, S.V.F., Souza, A.M.C., Maynard, R.: Statistical-mechanical foundation of the ubiquity of Lévy distributions in nature. Phys. Rev. Lett. **75**, 3589 (1995)
45. Silva Jr, R., Plastino, A.R., Lima, J.A.S.: A Maxwellian path to the q-nonextensive velocity distribution function. Phys. Lett. A **249**, 401 (1998)
46. Lima, J.A.S., Silva, R., Plastino, A.R.: Nonextensive thermostatistics and theH-theorem. Phys. Rev. Lett. **86**, 2938 (2001)
47. Abe, S., Martinez, S., Pennini, F., Plastino, A.: Nonextensive thermodynamic relations. Phys. Lett. A **281**, 126 (2001)
48. Tribeche, M., Merriche, A.: Nonextensive dust-acoustic solitary waves. Phys. Plasma **18**, 034502 (2011)
49. Ghosh, U.N., Chatterjee, P., Roychoudhury, R.: The effect of q-distributed electrons on the head-on collision of ion acoustic solitary waves. Phys. Plasma **19**, 012113 (2012)
50. Ghosh, U.N., Chatterjee, P., Kundu, S.K.: The effect of q-distributed ions during the head-on collision of dust acoustic solitary waves. Astrophys. Space Sci. **339**, 255 (2012)
51. Pakzad, H.R.: Cylindrical and spherical electron acoustic solitary waves with nonextensive hot electrons. Phys. Plasma **18**, 082105 (2011)
52. Leubner, M.P.: Consequences of entropy bifurcation in non-Maxwellian astrophysical environments. Nonlinear Process. Geophys. **15**, 531 (2008)
53. Shahmansouri, M., Alinejad, H.: Effect of electron nonextensivity on oblique propagation of arbitrary ion acoustic waves in a magnetized plasma. Astrophys. Space Sci. **344**, 463 (2013)
54. Shahmansouri, M., Astaraki, E.: Transverse perturbation on three-dimensional ion acoustic waves in electron–positron–ion plasma with high-energy tail electron and positron distribution. J. Theor. Appl. Phys. **8**, 189 (2014)
55. Shahmansouri, M., Alinejad, H.: Arbitrary amplitude electron acoustic waves in a magnetized nonextensive plasma. Astrophys. Space Sci. **347**, 305 (2013)
56. Sabetkar, A., Dorranian, D.: Non-extensive effects on the characteristics of dust-acoustic solitary waves in magnetized dusty plasma with two-temperature isothermal ions. J. Plasma Phys. **80**, 565 (2014)
57. Samanta, U.K., Saha, A., Chatterjee, P.: Bifurcations of dust ion acoustic travelling waves in a magnetized dusty plasma with a q-nonextensive electron velocity distribution. Phys. Plasma **20**, 022111 (2013)
58. Samanta, U.K., Saha, A., Chatterjee, P.: Bifurcations of nonlinear ion acoustic travelling waves in the frame of a Zakharov-Kuznetsov equation in magnetized plasma with a kappa distributed electron. Phys. Plasma **20**, 052111 (2013)
59. Samanta, U.K., Saha, A., Chatterjee, P.: Bifurcations of dust ion acoustic travelling waves in a magnetized quantum dusty plasma. Astrophys. Space Sci. **347**, 293 (2013)
60. Saha, A., Chatterjee, P.: Bifurcations of electron acoustic traveling waves in an unmagnetized quantum plasma with cold and hot electrons. Astrophys. Space Sci. **349**, 239 (2014)
61. Saha, A., Chatterjee, P.: Bifurcations of ion acoustic solitary waves and periodic waves in an unmagnetized plasma with kappa distributed multi-temperature electrons. Astrophys. Space Sci. **350**, 631 (2014b)
62. Saha, A., Chatterjee, P.: Bifurcations of ion acoustic solitary and periodic waves in an electron–positron–ion plasma through non-perturbative approach. J. Plasma Phys. **80**, 553 (2014)
63. Saha, A., Chatterjee, P.: Bifurcations of dust acoustic solitary waves and periodic waves in an unmagnetized plasma with nonextensive ions. Astrophys. Space Sci. **351**, 533 (2014)
64. Saha, A., Chatterjee, P.: New analytical solutions for dust acoustic solitary and periodic waves in an unmagnetized dusty plasma with kappa distributed electrons and ions. Phys. Plasma **21**, 022111 (2014)
65. Saha, A., Chatterjee, P.: Dust ion acoustic travelling waves in the framework of a modified Kadomtsev-Petviashvili equation in a magnetized dusty plasma with superthermal electrons. Astrophys. Space Sci. **349**, 813 (2014)
66. Saha, A., Chatterjee, P.: Electron acoustic blow up solitary waves and periodic waves in an unmagnetized plasma with kappa distributed hot electrons. Astrophys. Space Sci. **353**, 163 (2014)
67. Saha, A., Chatterjee, P.: Propagation and interaction of dust acoustic multi-soliton in dusty plasmas with q-nonextensive electrons and ions. Astrophys. Space Sci. **353**, 169 (2014)
68. Ghosh, D.K., Mandal, G., Chatterjee, P., Ghosh, U.N.: Nonplanar ion acoustic solitary waves in electron-positron-ion plasma with warm ions, and electron and positron following q-nonextensive velocity distribution. IEEE Trans. Plasma Sci. **41**, 1600 (2013)
69. Du, J.L.: Nonextensivity in nonequilibrium plasma systems with Coulombian long-range interactions. Phys. Lett. A **329**, 262 (2004)
70. Saha, A.: Bifurcation of travelling wave solutions for the generalized KP-MEW equations. Commun. Nonlinear Sci. Numer. Simulat. **17**, 3539 (2012)
71. Guckenheimer, J., Holmes, P.J.: Nonlinear Oscillations. Dynamical Systems and Bifurcations of Vector Fields. Springer, New York (1983)
72. Byrd, P.F., Friedman, M.D.: Handbook of Elliptic Integrals for Engineer and Scientists. Springer, New York (1971)

20

Linear and nonlinear analysis of dust acoustic waves in dissipative space dusty plasmas with trapped ions

A. M. El-Hanbaly[1] · E. K. El-Shewy[2] · M. Sallah[1] · H. F. Darweesh[1]

Abstract The propagation of linear and nonlinear dust acoustic waves in a homogeneous unmagnetized, collisionless and dissipative dusty plasma consisted of extremely massive, micron-sized, negative dust grains has been investigated. The Boltzmann distribution is suggested for electrons whereas vortex-like distribution for ions. In the linear analysis, the dispersion relation is obtained, and the dependence of damping rate of the waves on the carrier wave number k, the dust kinematic viscosity coefficient η_d and the ratio of the ions to the electrons temperatures σ_i is discussed. In the nonlinear analysis, the modified Korteweg–de Vries–Burgers (mKdV–Burgers) equation is derived via the reductive perturbation method. Bifurcation analysis is discussed for non-dissipative system in the absence of Burgers term. In the case of dissipative system, the tangent hyperbolic method is used to solve mKdV–Burgers equation, and yield the shock wave solution. The obtained results may be helpful in better understanding of waves propagation in the astrophysical plasmas as well as in inertial confinement fusion laboratory plasmas.

Keywords mKdV–Burgers equation · Solitary and shock waves · Reductive perturbation method · Bifurcations · Vortex-like ion distribution

✉ A. M. El-Hanbaly
a_elhanbaly@yahoo.com

M. Sallah
msallahd@mans.edu.eg

1 Theoretical Physics Research Group, Physics Department, Faculty of Science, Mansoura University, P.O. Box 35516, Mansoura, Egypt

2 Department of Physics, Taibah University, Al-Madinah Al-Munawarrah, Kingdom of Saudi Arabia

Introduction

Physics of dusty plasma studies the properties of charged dust in the presence of electrons and ions [31]. These media have been observed in lower and upper mesosphere, planetary magnetosphere, cometary tails, planetary rings, interplanetary spaces, interstellar media, etc. [17, 18, 26, 39]. The importance of studying such task appears also in the laboratory and plasmas technology such as low temperature physics (radio frequency), plasma discharge [5], and fabrication of modern materials such as semi conductors, optical fibres, nanostructural materials, and dusty crystals [6, 45].

Most of space and laboratory dusty plasma environments' dust are negatively charged [26]. However, there are also some space (i.e., upper part of ionosphere or lower part of magnetosphere) and some laboratory environments' dust are positively charged [13, 16, 32]. It has been noticed, theoretically and experimentally, that dust charge dynamics in dusty plasma modifies the existing plasma wave features and introduces different types of new wave modes [3]. For examples, dust ion acoustic (DIA) mode [27, 40], dust acoustic (DA) mode [23, 31], dust-drift mode [42], dust lattice (DL) mode [25] and Shukla–Varma mode [41], etc. Therefore, it is paramount to study linear and nonlinear waves in various configurations of laboratory and space plasma [4, 14, 19, 46]. Nonlinear dust acoustic solitary and shock waves in plasmas have a considerable attention to study the propagation behaviour of the nonlinear waves. Among such theoretical studies on dusty plasma, many authors studied solitary and shock waves in the frame of KdV–Burgers, mKdV–Burgers, KP–Burgers and mKP–Burgers-type equations, which are usually deduced from a dissipative plasma system. The dust fluid dissipation could be caused by dust fluid viscosity, dust–dust collision, dust charge fluctuation, and Landau damping, which would modify the wave structures [15, 21, 28, 47], and shock waves can be excited in the system under appropriate conditions.

Shukla and Mamun [38] derived KdV–Burgers equation by reductive perturbation method and they have analysed the properties of nonlinear waves (solitons and shock waves) for strongly coupled unmagnetized dusty plasmas. Dusty plasma system with an iso-nonthermal ion distribution has been studied by [24], and they have studied the properties of shock waves via the obtained mKdV–Burgers equation that was derived using a set of generalized hydrodynamic equations. Pakzad and Javidan [29] have interpreted the dust acoustic solitary and shock waves in strongly coupled dusty plasmas with nonthermal ions. Recently, [35] have investigated arbitrary amplitude DA waves in a nonextensive dusty plasma, where they have examined the effects of electrons and ions nonextensivity on the DA soliton profile. Shahmansouri and Tribeche [36] have also investigated the properties and formation of DA shock waves in complex plasmas. They have found that the influence of electrons and ions nonextensivity and dust charge fluctuation affect the basic properties of the collisionless DA shock waves drastically.

It is well known that the presence of trapped particles can significantly modify the wave propagation characteristics in collisionless plasmas [33]. In most laboratory dusty plasmas, ions are not always isothermal because they are trapped by the large amplitude DA wave potential. Accordingly, they may follow a vortex-like distribution [34], which affect drastically on the nonlinear propagation mode. Some recent theoretical studies focused on the effects of ion and electron trapping which is common not only in space plasmas but also in laboratory experiments [1, 30, 37]. El-Wakil et al. [11] theoretically investigated higher order contributions to nonlinear DA waves that propagate in a mesospheric dusty plasma with a complete depletion of the background (electrons and ions). Duha et al. [7] studied small amplitude DIA solitary and shock waves due to dust charge fluctuation with trapped electrons by reductive perturbation method. They showed that the effects of vortex-like/trapped electron distribution modify the properties of the DIA solitary and shock waves.

The major topic of this work is to study the linear (dispersion relation) and nonlinear (solitons, monotonic as well as oscillatory shocks) properties of the DA waves in a homogeneous system of an unmagnetized, collisionless and dissipative dusty space plasma. Here, the study is devoted for a plasma system consisting of extremely massive, micron-sized, negative dust grains, electrons that obey a Boltzmann distribution and the trapped ions described by the vortex-like distribution (Sect. 2). The normal mode method is used to investigate the obtained dispersion relation (Sect. 3). The reductive perturbation technique is used to obtain the one-dimensional mKdV–Burgers equation. Moreover, the bifurcation analysis has been illustrated to recognize different classes of admissible solutions (Sect. 4). Finally, some concluding remarks are given in Sect. 5.

Governing equations

Let us consider a homogeneous system of an unmagnetized, collisionless and dissipative dusty plasma. The system consists of extremely massive, micron-sized, negatively dust grains, electrons obey a Boltzmann distribution and ions follow the vortex-like distribution representing the ions trapped in the DA wave potential. The dynamics of the DA waves are governed by the following basic set of equations

$$\frac{\partial n_d}{\partial t}+\frac{\partial(n_d u_d)}{\partial x}=0, \tag{1}$$

$$\frac{\partial u_d}{\partial t}+u_d\frac{\partial u_d}{\partial x}+q\frac{\partial \phi}{\partial x}-\eta_d\frac{\partial^2 u_d}{\partial x^2}=0, \tag{2}$$

$$\frac{\partial^2 \phi}{\partial x^2}+qn_d-n_e+n_i=0. \tag{3}$$

In the above equations, n_d is the dust grain number density, u_d is the dust fluid velocity, q is the number of charges residing on the dust grain surface, ϕ is the electrostatic potential and η_d is the corresponding viscosity coefficient. The density of the Boltzmann distributed electrons, n_e, is given by

$$n_e=\mu_e \exp(\sigma_i\phi), \tag{4}$$

where μ_e is the initial equilibrium density of electrons and $\sigma_i=T_i/T_e$, with T_e is the temperature of electrons and T_i is the temperature of ions. The ion density, n_i, obeys the vortex-like distribution function [34] is given by

$$n_i=\mu_i\left[\exp(-\phi)-\frac{4}{3}b(-\phi)^{\frac{3}{2}}\right], \tag{5}$$

where μ_i is the ions initial equilibrium density and b is a constant representing the vortex parameter which depends on the temperature parameters of resonant ions (both free and trapped) and is given by

$$b=\frac{1}{\sqrt{\pi}}(1-\beta), \tag{6}$$

where β is a trapping parameter that determines the number of the trapped ions, and defined as the ratio of the free-ion temperature to the trapped-ion temperature. The case of $\beta<0$ refers to a vortex-like excavated trapped-ion distribution, and if $\beta=1$ (or $\beta=0$) we have Maxwellian (or flat-topped) ion distribution. This work focus on the case of $\beta<0$.

Linear analysis

To derive a dynamical equation for the dispersion relation of the DA waves from the basic equations (1, 2, 3, 4, 5), we use the normal mode method. With such method, we expand the dependent variables (n_d, u_d and ϕ) in terms of their equilibrium and perturbed parts as ($n_d = 1 + \tilde{n}_d$, $u_d = 0 + \tilde{u}_d$ and $\phi = 0 + \tilde{\phi}$). The perturbed quantities are proportional to $\exp[i(kx - \omega t)]$, and then the basic equations (1, 2, 3, 4, 5) are linearized and the corresponding first-order approximation yield

$$\tilde{n}_d = \frac{qk^2}{\omega^2 + i\eta_d k^2 \omega}\tilde{\phi}, \tag{7}$$

$$\tilde{u}_d = \frac{qk}{\omega + i\eta_d k^2}\tilde{\phi}. \tag{8}$$

In connection with Poisson equation (3), the linear dispersion relation follows

$$(k^2 + \mu_i + \mu_e\sigma_i)\omega^2 + i\eta_d k^2(k^2 + \mu_i + \mu_e\sigma_i)\omega - q^2k^2 = 0. \tag{9}$$

Let us consider

$$\omega = \omega_r + i\omega_i, \tag{10}$$

where ω_r and ω_i are the real and imaginary parts of the frequency ω.

By inserting Eq. (10) into Eq. (9), one obtains

$$2(k^2 + \mu_i + \mu_e\sigma_i)\omega_r\omega_i + \eta_d k^2(k^2 + \mu_i + \mu_e\sigma_i)\omega_r = 0, \tag{11a}$$

$$(k^2 + \mu_i + \mu_e\sigma_i)\omega_i^2 - (k^2 + \mu_i + \mu_e\sigma_i)\omega_r^2 + \eta_d k^2(k^2 + \mu_i + \mu_e\sigma_i)\omega_i + q^2k^2 = 0. \tag{11b}$$

By solving Eq. (11), one gets $\omega_r = 0$, whereas ω_i is given by

$$\omega_i = \frac{-k^2\eta_d}{2} + \frac{\sqrt{-4q^2k^2 + k^4\eta_d^2(k^2 + \mu_i + \mu_e\sigma_i)}}{2\sqrt{k^2 + \mu_i + \mu_e\sigma_i}}, \tag{12}$$

which depends mainly on the plasma parameters k, η_d and σ_i. The behaviour of the damping rate ω_i with such parameters is illustrated as shown in Figs. 1, and 2. From these figures, one can see that, the instability damping rate ω_i decreases by increasing the plasma parameters (carrier wave number k, the dust kinematic viscosity coefficient η_d and the ratio of the ion to the electron temperatures σ_i).

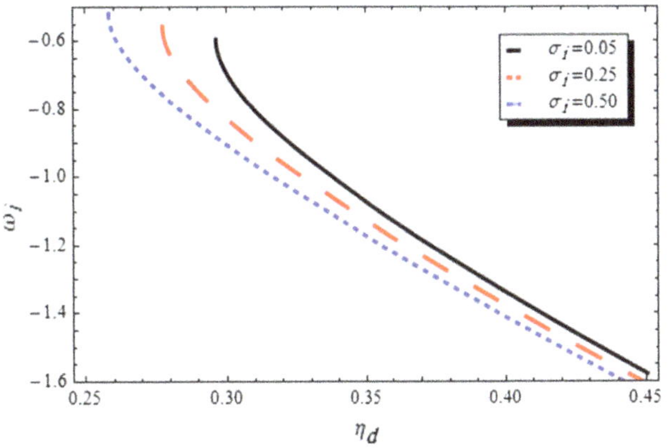

Fig. 1 The variation of ω_i vs. η_d at different values of σ_i for $\mu_e = 8$, $\mu_i = 7$, $k = 2, q = 1$

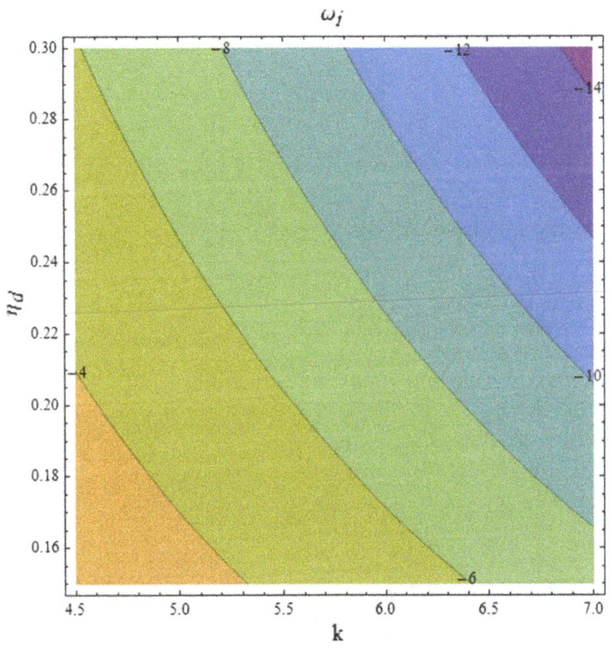

Fig. 2 Contour plot of ω_i vs. k and η_d for $\mu_e = 1.1$, $\mu_i = 0.1$, $\sigma_i = 0.2, q = 1$

Nonlinear analysis

Derivation of the mKdV–Burgers equation

To extend our analysis, the nonlinear dynamical features of DA waves are treated. This can be done using the standard reductive perturbation method. The reductive perturbation method is mostly applied to small amplitude nonlinear waves [43, 44]. This method rescales both space and time in the governing equations of the system to introduce space and time variables, which are appropriate for the description of long-wavelength phenomena. Accordingly, the independent variables are stretched as

$$\tau = \varepsilon^{\frac{3}{4}} t, \quad \xi = \varepsilon^{\frac{1}{4}}(x - \lambda t), \tag{13}$$

where ε is a small dimensionless expansion parameter measuring the strength of nonlinearity and λ is the wave speed. The value of η_d is assumed to be small, so that we may set its value as $\eta_d = \varepsilon^{\frac{1}{4}}\eta$, with η is a definite quantity. Furthermore, the physical quantities appearing in the basic equations (1, 2, 3, 4, 5) are expanded as power series in ε about their equilibrium values, viz

$$n_d = 1 + \varepsilon n_{d1} + \varepsilon^{\frac{3}{2}} n_{d2} + \varepsilon^2 n_{d3} + \cdots, \tag{14a}$$

$$u_d = \varepsilon u_{d1} + \varepsilon^{\frac{3}{2}} u_{d2} + \varepsilon^2 u_{d3} + \cdots, \tag{14b}$$

$$\phi = \varepsilon\phi_1 + \varepsilon^{\frac{3}{2}}\phi_2 + \varepsilon^2\phi_3 + \cdots \tag{14c}$$

Substituting Eqs. (13, 14) into basic equations (1, 2, 3, 4, 5) and equating coefficients of powers of ε, the lowest order of ε gives

$$n_{d1} = \frac{q}{\lambda^2}\phi_1, \tag{15}$$

$$u_{d1} = \frac{q}{\lambda}\phi_1. \tag{16}$$

and the phase velocity is given by

$$\lambda = \sqrt{\frac{q^2}{\mu_i + \mu_e\sigma_i}}. \tag{17}$$

The next order in ε yields the following

$$\frac{\partial n_{d1}}{\partial \tau} - \lambda\frac{\partial n_{d2}}{\partial \xi} + \frac{\partial u_{d2}}{\partial \xi} = 0, \tag{18a}$$

$$\frac{\partial u_{d1}}{\partial \tau} - \lambda\frac{\partial u_{d2}}{\partial \xi} + q\frac{\partial \phi_2}{\partial \xi} - \eta\frac{\partial^2 u_{d1}}{\partial \xi^2} = 0, \tag{18b}$$

$$\frac{\partial^3 \phi_1}{\partial \xi^3} + q\frac{\partial n_{d2}}{\partial \xi} - \frac{q^2}{\lambda^2}\frac{\partial \phi_2}{\partial \xi} + 2\mu_i b(-\phi_1)^{\frac{1}{2}}\frac{\partial \phi_1}{\partial \xi} = 0. \tag{18c}$$

Solving this system with the aid of Eqs. (15, 16, 17) and eliminating the second-order perturbed quantities (n_{d2}, u_{d2} and ϕ_2), we finally get the mKdV–Burgers equation as

$$\frac{\partial \phi_1}{\partial \tau} + A(-\phi_1)^{\frac{1}{2}}\frac{\partial \phi_1}{\partial \xi} + B\frac{\partial^3 \phi_1}{\partial \xi^3} + C\frac{\partial^2 \phi_1}{\partial \xi^2} = 0, \tag{19}$$

where

$$A = \frac{\mu_i b\lambda^3}{q^2}, B = \frac{\lambda^3}{2q^2}, C = -\frac{\eta}{2}. \tag{20}$$

Let us introduce the variable $\chi = \xi - u_0\tau$ where χ is the transformed coordinate relative to a frame which moves with the velocity u_0. Integrating Eq. (19) with respect to the variable χ, the reduced mKdV–Burgers equation leads to

$$\frac{d^2\phi_1}{d\chi^2} + \frac{C}{B}\frac{d\phi_1}{d\chi} - \frac{2A}{3B}(-\phi_1)^{\frac{3}{2}} - \frac{u_0}{B}\phi_1 = 0. \tag{21}$$

Owing to the presence of the Burgers term $\frac{C}{B}\frac{d\phi_1}{d\chi}$, Eq. (21) describes homogeneous and dissipative dusty plasmas. Therefore, the phase paths of such equation are, in general, no longer level curves of an energy $H(\phi_1, \frac{d\phi_1}{d\chi})$. So, in the dissipative case, it is reasonable to deal with $\frac{dH}{d\chi}$ rather than H. The mKdV–Burgers Eq. (21) can be written in the general form as

$$\frac{d^2\phi_1}{d\chi^2} + h\left(\phi_1, \frac{d\phi_1}{d\chi}\right)\frac{d\phi_1}{d\chi} + G(\phi_1) = 0, \tag{22}$$

where h and G are two functions that can be determined by comparing the Eqs. (21, 22).

In the conservative case ($h = 0$), the total energy associated with Eq. (22) is

$$H = \frac{1}{2}\left(\frac{d\phi_1}{d\chi}\right)^2 + V(\phi_1), \tag{23}$$

where $V(\phi_1)$ is the potential function and then

$$\frac{dH}{d\chi} = \frac{d\phi_1}{d\chi}\left(\frac{d^2\phi_1}{d\chi^2} + \frac{dV}{d\phi_1}\right). \tag{24}$$

If $\frac{dV}{d\phi_1} = G(\phi_1)$ in addition to Eq. (22), the total derivative of H will be

$$\frac{dH}{d\chi} = -h\left(\phi_1, \frac{d\phi_1}{d\chi}\right)\left(\frac{d\phi_1}{d\chi}\right)^2, \tag{25}$$

which is a decreasing function for the variable χ if $h > 0$. This equation is very important for studying the stability of the system.

In the present study, the function $\frac{dH}{d\chi}$ corresponds to Burgers term in the mKdV–Burgers Eq. (21), as

$$\frac{dH}{d\chi} = \frac{C}{B}\left(\frac{d\phi_1}{d\chi}\right)^2, \tag{26}$$

which shows that the energy of the plasma system is not conserved and hence it is not easy to find out exact analytical solution for mKdV–Burgers equation.

In terms of the viscosity coefficient η, Eq. (26) can be written as

$$\frac{dH}{d\chi} = -\frac{q^2\eta}{\lambda^3}\left(\frac{d\phi_1}{d\chi}\right)^2, \tag{27}$$

which is always a decreasing function since q^2, η and λ are positive quantities.

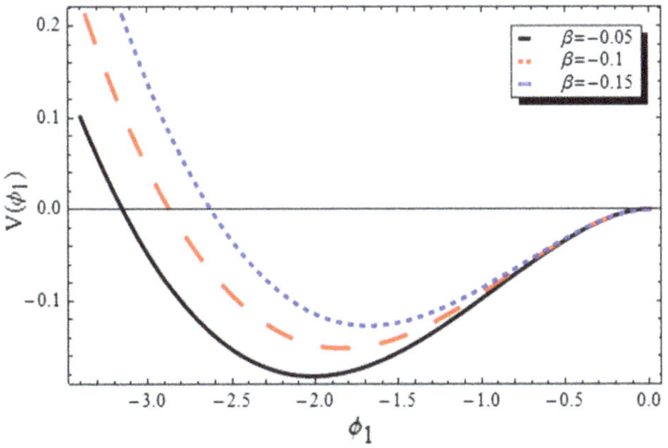

Fig. 3 The variation of $V(\phi_1)$ vs. ϕ_1 for different values of β for $\mu_e = 1.4, \mu_i = 0.4, \sigma_i = 0.2, u_0 = 0.4, q = 1$

Bifurcation analysis and soliton solution

In the absence of the Burgers coefficient $(C = 0)$, Eq. (21) becomes conservative and consequently $\frac{dH}{d\chi} = 0$. Then, the total energy has the form

$$H = \frac{1}{2}\left(\frac{d\phi_1}{d\chi}\right)^2 - \frac{u_0}{2B}\phi_1^2 + \frac{4A}{15B}(-\phi_1)^{\frac{5}{2}}, \tag{28}$$

where the potential function $V(\phi_1)$ is

$$V(\phi_1) = -\frac{u_0}{2B}\phi_1^2 + \frac{4A}{15B}(-\phi_1)^{\frac{5}{2}}. \tag{29}$$

The bifurcation analysis is investigated via Eqs. (28, 29) under the condition $A > 0, B > 0$ and $u_0 > 0$. The potential function $V(\phi_1)$ versus ϕ_1 for different values of β is shown graphically, Fig. 3, and it is clear that the potential well has one hump and a pit. These two points correspond to a saddle point at $(0, 0)$ and a center point at $(-[3u_0/2A]^2,\ 0)$ in the phase portrait. From the topology of the phase portrait diagram, Fig. 4, one can see a family of periodic orbits at $(-[3u_0/2A]^2,\ 0)$, which refer to a family of periodic wave solutions. On the other hand, the homoclinic orbit at $(0, 0)$ refers to one solitary wave solution. Moreover, Fig. 4 shows a series of bounded open orbits that corresponds to a series of breaking wave solutions.

These trajectories that are shown in Fig. 4 refer to the existence of stable solitonic solution that should satisfy the following condition [2]

$$\left[\frac{d^2V}{d\phi_1^2}\right]_{\phi_1=0} < 0$$

which explains that there must exist a nonzero crossing point $\phi_1 = \phi_0$ that $V(\phi_1 = \phi_0) = 0$. In addition, there must exist a ϕ_1 between $\phi_1 = 0$ and $\phi_1 = \phi_0$ to make $V(\phi_1) < 0$. Obviously, from Eq. (28), the condition of existence of stable solitonic solution is satisfied, since

$$\left[\frac{d^2V}{d\phi_1^2}\right]_{\phi_1=0} = -\frac{u_0}{B} < 0,$$

where the parameters u_0 and B are positive. However, the corresponding stable solitonic solution is given by

$$\phi_1 = -\phi_0 \text{sech}^4\left(\frac{\chi}{W}\right), \tag{30}$$

where $\phi_0 = \left(\frac{15u_0}{8A}\right)^2$ is the amplitude and $W = 4\sqrt{\frac{B}{u_0}}$ is the width of the soliton solution in the absence of Burgers term. The behaviours of the obtained solution and its characterizing parameters (DA wave amplitude and its width) are presented in Figs. 5, 6, 7, and 8. Figure 5 shows the variation in ϕ_1 with χ for different values of β (the ratio of the free-ion temperature to the trapped-ion temperature). It emphasizes that the amplitude of the soliton wave increases with increasing of β. The single-pulse soliton solution ϕ_1 is plotted versus ξ, and its propagation is shown at different time scales τ (Fig. 6).

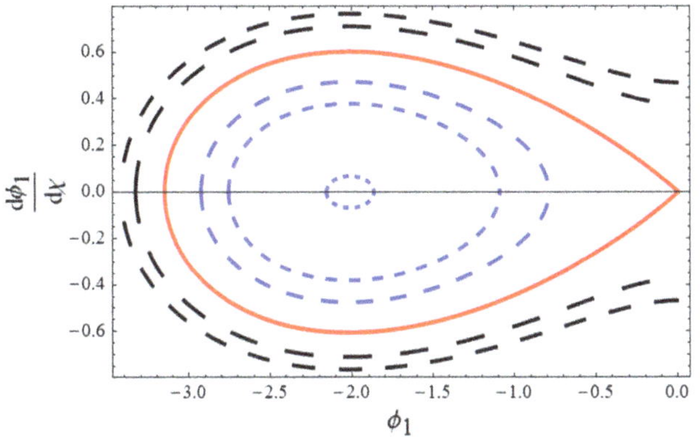

Fig. 4 Phase portrait $d\phi_1/d\chi$ vs. ϕ_1 for different H for $\mu_e = 1.4, \mu_i = 0.4, \sigma_i = 0.2, \beta = -0.05, u_0 = 0.4, q = 1$

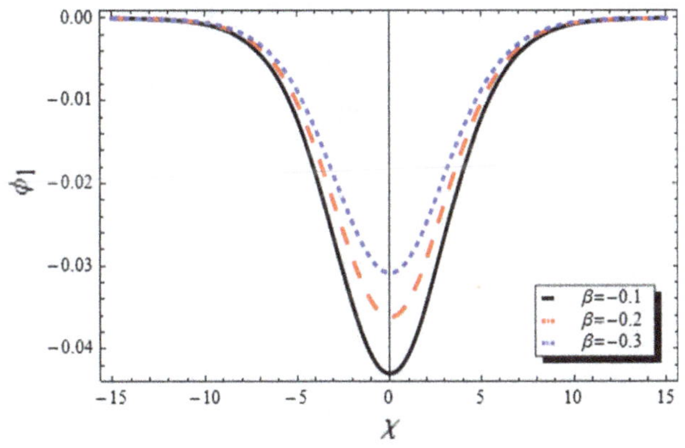

Fig. 5 Evolution of ϕ_1 vs. χ for different values of β for $\mu_e = 1.4, \mu_i = 0.4, \sigma_i = 0.2, u_0 = 0.4, q = 1$

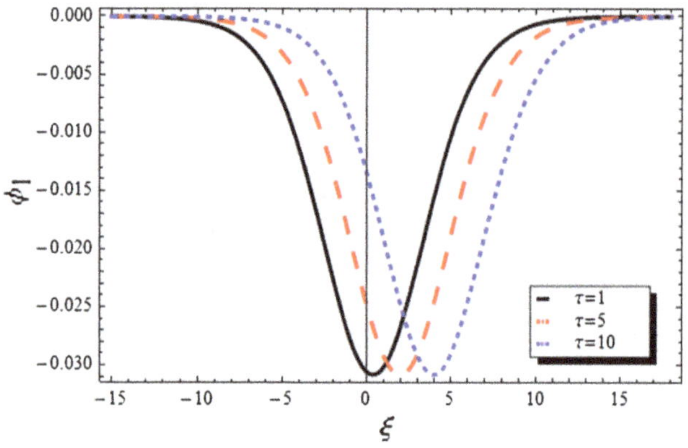

Fig. 6 Evolution of ϕ_1 vs. ξ for different values of τ for $\mu_e = 1.4, \mu_i = 0.4, \sigma_i = 0.2, \beta = -0.3, u_0 = 0.4, q = 1$

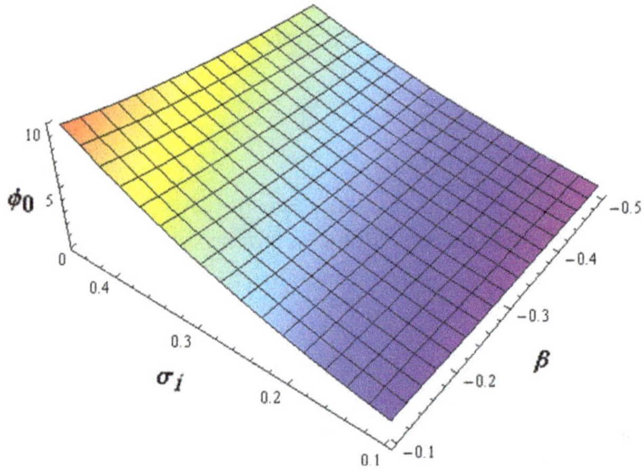

Fig. 7 Evolution of ϕ_0 vs. (β, σ_i) for $\mu_e = 1.4, \mu_i = 0.4, u_0 = 0.4, q = 1$

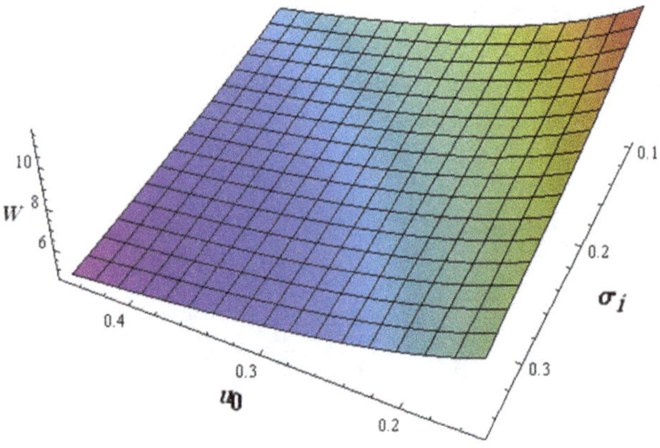

Fig. 8 Evolution of W vs. (σ_i, u_0) for $\mu_e = 1.4, \mu_i = 0.4, q = 1$

The evolution of the amplitude ϕ_0 with β and σ_i is presented in Fig. 7. It can be seen that the amplitude of the soliton wave increases with increasing the values of both β and σ_i. However, the variation of the width W with u_0 and

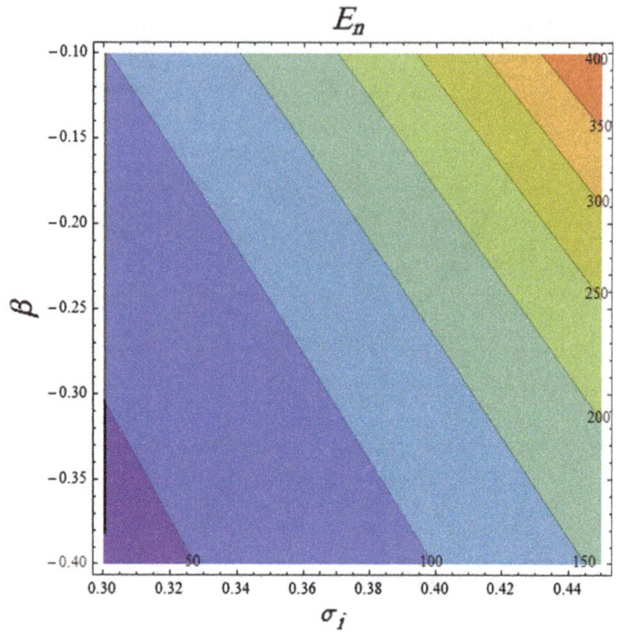

Fig. 9 *Contour plot* of E_n vs. (β,σ_i) for $\mu_e = 1.4, \mu_i = 0.4, u_0 = 0.4, q = 1$

σ_i is graphed in Fig. 8, which shows that the soliton wave width increases with decreasing u_0 and σ_i.

The soliton energy E_n is obtained according to the integral

$$E_n = \int_{-\infty}^{\infty} u_{d1}^2(\chi)\mathrm{d}\chi. \tag{31}$$

Upon integrating (31), yields

$$E_n = \frac{45.2\, q^2 u_0^4}{A^4 \lambda^2 \sqrt{u_0/B}}. \tag{32}$$

which, clearly, depends mainly on the plasma parameters via the coefficients A and B. The variation of the soliton energy E_n with σ_i and β is shown graphically in Fig. 9, which shows that the soliton energy increases with increasing the values of both σ_i and β.

Moreover, the associated electric field is obtained according to the relation

$$\mathbf{E} = -\nabla\phi_1, \tag{33}$$

and then, the electric field has the expression

$$E = -\frac{225 u_0^2 \sqrt{u_0/B}}{64A^2} \operatorname{sech}^4\left(\frac{1}{4}\sqrt{u_0/B}\chi\right) \tanh\left(\frac{1}{4}\sqrt{u_0/B}\chi\right). \tag{34}$$

The behaviour of the electric field E is presented graphically as shown in Figs. 10, and 11. Figure 10 shows the variation of the electric field E against χ for different values of β. Obviously, the electric field increases with the

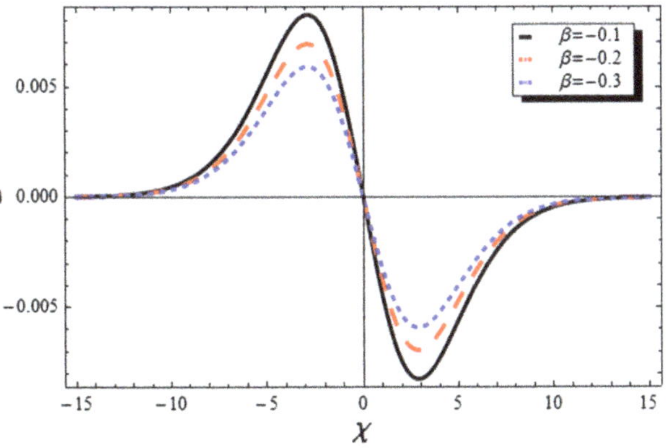

Fig. 10 Variation of E vs. χ for different values of β for $\mu_e = 1.4, \mu_i = 0.4, \sigma_i = 0.2, u_0 = 0.4, q = 1$

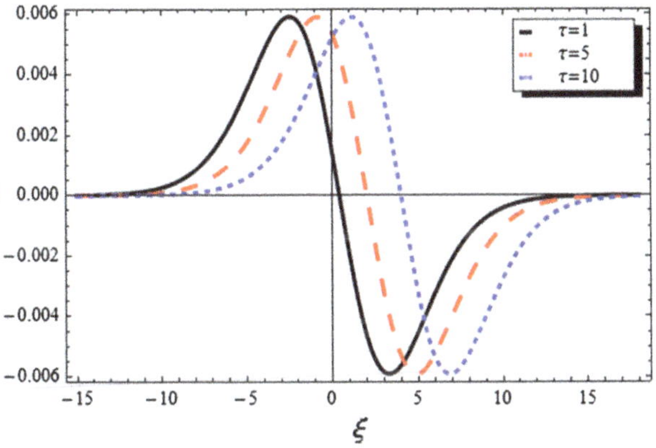

Fig. 11 Evolution of E vs. ξ for different values of τ for $\mu_e = 1.4, \mu_i = 0.4, \sigma_i = 0.2, \beta = -0.3, u_0 = 0.4, q = 1$

parameter β, whereas Fig. 11 represents the evolution of the electric field E versus ξ for different values of time τ.

Shock wave solutions

In the presence of Burgers term, $C \neq 0$, the system of equations becomes dissipative and the total energy H is not conservative. In this case, the exact solution of Eq. (21) cannot be easily constructed. Therefore, different mathematical methods [8–10, 12, 20] have arose to solve such types of these nonlinear differential equations. Among those, the tangent hyperbolic (tanh) method has been proved to be a powerful mathematical technique for solving nonlinear partial differential equations. Following the procedure of the tanh method [22], we consider the solution in the following form

$$\phi_1 = \sum_{n=0}^{N} a_n \tanh^n(\chi), \tag{35}$$

where the coefficients a_n and N should be determined. Balancing the nonlinear and dispersion terms in Eq. (21), we obtain $N = 4$. Substituting Eq. (35) into Eq. (21) and equating to zero the different coefficients of different powers of tanh(χ) functions, one can obtain the following set of algebraic equations

$$-\frac{u_0}{B}a_0 - \frac{2A}{3B}(-a_0)^{\frac{3}{2}} + \frac{C}{B}a_1 + 2a_2 = 0, \tag{36a}$$

$$-2a_1 - \frac{u_0}{B}a_1 + \frac{A}{B}\sqrt{-a_0}a_1 + \frac{2C}{B}a_2 = 0, \tag{36b}$$

$$-\frac{C}{B}a_1 - \frac{A}{3B}\frac{a_1^2}{\sqrt{-a_0}} - 8a_2 - \frac{u_0}{B}a_2 + \frac{A}{B}\sqrt{-a_0}a_2 = 0, \tag{36c}$$

$$2a_1 - \frac{2C}{B}a_2 - \frac{2A}{3B}\frac{a_1 a_2}{\sqrt{-a_0}} = 0, \text{ and } 6a_2 - \frac{A}{3B}\frac{a_2^2}{\sqrt{-a_0}} = 0. \tag{36d}$$

Solving these set of algebraic equations, one gets

$$a_0 = -\frac{(12B + u_0)^2}{A^2}, \tag{37a}$$

$$a_1 = -\frac{18C(12B + u_0)}{5A^2}, \tag{37b}$$

$$a_2 = \frac{18B(12B + u_0)}{A^2}, \tag{37c}$$

$$C = 10B, \quad \text{and} \quad u_0 = 42B. \tag{37d}$$

With the knowledge of the above coefficients, one can write down the explicit solution of the mKdV–Burgers equation (21) in terms of tanh function

$$\phi_1 = \frac{(12B + u_0)}{A^2}\left(6B - u_0 - \frac{18C}{5}\tanh(\chi) - 18B\text{sech}^2(\chi)\right). \tag{38}$$

This class of solution represents a particular combination of a solitary wave [(sech$^2(\chi)$ term on the right-hand side of Eq. (38)] with a Burgers shock wave [(tanh(χ) term]. The behaviour of this solution is shown graphically as in Figs. 12, 13, and 14. Figures 12, and 13 show the variation of ϕ_1 with χ for different values of β and σ_i, respectively, and indicate that the amplitude of the wave increases with increasing the values of both β and σ_i. The variation of ϕ_1 against ξ due to the change of the time τ is plotted in Fig. 14. One can see from these figures that both soliton and shock structures are obtained due to the presence of both dispersive and dissipative coefficients.

Another type of solution can be obtained when the dissipation term is dominant over the dispersion term. In this case, Eq. (21) reduces to the following nonlinear first-order differential equation

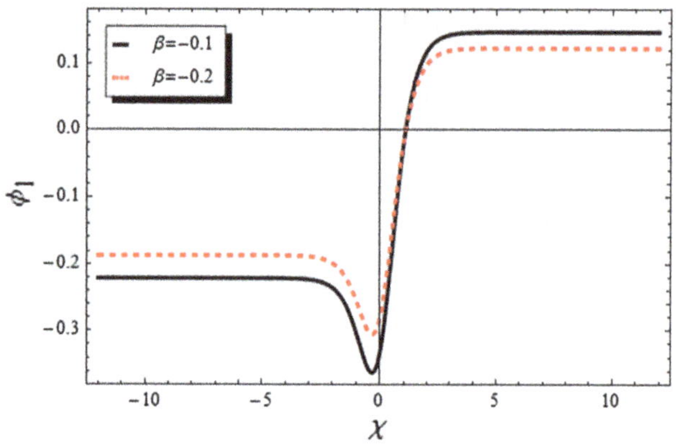

Fig. 12 Variation of ϕ_1 vs. χ for different values of β for $\mu_e = 5, \mu_i = 4, \sigma_i = 0.15, \eta = 0.3, u_0 = 0.4, q = 1$

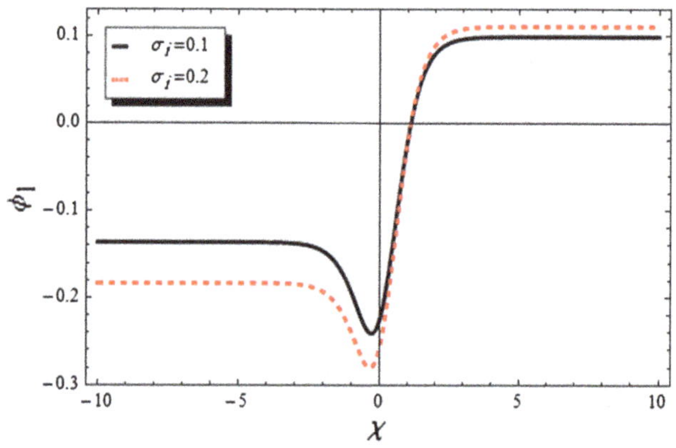

Fig. 13 Variation of ϕ_1 vs. χ for different values of σ_i for $\mu_e = 5, \mu_i = 4, \beta = -0.3, \eta = 0.3, u_0 = 0.4, q = 1$

$$\frac{\mathrm{d}\phi_1}{\mathrm{d}\chi} = \frac{u_0}{C}\phi_1 + \frac{2A}{3C}(-\phi_1)^{\frac{3}{2}}, \tag{39}$$

that admits the following solution

$$\phi_1 = -\frac{9u_0^2 \exp(\frac{u_0}{C}\chi)}{\left[1 + 2A\exp(\frac{u_0}{2C}\chi)\right]^2}. \tag{40}$$

This type of solution actually describes monotonic shock wave.

On the other hand, another type of solution of special interest can be obtained if one considers the boundary condition $\chi \to \pm\infty \Rightarrow \frac{\mathrm{d}^2\phi_1}{\mathrm{d}\chi^2} = \frac{\mathrm{d}\phi_1}{\mathrm{d}\chi} = 0$. With this condition, one obtains the asymptotic solution

$$\phi_c = -\left(\frac{3u_0}{2A}\right)^2, \tag{41}$$

of the nonlinear mKdV–Burgers differential equation.

Using $\phi_1 = \phi_c + \tilde{\phi}$ for $|\phi_c| \gg |\tilde{\phi}|$, Eq. (21) can be linearized to the second-order linear differential equation

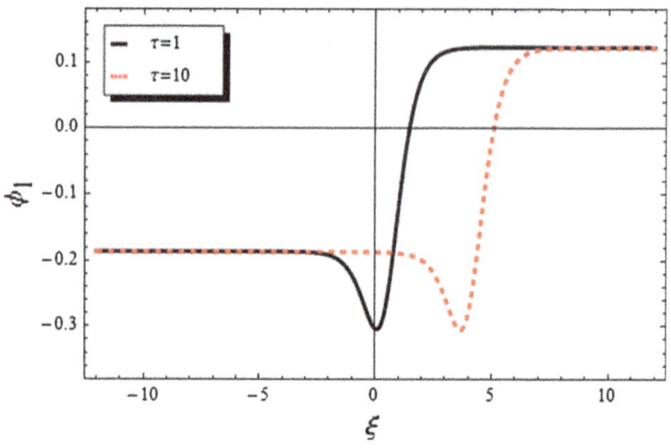

Fig. 14 Evolution of ϕ_1 vs. ξ for different values of τ for $\mu_e = 5, \mu_i = 4, \sigma_i = 0.15, \beta = -0.2, \eta = 0.3, u_0 = 0.4, q = 1$

$$\frac{\mathrm{d}^2\tilde{\phi}}{\mathrm{d}\chi^2} + \frac{C}{B}\frac{\mathrm{d}\tilde{\phi}}{\mathrm{d}\chi} + \frac{u_0}{2B}\tilde{\phi} = 0. \tag{42}$$

The solution of the linear differential equation (42) can be expressed in the exponential form $\tilde{\phi} = \exp(M\chi)$, where M is defined by

$$M = \frac{C}{2B}\left[-1 \pm \sqrt{\left(1 - \frac{2u_0B}{C^2}\right)}\right]. \tag{43}$$

For $C^2 \ll 2u_0B$, the oscillatory shock wave solution is given by

$$\phi_1 = \phi_c + \tilde{K}\exp\left(-\frac{C}{2B}\chi\right)\cos\left(\sqrt{\frac{u_0}{2B}}\chi\right), \tag{44}$$

where $\tilde{K}$ is an arbitrary constant. The behaviour of the obtained solution in terms of the coordinates ξ and τ is shown graphically in Fig. 15. In addition to oscillatory shock wave, the mKdV–Burgers equation exhibits solitonic

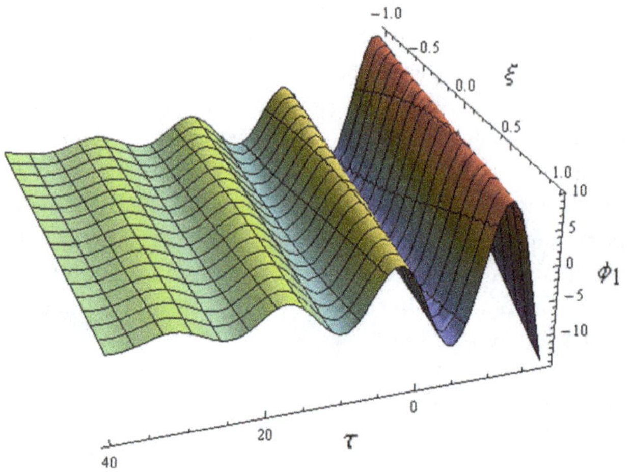

Fig. 15 Evolution of ϕ_1 vs. ξ for different values of τ for $\mu_e = 2.5,\ \mu_i = 1.5,\ \sigma_i = 0.2,\ \beta = -0.4, \eta = 0.1, u_0 = 0.4, q = 1$

and monotonic shock wave due to the Burgers term that arises from the fluid viscosity.

Conclusions

The linear and nonlinear properties of the DA waves in a system of homogeneous, unmagnetized, collisionless and dissipative dusty plasma consisted of extremely massive, micron-sized, negative dust grains, electrons obey a Boltzmann distribution and traped ions have been investigated. In the linear analysis, the normal mode method is used to reduce the basic set of fluid equations to a linear dispersion relation. Assuming $\omega = \omega_r + i\omega_i$, an expression for the damping rate ω_i is obtained and its behaviour with the plasma parameter (carrier wave number k, dust kinematic viscosity coefficient η_d, ratio of the ion to the electron temperatures σ_i) is shown graphically. Such graphs show that the instability damping rate decreases as the plasma parameters (k, η_d,σ_i) increase.

In the nonlinear analysis, the reductive perturbation method is used to reduce the basic set of fluid equations to the nonlinear partial differential mKdV–Burgers equation (19). Such nonlinear equation is not an integrable Hamiltonian system because the energy of plasma system is not conserved due to the dissipative Burgers term. This implies that finding exact solutions of mKdV–Burgers equation is a complicated task in general case. Therefore, in the absence of Burgers term, the equation becomes integrable and one can easily write down the explicit analytical solution. By solving the obtained mKdV–Burgers equation, we have succeeded to distinguish some interesting classes of analytical solutions due to Burgers coefficient C.

In the absence of the Burgers term ($C = 0$), i.e., the dissipation effect is negligible in comparison with that of the nonlinearity and dispersion, the topology of phase portrait and potential diagram is investigated as shown in Figs. 3, and 4. The advantage of using the phase portrait is that one may predict wide classes of the travelling wave solutions of mKdV–Burgers equation. One of these solutions is the soliton solution (30), whose behaviour is discussed in terms of plasma parameters (β, σ_i , u_0, τ) as shown in Figs. 5, 6, 7, and 8. In addition, the solitonic energy E_n and the associated electric field E are calculated and shown in Figs. 9, 10, and 11. It is obvious that the behaviour of the solitonic energy and the electric field depends crucially on the values of the parameters σ_i , β and τ.

In the presence of Burgers term ($C \neq 0$), the mKdV–Burgers equation admits some solutions of physical interest which are related to a combination between shock and soliton waves, monotonic and oscillatory shocks. The combination solution between shock and soliton waves is obtained explicitly using tanh method. In this case, both the dispersion and dissipation coefficients remain finite in comparison with each other and the profile of such case exhibits both shock and soliton waves for different values of β, σ_i and τ as shown in Figs. 12, 13, and 14. The monotonic shock wave can exist when the dissipation term is dominant over the dispersion term. The oscillatory shock wave can exist when dispersion term is dominant over the dissipation term and its behaviour is presented in Fig. 15. However, the formation of these solutions depends crucially on the value of the Burgers term as well as the plasma parameters.

Finally, it is concluded that a plasma with dissipative and dispersive properties supports the existence of shock waves instead of solitons. Under appropriate conditions, shock waves can be propagated in the plasma medium. It is emphasized that the present investigation may be helpful in better understanding of waves propagation in the astrophysical plasmas as well as in inertial confinement fusion laboratory plasmas.

References

1. Alinejad, H.: Influence of arbitrarily charged dust and trapped electrons on propagation of localized dust ion-acoustic waves. Astrophys. Space Sci. **337**, 223 (2012)
2. Annou, R.: Current-driven dust ion-acoustic instability in a collisional dusty plasma with a variable charge. Phys. Plasmas **5**, 2813 (1998)
3. Barkan, A., Merlino, R.L., D'Angelo, N.: Laboratory observation of the dust ion acoustic wave mode. Phys. Plasmas **2**, 3563 (1997)
4. Chatterjee, P., Ghosh, U.N.: Head-on collision of ion acoustic solitary waves in electron-positron-ion plasma with superthermal electrons and positrons. Eur. Phys. J. D **64**, 413 (2011)
5. Chu, J.H., Du, J.B., Lin, I.: Coulomb solids and low-frequency fluctuations in RF dusty plasmas. J. Phys. D **27**, 296 (1994)
6. Chu, J.H., Du, J.B., Lin, I.: Direct observation of coulomb crystals and liquids in strongly coupled rf dusty plasmas. Phys. Rev. Lett. **72**, 4009 (1994)
7. Duha, S.S., Anowar, M.G.M., Mamun, A.A.: Dust ion-acoustic solitary and shock waves due to dust charge fluctuation with vortexlike electrons. Phys. Plasmas **17**, 103711 (2010)
8. Dutta, M., Ghosh, S., Chakrabarti, N.: Electron acoustic shock waves in a collisional plasma. Phys. Rev. E **86**, 066408 (2012)
9. El-Hanbaly, A.M.: On the solution of the integro-differential fragmentation equation with continuous mass loss. J. Phys. A **36**, 8311 (2003)
10. El-Hanbaly, A.M., Abdou, M.: New application of Adomian decomposition method on Fokker-Planck equation. J. Appl. Math. Comput. **182**, 301 (2006)
11. El-Wakil, S.A., Attia, M.T., Zahran M.A., El-Shewy E.K., Abdelwahed H.G.: Effect of higher order corrections on the propagation of nonlinear dust-acoustic solitary waves in mesospheric dusty plasmas. Z. Naturforsch. **61a**, 316 (2006)
12. El-Wakil, S.A., El-Hanbaly, A.M., El-Shewy, E.K., El-Kamash, I.S.: Symmetries and exact solutions of KP equation with an arbitrary nonlinear term. J. Theor. Appl. Phys. **8**, 130 (2014)
13. Fortov, V.E., Nefedov, A.P., Vaulina, O.S., Lipaev, A.M., Molotkov, V.I., Samaryan, A.A., Nikitskii, V.P., Ivanov, A.I.,

Savin, S.F., Kalmykov, A.V., Solov'ev, A.Y., Vinogradov, P.V.: Dusty plasma induced by solar radiation under microgravitational conditions: an experiment on board the Mir orbiting space station. J. Exp. Theor. Phys. **87**, 1087 (1998)
14. Ghosh, U.N., Chatterjee, P.: Interaction of cylindrical and spherical ion acoustic solitary waves with superthermal electrons and positrons. Astrophys. Space Sci. **344**, 127 (2013)
15. Gupta, M.R., Sarkar, S., Ghosh, S., Debnath, M., Khan, M.: Effect of nonadiabaticity of dust charge variation on dust acoustic waves: generation of dust acoustic shock waves. Phys. Rev. E **63**, 046406 (2001)
16. Havnes, O., Troim, J., Blix, T., Mortensen, W., Naesheim, L.I., Thrane, E., Tonnesen, T.: First detection of charged dust particles in the Earth's mesosphere. J. Geophys. Res. **101**, 10839 (1996)
17. Horányi, M.: Charged dust dynamics in the solar system. Annu. Rev. Astrophys **34**, 383 (1996)
18. Horányi, M., Mendis, D.A.: The dynamics of charged dust in the tail of comet Giacobini-Zinner. J. Geophys. Res **91**, 355 (1986)
19. Ikezi, H., Taylor, R.J., Baker, D.R.: Formation and interaction of ion-acoustic solitions. Phys. Rev. Lett. **25**, 11 (1970)
20. Mahmood, S., Ur-Rehman, H.: Formation of electrostatic solitons, monotonic, and oscillatory shocks in pair-ion plasmas. Phys. Plasmas **17**, 072305 (2010)
21. Maitra, S., Roychoudhury, R.: Sagdeev's approach to study the effect of the kinematic viscosity on the dust ion-acoustic solitary waves in dusty plasma. Phys. Plasmas **12**, 054502 (2005)
22. Malfliet, W., Hereman, W.: The tanh method: II, Perturbation technique for conservative systems. Phys. Scr. **54**, 563 (1996)
23. Mamun, A.A.: Arbitrary amplitude dust-acoustic solitary structures in a three-component dusty plasma. Astrophys. Space Sci. **268**, 443 (1999)
24. Mamun, A.A., Eliasson, B., Shukla, P.K.: Dust-acoustic solitary and shock waves in a strongly coupled liquid state dusty plasma with a vortex-like ion distribution. Phys. Lett. A **332**, 412 (2004)
25. Melandso, F.: Lattice waves in dust plasma crystals. Phys. Plasmas **3**, 3890 (1996)
26. Mendis, D.A., Rosenberg, M.: Cosmic dusty plasma. Annu. Rev. Astron. Astrophys. **32**, 418 (1994)
27. Merlino, R.L., Barkan, A., Thompson, C., D'Angelo, N.: Laboratory studies of waves and instabilities in dusty plasmas. Phys. Plasmas **5**, 1607 (1998)
28. Nakamura, Y., Sarma, A.: Observation of ion-acoustic solitary waves in a dusty plasma. Phys. Plasmas **8**, 3921 (2001)
29. Pakzad, H.R., Javidan, K.: Dust acoustic solitary and shock waves in strongly coupled dusty plasmas with nonthermal ions. Pramana J. Phys. **73**(5), 913 (2009)
30. Rahman, O., Mamun, A.A., Ashrafi, K.S.: Dust-ion-acoustic solitary waves and their multi-dimensional instability in a magnetized dusty electronegative plasma with trapped negative ions. Astrophys. Space Sci. **335**, 425 (2011)
31. Rao, N.N., Shukla, P.K., Yu, M.: Dust acoustic waves in dusty plasmas. Planet. Space Sci. **38**, 543 (1999)
32. Rosenberg, M., Mendis, D.A., Sheehan, D.P.: Positively charged dust crystals induced by radiative heating. IEEE Trans. Plasma Sci. **27**, 239 (1999)
33. Schamel, H.: Snoidal and sinusoidal ion acoustic waves. Plasma Phys. **14**, 905 (1972)
34. Schamel, H., Bujarbarua, S.: Solitary plasma hole via ionvortex distribution. Phys. Fluids **23**, 2498 (1980)
35. Shahmansouri, M., Tribeche, M.: Arbitrary amplitude dust acoustic waves in a nonextensive dusty plasma. Astrophys. Space Sci. **344**, 99 (2013)
36. Shahmansouri, M., Tribeche, M.: Nonextensive dust acoustic shock structures in complex plasmas. Astrophys. Space Sci. **346**, 165 (2013)
37. Shchekinov, Y.A.: Dust-acoustic soliton with nonisothermal ions. Phys. Lett. A **225**, 117 (1997)
38. Shukla, P.K., Mamun, A.A.: Dust-acoustic shocks in a strongly coupled dusty plasma. IEEE Trans. Plasma Sci. **29**, 221 (2001)
39. Shukla, P.K., Mamun, A.A.: Introduction to Dusty Plasma Physics. Institute of Physics Publishing Ltd, Bristol, UK (2002)
40. Shukla, P.K., Slin, V.P.: Dust ion-acoustic wave. Phys. Scr. **45**, 508 (1992)
41. Shukla, P.K., Varma, R.K.: Convective cells in nonuniform dusty plasmas. Phys. Fluids B **5**, 236 (1993)
42. Shukla, P.K., Yu, M.Y., Bharuthram, R.: Linear and nonlinear dust drift waves. J. Geophys. Res. **96**, 21343 (1992)
43. Taniuti, T., Wei, C.C.: Reductive perturbation method in nonlinear wave propagation I. J. Phys. Soc. Jpn. **24**, 941 (1968)
44. Taniuti, T., Yajima, N.: Perturbation method for a nonlinear wave modulation. J. Math. Phys. **10**, 1369 (1969)
45. Thomas, H., Morfill, G.E., Dammel, V.: Plasma crystal: coulomb crystallization in a dusty plasma. Phys. Rev. Lett. **73**, 652 (1994)
46. Washimi, H., Taniuti, T.: Propagation of ion-acoustic solitary waves of small amplitude. Phys. Rev. Lett. **17**, 996 (1966)
47. Xue, J.K.: Cylindrical and spherical ion-acoustic solitary waves with dissipative effect. Phys. Lett. A **322**, 225 (2004)

21

Multifocal terahertz radiation by intense lasers in rippled plasma

Reenu Gill[1] · Divya Singh[2] · Hitendra K. Malik[1]

Abstract This paper presents a theoretical model for the generation of terahertz radiation by cosh-Gaussian laser beams of high intensity, which are capable of creating relativistic–ponderomotive nonlinearity. We find the components of the terahertz radiation for the relativistic laser plasma interaction, i.e. beating of the two lasers of same amplitude and different frequency in under dense plasma. We plot the electric field profile of the emitted radiation under the effect of lasers index. By creating a dip in peak of the incident lasers' fields, we can achieve multifocal terahertz radiation.

Keywords Coshyperbolic-Gaussian lasers · Laser index · Relativistic effect · Rippled plasma · Multifocal terahertz radiation

Introduction

Recently, the high-field THz pulses have been generated from large-scale accelerators using ultrashort electron bunches. For example, peak field of 44 MV/cm has been obtained via coherent transition radiation in the Linac Coherent Light Source [1]. In the laser-based schemes, THz pulses with field strength >8 MV/cm at 1 kHz repetition rate have been generated via two-color laser filamentation [2]. Single-cycle THz pulse with field strength up to 36 MV/cm can be generated by optical rectification of a mid infrared laser in a large-size nonlinear organic crystals assembly [3]. Electromagnetic radiations have been generated by the interaction of ultrashort lasers with gas-jet plasmas [4, 5] and water vapor [6]. In addition to these, other methods that have produced high-field THz pulses (but only with components <20 THz) include difference frequency mixing process of two near-infrared lasers in second-order nonlinear crystals [7], and optical rectification (with component limited to <1.5 THz) in lithium niobate ($LiNbO_3$) crystals with tilted laser front [8]. Besides, high-field THz radiations can be generated from relativistic laser irradiated plasmas via various mechanisms [9–12].

Several experiments employ plasma as a nonlinear medium for the generation of THz radiation by using sub picosecond laser pulses and energetic electron beams [13], since the plasma has the advantage of handling very high field with a remarkable property of being not damaged and having strong nonlinearity [14]. Malik et al. [15] have analytically investigated the THz generation by tunnel ionization of a gas jet with superposed femtosecond laser pulses shone onto it after passing through an axicon. In this mechanism, the temporal evolution of plasma density and hence, oscillatory current density via oscillating dipoles gives rise to the emission of THz radiation. Electron positron plasma has been suggested for high-efficient THz generation by using laser pulse [16]. All the methods suggest that the intense THz radiation can be achieved when very high-intensity lasers are employed in the plasma. However, under that situation, the electron motion will become relativistic and hence, their mass will be varied. This situation is considered in the present article along with the use of coshyperbolic-Gaussian lasers.

✉ Hitendra K. Malik
hkmalik@physics.iitd.ac.in

1 Plasma Waves and Particle Acceleration (PWAPA) Laboratory, Department of Physics, Indian Institute of Technology Delhi, New Delhi 110016, India

2 Rajdhani College, Delhi University, New Delhi, India

Coshyperbolic-Gaussian lasers

We consider linearly polarized cosh-Gaussian electromagnetic beams propagating along the z-axis. Their electric fields are along the y-axis, given by

$$\vec{E}_j = E_0 \cosh\left(\frac{ys}{b}\right) \exp\left(-\left(\frac{ys}{b}\right)^p\right) \exp\left(i\left(k_j z - \omega_j t\right)\right)\hat{y},$$

together with $j = 1, 2$ and p as the index of the lasers. Here E_0 is the amplitude of the fields; b is the beam width of the lasers. Figure 1 shows the profile of the incident laser beam when $p \geq 2$ and for different values of the skewness parameter s.

Relativistic ponderomotive force

We use the cold fluid equations for electrons with their number density n_0 and velocity υ under the impact of lasers fields. Hence, the electrons oscillatory velocity is obtained as

$$\vec{\upsilon}_j = \frac{e\vec{E}_j}{mi\omega_j}. \tag{1}$$

Here we neglect the effect of collisions in the plasma because collision effect is not significant for high-intensity laser beam. The density n obeys the continuity equation $\frac{\partial n}{\partial t} + \vec{\nabla} \cdot (n\vec{\upsilon}) = 0$. In view of the high-intensity lasers, we consider relativistic motion of the electrons. The relativistic ponderomotive force in the presence of an intense electromagnetic beam can be represented [17, 18] as

$\vec{F}_p = -mc^2\vec{\nabla}(\gamma - 1)$ together with

$$\gamma = \left(1 - \frac{|\upsilon_1 + \upsilon_2|^2}{c^2}\right)^{-\frac{1}{2}}. \tag{2}$$

Here m is the rest mass of the electron, c is the velocity of the light, and υ_1 and υ_2 are the oscillatory drift velocities of the electrons. The use of Eq. (1) in Eq. (2) yields the ponderomotive force as

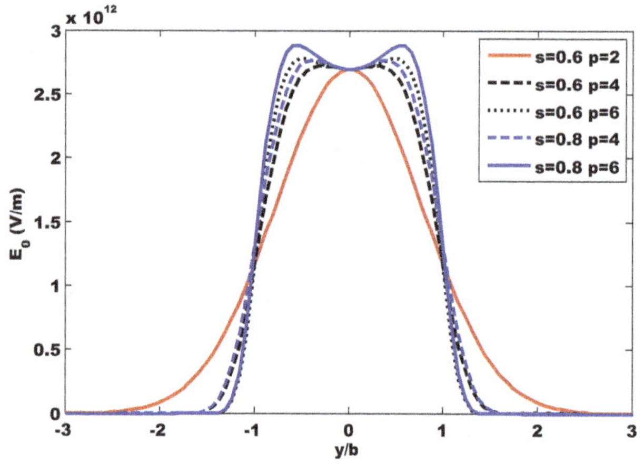

Fig. 1 Electric field distribution of the incident laser profile

$$\vec{F}_p = -mc^2\vec{\nabla}\left(1 - \frac{2A}{c^2\omega_1\omega_2}\cos(kz - \omega t)\right)^{-\frac{1}{2}}, \tag{3}$$

where $A = \frac{e^2E_0^2}{m^2}\cosh^2\left(\frac{ys}{b}\right)e^{-2\left(\frac{ys}{b}\right)^p}$, $k = k_1 - k_2$ and $\omega = \omega_1 - \omega_2$.

We obtain the following expression for the ponderomotive force:

$$\vec{F}_p^{NL} = \frac{imc^2A}{\omega_1\omega_2}\left(1 - \frac{2Ac^2}{\omega_1\omega_2}\cos(kz - \omega t)\right)^{-\frac{3}{2}} \left[\left(2p\left(\frac{ys}{b}\right)^{p-1}\frac{s}{b} - \tanh\left(\frac{ys}{b}\right)\frac{s}{b}\right)\cos(kz - \omega t)\hat{y} + k\sin(kz - \omega t)\hat{z}\right]. \tag{4}$$

This nonlinear force causes perturbations in the density of the plasma. These perturbations generate space charge potential that leads to the linear density perturbations. The nonlinear density perturbations are obtained as

$$n_1^{NL} = \frac{n_0\vec{\nabla} \cdot \vec{F}_p^{NL}}{m\omega^2}. \tag{5}$$

Finally, the nonlinear velocity of the electrons is obtained from equation of motion as

$$\vec{V}^{NL} = -\left(\frac{\omega^2\left(\omega^2 - \omega_p^2\right) + \omega_p^2\omega^2}{mi\omega^3\left(\omega^2 - \omega_p^2\right)}\right)\vec{F}_p. \tag{6}$$

Here, ω_p is the plasma frequency given by $\omega_p = \sqrt{\frac{4\pi n_0 e^2}{m}}$.

The nonlinear current density can be obtained from $\vec{J}^{NL} = -\frac{1}{2}ne\vec{V}^{NL}$, where $n = n_0 + n_2$. Here n_2 is the ripple density in the plasma given as $n_2 = n_\alpha e^{i\alpha z}$ together with α as the wave number of the ripples produced in the plasma and n_α as the amplitude of the density ripples. These density ripples are generally used to achieve the resonance condition. These ripples in density may be produced using various techniques, which involve transmissive ring grating and a patterned mask where the control of ripple parameters might be possible by changing the groove structure, groove period, and duty cycle in such a grating and by adjusting the period and size of the mask [19–23]. Finally, the nonlinear current density is obtained as

$$\vec{J}_y^{NL} = -\frac{1}{2}n_\alpha e^{i\alpha z}e\left[-\frac{\omega^2\left(\omega^2 - \omega_p^2\right) + \omega_p^2\omega^2}{mi\omega^3\left(\omega^2 - \omega_p^2\right)}\vec{F}_{py}\right]. \tag{7}$$

Calculation of electric field of terahertz radiation

The wave equation that governs the emitted THz radiation is given by

$$-\nabla^2 \vec{E}_{\mathrm{THz}} + \vec{\nabla}(\vec{\nabla} \cdot \vec{E}_{\mathrm{THz}}) = \frac{4\pi i \omega}{c^2} \vec{J}^{\mathrm{NL}} + \frac{\omega^2}{c^2} \varepsilon \vec{E}_{\mathrm{THz}}. \quad (8)$$

Here $\vec{E}_{\mathrm{THz}}$ is the electric field associated with the THz radiation and ε is dielectric constant of the plasma amounting to $1 - \frac{\omega_p^2}{\omega^2}$. From Eq. (8), we get the amplitude of the THz radiation as

$$E_{THzy} = \begin{bmatrix} -\frac{4\pi i \omega m z}{c^2 2k'} \frac{e n_\alpha e^{i\alpha z}}{2} \left(\frac{Ac^2}{\omega_1 \omega_2}\right)^{-\frac{1}{2}} \frac{1}{2\sqrt{2}} [\cos(kz - \omega t)]^{-\frac{3}{2}} - \frac{4\pi i \omega z}{c^2 2k'} \frac{e m n_\alpha e^{i\alpha z}}{2} \\ \left(\frac{Ac^2}{\omega_1 \omega_2}\right)^{-\frac{1}{2}} \frac{3}{4} \left(\frac{Ac^2}{\omega_1 \omega_2}\right)^{-1} \frac{1}{2\sqrt{2}} [\cos(kz - \omega t)]^{-\frac{5}{2}} \end{bmatrix} \times [CD \cos(kz - \omega t)] \quad (9)$$

Here C and D are frequency-dependent terms defined as $C = 2p\left(\frac{ys}{b}\right)^{p-1} \frac{s}{b} - \frac{s}{b} \tanh\left(\frac{ys}{b}\right)$ and $D = \frac{i\omega}{m\left(\omega^2 - \omega_p^2\right)}$ and $k' = k + \alpha$.

In order to find the amplitude of the THz radiation, we analyze the expression (9). This relation can be bifurcated into two different expressions as shown in Eq. (10), indicating that it is superposition of two different waves. This is the consequence of the different components of the ponderomotive force. Finally, we achieve the following expression for E_{THz}:

$$E_{\mathrm{THz}} = J \cos(kz - \omega t) + R \cos^2(kz - \omega t). \quad (10)$$

Here J and R are the amplitudes of the different wave components that are generated due to the relativistic ponderomotive force nonlinearity in the plasma. The symbols J, R and G are defined as

$$J = -\mathrm{GCD},$$

$$R = -\frac{\mathrm{GADC}}{3\omega_1 \omega_2 c^2} \text{ and } G = -\frac{\pi i \omega z n_\alpha e m}{2\sqrt{2} c^2 k'} \left(\frac{Ac^2}{\omega_1 \omega_2}\right)^{-\frac{1}{2}}.$$

For the high-intensity laser, where a part of its energy is used in heating of the electrons in the plasma, the resonance condition is expected to be modified as $\omega = \sqrt{\omega_p^2 + k^2 v_{\mathrm{th}}^2}$, where v_{th} is the thermal velocity of the electrons. However, the second term would play a significant role only when k is very large, i.e. for the short wavelength laser pulses. So it is clear that the resonance condition is departed from $\omega = \omega_p$ when one considers the heating in the plasma and under this situation reduced THz field would be realized (clear from Fig. 3, later).

Results and discussion

This section deals with the results obtained from Eq. (10) for the electric field of THz radiation. The following set of parameters has been used in the numerical analysis: $E_0 = 2.7 \times 10^{12} V/m$, $b = 0.01 \times 10^{-2} m$, $s = 0.6$, $\omega = 1.15\omega_p$, $\omega_1 = 1.65 \times 10^{15}$rad/s and $\omega_p = 2.0 \times 10^{13}$ rad/s. This is seen that out of two components of fields, given by Eq. (10), only one component (second term) dominates and is solely responsible for the THz amplitude.

Figure 1 shows the electric field distribution of the incident lasers. This is clear that a dip occurs in the peak of the fields when $p > 2$. As the skewness parameter s is increased this dip takes a prominent form in the incident laser profile. Figure 2 shows the output profile (normalized amplitude) of the THz radiation with the normalized distance, when $n_\alpha = 0.3n_0$. The maximum amplitude occurs at particular values of y/b. Peak positions are symmetric in nature because of skewness parameter s. Maximum amplitudes correspond to the higher SG index lasers. A comparison of Figs. 1 and 2 shows that in general two peaks of the emitted THz radiation are obtained for the lasers having one peak. However, when there is a dip in the peak value of the lasers field, small peaks start generating in the produced radiations. Also the higher peak is further enhanced for the larger dip in the incident laser profile. This is true for the lasers having higher skewness in their profiles ($s > 0.6$).

The dependence of THz field on the ripple amplitude n_α and beating frequency ω is shown in Fig. 3. It is evident that the amplitude E_{THz} is high when n_α carries higher values, but it decreases sharply when the resonance condition $\omega \sim \omega_p$ is departed. Ripples in plasma density play a positive role to enhance the THz amplitude because large

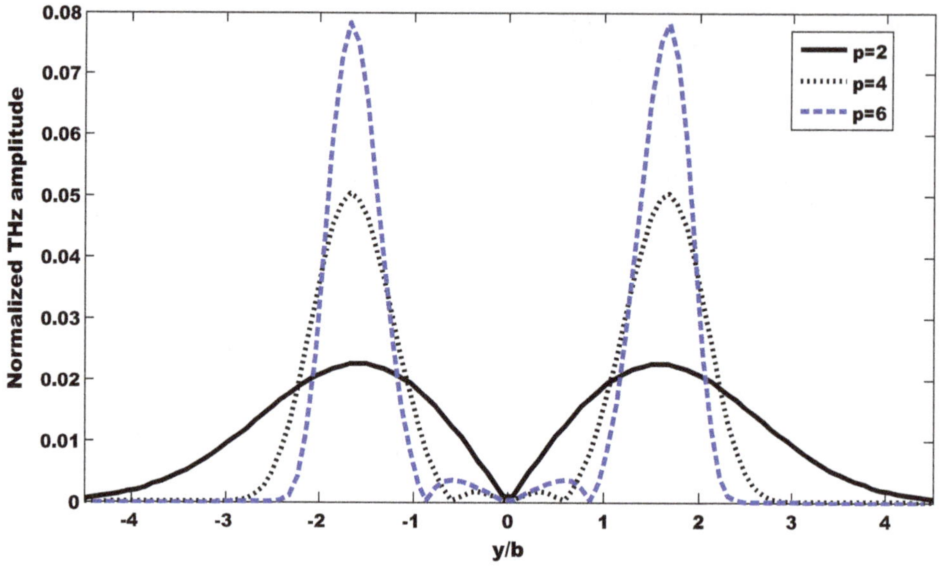

Fig. 2 The profile of electric field of emitted THz radiation

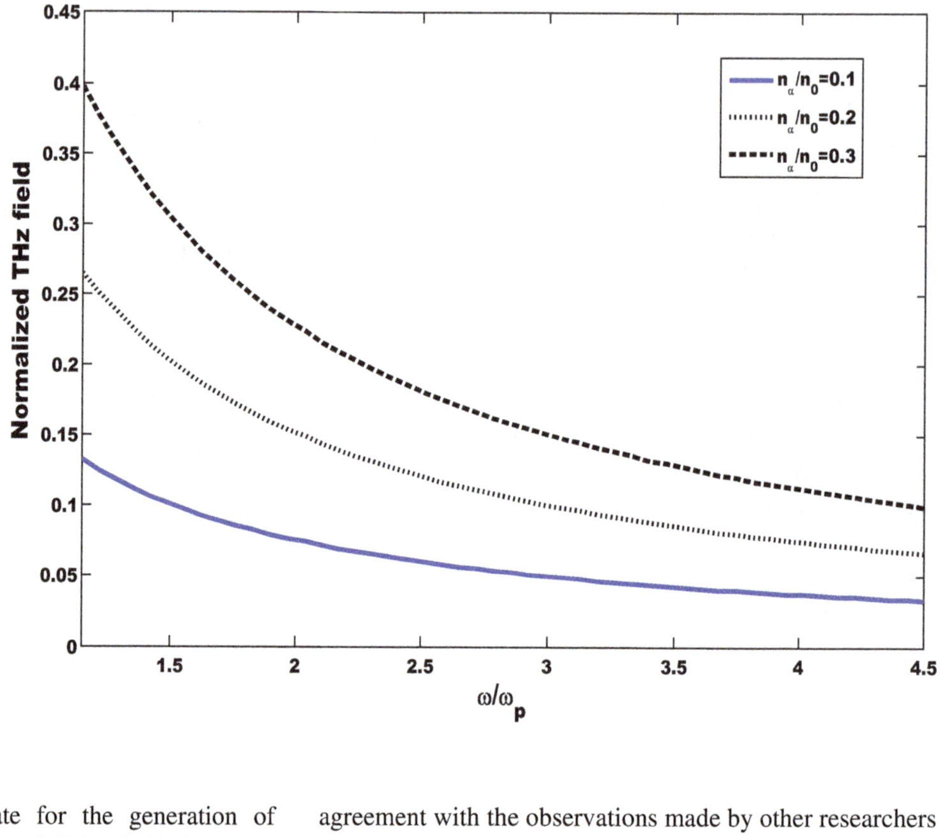

Fig. 3 Variation of THz radiation normalized field with beating frequency ω and density ripple amplitude n_α, for the same parameters as in Fig. 2

numbers of electrons participate for the generation of nonlinear current so amplitude of THz field increases accordingly.

In order to find the effect of beam width of the lasers on the THz radiation, we have plotted the intensity (normalized by $|E_0|^2$, i.e. the intensity of lasers) with b and p in Fig. 4. Here, we clearly observe that the lasers with lower beam width produce intense radiation and the same is the case with higher index lasers (p). These results are in agreement with the observations made by other researchers [24–26].

Comparative study

Kumar et al. [27] have studied the Gaussian lasers beam beating for THz generation. Our results are comparable to their observations when we consider the relativistic case

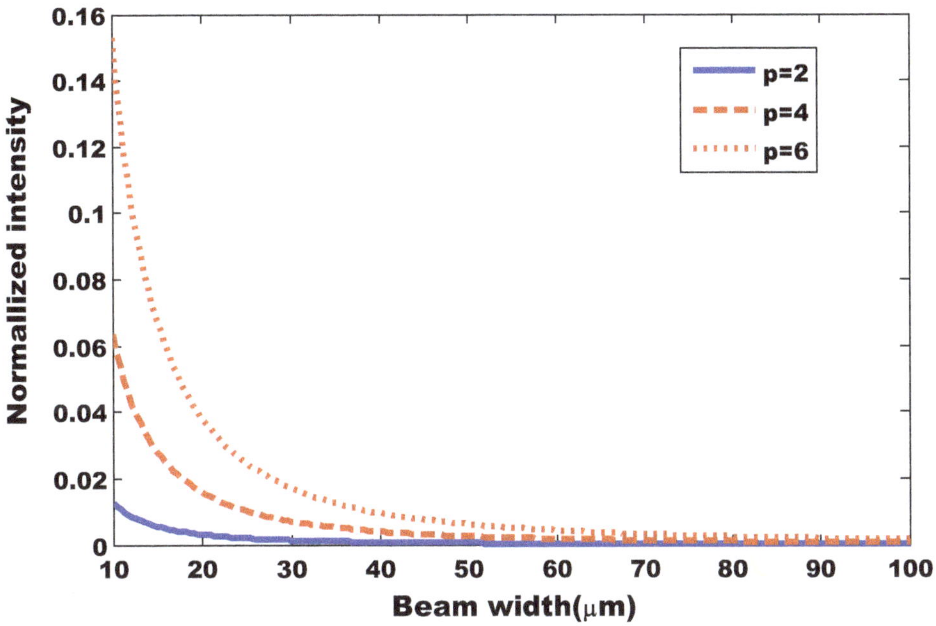

Fig. 4 Variation of normalized intensity of the generated THz radiation with beam width of the lasers

and use coshyperbolic-Gaussian lasers. Flat top or super-Gaussian like lasers destabilize as these propagate in the medium leading to two separate peaks in THz profile, whereas Gaussian-like beam retains its shape as it propagates through the medium. Therefore, the THz field in Fig. 2 shows small intensity peak at the center. In nonrelativistic cases the SG index plays important role to enhance the amplitude of the THz [26], but in case of relativistic ponderomotive force the density ripples are found to enhance the amplitude significantly, whereas SG index does not play that much significant role. Main difference between the results obtained in relativistic and nonrelativistic cases is that multiple components appear in Eq. [10] as obtained by Bituk and Fedorov [28] also.

Conclusions

When we consider the relativistic effect in plasma due to the higher intensity lasers, the emitted THz radiation is not found to be focused at a single place, rather it is focused at least at two places. These amplitudes carry different magnitudes. We can create multiple focal THz radiation by creating a dip in the electric field peak of the beating lasers. Other effects of the beam width and density ripples are found to be the similar as in non-relativistic case.

Acknowledgements The authors thank DRDO, Government of India, for providing financial support.

References

1. Wu, Z., Fisher, A.S., Goodfellow, J., Fuchs, M., Wen, H., Ghimire, S., Reis, D.A., Loos, H., Lindenberg, A.: Rev. Sci. Instrum. **84**, 022701 (2013)
2. Oh, T.I., Yoo, Y.J., You, Y.S., Kim, K.Y.: Appl. Phys. Lett. **105**, 041103 (2014)
3. Vicario, C., Monoszlai, B., Hauri, C.P.: Phys. Rev. Lett. **112**, 213901 (2014)
4. Dorranian, D., Starodubtsev, M., Kawakami, H., Ito, H., Yugami, N., Nishida, Y.: Phys. Rev. E **68**, 026409 (2003)
5. Dorranian D., Ghoranneviss M., Starodubtsev M., Ito H., Yugami N., Nishida Y.: Phys. Lett. A **331**, 77 (2004)
6. Johnson, K., Price-Gallagher, M., Mamerc, O., Lesimple, A., Fletcherb, C., Chend, Y., Lu, X., Yamaguchi, M., Zhang, X.-C.: Phys. Lett. A **372**, 6037 (2008)
7. Sell, A., Leitenstorfer, A., Huber, R.: Opt. Lett. **33**, 2767 (2008)
8. Stepanov, A.G., Henin, S., Petit, Y., Bonacina, L., Kasparian, J., Wolf, J.-P.: Appl. Phys. B **115**, 293 (2014)
9. Sheng, Z.-M., Mima, K., Zhang, J., Sanuki, H.: Phys. Rev. Lett. **94**, 095003 (2005)
10. Li, Y.T., Li, C., Zhou, M.L., Wang, W.M., Du, F., Ding, W.J., Lin, X.X., Liu, F., Sheng, Z.M., Peng, X.Y., Chen, L.M., Ma, J.L., Lu, X., Wang, Z.H., Wei, Z.Y., Zhang, J.: Appl. Phys. Lett. **100**, 254101 (2012)
11. Gopal, A., Herzer, S., Schmidt, A., Singh, P., Reinhard, A., Ziegler, W., Brommel, D., Karmakar, A., Gibbon, P., Dillner, U., May, T., Meyer, H.-G., Paulus, G.G.: Phys. Rev. Lett. **111**, 074802 (2013)
12. Chen, Z.-Y., Li, X.-Y., Yu, W.: Phys. Plasmas **20**, 103115 (2013)
13. Leemans, W.P., van Tilborg, J., Faure, J., Geddes, C.G.R., Toth, C., Schroeder, C.B., Esarey, E., Fubioni, G., Dugan, G.: Phys. Plasmas **11**, 2899 (2004)
14. Pukhov, A.: Rep. Prog. Phys. **66**, 47 (2003)
15. Malik, A.K., Malik, H.K., Kawata, S.: J. Appl. Phys. **107**, 113105 (2010)
16. Malik, H.K.: Phys. Lett. A **379**, 2826 (2015)
17. Brandi, H.S., Manus, C., Mainfray, G., Lehner, T., Bonnaud, G.: Phys. Fluids **5**, 3539 (1993)
18. Gupta, M.K., Sharma, R.P., Gupta, V.L.: Phys. Plasmas **12**, 1231011 (2005)

19. Kim K.Y., Taylor A.J., Glownia J.H., Rodriguez G.: Nat. Photonics **2**, 605 (2008)
20. Kuo, C.C., Pai, C.H., Lin, M.W., Lee, K.H., Lin, J.Y., Wang, J., Chen, S.Y.: Phys. Rev. Lett. **98**, 033901 (2007)
21. Hazra, S., Chini, T.K., Sanyal, M.K., Grenzer, J.: Phys. Rev. B **70**, 121307(R) (2004)
22. Layer, B.D., York, A., Antonson, T.M., Varma, S., Chen, Y.-H., Leng, Y., Milchberg, H.M.: Phys. Rev. Lett. **99**, 035001 (2007)
23. Malik, H.K.: Europhys. Lett. **106**, 55002 (2014)
24. Malik, A.K., Malik, H.K., Stroth, U.: Phys. Rev. E **85**, 016401 (2012)
25. Malik, A.K., Malik, H.K., Nishida, Y.: Phys. Lett. A **375**, 1191 (2011)
26. Singh, D., Malik, H.K.: Plasma Sour. Sci. Technol. **24**, 045001 (2015)
27. Kumar, S., Singh, R.K., Sharma, R.P.: Phys. Plasmas **22**, 103101 (2015)
28. Bituk, D.R., Fedorov, M.V.: JETP **89**(4), 640 (1999)

22

Dynamical and transport properties in plasmas including three-particle spatial correlations

Hakima Ababsa[1] · Med Tayeb Meftah[1] · Thouria Chohra[1]

Abstract In this work, we study the two and triplet static correlation functions in plasma when the ions interact via the Debye screened potential and via the Deutsch screened potential. The latter takes into consideration the possible quantum effects at short distances. The ratio of the mean distance between two ions and the thermal De Broglie wavelength r_i/λ_T gives the measure of these effects. Our investigation is developed in the conditions of weak coupling parameter ($\Gamma < 1$). The pair and the triplet correlation functions are calculated numerically and compared to the correlation functions due to the Kirkwood superposition approximation (KSA). Some applications to the ion velocity auto-correlation function *D*(*t*) and the electric field auto-correlation function *C*(*t*) at an ion (assumed to be an impurity) and the diffusion coefficient *D* are calculated for the two kinds of potentials in different plasma conditions. The comparison with other results found in the literature shows a well satisfactory agreement, for the static as well as the dynamic properties.

Keywords Static pair correlation · Triplet correlation · Time auto-correlation function · Velocity · Electric field · Diffusion coefficient

✉ Med Tayeb Meftah
mewalid@yahoo.com

[1] LRPPS Laboratory, Department of Physics, Faculty of Mathematics, Kasdi Merbah University, 30000 Ouargla, Algeria

Introduction

The equilibrium properties of a plasma considered as a liquid of charged particles (in the hydrodynamical description) are fully described by a set of probability density functions $g(\mathbf{r}_1, \ldots, \mathbf{r}_n)$ of location of particles at points $\mathbf{r}_1, \ldots, \mathbf{r}_n$, when the total potential energy of a liquid is given by a sum of isotropic pair potentials [1], and the physical properties (pressure, energy density, etc.) [2] are defined by the pair correlation function $g(\mathbf{r}_1, \mathbf{r}_2)$. However, even in the approximation of pair interaction, higher order correlation functions are of interest. Information on the triplet correlation function $g(\mathbf{r}_1, \mathbf{r}_2, \mathbf{r}_3)$ is of importance in calculating the properties of the medium (entropy, thermal expansion coefficients, etc.) [3]. Explicit knowledge of triplet correlations is also required in perturbation theories for static fluid properties [4] and in the theories of transport properties [5]. The triple distribution function can be computed either via computer simulation methods like (the Monte Carlo method, molecular dynamics method) [6–8], or via the theoretical methods [9]. It can also be determined experimentally [3, 4, 10].

In this paper the obtained results are based on theoretical studies of the structure functions and the transfer phenomena in the plasmas in terms of two-particle and three-particle correlation. We have theoretically obtained three-particle correlation functions and analyzed them and compared with the superposition approximation [11]. Overall in this paper we use two kinds of screened potentials: the screened Debye potential $V(r) = q\exp(-r/\lambda_D)/r$ and the screened Deutsch potential $V^{SD}(r) = q\exp(-r/\lambda_D)(1 - \exp(-r/\lambda_T))/r$ [12], where r is the interparticle spacing, λ_D is the screening Debye length, λ_T the thermal Broglie wave length and q is the ion charge. $\Gamma = q^2/k_B T r_i$ is the coupling parameter, where $r_i = (3/4\pi\rho)^{1/3}$ is the mean interparticle spacing, ρ the particles density whereas k_B is the Boltzmann constant and T is the equilibrium temperature of the system. Our paper is organized as follows: in the next section, we present some theoretical implementations describing the pair correlation functions, following which the theoretical derivation of the three correlation functions for different plasma conditions for both kinds of potential is presented. In the subsequent section, we present some applications of the static structure functions on the calculation of the dynamical properties as the time auto-correlation functions of the ions velocity and of the electrical micro-field on an ion (considered as an impurity), and the related transport quantities. The results with discussion are reported before the concluding section. The final section contains conclusion with some perspectives.

Static pair correlation functions

To define the pair correlation functions $g(r)$ that represents a static structure function, we introduce the probability of a configuration of N particles in a volume V in equilibrium with a thermal bath at temperature T,

$$g(r_1, r_2) = V^2\left(1 - \frac{1}{N}\right)\frac{\int \exp(-\beta V_N(r^N))\mathbf{dr}_3 \ldots \mathbf{dr}_N}{Z_N} \tag{1}$$

For the case where $N \to \infty$ such as $N/V = \rho$ is finite (thermodynamical limit):

$$g(r_1, r_2) = V^2\frac{\int \exp(-\beta V_N(r^N))\mathbf{dr}_3 \ldots \mathbf{dr}_N}{Z_N} \tag{2}$$

where $\beta = 1/k_B T$, Z_N and $V_N(r^N)$ are the configuration integral and the total potential energy, respectively, given by

$$Z_N = \int \exp(-\beta V_N(r^N))\mathbf{dr}_1 \ldots \mathbf{dr}_N \tag{3}$$

$$V_N(r^N) = \sum_{1 \le i \langle j \le N} v(r_{ij}) \tag{4}$$

where $v(r_{ij})$ is the binary interaction between the ith and jth ions. By substituting (4) in (2), we find

$$g(r_1, r_2) = V^2\frac{\int \exp\left(-\beta\sum_{1 \le i < j \le N} v(r_{ij})\right)\mathbf{dr}_3 \ldots \mathbf{dr}_N}{Z_N} \tag{5}$$

When we extract away the first term $v(r_{12})$ from the sum present at the exponential, because it is not concerned by the integration, we find

$$g(r_1, r_2) = \frac{V^2}{Z_N}\exp(-\beta v(r_{12})) \times \int \exp\left(-\beta \sum_{i \ge 1, j \ge 3(i \langle j)} v(r_{ij})\right)\mathbf{dr}_3 \ldots \mathbf{dr}_N \tag{6}$$

$$= \frac{V^2}{Z_N}\exp(-\beta v(r_{12})) \int \prod_{i \ge 1, j \ge 3(i \langle j)} \exp(-\beta v(r_{ij}))\mathbf{dr}_3 \ldots \mathbf{dr}_N \tag{7}$$

then, the pair correlation function becomes

$$g(r_1, r_2) = \frac{V^2}{Z_N}\exp(-\beta v(r_{12})) \int \prod_{i \ge 1, j \ge 3(i<j)} (1 + f_{ij})\mathbf{dr}_3 \ldots \mathbf{dr}_N \tag{8}$$

where

$$f_{ij} = \exp(-\beta v(r_{ij})) - 1. \tag{9}$$

As we have the product expansion

$$\prod_{ij}(1 + f_{ij}) = 1 + \sum_{i<j} f_{ij} + \sum_{i<j,k<l} f_{ij}f_{kl} + \cdots \tag{10}$$

$$\approx 1 + \sum_{i \prec j} f_{ij} + \sum_{i<j,k<l} f_{ij}f_{kl} \tag{11}$$

we can develop the pair correlation as the following

$$g(r_1, r_2) = \frac{V^2}{Z_N}\exp(-\beta v(r_{12})) \times \int \left(1 + \sum_{i \ge 1, j \ge 3(i<j)} f_{ij} + \sum_{i \ge 1, j \ge 3(i<j), k \ge 1, l \ge 4(k<l)} f_{ij}f_{kl}\right) \mathbf{dr}_3 \ldots \mathbf{dr}_N \tag{12}$$

The integrals in Eq. (12) can be formally written as

$$\int \mathbf{dr}_3 \ldots \mathbf{dr}_N = V^{N-2} \tag{13}$$

$$\int \sum_{i \ge 1, j \ge 3(i<j)} f_{ij}\mathbf{dr}_3 \ldots \mathbf{dr}_N = V^{N-2}\left(\rho\frac{N}{2}\beta_1\right) \tag{14}$$

$$\int \sum_{i \ge 1, j \ge 3(i<j), k \ge 1, l \ge 4(k<l)} f_{ij}f_{kl}\mathbf{dr}_3 \ldots \mathbf{dr}_N = V^{N-2}\left(\frac{N^2}{8}\rho^2\beta_1^2 + \rho b(r_1, r_2)\right) \tag{15}$$

where ρ is the ion density and

$$\beta_1 = \int (\exp(-\beta v(r)) - 1)\mathbf{dr} = \int f(r)\mathbf{dr} \tag{16}$$

$$b(r_1, r_2) = \int f_{13}f_{23}\mathbf{dr}_3 = \frac{1}{(2\pi)^3 r_{12}}\int k\mathbf{dk}|f(k)|^2 \sin(kr_{12}) \tag{17}$$

where $f(k)$ is the Fourier transform of $f(r)$

$$f(k) = \int f(r)\exp(i\mathbf{k}.\mathbf{r})d\mathbf{r} \quad (18)$$

When we replace formula (13)–(15) in Eq. (12), we obtain $g(r_1, r_2)$ by the following result

$$g(r_1, r_2) = \frac{\exp(-\beta v(r_{12}))V^N}{Z_N} \left(1 + \rho b(r_1, r_2) + \frac{N}{2}\rho\beta_1 + \frac{N^2}{8}\rho^2\beta_1^2\right) \quad (19)$$

Using the condition of the weakness coupling ($1 \gg \frac{\rho\beta_1}{2}$), we estimate in the same manner the partition function Z_N which appears in (19):

$$Z_N = V^N\left(1 + \rho\beta_1\frac{N}{2} + \frac{N^2}{8}\beta_1^2\rho^2\right) = V^N\left(1 + \frac{\rho\beta_1}{2}\right)^N \simeq V^N \quad (20)$$

When we drop the last two terms in (19), we can write the pair correlation function $g(r_1, r_2)$ as

$$g(r_1, r_2) = \exp(-\beta v(r_{12}))(1 + \rho b(r_1, r_2)) \quad (21)$$

If the interparticle interaction is spherically symmetric and if the system is treated as an isotropic fluid, then $g(r_1, r_2)$ depends only on the distance $r_{12} = |\mathbf{r}_1 - \mathbf{r}_2|$ between ions 1 and 2. We adopt the notation $g(r) = g(r_{12})$ and define the radial distribution function $g(r)$ as

$$g(r) = \exp(-\beta v(r))(1 + \rho b(r)) \quad (22)$$

The last formula gives a correction to the commonly known formula $g(r) = \exp(-\beta v(r))$ (Boltzmann factor formula), usually used by many authors for representing an acceptable approximation of the radial distribution function.

Three correlation functions

In the canonical ensemble in which particles interact by a pair potential $v(r)$, the triplet correlation function is defined as [1]

$$g^{(3)}(\mathbf{r}_1, \mathbf{r}_2, \mathbf{r}_3) = V^3\frac{\int\exp(-\beta V_N(r^N))d\mathbf{r}_4\ldots d\mathbf{r}_N}{Z_N} \quad (23)$$

where $\mathbf{r}_i$ is the position vector of the ith particle. The function $g^{(3)}(\mathbf{r}_1, \mathbf{r}_2, \mathbf{r}_3)$ defines the probability of simultaneous detection of three particles in the vicinity of points $\mathbf{r}_1$, $\mathbf{r}_2$, $\mathbf{r}_3$. Unlike the binary function $g(r)$, the $g^{(3)}(\mathbf{r}_1, \mathbf{r}_2, \mathbf{r}_3)$ depends on three space coordinates and, accordingly, enables one to obtain an additional information about the structure of particles. the Kirkwood superposition approximation (KSA) [11] is most frequently employed to approximate the three-particle correlation function

$$g^{(3)}(\mathbf{r}_1, \mathbf{r}_2, \mathbf{r}_3) \simeq g^{(3)}_{SA}(\mathbf{r}_1, \mathbf{r}_2, \mathbf{r}_3) = g(\mathbf{r}_1, \mathbf{r}_2)g(\mathbf{r}_2, \mathbf{r}_3)g(\mathbf{r}_1, \mathbf{r}_3) \quad (24)$$

We calculate the function $g(\mathbf{r}_1, \mathbf{r}_2, \mathbf{r}_3)$ in the same way for computing $g(r)$ (see Eq. (21)), and we have acquired the result

$$g^{(3)}(\mathbf{r}_1, \mathbf{r}_2, \mathbf{r}_3) = e^{-\beta(v(r_{12})+v(r_{23})+v(r_{13}))} (1 + \rho(b_1(r_{12}) + b_2(r_{23}) + b_3(r_{13})) \quad (25)$$

where

$$b_1(r_{12}) = \int f_{14}f_{24}d\mathbf{r}_4 = \frac{1}{(2\pi)^3 r_{12}}\int k|f(k)|^2\sin(kr_{12})d\mathbf{k} \quad (26)$$

$$b_2(r_{23}) = \int f_{24}f_{34}d\mathbf{r}_4 = \frac{1}{(2\pi)^3 r_{23}}\int k|f(k)|^2\sin(kr_{23})d\mathbf{k} \quad (27)$$

$$b_3(r_{13}) = \int f_{14}f_{34}d\mathbf{r}_4 = \frac{1}{(2\pi)^3 r_{13}}\int k|f(k)|^2\sin(kr_{13})d\mathbf{k} \quad (28)$$

and

$$f(k) = \frac{1}{k}\int rf(r)\sin(kr)d\mathbf{r} \quad (29)$$

Numerical calculations of the function b [formula (17)], $b1$, $b2$ and $b3$ [formula (26)–(28)], for both kinds of potentials, allow us to compute the static structure functions as shown in Figs. 1 and 2. Now, we are able to apply the above results to compute the dynamic properties and the transport coefficients.

Application to dynamic and transport properties

By following the theory presented in [13], we have obtained the velocity and the micro-field auto-correlation functions $D(t)$ and $C(t)$ as the sum of three exponentials

$$D(t) = \sum D_i e^{Z_i t}, \quad C(t) = \sum C_i e^{Z_i t} \quad (30)$$

where the coefficients D_i and C_i are given by

$$D_i = -(\omega_0/Z_i)C_i \quad (31)$$

and

$$C_1 = (\Lambda + Z_1)Z_1(Z_3 - Z_2)/\Delta \quad (32)$$

$$C_2 = (\Lambda + Z_2)Z_2(Z_1 - Z_3)/\Delta \quad (33)$$

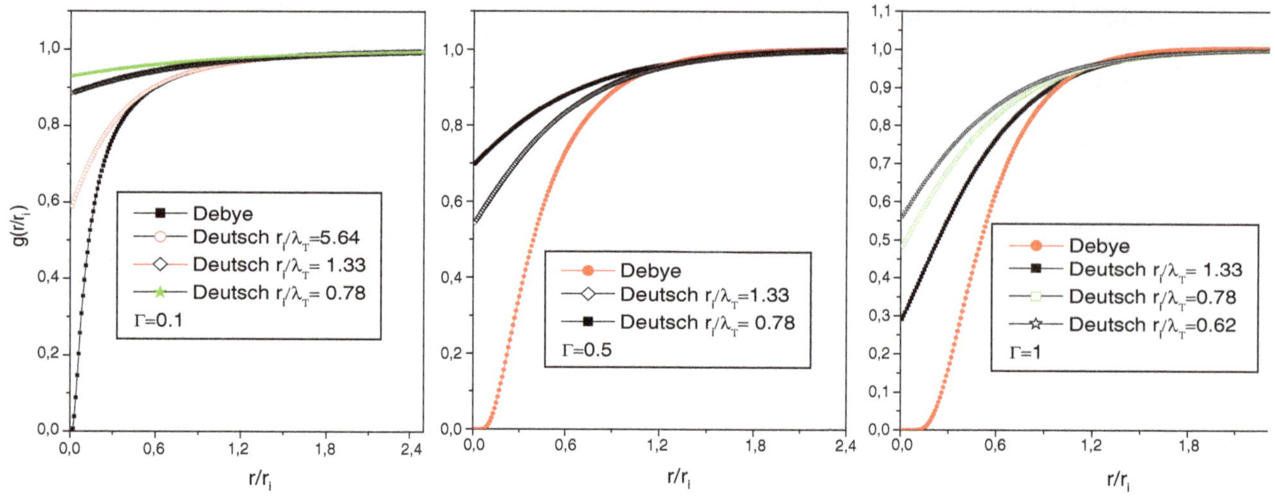

Fig. 1 Pair correlation function $g(r)$ calculated by our model for different coupling parameters $\Gamma = 0.1, 0.5, 1.$ and different effect quantum parameters r_i/λ_T and for two interaction potentials (Debye and Deutsch)

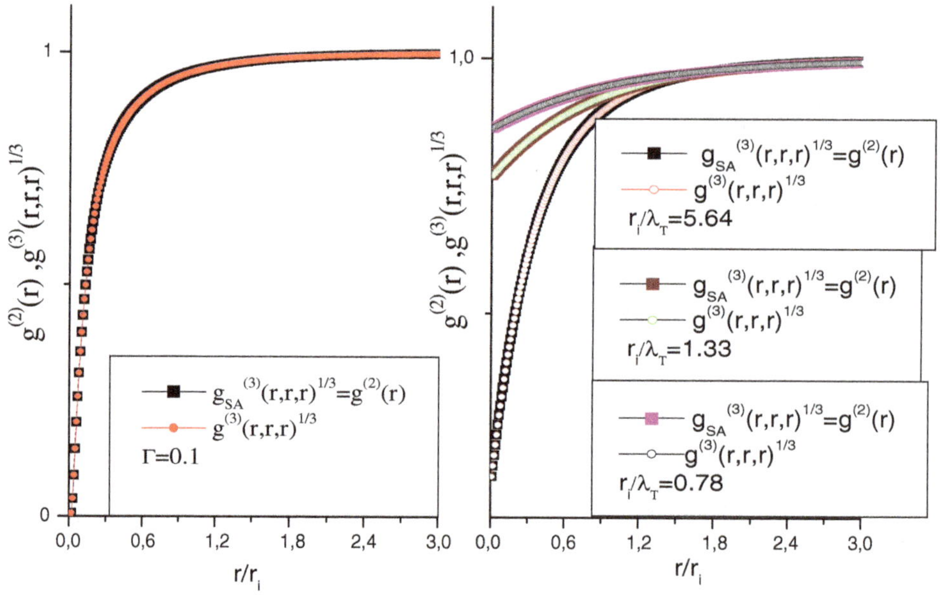

Fig. 2 Comparison between triplet correlation function $g^{(3)}(r,r,r)^{1/3}$ and the triplet correlation function in the Kirkwood superposition approximation for: Debye potential ($\Gamma = 0.1$) at the *left*, Deutsch potential for different r_i/λ_T at the *right*

$$C_3 = (\Lambda + Z_3)Z_3(Z_2 - Z_1)/\Delta \tag{34}$$

$$\Delta = (Z_1 - Z_2)(Z_2 - Z_3)(Z_3 - Z_2) \tag{35}$$

and the $\{Z_i\}$ are solutions to the cubic equations (relative to Debye and Deutsch potentials respectively)

$$\begin{aligned} Z^3 + \Lambda_{\text{Debye}} Z^2 + \omega_1^2 Z + \Lambda_{\text{Debye}} \omega_0^2 = 0 \\ Z^3 + \Lambda_{\text{Deutsch}} Z^2 + \Omega_1^2 Z + \Lambda_{\text{Deutsch}} \Omega_0^2 = 0 \end{aligned} \tag{36}$$

The analysis made up to this point applies for arbitrary interaction potentials and plasma composition. To illustrate the physical content of our investigation, we consider the special case of a OCP one-component plasma with an impurity ion of the same mass and the same charge (m, q) for one component of the plasma ions. Before starting the evaluation of the three necessary parameters to solve the above cubic equation, remember the well-known useful Green–Kubo relation that gives the self-diffusion coefficient D from the velocity auto-correlation function $D(t)$

$$D = \frac{1}{m\beta} \int_0^\infty D(t)\mathrm{d}t \tag{37}$$

Determination of parameters

To calculate the auto-correlation functions $D(t)$ and $C(t)$ we need to know the three Z_i roots which are connected to parameters ω_0, ω_1, Λ which in turn must be calculated.

We attack now to determine all the parameters necessary in the description of the dynamics. These parameters allow us, first, to solve the cubic algebraic equation (36).

Evaluation of ω_0, Ω_0

Consider first ω_0 defined in [14]

$$\omega_0^2 = (\beta q^2/3m)\langle E^2\rangle = -(\beta q/3m)\langle \mathbf{E}\cdot\nabla v(r)\rangle \tag{38}$$

where $v(r)$ is the potential of the interaction between the impurity ion and the surrounding plasma. This expression (38) can be rewritten, after making a part integral

$$\omega_0^2 = -(q/3m)\langle \nabla\cdot\mathbf{E}\rangle \tag{39}$$

It is noted here that the background does not contribute to ω_0^2

$$\nabla_0\cdot\mathbf{E}_b = 0 \tag{40}$$

Then (38) leads to the two results

$$\omega_0^2 = -\frac{1}{3}(nq/m)\int \mathrm{d}\mathbf{r}\nabla.\mathbf{e}(\mathbf{r})g(r) \tag{41}$$

where $g(r)$ is the pair correlation functions for the probability to find a plasma ion of species a at a distance r from the impurity ion of mass m and charge q, n is the density of the plasma, and

$$\nabla\cdot e(\mathbf{r}) = -\Delta v(r) = -\left(\frac{\partial^2}{\partial r^2}+\frac{2}{r}\frac{\partial}{\partial r}\right)v(r) \tag{42}$$

Here $v(r)$ must be either the screened Debye potential or the screened Deutsch potential.

Debye potential case

$$\omega_0^2 = \frac{1}{3}\omega_p^2\int_0^\infty \mathrm{d}rrk_\mathrm{D}^2\mathrm{e}^{-k_\mathrm{D}r}g(r),\quad \omega_p^2 = 4\pi\rho q^2/m \tag{43}$$

Deutsch potential case

$$\Omega_0^2 = \frac{1}{3}\left(\frac{mq}{m_0q}\right)\omega_p^2\int_0^\infty \mathrm{d}rr(k_\mathrm{D}^2\mathrm{e}^{-k_\mathrm{D}r} - (k_\mathrm{D}+k_\mathrm{T})^2\mathrm{e}^{-(k_\mathrm{D}+k_\mathrm{T})r})g(r) \tag{44}$$

Evaluation of ω_1, Ω_1

$$\omega_1^2 = (q^2/3m\omega_0^2)(nm/\mu)\int \mathrm{d}\mathbf{r}[\partial e_\alpha(\mathbf{r})/\partial r_j]^2 g(r) \tag{45}$$

$$+ (9\omega_0^2)^{-1}\left\{(nq/m)\int \mathrm{d}\mathbf{r}\nabla\cdot\mathbf{e}_\alpha(\mathbf{r})g_\alpha(r)\right\}^2 \tag{46}$$

$$+ (q^2/3m\omega_0^2)\sum_\sigma(n^2)\int \mathrm{d}\mathbf{r}\mathrm{d}\mathbf{r}'[\partial e(\mathbf{r})/\partial r_j'][\partial e(\mathbf{r})/\partial r] \tag{47}$$

$$\times\ \{g^{(3)}(\mathbf{r},\mathbf{r}') - g(r)g(r')\} \tag{48}$$

where $\mu = m/2$ is the reduced mass. The second term of the last formula can be simplified to give the form

$$\omega_1^2 = \omega_0^2 + (q^2/3m\omega_0^2)(nm/\mu)\int \mathrm{d}\mathbf{r}[\partial e(\mathbf{r})/\partial r_j]^2 g(r) \tag{49}$$

$$+ (q^2/3m\omega_0^2)(n^2)\int \mathrm{d}\mathbf{r}\mathrm{d}\mathbf{r}'[\partial e(\mathbf{r})/\partial r_j'][\partial e(\mathbf{r})/\partial r] \tag{50}$$

$$\times\ \{g^{(3)}(\mathbf{r},\mathbf{r}') - g(r)g(r')\} \tag{51}$$

To be able to make the last integrals, we need two and three correlation functions (static structure functions) g^2, g^3 that are given in "Static pair correlation functions" and "Three correlation functions".

Debye potential case

$$\omega_1^2 = \omega_0^2(1 + (m/\mu I_0^2)I_1 + (3/I_0^2)I_2) \tag{52}$$

where

$$I_0 = \int_0^\infty \mathrm{d}rrk_\mathrm{D}^2 e^{-k_\mathrm{D}r}g(r) \tag{53}$$

and

$$I_1 = \int_0^\infty \mathrm{d}rr^{-4}\mathrm{e}^{-2k_\mathrm{D}r}[6 + 12k_\mathrm{D}r + 10(k_\mathrm{D}r)^2 + 4(k_\mathrm{D}r)^3 + (k_\mathrm{D}r)^4]g(r) \tag{54}$$

$$I_2 = \int_0^\infty \mathrm{d}r\mathrm{d}r'rr'k_\mathrm{D}^4\mathrm{e}^{-k_\mathrm{D}r}\mathrm{e}^{-k_\mathrm{D}r'}\{g^{(3)}(\mathbf{r},\mathbf{r}') - g(r)g(r')\} \tag{55}$$

Deutsch potential case

$$\Omega_1^2 = \Omega_0^2(1 + (m/\mu I_0'^2)I_1' + (3/I_0'^2)I_2') \tag{56}$$

where

$$I_0' = \int_0^\infty \mathrm{d}rr(k_\mathrm{D}^2\mathrm{e}^{-k_\mathrm{D}r} - (k_\mathrm{D}+k_\mathrm{T})^2\mathrm{e}^{-(k_\mathrm{D}+k_\mathrm{T})r})g(r) \tag{57}$$

$$I_1' = \int_0^\infty \mathrm{d}rr^{-4}\mathrm{e}^{-2k_\mathrm{D}r}[A - 2\mathrm{e}^{-k_\mathrm{T}r}B + \mathrm{e}^{-2k_\mathrm{T}r}C]g(r) \tag{58}$$

and

$$I_2' = \int_0^\infty \mathrm{d}r\mathrm{d}r'rr'(k_\mathrm{D}^2\mathrm{e}^{-k_\mathrm{D}r} - k_\mathrm{DT}^2\mathrm{e}^{-k_\mathrm{DT}r})(k_\mathrm{D}^2\mathrm{e}^{-k_\mathrm{D}r'} - k_\mathrm{DT}^2\mathrm{e}^{-k_\mathrm{DT}r'}) \times\{g^{(3)}(\mathbf{r},\mathbf{r}') - g(r)g(r')\} \tag{59}$$

where

$$A = (k_D r)^4 + 4(k_D r)^3 + 10(k_D r)^2 + 12k_D r + 6 \quad (60)$$

$$B = (k_D k_{DT})^2 r^4 + 2(k_D k_{DT}^2 + k_D^2 k_{DT}) r^3 + 2(3k_D k_{DT} + k_D^2 + k_{DT}^2) r^2 + 6(k_D + k_{DT}) r + 6 \quad (61)$$

$$C = (k_{DT} r)^4 + 4(k_{DT} r)^3 + 10(k_{DT} r)^2 + 12k_{DT} r + 6 \quad (62)$$

$$k_{DT} = k_D + k_T \quad (63)$$

It is understood that the integration variables and screening length k_D^{-1}, k_T^{-1} are in units of the ion sphere radius, $r_i = (3/4\pi\rho)^{1/3}$.

Evaluation of $\Lambda_{Debye}, \Lambda_{Deutsch}$

We use the following expression [13]

$$\Lambda_{Debye} = ((m/\mu)I_1 + 3I_2)/(I_0^2 m\beta D) \quad (64)$$

and the same formula for Deutsch case using $I'_{0,1,2}$ defined above (56)–(59). At this stage, we need the self coefficient diffusion D. To proceed, we calculate D with self-consistent method: we give at first an initial value to D, and compute Λ with respect the last formula. Then we have the three coefficients of the cubic equation (ω_0, ω_1 and Λ). We are able to solve this equation and then to have the velocity auto-correlation function $D(t)$. Using the Eq. (37), we get the self-diffusion coefficient D. Using the obtained value of D in the equation to have a new value of Λ, and by solving the cubic equation, we get the velocity auto-correlation function $D(t)$. Integrated, the latter gives a new self-diffusion coefficient D. This procedure must be repeated till we get the convergence.

Results and discussions

We have presented a model for the pair correlation function and the triplet correlation function given by the Eqs. (22) and (25) for different coupling parameters Γ, using the screened Debye potential and the screened Deutsch potential

In Fig. 1 we have presented and compared the pair correlation function $g(r)$ calculated by our model for different coupling parameters Γ and different ratios r_i/λ_T for the two potentials. We note, in the case of the Debye potential, that the curve starts from zero: when the distance r (between two ions) goes to zero; $g(r)$ goes to zero too. Whereas when the distance goes to infinity, the pair correlation function $g(r)$ goes to one. For the screened Deutsch potential, we note that the curve not start from zero (contrary to Debye case). This means that the correlation between the ions is more significant in the weak range of the ratio r_i/λ_T. This indicates that, in this range, the probability of interaction between two ions is more important. Readibly, this phenomenon is due to the quantum effects.

Figure 2 shows the triplet correlation function in the equilateral triangle geometry for the coupling parameter $\Gamma = 0.1$ for the two potentials. To allow a comparison, we have taken the cubic root $\sqrt[3]{g^{(3)}(r,r,r)}$ and the cubic root of the triplet correlation function in the Kirkwood superposition approximation $\sqrt[3]{g_{SA}^{(3)}(r,r,r)} = g^{(2)}(r)$ [11]. We have obtained a very good agreement as it is shown in Fig. 2.

To calculate the time auto-correlation functions $D(t)$ and $C(t)$ we need to know the frequencies Z_i (roots of the algebraic cubic equations (36)) and the coefficients C_i (or D_i, respectively). They are expressed by three parameters—the diffusion constant D, and the frequencies ω_0, ω_1 for the screened Debye potential and Ω_0, Ω_1 for the screened Deutsch potential. The frequencies ω_0, Ω_0 are functions of the pair correlation function g^2, whereas the frequencies ω_1, Ω_1 are functions of the triplet correlation function g^3. To calculate the frequencies ω_1 and Ω_1 we need the knowledge of the triplet correlation functions.Table 1 shows the values of the diffusion coefficients (D^*_{Debye} and $D^*_{Deutsch}$) (given in units of $\omega_p r_i^2$), for different values of the coupling parameter Γ and different values of the ratio r_i/λ_T. We have compared D^*_{Debye} and $D^*_{Deutsch}$ with the diffusion coefficient computed by the simulation technique earlier by (M. A. Berkovsky) [14]. We note a good agreement between Debye case and M. A. Berkovsky simulation. So, there is a difference with the Deutsch potential case. We also note that the coefficient $D^*_{Deutsch}$ increases when r_i/λ_T decreases in each coupling category because it is related to the temperature contrary to the case of Debye potential.

Table 1 Γ Dependence of the diffusion coefficients D^* is given in units of $\omega_p r_i^2$

Γ	r_i/λ_T	ρ	$T \times 10^4$	$D^*_{Deutsch}$	D^*_{Debye}	D^* [14]
0.1	5.64	2×10^{20}	15.73	277.59	153.29	–
	1.33	1.14×10^{24}	281	705.54		
	0.78	2.85×10^{25}	821	1659.04		
0.5	1.33	9.12×10^{21}	11.26	44.83	7.80	8.71
	0.78	2.28×10^{23}	32.92	92.18		
1.0	1.33	1.14×10^{21}	2.81	13.89	2.787	2.64
	0.78	2.85×10^{22}	8.21	29.24		
	0.62	1.14×10^{23}	13.05	41.48		

Table 2 Γ dependence of ω_0, ω_1, are given in units of ion plasma frequency ω_p

Γ	r_i/λ_T	$\Omega_{0\text{Deutsch}}$	$\omega_{0\text{Debye}}$	$\Omega_{1\text{Deutsch}}$	$\omega_{1\text{Debye}}$	Λ_{Deutsch}	Λ_{Debye}
0.1	5.64	0.254	0.553	16.416	92.92	49.89	573.63
	1.33	0.102		7.39		15.74	
	0.78	0.067		6.008		19.44	
0.5	1.33	0.172	0.538	4.65	11.58	10.79	39.46
	0.78	0.113		4.28		10.29	
1.0	1.33	0.202	0.498	3.75	6.04	8.19	17.49
	0.78	0.134		3.63		8.27	
	0.62	0.111		3.81		9.38	

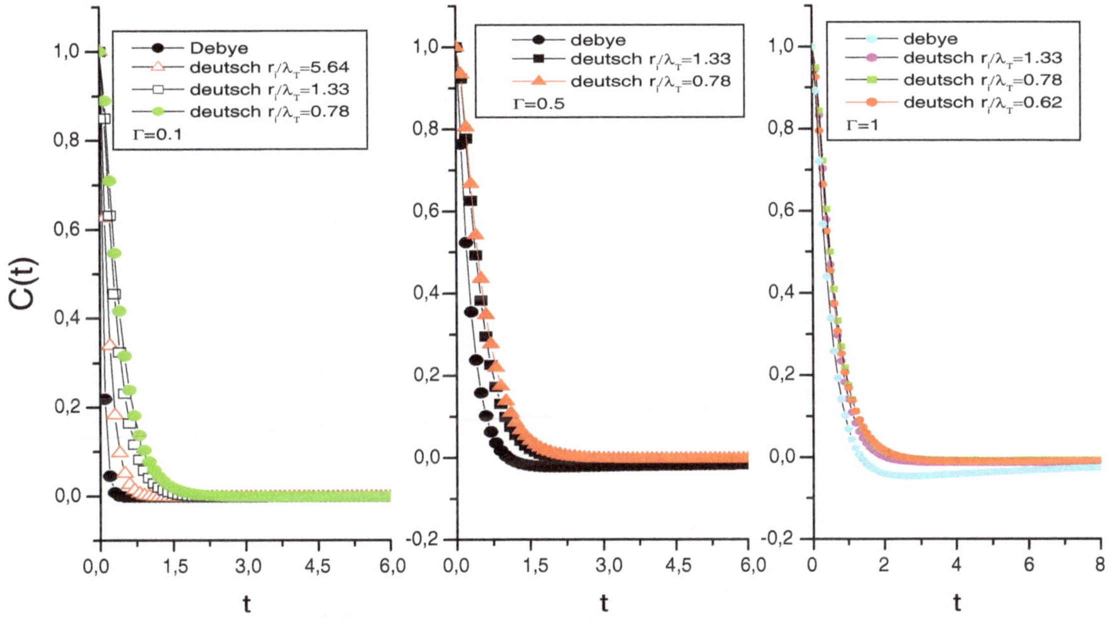

Fig. 3 Time dependence of the electric field correlation function for different coupling parameters $\Gamma = 0.1, 0.5, 1.$ and for different r_i/λ_T and different potentials. The time t is given in units of ω_p^{-1}

Table 2 shows the Γ dependence of ω_0, Ω_0, ω_1 and Ω_1 (in units of the OCP plasma frequency ω_p) and Λ. We show that the results for the case of the Debye potential are different from those obtained in the case of the screened Deutsch potential. The comparison was made for different values of the coupling parameter Γ and for different values of the ratio r_i/λ_T.

Figure 3 shows the time auto-correlation function of the electric micro-field $C(t)$ for different coupling parameter Γ and different values of the ration r_i/λ_T. We notice that the time auto-correlation function of the electric micro-field $C(t)$ increases when the ratio r_i/λ_T decreases. Furthermore, we note that $C(t)$ decreases up to zero in small time. This indicates that the ratio r_i/λ_T is small when the correlation between the ions is very strong. Figure 4 shows the velocity auto-correlation function $D(t)$ for different coupling parameters Γ and different ratios r_i/λ_T. As it is clear, $D(t)$ increases when the ratio r_i/λ_T decreases. Here we note that the time auto-correlation function of the velocity $D(t)$ decreases up to zero in a long time. This indicates that the ratio r_i/λ_T is small when the correlation between the ions is very strong. Therefore, the diffusion coefficient increases when the time auto-correlation function of the velocity $D(t)$ is larger, because the diffusion coefficient D is equal to the integral of $D(t)$ [see Green–Kubo formula (37)].

Conclusion and perspectives

In this work, we have presented a model for the pair and the triplet correlation function theoretically of a one-component plasma at for weak and intermediate coupling $\Gamma \leq 1$ and for $r_i/\lambda_T \leq 1$. This is done using the screened Debye potential and the screened Deutsch potential. We have

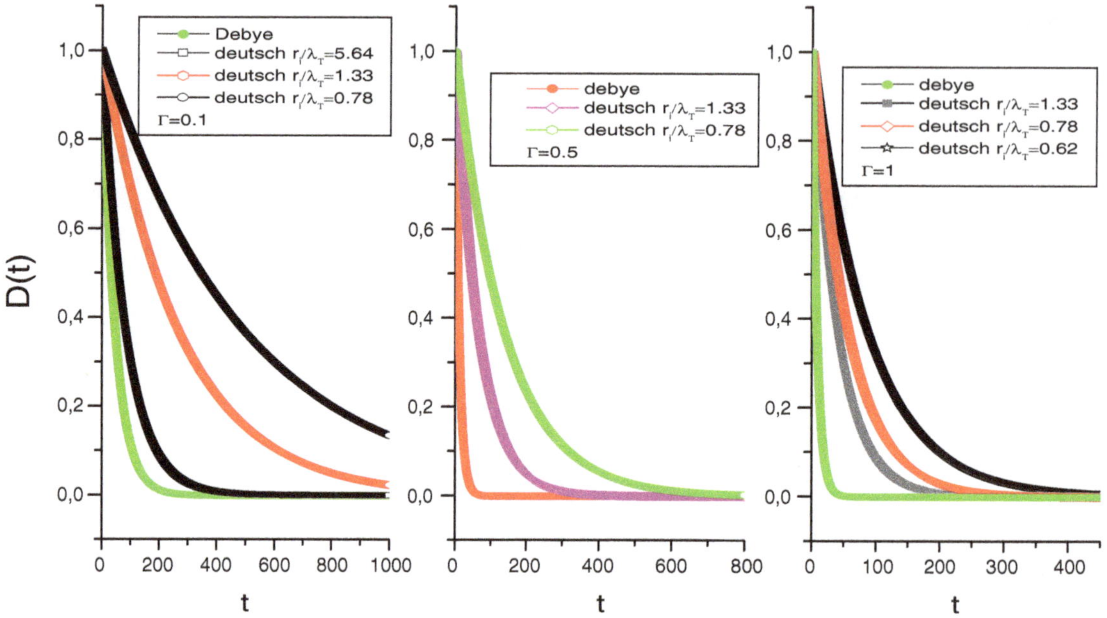

Fig. 4 Time dependence of the velocity auto-correlation function for coupling parameters $\Gamma = 0.1, 0.5, 1.$ and for different r_i/λ_T and different potentials. The time t is given in units of ω_p^{-1}

found that the pair correlation function $g^2(r)$, for both potentials, are rather close, but start to deviate from each other when the coupling parameter Γ increases with a decreasing r_i/λ_T. Nevertheless, the use of the Deutsch potential is preferable because it considers in more realistic physics the short distance collisions, i.e. the collisions occurring at distances smaller than the De Broglie wave-length. In other hand, we have computed the dynamical properties: the time auto-correlation function of the velocity $D(t)$ and the time auto-correlation function of the electric micro-field $C(t)$. We have noticed, when the ratio r_i/λ_T is small the correlation between the ions is very strong. Therefore, the velocity auto-correlation function $D(t)$ and the electric field auto-correlation function $C(t)$ are large for a weak value of r_i/λ_T. As a consequence, the self-diffusion coefficient D is more important for the strongly coupled plasmas.

References

1. Hansen, J.P., McDonald, I.R.: Theory of Simple Liquids. Academic, London (1986)
2. Frenkel, Ya.I.: Kinetic Theory of Liquids. Clarendon Press, Oxford (1946)
3. Fortov, V.E., Vaulina, O.S., Petrov, O.F.: Dusty plasma liquid: structure and transfer phenomena. Plasma Phys. Control. Fusion **47**(12B), B551–B563 (2005)
4. Zahn, K., Maret, G., Ruß, C., von Grunberg, H.H.: Three-particle correlations in simple liquids. Phys. Rev. Lett. **91**(11), 115502-1–115502-4 (2003)
5. Wang, H., Lettinga, M.P., Dhont, J.K.G.: Microstructure of a near-critical colloidal dispersion under stationary shear flow. J. Phys. Condens. Matter **14**, 7599–7615 (2002)
6. Roman, F.L., White, J.A., Gonzalez, A., Velasco, S.: Fluctuations in a small hard disk system: implicit finite size effects. J. Phys. **11**, 3789–3798 (1999)
7. Block, R., Schommers, W.: Triplet correlations in disordered systems: a study for liquid rubidium. J. Phys. C Solid State Phys. **8**, 1997–2002 (1975)
8. Linse, P.: Highly asymmetric electrolyte: triplet correlation functions from simulation in one and two component model systems. J. Chem. Phys. **94**(12), 8227–8233 (1991)
9. Taylor, M.P., Lipson, J.E.G.: On the Born–Green–Yvon equation and triplet distributions for hard spheres. J. Chem. Phys. **97**(6), 4301–4308 (1992)
10. Ruß, C., Zahn, K., von Grunberg, H.H.: Triplet correlations in two-dimensional colloidal model liquids. J. Phys. Condens. Matter **15**, S3509–S3522 (2003)
11. Kirkwood, J.G.: Statistical mechanics of fluid mixtures. J. Chem. Phys. **3**, 300–313 (1935)
12. Deutsch, C.: Nodal expansion in a real matter plasma. Phys. Lett. A **60A**(4), 317–318 (1977)
13. Meftah, M.T., Chohra, T., Bouguettaia, H., Khelfaoui, F., Talin, B., Calisti, A., Dufty, J.W.: Electric field dynamics at charged point in two component plasma. Eur. Phys. J. B **37**, 39–46 (2004)
14. Berkovsky, M., Dufty, J.W., Calisti, A., Stamm, R., Talin, B.: Electric field dynamics at a charged point. Phys. Rev. E **54**(4), 4087–4097 (1996)

23

Experimental investigation of the effect of insulator sleeve length on the time to pinch and multipinch formation in the plasma focus facility

M. Momenei[1] · Z. Khodabakhshei[1] · N. Panahi[2] · M. A. Mohammadi[3]

Abstract The length of insulator sleeve is varied to investigate its effect on the pinch formation in the plasma focus facility. In this paper, the effect of insulator length on the time to pinch at various pressures and working voltages in the 1.15 kJ Mather type plasma focus is investigated. The results show that with 4.5 cm insulator length the time to pinch at all pressures is minimum. Other results also confirm that with increasing of pressure the time to pinch is increased. Moreover, with increasing working voltage the time to pinch is decreased. Pictures, captured using a digital single lens reflex (DSLR) Canon EOS 7D system, show that multipinch phenomenon is formed.

Keywords Plasma focus · Insulator sleeve · Multipinch

Introduction

Plasma focus device (PFD) is an effective device in laboratory for production of high-temperature ($\sim$1 keV) and high-density ($\approx 10^{25}$–10^{26} m^{-3}) plasma. PFD was developed in the early 1960s in the former Soviet Union (Filippov type) [1] and in the USA (Mather type) [3] independently. The PFD was initially considered as a fast neutron source [3, 4]. It is also a rich source of soft and hard X-rays [5, 6], highly energetic ions [7, 8] and relativistic electrons [9]. The X-ray emission from PFD has been used for defectoscopy, X-ray lithography activation of enzymes, micro-machining and radiography [10–13]. The energetic ions have been used for material processing such as ion implantation and thin films [14–16]. PFD has also been used as a pump source for lasers [17]. Plasma produced in PFD can be affected with the plasma focus insulator sleeve length and PFD working voltage and pressure. Recently, experimental studies have been carried out on the effects of insulator sleeve length and pressure on time to pinch and current sheath structure [18, 19]. The effect of insulator sleeve length on X-ray emission has also been reported by Rawat et al. [20]. In [21, 22] Zhang et al. and Zakaullah et al. show that the X-ray and neutron yield is affected by insulator sleeve length. The current sheath formation dynamics and its structure for different insulator lengths in plasma focus device are investigated by Seng et al. [23]. The current sheath dynamics and multipinch phenomena have also been reported by Mohammadi et al. [24].

In this study, we have shown that the change in insulator sleeve length, pressure and working voltage affects the pinch time. The formation of multipinch phenomena is also reported.

Experimental setup

The present investigation was performed on a simple single capacitor DPF device designated at the Shahrood University (SHUPF). It is a Mathertype focus device, energized by a single 16 μF, 12 kV fast discharging capacitor, with a maximum energy storage of 1.15 kJ. In our investigation, the device was operated at a charging voltage ranging between 7 and 9 kV. In our plasma focus, SHUPF,

✉ M. A. Mohammadi
mohammadidorbash@yahoo.com

[1] Faculty of Physics, University of Shahrood, Shahrood, Iran

[2] Department of Physics, Bandar Abbas Branch, Islamic Azad University, Bandar Abbas, Iran

[3] Department of Atomic and Molecular Physics, Faculty of Physics, University of Tabriz, Tabriz, Iran

cylindrical anode made of copper has 60 mm in length and 20 mm in diameter. The cathode was built in six brass rods each of 10 mm diameter and 60 mm length and symmetrically located around the anode. The device was evacuated to a vacuum (~0.005 Torr) by a rotary pump and was filled with argon gas to a different pressure (1–1.6 Torr) before the operation.

An insulator sleeve of Pyrex glass with fixed 24 mm outer diameter, in different effective lengths 35, 40, 45 and 50 mm, separates anode and cathode, as shown in Fig. 1. For determination of the temporal pinch zone, high-voltage probe is used. All data are captured with the GPS 200 MHz digital oscilloscope. A digital single lens reflex (DSLR) Canon EOS 7D was used for the time-integrated plasma column photography. The open shutter camera with an analyzer and polarizer was fixed at a distance of 15 cm in front of the window.

Results and discussion

In Fig. 2 typical signal of voltage probe is shown. The first peak (I) of signal coincides with breakdown phase and the second and third peaks (II, III) are pinch signals. Time distance between I and II (the time from start of discharge to the pinch) is time to pinch. In Fig. 3 variation of time to pinch versus insulator sleeve for different pressure is shown. As seen, the minimum and maximum time to pinch are 4.99 ± 0.04 (μs) to 4.5 cm at 1 Torr and 7.36 ± 0.06 (μs) for 3.5 cm at 1.4 Torr, respectively. At 1.6 Torr pressure with 3.5 cm insulator sleeve length pinch is not formed. At all pressure, the time to pinch for insulator with 4.5 cm is minimum. This result confirms that at all insulators, there is one pressure, which time to pinch is minimum. At all pressures when we deviate from an insulator sleeve with 4.5 cm length the time to pinch increases. This means that this insulator sleeve length is the optimum length of this plasma focus facility. In Fig. 4 variation of time to pinch versus insulator sleeve length for different working voltage is shown. This figure shows that with increasing of voltage for all insulator sleeve length time to pinch is decreased. With increasing of voltage the current is increased, and then the Lorentz force for driving current sheath is increased. Also the result of this figure shows that time to pinch at insulator with 4.5 cm length for all voltages is minimum. This means that the optimum length is independent of working voltage and pressure. When the insulator length is increased or decreased from the optimum value (4.5 cm length) the time to pinch is increased. The modification factor is defined as follows [25]:

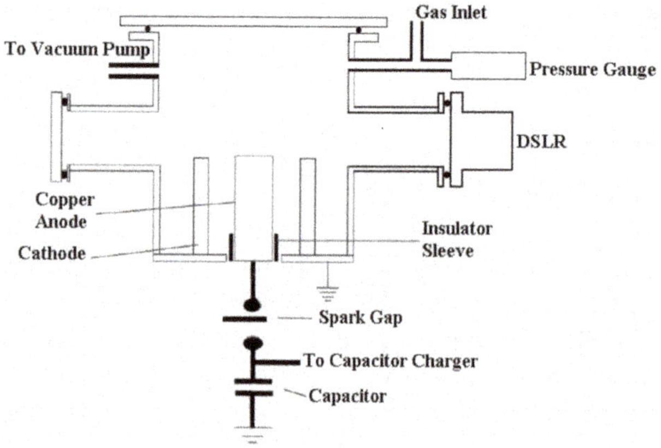

Fig. 1 The schematic view of SHUPF

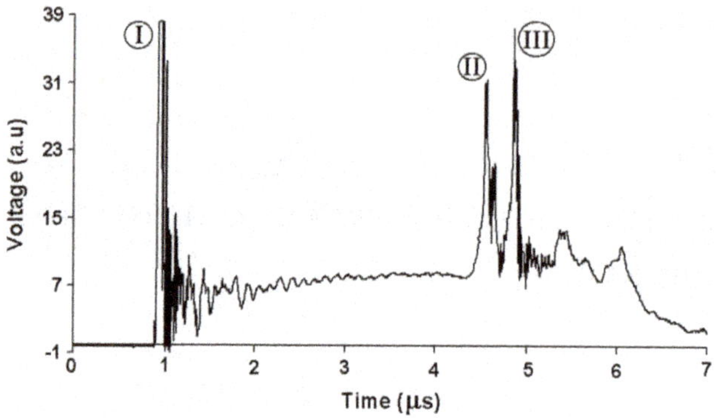

Fig. 2 Typical signal of voltage probe

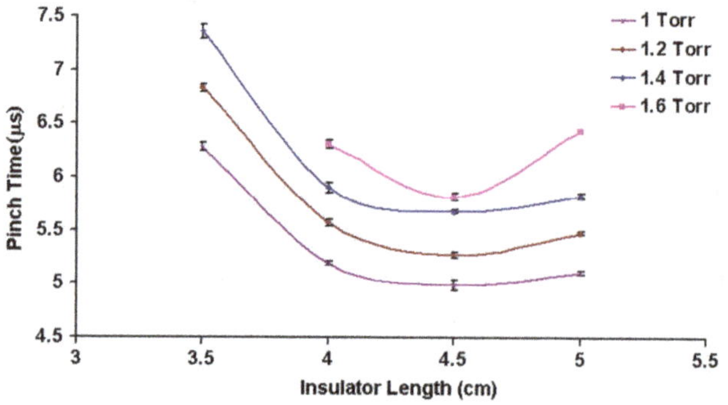

Fig. 3 Variation of time to pinch with insulator sleeve length for different pressure

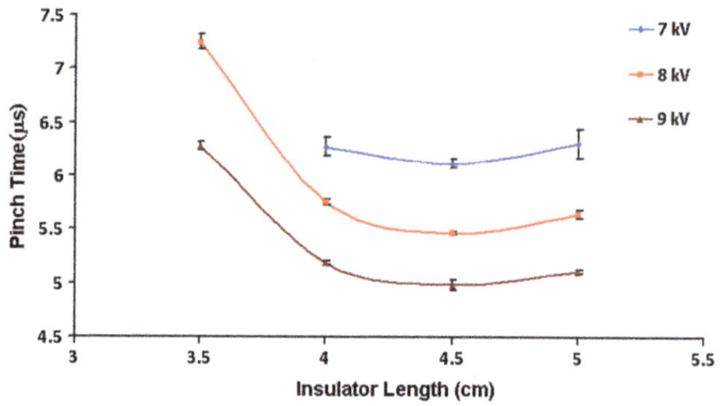

Fig. 4 Variation of time to pinch with insulator sleeve length for different voltage

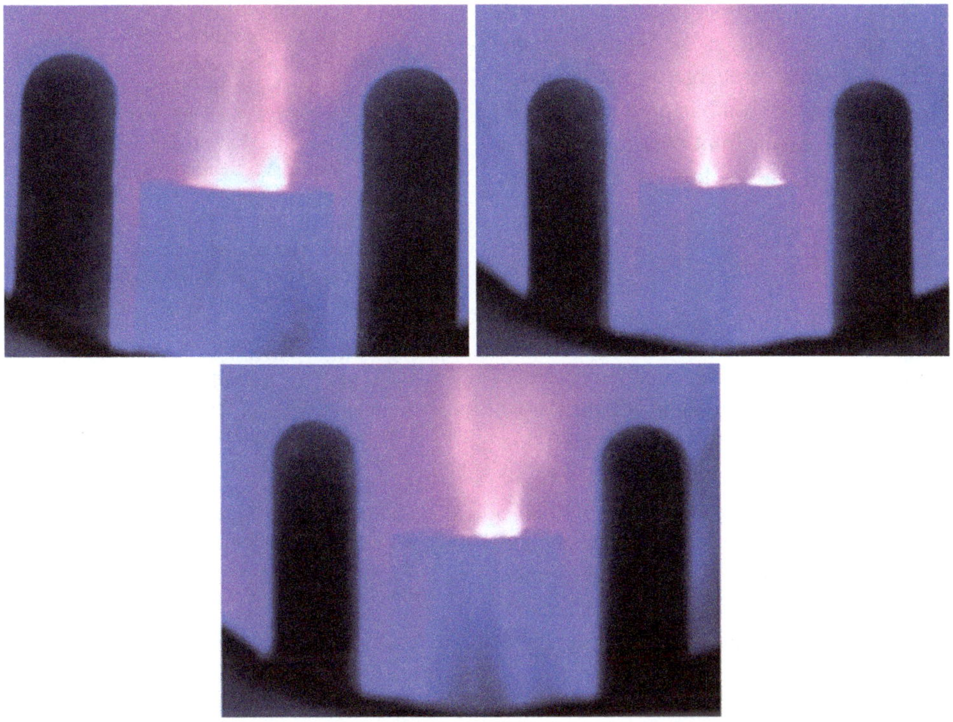

Fig. 5 Time integrate picture of pinch zone

$$F = f_c / \sqrt{f_m}. \quad (1)$$

where f_c and f_m are fraction of current and mass swept factor driving the plasma sheath, respectively. With the deviating of optimum value of insulator sleeve length, the leakage current will be increased and the modification factor is decreased. With considering of modification factor the current sheath velocity equation at axial velocity is defined as [26]

$$U_a = F\left[\frac{\mu \ln C}{4\pi^2(C^2-1)}\right]^{1/2}\frac{I_0}{a\sqrt{\rho}}. \quad (2)$$

This equation shows that with decreasing the modification factor the axial velocity is decreased, so the time to pinch is increased. The optimum insulator sleeve length corresponds to the conditions for uniform discharge development and its take off across the insulator sleeve surface. When the sleeve is too long the increased inductance may cause the current sheath to remain at the sleeve surface for longer period of time. When the sleeve is too short, the rapid current sheath development may cause spoke formation. As a result, when the insulating sleeve is not of appropriate length, the current sheath no longer remains uniform, and the so-called filaments or spokes are developed [26].

The sheath formation time is given as [27]

$$t_f = \left(\frac{2\pi r_s l_s d_s L_i}{\eta} w_{is}\right)^{1/2} \cdot \frac{1}{U}. \quad (3)$$

where U is the working voltage, r_s and l_s are the radius and the length of insulator, respectively; L_i is the inductance; d_s is the sheath thickness; η is the efficiency of energy fed to the discharge and w_{is} is the energy density. When the insulator sleeve length is longer than the optimum value, the sheath formation time is increased and then the time to pinch is increased. Equation 3 shows that with increasing of working voltage the sheath formation time decreased, which is confirmed by Fig. 4.

For the time-integrated study of plasma column a digital single lens reflex (DSLR) Canon EOS 7D is used. Time-integrated picture of pinch zone is shown in Fig. 5. This figure shows that the multipinch is formed on top of the anode surface. The II and III peaks of the voltage probe signal, which is shown in Fig. 2, also confirm that the multipinch is formed. The multipinch formation can be explained as follows: in the Lee model the mass carried by current sheath at the position z is [28]

$$\rho\pi(b^2 - a^2)m_f z. \quad (4)$$

In this equation ρ, b and a are gas density, cathode radius and anode radius, respectively. m_f is the mass factor being less than one. This equation explains that the current sheath cannot carry 100% of gas, but some gas is left back near the insulator sleeve. After the first pinch/compression phase, indicated by the first peak in the voltage probe signal, shown in Fig. 2, another discharge on the insulator is produced and a second current sheath is produced which, owing to the low density of gas in front of it moves much faster and collapses at the anode top as the second pinch/compression phase.

Conclusion

Pinch formation time with the various insulator sleeve length is investigated. It was found that at all insulators we have one minimum time to pinch. It was obtained that with insulator with 4.5 cm length the time to pinch at all pressure is minimum. The average pinch time with different pressure shows that with increasing of pressure the time to pinch increased. Experimentally, it was shown that at a higher voltage the time to pinch is decreased. Experiments demonstrate that the multipinch phenomenon is formed.

References

1. Filippov, N.V., Filippova, T.I., Vinogradov, V.P., Dense, high-temperature plasma in a non-cylinderical Z-pinch compression. Nucl. Fusion Supl. **2**, 577 (1962)
2. Mather, J.W., Investigation of the high energy acceleration mode in coaxial gun. Phys. Fluids **S28**, (1964)
3. Springham, S.V., Lee, S., Rafique, M.S., Correlated deuteron energy spectra and neutron yield for a 3 kJ plasma focus, Plasma Phys. Control Fusion **42**, 1023 (2000)
4. Mohammadi, M.A., Sobhanian, S., Rawat, R.S., Neutron production with mixture of deuterium and krypton in Sahand Filippov type plasma focus facility. Phys. Lett. A **375**, 3002 (2011)
5. Mohammadi, M.A., Verma, R., Sobhanian, S., Wong, C.S., Lee, S., Springham, S.V., Tan, T.L., Lee, P., Rawat, R.S., Neon soft X-ray emission studies from UNU-ICTP plasma focus operated with longer than optimal anode length. Plasma Sour. Sci. Tech. **16**, 785 (2007)
6. Zakaullah, M., Alamgir, A., Shafiq, M., Sharif, M., Waheed, A., Scope of plasma focus with argon as a soft X-Ray source. IEEE Trans. Plasma Sci. **30**, 2089 (2002)
7. Valipour, M., Mohammadi, M.A., Sobhanian, S., Rawat, R.S., Increasing of hardness of titanium using energetic nitrogen ions from Sahand as a Filippov Type plasma focus facility. J Fusion Energ. **31**, 65 (2012)
8. Ghareshabani, E., Mohammadi, M.A., Measurement of the energy of nitrogen ions produced in Filippov Type plasma focus used for the nitriding of titanium. J Fusion Energ. **31**, 595 (2012)
9. Patran, A., Stoenescu, D., Rawat, R.S., Springham, S.V., Tan, T.L., Tan, L.C., Rafique, M.S., Lee, P., Lee, S., A magnetic electron analyzer for plasma focus electron energy distribution studies. J. Fusion Energy **25**, 57 (2006)
10. Lee, S., et al., Application of plasma focus as a source of high energy electron. Singap. J. Phys. **173**, 276 (2003)
11. Kato, Y., Be, S.H., Generation of soft x rays using a rare gas-hydrogen plasma focus and its application to x-ray lithography. Appl. Phys. Lett. **48**, 686 (1986)
12. Castillo, F., Gamboa-deBuen, I., Herrera, J.J.E., Rangel, J., Villalobos, S., High contrast radiography using a small dense plasma focus. Appl. Phys. Lett. **92**, 051502 (2008)
13. Ghareshabani, E., Rawat, R.S., Sobhanian, S., Verma, R., Karamat, S., Pan, Z.Y., Synthesis of nanostructured multiphase Ti(C, N)/aC films by a plasma focus device. Nucl. Instrum. Methods Physi. Res. B **268**, 2777–2784 (2010)
14. Khan, I.A., Hassan, M., Ahmad, R., Qayyum, A., Murtaza, G., Zakaullah, M., et al., Nitridation of zirconium using energetic ions from plasma focus device. Thin Solid Films **516**, 8255–8263 (2008)
15. Gupta, Ruby, Srivastava, M.P., Carbon ion implantation on titanium for TiC formation using a dense plasma focus device. Plasma Sources Sci. Technol. **13**, 371–374 (2004)
16. Ruby Gupta, Srivastava, M.P., Balakrishnan, V.R., Kodama, R., Peterson, M.C., Deposition of nanosized grains of ferroelectric lead zirconate titanate on thin films using dense plasma focus. J. Phys. D Appl. Phys. **37**, 1091–1094 (2004)
17. Kozlov. N. P., Aleksev. V. A., Protsov. Y. S. and Rubinov. A. B., High-power ultraviolet paraterphenyl-solution laser excited by the plasma focus of a magnetoplasma compressor. JEPT Lett. **20**, 331 (1974)
18. Koohestani, S., Habibi, M., Amrollahi, R., Baghdadi, R., Roomi, A., Effect of quartz and pyrex insulators length on hard-X ray signals in APF plasma focus device. J. Fusion Energy **30**, 68–71 (2011)
19. Feugeas, J.N., The influence of the insulator surface in the plasma focus behavior. J. Appl. Phys. **66**, 3467 (1989)
20. Rawat, R.S., Zhang, T., Phua, C.B.L., Then, J.X.Y., Chandra, K.A., Lin, X., Patran, A., Lee, P., Effect of insulator sleeve length on soft x-ray emission from a neon-filled plasma focus device. Plasma Sour. Sci. Technol. **13**, 569–575 (2004)
21. Zhang, T., Lin, X., Chandra, K.A., Tan, T.L., Springham, S.V., Patran, A., Lee, P., Lee, S., Rawat, R.S. Current sheath curvature correlation with the neon soft x-ray emission from plasma focus device. Plasma Sources Sci. Technol. **14**, 368–374 (2005)
22. Zakaullah, M., et al., Effect of insulator sleeve length on neutron emission in a plasma focus. Phys. Lett. A **137**, 39 (1989)
23. Seng, Y.S., Lee, P., Rawat, R.S., Current sheath formation dynamics and structure for different insulator lengths of plasma focus device. Phys. Plasmas **21**, 113508 (2014)
24. Mohammadi, M.A., Sobhanian, S., Wong, C.S., Lee, S., Lee, P., Rawat, R.S., The effect of anode shape on neon soft x-ray emissions and current sheath configuration in plasma focus device. J. Phys. D Appl. Phys. **42**, 045203 (2009)
25. Yousefi, H.R., Aghamir, F.M., Masugata, K., Effect of the insulator length on Mather-type plasma focus devices. Phys. Lett. A **361**, 360–363 (2007)
26. Zakaullah, M., Mrtaza, G., Ahmad, I., Beg, F.N., Beg, M.M., Shabbir, M., Comparative study of low energy Mather-type plasma focus devices. Plasma Sour. Sci. Technol. **4**, 117–124 (1995)
27. Kies, W., Power limits for dynamical pinch discharges. Plasma Phys. Controll Fusion **28**, 1645–1657 (1986)
28. Serban, A., Anode geometry and focus characteristics. PhD thesis, Nanyang Technological University (1995)

24

Solitary and double-layer structures in quantum bi-ion plasma

Mehran Shahmansouri[1] · Mouloud Tribeche[2]

Abstract Weak ion-acoustic solitary waves (IASWs) in an unmagnetized quantum plasmas having two-fluid ions and fluid electrons are considered. Using the one-dimensional quantum hydrodynamics model and then the reductive perturbation technique, a generalized form of nonlinear quantum Korteweg-de Vries (KdV) equation governing the dynamics of weak ion acoustic solitary waves is derived. The effects of ion population, warm ion temperature, quantum diffraction, and polarity of ions on the nonlinear properties of these IASWs are analyzed. It is found that our present plasma model may support compressive as well as rarefactive solitary structures. Furthermore, formation and characteristics properties of IA double layers in the present bi-ion plasma model are investigated. The results of this work should be useful and applicable in understanding the wide relevance of nonlinear features of localized electro-acoustic structures in laboratory and space plasma, such as in super-dense astrophysical objects [24] and in the Earth's magnetotail region (Parks [43]. The implications of our results in some space plasma situations are discussed.

Keywords Bi-ion plasma · Ion acoustic wave · Quantum hydrodynamics

✉ Mehran Shahmansouri
mshmansouri@gmail.com

[1] Department of Physics, Faculty of Science, Arak University, Arak 38156-8 8349, Iran

[2] Theoretical Physics Laboratory, Faculty of Physics, Plasma Physics Group, University of Bab-Ezzouar, USTHB, B.P. 32, El Alia, Algiers 16111, Algeria

Introduction

Ion acoustic solitary waves (IASWs), as one of the most important nonlinear excitations in plasma physics, have been studied by a number of authors in various models of plasma physics in the last few decades [2, 3, 8, 23, 30, 40, 78, 82, 86]. They arise due to a delicate balance between the dispersive and nonlinear properties of the medium. In the absence of dissipative effects and geometry distortion, a Korteweg-de Vries (KdV) equation describes the propagation of IA waves in a weakly nonlinear regime as shown by Washimi et al. [82]. The IA solitons have been observed experimentally by Ikezi et al. [23] (also see Refs. [3, 8, 86]). The IA waves are highly Landau damped unless one considers the case $T_i \ll T_e$ [2]. Properties of IA waves may be useful to explain and understand the wave characteristics observed in the Earth's ionosphere [44, 70, 71], interplanetary space [12] and transport phenomena in the solar wind, corona, and chromospheres [50]. Formation of IA solitons [1, 10, 16, 23, 54, 56, 57, 60, 63, 78], double layers or shocks [8, 12, 59, 79, 80, 86] in plasma have been studied by a number of authors.

In the presence of a second ion species in a plasma, the basic properties of ion acoustic wave would be significantly modified in the linear [17, 39, 77] and nonlinear regime [78, 85] (see also Refs. [9, 13, 14, 17, 20, 35, 36, 45, 47–49, 61, 62, 64, 72] studied linear IA waves in a plasma composed of two ion species. They found that their model supports the existence of two ion acoustic modes, called the light and heavy ion modes. The experimental investigations [39, 77] verify the existence of these two ion wave modes. Song et al. [72] investigated propagation and damping of ion acoustic waves in a Q-machine plasma consisting of K^+ positive ions, SF_6^- negative ions, and electrons. McKenzie et al. [35] have investigated propagation properties of IASWs in a bi-ion plasma. They found that the compressive as well as rarefactive IA solitons may be excited in their plasma model. Drift instability has been studied by Rosenberg and Merlino [45] in a magnetized plasma comprising of positive ions and negative ions.

On the other hand, investigation of the propagation properties of waves in quantum plasmas is of fundamental importance due to their vital role for understanding the collective behaviors in super-dense astrophysical objects [24], in intense laser-solid density plasma experiments [6, 33], in ultra-cold plasma [27], in microelectronic devices [34], in nanowires [67], carbon nanotubes [84], and quantum diodes [7]. New pressure laws and new quantum forces are included in a quantum plasma, such that in the presence of these new forces the collective behavior of quantum plasmas affects significantly. During the last few decades, the different electrostatics and electromagnetic plasma modes [19, 68, 69, 73]; Ali and Shukla [4, 43, 53] have investigated in quantum plasmas using the QHD and quantum magneto-hydrodynamic models. Haas et al. [19] have presented a detailed study of linear and nonlinear IA waves in an unmagnetized quantum plasma using QHD model. They reported several new quantum features of the IA waves referred to as quantum IA waves. Ion-acoustic shocks in quantum dusty pair-ion plasma were investigated by Misra [38]. The linear and nonlinear quantum IA waves were investigated by Rehman [51] in a plasma with positive and negative ions and Fermi electron gas. A theoretical study on the non-planar electrostatic solitary waves in a negatiove ion degenerate plasma is presented by Hussain et al. [22]. Wang et al. [83] investigated dressed IA solitons in a quantum pair-ion plasma in the presence of a relativistic electron beam. Sahu and Singha [52] investigated the arbitrary amplitude ion acoustic solitary waves as well as double-layer structures for an ultra-relativistic degenerate dense plasma comprising cold and hot electrons and inertial ultra-cold ions. Formation of IA solitons [5], double layers or shocks [46] and vortices [21] has been studied in quantum plasmas.

The spacecraft observations indicate that the Earth's plasma sheet boundary layer is consist of the back ground electrons, the cold electron beams (with energies of the order of a few eV—a few 100 s of eV), the back ground cold ions and the ion beams (having energies of the order of a few eV −10 s of keV) [42, 25, 26, 41]. Here, our aim is the study of solitary as well as double-layer structures in a bi-ion plasma, consisting of warm ions, cool ions and Maxwellian electrons.

The manuscript is organized as follows. In the next section, we present the basic equations of our theoretical model followed by discussion on Soliton solution; The m-KdV Equation is derived in the subsequent section and our findings and conclusion are provided in the last section.

Theoretical Model

Let us consider an unmagnetized plasma consisting of electrons with quantum effects and two-temperature fluid ions. In one dimension, the quantum hydrodynamic (QHD) governing the dynamics of nonlinear IASWs, in normalized form, reads

$$\frac{\partial n_{ic}}{\partial t}+\frac{\partial(n_{ic}u_{ic})}{\partial x}=0, \tag{1}$$

$$\frac{\partial u_{ic}}{\partial t}+u_{ic}\frac{\partial u_{ic}}{\partial x}=-s_c\frac{\partial \phi}{\partial x}-3\sigma_c n_{ic}\frac{\partial n_{ic}}{\partial x}, \tag{2}$$

$$\frac{\partial n_{ih}}{\partial t}+\frac{\partial(n_{ih}u_{ih})}{\partial x}=0, \tag{3}$$

$$\frac{\partial u_{ih}}{\partial t}+u_{ih}\frac{\partial u_{ih}}{\partial x}=-s_h\frac{\partial \phi}{\partial x}-3\sigma_h n_{ih}\frac{\partial n_{ih}}{\partial x}, \tag{4}$$

$$\frac{\partial n_{e}}{\partial t}+\frac{\partial(n_{e}u_{e})}{\partial x}=0, \tag{5}$$

$$\frac{m_e}{m_c}\frac{\partial u_e}{\partial t}+\frac{m_e}{m_c}u_e\frac{\partial u_e}{\partial x}=\frac{\partial \phi}{\partial x}-n_e\frac{\partial n_e}{\partial x}+\frac{H^2}{2}\frac{\partial}{\partial x}\left(\frac{1}{\sqrt{n_e}}\frac{\partial^2\sqrt{n_e}}{\partial x^2}\right), \tag{6}$$

$$\frac{\partial^2 \phi}{\partial x}=n_e-s_c f n_{ic}-(1-s_c f)n_{ih}, \tag{7}$$

where the following normalized variables $x \rightarrow \omega_{pi}x/c_s$, $t \rightarrow \omega_{pi}t$, $n_j \rightarrow n_j/n_{j0}$, $u_{ic} \rightarrow u_{ic}/c_s$, $u_{ih} \rightarrow u_{ih}/c_s$, $\phi \rightarrow e\phi/2T_{Fe}$, have been used. Here, $\omega_{pi}=\sqrt{4\pi n_0 e^2/m_c}$ is the ion plasma frequency, $c_s=\sqrt{2T_{Fe}/m_c}$ is the acoustic speed, $H=\hbar\omega_{pe}/2T_{Fe}$ is quantum diffraction parameter, T_{Fe} is the electron Fermi temperature, $\sigma_c=T_c/T_{Fe}$, $\sigma_h=T_h/T_{Fe}$, and the subscript c (h) stands for the cold (hot) ions. At equilibrium, the charge neutrality reads as $n_{e0}=s_c n_{ic0}+s_h n_{ih0}=n_0$, from which we define $f=n_{ic0}/n_0=(1-s_h n_{ih0}/n_0)/s_c$. In the following sections we neglect the left hand side of Eq. (6), due to $m_e/m_c \ll 1$.

Linear dispersion relation

To obtain the linear dispersion relation of the IA waves in quantum bi-ion plasma we linearize Eqs. (1)–(7) to a first-order approximation. The zeroth-order fields and velocities are also neglected and linear dispersion relation takes the form

$$\frac{s_c^2 f}{\omega^2-3\sigma_c k^2}+\frac{s_h(1-s_c f)}{\omega^2-3\sigma_h k^2}=\frac{1+k^2(1+H^2k^2/4)}{k^2(1+H^2k^2/4)} \tag{8}$$

In the above linear dispersion relation of IA waves the quantum effects, ion polarity and ions temperature are included. This equation gives rise to two IA modes propagating with different phase velocities, similar to that observed theoretically [17] and experimentally [39, 77]. The mode with larger phase velocity is called the fast mode whereas the mode with smaller phase velocity is known as the slow mode. Note that in the absence of thermal effects the slow mode does not appear in the present plasma model.

Substituting Eqs. (10a)–(10c) into Eq. (10d), one can obtain the following relation

$$\frac{s_c^2 f}{U_0^2 - 3\sigma_c} + \frac{(1 - s_c f)s_h}{U_0^2 - 3\sigma_h} - 1 = 0. \tag{11}$$

The above equation is quadratic in U_0^2, and shows that the present plasma system supports two distinct IA modes. In this case, the phase speeds of the fast (positive sign) and slow (minus sign) IA modes are given by

$$\begin{aligned} U_{0\pm}^2 &= \frac{3}{2}(\sigma_c + \sigma_h) + \frac{s_c^2 f + (1 - s_c f)s_h}{2} \\ &\pm \sqrt{\left(\frac{3}{2}(\sigma_c + \sigma_h) + \frac{s_c^2 f + (1 - s_c f)s_h}{2}\right)^2 - 9\sigma_c\sigma_h - 3\sigma_h s_c^2 f - 3\sigma_c(1 - s_c f)s_h}. \end{aligned} \tag{12}$$

Soliton solution

To study weakly nonlinear IASWs that may develop in our present plasma model, we employ the standard reductive perturbation technique. The stretched coordinates are defined by $\xi = \varepsilon^{1/2}(x - U_0 t)$, $\tau = \varepsilon^{3/2}t$, where ε is a small parameter measuring the strength of nonlinearity or dispersion, and U_0 the normalized phase velocity. The variables in (1)–(7) are expanded in power series of ε as [75, 76]

$$n_{ic} = 1 + \varepsilon n_{ic1} + \varepsilon^2 n_{ic2} + \cdots, \tag{9a}$$

$$u_{ic} = \varepsilon u_{ic1} + \varepsilon^2 u_{ic2} + \cdots, \tag{9b}$$

$$n_{ih} = 1 + \varepsilon n_{ih1} + \varepsilon^2 n_{ih2} + \cdots, \tag{9c}$$

$$u_{ih} = \varepsilon u_{ih1} + \varepsilon^2 u_{ih2} + \cdots, \tag{9d}$$

$$n_e = 1 + \varepsilon n_{e1} + \varepsilon^2 n_{e2} + \cdots, \tag{9e}$$

$$u_e = \varepsilon u_{e1} + \varepsilon^2 u_{e2} + \cdots, \tag{9f}$$

$$\phi = \varepsilon\phi_1 + \varepsilon^2\phi_2 + \cdots. \tag{9g}$$

Substituting the power series (9a–9g) along with the stretching coordinates into Eqs. (1)–(7), and collecting the terms in different powers of ε, we obtain at the lowest order

$$n_{ic1} = \frac{u_{ic1}}{U_0} = \frac{s_c\phi_1}{U_0^2 - 3\sigma_c}, \tag{10a}$$

$$n_{ih1} = \frac{u_{ih1}}{U_0} = \frac{s_h\phi_1}{U_0^2 - 3\sigma_h}, \tag{10b}$$

$$n_{e1} = \phi_1, \tag{10c}$$

$$n_{e1} - s_c f n_{ic1} - (1 - s_c f)n_{ih1} = 0. \tag{10d}$$

Gill et al. [18] reported the similar results in their study. For $\sigma_c \to 0$ and $\sigma_h \to 0$, the present plasma model does not support slow IA modes, as already reported by Mishra and Chhabra [37].

At the next higher order of ε, the condition for annihilation of secular terms Eq. (11), leads to the following Korteweg de Vries (KdV) equation for IA waves

$$\frac{\partial\phi_1}{\partial\tau} + A\phi_1\frac{\partial\phi_1}{\partial\xi} + B\frac{\partial^3\phi_1}{\partial\xi^3} = 0 \tag{13}$$

where

$$A = \left[1 + \frac{3s_c^3 f(U_0^2 + \sigma_c)}{(U_0^2 - 3\sigma_c)^3} + \frac{3(1 - s_c f)s_h^2(U_0^2 + \sigma_h)}{(U_0^2 - 3\sigma_h)^3}\right] \Bigg/ \left[\frac{2fU_0 s_c^2}{(U_0^2 - 3\sigma_c)^2} + \frac{2(1 - s_c f)s_h U_0}{(U_0^2 - 3\sigma_h)^2}\right],$$

$$B = (1 - H^2/4) \Bigg/ \left[\frac{2fU_0 s_c^2}{(U_0^2 - 3\sigma_c)^2} + \frac{2(1 - s_c f)s_h U_0}{(U_0^2 - 3\sigma_h)^2}\right].$$

The nonlinear and dispersive coefficients in Eq. (13), both are depend on the relevant parameters such as ions population, ions temperature, and polarity of ions. The latter is also depends on the quantum diffraction.

To solve the KdV Eq. (13), we use the transformation $\eta = \varepsilon^{1/2}[x - (U_0 + \varepsilon V_0)t]$, where V_0 is the normalized solitary wave speed in the stationary frame. We can then define the Mach number as $M = U_0 + \varepsilon V_0$ (or normalized solitary wave speed in laboratory frame). The stationary solution of Eq. (13) can then be obtain as [32, 81]

$$\phi_1(\eta) = \Phi_{\max}\sec h^2\left(\frac{\eta}{\Delta}\right) \tag{14}$$

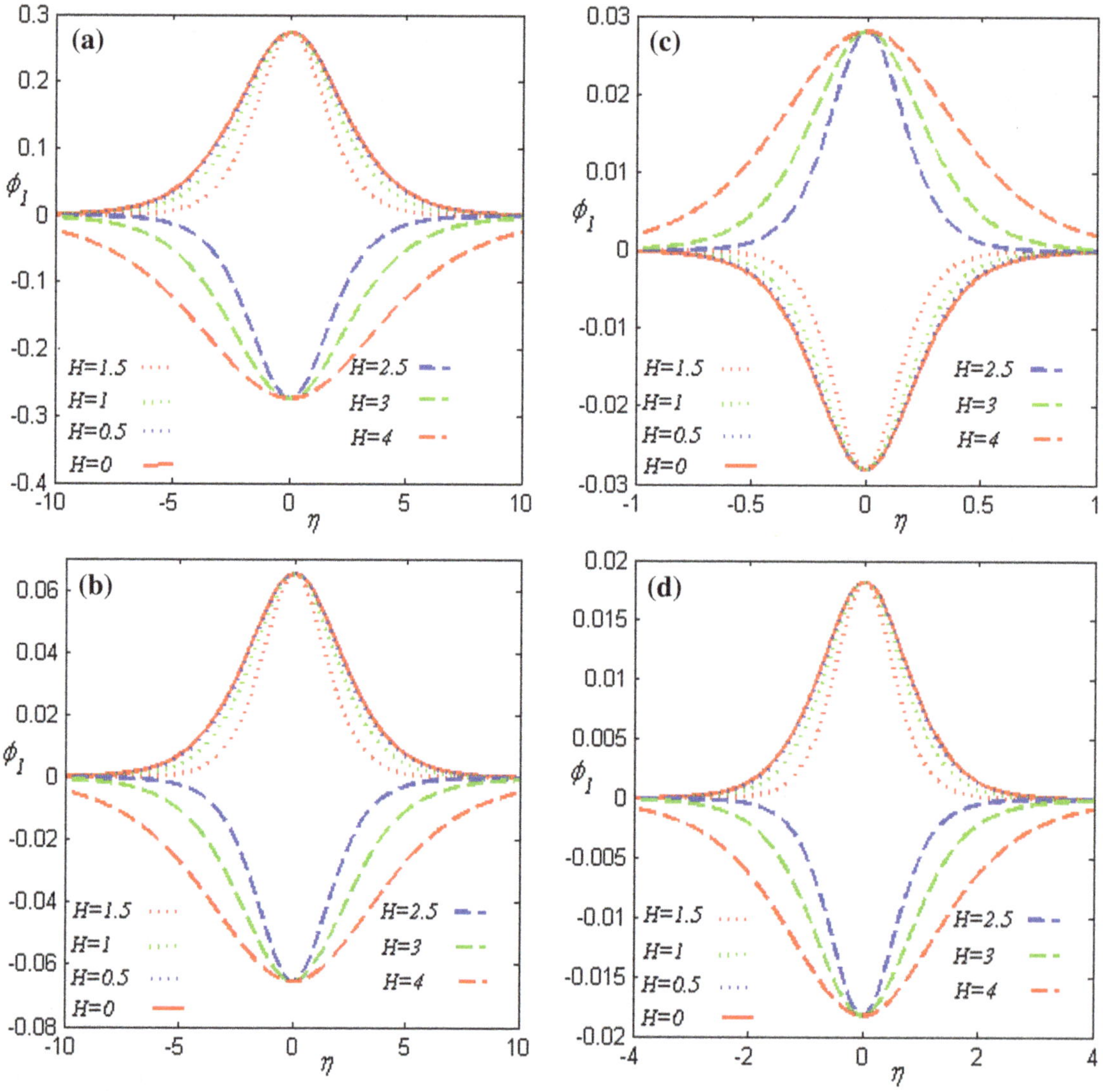

Fig. 1 The IA solitary wave profile for different values of quantum parameter in the cases of fast mode with **a** $s_h = +1$ and **b** $s_h = -1$, and slow mode with **c** $s_h = +1$ and **d** $s_h = -1$, where the other parameters are: $f = 0.8$, $\sigma_c = 0.05$, $\sigma_h = 0.1$, and $s_c = +1$

where $\Phi_{\max} = 3V_0/A$ and $\Delta = (4B/V_0)^{1/2}$ represent, respectively, the amplitude and width of the solitary wave. The IA solitary structures are depicted in Figs. 1, 2 and 3 for different values of the plasma parameters. It can be seen from Fig. 1 that both the compressive and rarefactive IA solitons will be observe in this plasma model, due to the quantum parameter value. Also an increase of the quantum parameter leads to decrease (increase) of the fast compressive (rarefactive) IA solitons width whereas the amplitude of IA solitons remains unchanged. This means that the bi ion plasma in the presence of strong quantum effects supports narrower (wider) compressive (rarefactive) IA solitons. A comparison between Fig. 1a, b shows that the structure of IA solitary waves depends sensitively on the polarity of warm ions, as for negative warm ions IA solitons have smaller amplitudes. Similar figures for slow mode are depicted in Fig. 1c, d. Note that the effect of quantum parameter on the slow IA solitons width is reverse in comparison with the fast one. Coexistence of compressive and rarefactive solitary structures in the present model (in the presence as well as in the absence of quantum effects) may be applicable to describe such behavior in the magnetospheric regions, which is consisting of multi-ion plasma. Lui et al [31] observed the presence of earthward and anti-earthward streaming of ions having density $\leq 0.1\,\text{cm}^{-3}$ with temperatures of the order of $0.1 - 1\,\text{keV}$ (or 50 eV [65, 66]) in the plasma sheet boundary layer. Dusenbery and Lyons [15] discussed the effect of two population ion species on the formation of electrostatic structures in the plasma sheet boundary layer. Here, Fig. 1 is depicted for parameters $T_c/T_h = 0.5$, cold ion density of the order of 0.2 for different values of quantum diffraction

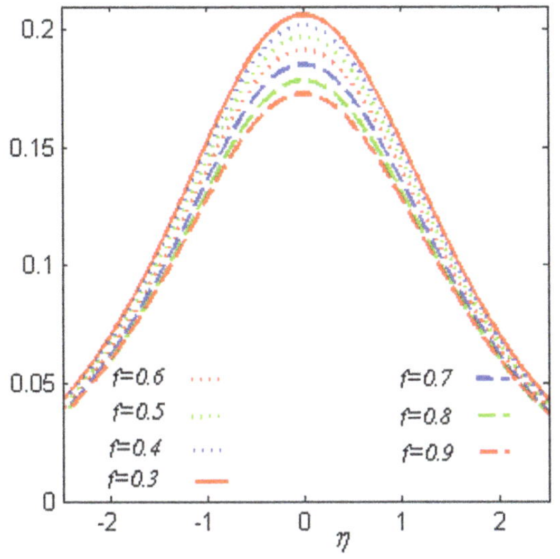

Fig. 2 The fast IA solitary wave profile for different values of cold ion concentration with parameters: $H = 1.5$, $\sigma_c = 0.05$, $\sigma_h = 0.1$ and $s_c = s_h = +1$

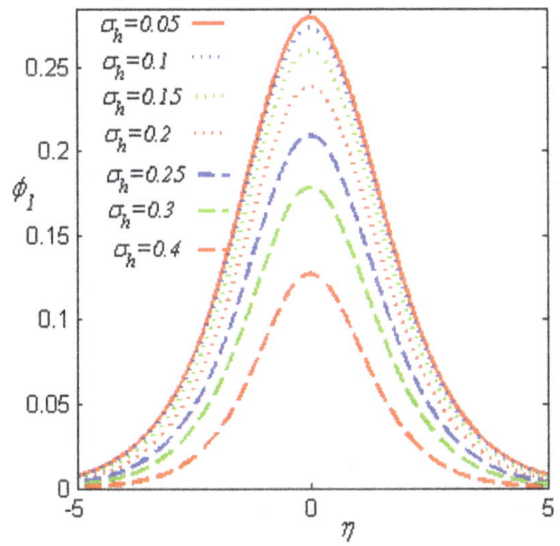

Fig. 3 The fast IA solitary wave profile for different values of σ_h with parameters: $f = 0.8$, $H = 1.5$, $\sigma_c = 0.05$ and $s_c = s_h = +1$

parameter $0 \leq H \leq 4$, in which we found the electrostatic potentials which support the coexistence of compressive and rarefactive solitons. It is easy to show that such compressive and rarefactive structures have bipolar electric fields. The effect of cold ions concentration f on the fast IA solitons is shown in Fig. 2. It can be seen that the amplitude of IA solitons shows a decreasing behavior with f while the width has an inverse behavior. Thus, a decrease in f leads to appearance of larger and narrower IA solitons, as in the presence of more warm ions the solitary structures become more spiky.

Figure 3 indicates the influence of warm ion temperature on the fast IA solitary wave. An increase of the warm ion temperature decreases the amplitude of fast IA solitons. This shows that the warm ion temperature has a destructive effect for the formation of IA solitary structures in such a quantum bi-ion plasma. Similar result was also observed for dust acoustic solitons in dusty plasma as shown in the work of Shahmanouri and Alinejad [58].

Derivation of m-KdV equation

In order to study the double-layer structures in the present plasma model, we consider the higher order effects to obtain the evolution equation. Thus, we introduce the new stretching coordinates as follows:

$$\xi = \varepsilon(x - U_0 t), \quad \tau = \varepsilon^3 t, \tag{15}$$

Then, we substitute the above stretching coordinates along with the set of Eqs. (9a–9g) into the Eqs. (1)–(7), and again collect the terms in different powers of ε. The second power of ε in Poisson's equation leads to

$$A\phi^{(1)^2} = 0, \tag{16}$$

We have assumed that $\phi^{(1)} \neq 0$ and to derive the modified KdV (m-KdV) equation we also consider $A \neq 0$, then the magnitude of A should be of the first order of ε [18, 55]. Therefore, $A\phi^{(1)^2}$ becomes of the order of ε^3, and will appear in the third order of the Poisson's equation. After a long algebraic but straightforward manipulation we eliminate the third-order variables to derive the following m-KdV equation

$$\frac{\partial \phi^{(1)}}{\partial \tau} + A\phi^{(1)}\frac{\partial \phi^{(1)}}{\partial \xi} + D\left(\phi^{(1)}\right)^2\frac{\partial \phi^{(1)}}{\partial \xi} + B\frac{\partial^3 \phi^{(1)}}{\partial \xi^3} = 0, \tag{17}$$

where

$$D = \left[-\frac{3}{2} + \frac{3s_c^4 f}{2\left(U_0^2 - 3\sigma_c\right)^5}\left(3U_0^2\left(U_0^2 + \sigma_c\right) + 2U_0^4 - 18\sigma_c^2 + 27\sigma_c\left(U_0^2 + \sigma_c\right)\right) + \frac{3(1 - s_c f)s_h^3}{2\left(U_0^2 - 3\sigma_h\right)^5} \times \left(2U_0^4 + 3U_0^2\left(U_0^2 + \sigma_h\right) + 2U_0^4 - 18\sigma_h^2 + 27\sigma_h\left(U_0^2 + \sigma_h\right)\right)\right] \Bigg/ \left[\frac{2fs_c^2 U_0}{\left(U_0^2 - 3\sigma_c\right)^2} + \frac{2(1 - s_c f)s_h U_0}{\left(U_0^2 - 3\sigma_h\right)^2}\right]$$

By introducing a new variable $\eta = \xi - u_0\tau$ in a stationary frame, we can write Eq. (17) in the form of an energy-like equation as follows

$$\frac{1}{2}\left(\frac{\partial \phi}{\partial \eta}\right)^2 + \psi(\phi) = 0, \tag{18}$$

where $\phi = \phi^{(1)}$, and the Sagdeev pseudo-potential is given by

$$\psi(\phi) = -\frac{u_0}{2B}\phi^2 + \frac{A}{6B}\phi^3 + \frac{D}{12B}\phi^4 \tag{19}$$

Equation (18) is similar to the "energy-integral" of an oscillating particle of a unit mass, with speed $d\phi/d\xi$ and position ϕ, in a potential $\psi(\phi)$. The first term of (18) can be considered as the kinetic energy of the unit mass, and $\psi(\phi)$ is the potential energy. Since the kinetic energy is a positive quantity, thus it requires that $\psi(\phi) \leq 0$ for the entire of motion. The existence condition of electrostatic structures could be determined by the behavior of Sagdeev potential. The existence of localized solution requires that the Sagdeev potential have a maximum value at the origin, viz. $\psi(\phi = 0) = 0$ and $(\partial\psi/\partial\xi)|_{\phi=0} = 0$, $(\partial^2\psi/\partial\xi^2)|_{\phi=0} < 0$. Also it needs that $\psi(\phi)$ be negative in the interval $0 < \phi < \phi_{\max}$ ($\phi_{\min} < \phi < 0$) for the compressive (rarefactive) solitary waves. Where $\phi_{\max}(\phi_{\min})$ is the maximum (minimum) value of ϕ for which $\psi(\phi) = 0$. Additional conditions that the double-layers need to existence are as follows $\psi(\phi = \phi_m) = 0$ and $(\partial\psi/\partial\xi)|_{\phi=\phi_m} = 0$, $(\partial^2\psi/\partial\xi^2)|_{\phi=\phi_m} < 0$. The first two boundary condition is required to satisfy the global charge neutrality, and lead to the following conditions

$$\phi_m = -A/D, \quad u_0 = -D\phi_m^2/6, \tag{20}$$

Substituting the above expressions into Eq. (18), we obtain the energy integral in the following form

$$\left(\frac{\partial\phi}{\partial\eta}\right)^2 + \frac{D\phi^2}{6B}(\phi - \phi_m)^2 = 0. \tag{21}$$

The above equation is analytically solvable, and thus one can obtain a stationary solution for ϕ of the form [32]

$$\phi(\xi, \tau) = \frac{\phi_m}{2}\left[1 - \tanh\left(\frac{\xi - u_0\tau}{L}\right)\right], \tag{22}$$

where L is the thickness of the IA double-layer, and it is given by $L = \sqrt{-24B/D}/|\phi_m|$. The definition of L provided that $B/D < 0$. It appears appropriate here to add that the nature of the IA double layer depends on the sign of A/D. Positive sign of A/D supports appearance of rarefactive double layers whereas negative sign of A/D supports compressive one.

To examine the possibility of double layer excitation in the present plasma model, we must check the necessary conditions, for which $D < 0$, $B > 0$ and $\sqrt{-6u_0/D} = -A/D$ are satisfied simultaneously. The result of a careful study of these cases is illustrated for different values of the relevant plasma parameters in Table 1. In this table, we have listed the corresponding values of B/D, $-6u_0/D$, $-A/D$ and also the double layer amplitude ϕ_m. Note that the essential conditions for the formation of double layers are confined to the simultaneous satisfaction of the following conditions: (1) $D < 0$, (2) $B > 0$ and (3) $-\sqrt{-6u_0/D} = -A/D$ (for rarefactive double layers) or $\sqrt{-6u_0/D} = -A/D$ (for compressive double layers). First, we have listed various values of the cold ion concentration

Table 1 The numerical results of the double layer conditions for different relevant physical parameters

	$H = 1$, $u_0 = 0.0013, \sigma_c = 0.05, \sigma_h = 0.2$				$f = 0.6, u_0 = 0.0025, \sigma_c = 0.05, \sigma_h = 0.2$			
	$f = 0.5$	$f = 0.6$	$f = 0.7$	$f = 0.8$	$H = 0$	$H = 1$	$H = 3$	$H = 4$
B/D	0.0002	−0.0001	−0.0001	−0.0001	−0.0013	−0.0010	0.0016	0.0039
$\frac{-6u_0}{D}$	−0.0001	0.0002	0.0000	0.0000	0.0005	0.0005	0.0005	0.0005
$\frac{-A}{D}$	−0.0205	0.0217	−0.0053	−0.0053	0.0217	0.0217	0.0217	0.0217
$\lvert\phi_m\rvert = \sqrt{\frac{-6u_0}{D}}$	0.008i	0.0157	0.0053	0.0053	0.0217	0.0217	0.0217	0.0217
$\phi_m = \frac{-A}{D}$	−0.0205	0.0217	−0.0053	−0.0053	0.0217	0.0217	0.0217	0.0217
	$f = 0.8,\ H = 1, \sigma_c = 0.05, \sigma_h = 0.2$				$f = 0.8,\ H = 1,\ u_0 = 0.0036, \sigma_c = 0.05$			
	$u_0 = 0.001$	$u_0 = 0.002$	$u_0 = 0.003$	$u_0 = 0.0036$	$\sigma_h = 0.1$	$\sigma_h = 0.15$	$\sigma_h = 0.2$	$\sigma_h = 0.25$
B/D	−0.0000	−0.0000	−0.0000	−0.0000	$-1e-7$	−0.0000	−0.0000	−0.0001
$\frac{-6u_0}{D}$	0.0001	0.0000	0.0000	0.0000	$6e-7$	0.0000	0.0000	0.0001
$\frac{-A}{D}$	−0.0043	−0.0043	−0.0043	−0.0043	$-6.24e-4$	−0.0021	−0.0043	−0.0066
$\lvert\phi_m\rvert = \sqrt{\frac{-6u_0}{D}}$	0.0023	0.0032	0.0039	0.0043	$-7.94e-4$	0.0022	0.0043	0.0076
$\phi_m = \frac{-A}{D}$	−0.0043	−0.0043	−0.0043	−0.0043	$-6.24e-4$	−0.0021	−0.0043	−0.0066

The columns which are included in highlight indicate the occurrence of slow double layer structures, where the essential conditions would be satisfied simultaneously

Table 2 The slow double layer conditions for different values of u_0, with $f = 0.7$, $H = 1$, $\sigma_c = 0.05$ and $\sigma_h = 0.2$

$\frac{d\psi}{d\phi}$									
$u_0 = 0.002$	0.00921	0.00161	0.00002	−0.00247	0.00003	0	−0.00066	−0.00247	−0.01169
$u_0 = 0.002005$	0.00923	0.00162	0.0000	−0.00248	0.0000	0	−0.00066	−0.00248	−0.01170
$u_0 = 0.00205$	0.00924	0.00164	0.00004	−0.00248	0.00003	0	−0.00066	−0.00248	−0.01171
ψ									
$u_0 = 0.002$	−5.21e−6	−4.7e−7	0.0000	−8.5e−7	−2e−8	0	−1.3e−7	−8.5e−7	−7.19e−6
$u_0 = 0.002005$	−5.22e−6	−4.8e−7	−1e−8	−8.5e−7	−2e−8	0	−1.3e−7	−8.5e−7	−7.10e−6
$u_0 = 0.00205$	−5.27e−6	−5e−7	−1e−8	−8.6e−7	−2e−8	0	−1.3e−7	−8.6e−7	−7.21e−6
ϕ									
	−0.003	−0.002	−0.00122	−0.001	−0.0005	0	0.0005	0.001	0.002

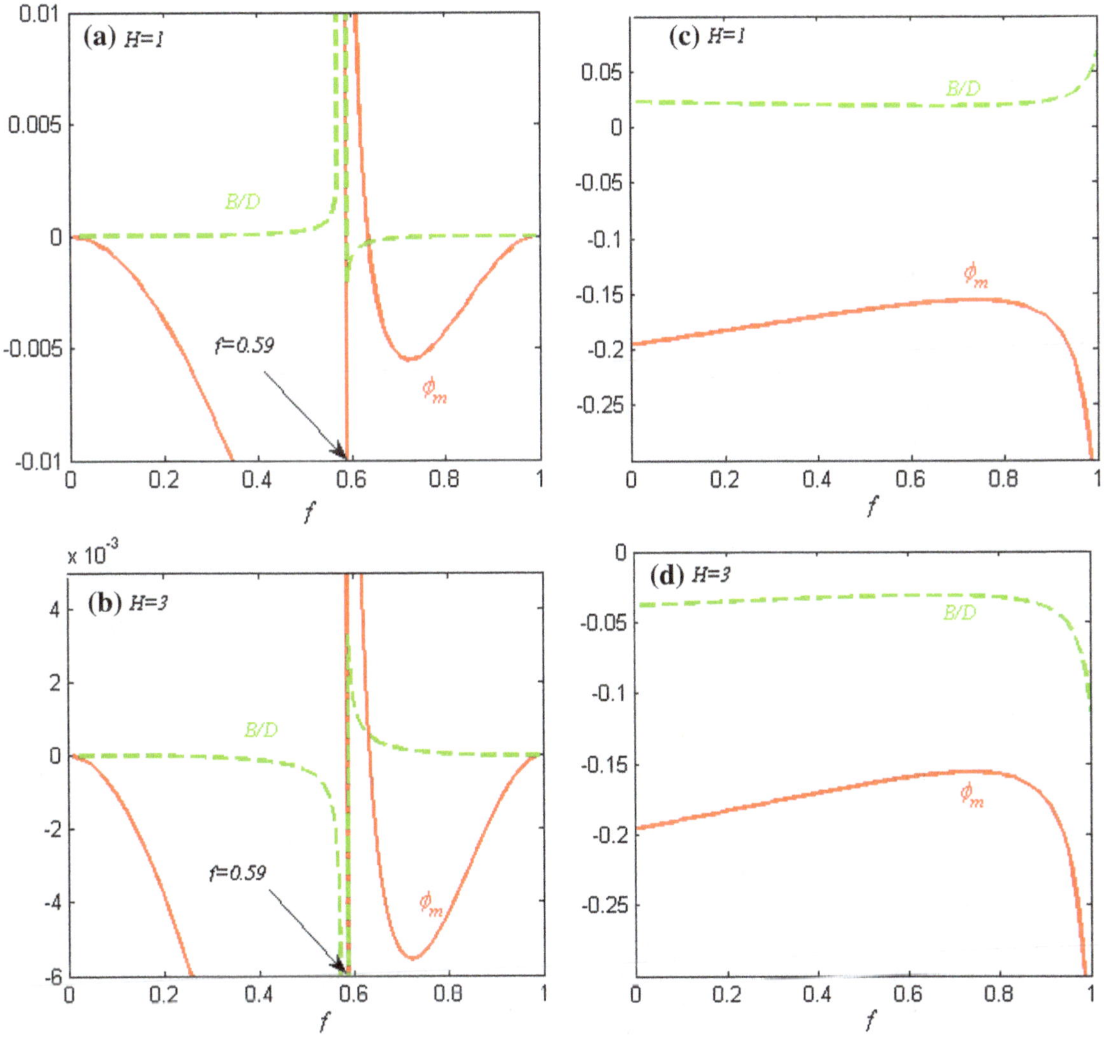

Fig. 4 Variation of B/D and ϕ_m with respect to cold ion concentration for slow mode with **a** $H = 1$ and **b** $H = 3$, and for fast mode with **c** $H = 1$ and **d** $H = 3$, where $\sigma_c = 0.05$, $\sigma_h = 0.2$

lying in the range $0.6 < f < 0.9$. For each such value, we have obtained the corresponding values of B/D, $-6u_0/D$, $-A/D$. The table shows that only for $f = 0.7$ (the third column) both double layer conditions satisfy and a rarefactive IA double layer can be form in the system. In the second four columns we have listed various values of the

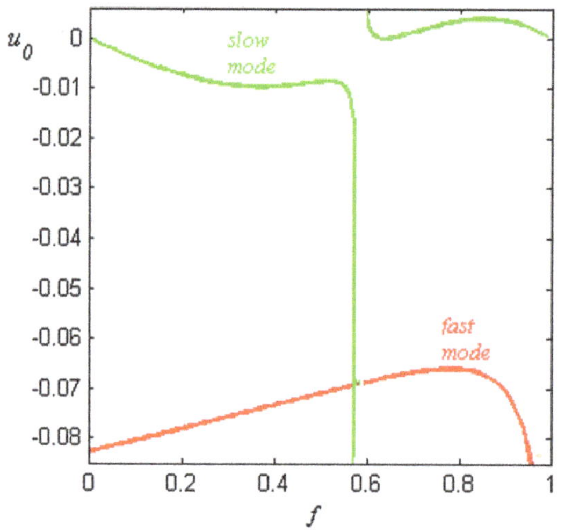

Fig. 5 The primitive values of u_0 which are corresponding to the essential condition for the formation of IA double layers, where $\sigma_c = 0.05$, $\sigma_h = 0.2$, $H = 3$

quantum parameter. In this case we see that only for $H = 1$(the 6th column) the double layer conditions satisfy, and system supports excitation of compressive IA double layer. A similar treatment for other two cases, i.e., different values of u_0 and σ_h, are shown in another columns as in the 12th and 15th columns double layers may occur.

The formation of double layers can be also investigated via another simple fashion, in which we have obtained the relevant value of $\psi(\phi)$ and $\mathrm{d}\psi(\phi)/\mathrm{d}\phi$ for various values of u_0 in Table 2. The existence of double layers need to simultaneous zero in $\psi(\phi)$ and $\mathrm{d}\psi(\phi)/\mathrm{d}\phi$ for $\phi_m \neq 0$. Such this condition satisfies at $\phi = -0.00122$, which is in accordance with a rarefactive IA double layer.

In order to study the nature of double layer (i.e., compressive or rarefactive double layers), we analyzed numerically the essential conditions for the existence of double layer, for typical parameters $\sigma_c = 0.05$, $\sigma_h = 0.2$ and $s_c = s_h = +1$. Variation of $\phi_m = -A/D$ and B/D as a function of cold ion concentration are plotted in Fig. 4. From the sign of $\phi_m = -A/D$ and B/D we can show that for $H = 1$ in slow mode, both compressive and rarefactive double layers are obtained respectively for $0.59 < f < 0.636$ and $0.636 < f < 1$ while for $H = 3$ only negative double layers are obtained for $f < 0.59$. The amplitude of double layers tend to zero when system goes to two-component plasma, i.e., in the cases of $f \to 0$ and $f \to 1$. On the other hand, for the fast mode with $H = 1$ the double layer structures do not exist in the system due to the positive sign of B/D. While for the fast mode with $H = 3$ only the rarefactive double layers will exist.

One of the necessary conditions for the formation of double layers bounds u_0, as it needs to satisfy $u_0 = -A^2/6D$. It can be seen that the primitive values of u_0

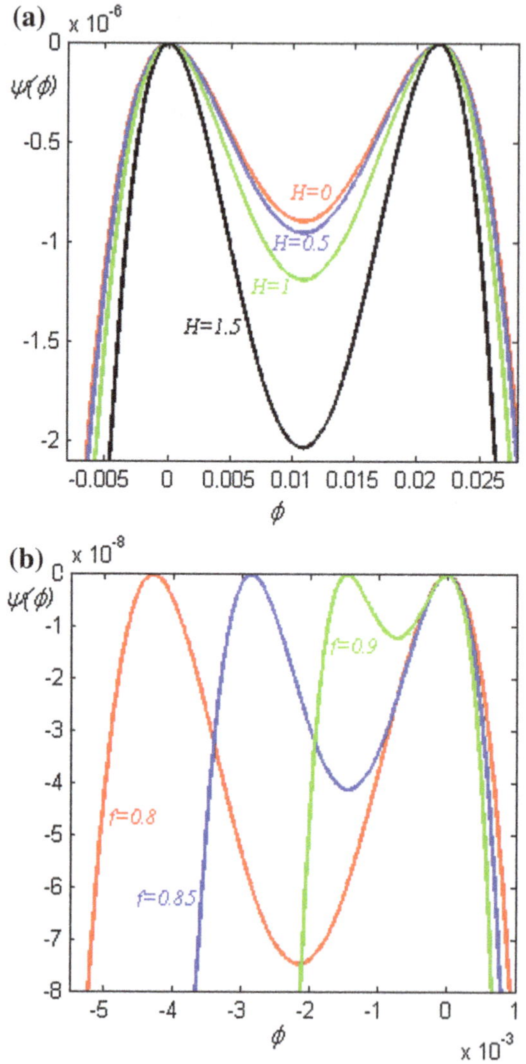

Fig. 6 The slow IA double layers for **a** different values of quantum parameter and $f = 0.6$, and **b** different values of cold ions concentration and $H = 1$, with parameters: $\sigma_c = 0.05$, $\sigma_h = 0.2$, $s_c = s_h = +1$

depends sensitively on the cold ion concentration. The behavior of u_0 with respect to f for both slow and fast modes is depicted in Fig. 5. We found that for the fast mode, u_0 is always negative whereas for the slow mode in the region of $0 < f < 0.58$, u_0 is negative and in the region of $0.58 < f < 1$, u_0 is positive.

The effect of quantum parameter on the slow IA double layers is depicted in Fig. 6a. It can be seen that the double layer structure is highly sensitive to the quantum parameter. This shows that an increase in the quantum parameter leads to increase of the depth of potential well. Indeed, this is equivalent with a decrease in the width of double layers, while the amplitude of the produced double layer remains unchanged. Furthermore, the double layer structure for different values of the cold ion concentration is depicted in Fig. 6b. Both characteristics properties of double layers, i.e., depth of potential well and amplitude of double layer ϕ_m, decrease with an increasing f. Similar to that observed in Fig. 4a, when f tends to the unit value the amplitude of double layer tends to zero.

Conclusions

Ion acoustic solitary waves and double layer structures are studied in a bi-ion plasma consisting of cold and warm fluid ions and quantum electrons. The existence of solitary waves and double layers is shown to be dependent on the different parameters such as quantum parameter, ion concentration and ion temperature in detail. The main results are as follows:

1. We found that the present bi-ion plasma model supports two distinct IA modes which propagates with different phase speed, these modes are knows as the fast and slow IA modes. When both ion species are cold, i.e., $\sigma_c = \sigma_h = 0$, the slow mode disappears and the fast mode reduces to the usual IA mode.
2. We found that the system supports the compressive as well as rarefactive excitations (solitary and double-layer structures), due to the particular set of plasma parameters.
3. We found that the IA solitary waves depend sensitively on the plasma parameters such as the quantum effects, ion concentration, ion polarity and ion temperature. The present quantum bi-ion plasma model supports formation of both the compressive and rarefactive IA solitons, due to the quantum parameter value. Also an increase of the quantum parameter leads to decrease (increase) of the fast compressive (rarefactive) IA solitons width whereas the amplitude of IA solitons remains unchanged. This means that the quantum bi-ion plasma in the presence of strong quantum effects supports narrower (wider) compressive (rarefactive) IA solitons. Note that the effect of quantum parameter on the slow IA solitons width is reverse in comparison with the fast one. The effect of cold ions concentration f on the fast IA solitons is decrease of their amplitude. Also we found that the warm ion temperature has a destructive effect on the fast IA solitary wave, as an increase of the warm ion temperature decreases the amplitude of fast IA solitons.
4. We determined numerically the existence domain of compressive and rarefactive IA double layers. As for $H = 1$ in slow mode, both compressive and rarefactive double layers are obtained respectively for $0.59 < f < 0.636$ and $0.636 < f < 1$ while for $H = 3$ only negative double layers are obtained for $f < 0.59$. The amplitude of double layers tend to zero when system goes to two-component plasma, i.e., in the cases of $f \to 0$ and $f \to 1$. On the other hand, for the fast mode with $H = 1$ the double layer structures do not exist in the system due to the positive sign of B/D. While for the fast mode with $H = 3$ only the rarefactive double layers will exist.

The numerical analysis in the present study is a systematic investigation in a wide range of physical parameters. Therefore, our results should be useful and applicable in understanding the wide relevance of nonlinear features of localized electro-acoustic structures in laboratory and space plasma, such as in super-dense astrophysical objects [24] and in the Earth's magnetotail region [42].

References

1. Alinejad, H.: Phys. Scr. **81**, 025501 (2010)
2. Agrimson, E., D'Angelo, N., Merlino, R.L.: Phys. Rev. Lett. **86**, 5282 (2001)
3. Akimoto, K., Papadopoulos, K., Winske, D.: J. Plasma Phys. **34**, 467 (1985)
4. Ali, S., Shukla, P.K.: Phys. Plasmas **13**, 052113 (2006)
5. Ali, S., Moslem, W.M., Shukla, P.K., Kourakis, I.: Phys. Lett. A **366**, 606 (2007)
6. Andreev, A.V.: JETP Lett. **72**, 238 (2000)
7. Ang, L.K., Kwan, T.J.T., Lau, Y.Y.: Phys. Rev. Lett. **91**, 208303 (2003)
8. Bharuthram, R., Shukla, P.K.: Phys. Fluids **20**, 3214 (1986)
9. Bharuthram, R., Shukla, P.K.: Planet. Space Sci. **40**, 973 (1992)
10. Boubakour, N., Tribeche, M., Aoutou, K.: Phys. Scr. **79**, 065503 (2009)
11. Cranmer, S.R., Ballegooijen, A.A., Edgar, R.J.: Astrophys. J. **171**, 520 (2007)
12. Das, G.C., Sen, K.M.: Earth Moon Planets **64**, 47 (1994)
13. Dezfuly, SGh, Dorranian, D.: Cont. Plasma Phys. **53**, 564 (2013)
14. Dorranian, D., Sabetkar, A.: Phys. Plasmas **19**, 013702 (2012)
15. Dusenbery, P.B., Lyons, L.R.: J. Geophys. Res. **90**, 10935 (1985)
16. El-Awady, E.I., El-Tantawy, S.A., Moslem, W.M., Shukla, P.K.: Phys. Lett. A **374**, 3216 (2010)
17. Fried, B.D., White, R.B., Samec, ThK: Phys. Fluids **14**, 2388 (1971)
18. Gill, T.S., Bala, P., Kaur, H., Saini, N.S., Bansal, S., Kaur, J.: Eur. Phys. J. D **31**, 91 (2004)
19. Haas, F., Garcia, L.G., Goedert, J., Manfredi, G.: Phys. Plasmas **10**, 3858 (2003)
20. Haidar, M.M., Frdous, T., Duha, S.S.: J. Theor. Appl. Phys. **9**, 159 (2015)
21. Haquea, Q., Mahmood, S.: Phys. Plasmas **15**, 034501 (2008)
22. Hussain, S., Akhtar, N., Saeed-ur-Rehman, Chin: Phys. Lett. **28**, 045202 (2011)
23. Ikezi, H., Taylor, R., Baker, D.: Phys. Rev. Lett. **25**, 11 (1970)
24. Jung, Y.D.: Phys. Plasmas **8**, 3842 (2001)
25. Kakad, A.P., Singh, S.V., Reddy, R.V., Lakhina, G.S., Tagare, S.G.: Adv. Space Res. **43**, 1945 (2009)
26. Kakad, A.P., Singh, S.V., Reddy, R.V., Lakhina, G.S., Tagare, S.G., Verheest, F.: Phys. Plasmas **14**, 052305 (2007)

27. Killian, T.C.: Nature (Lond.) **441**, 298 (2006)
28. Kim, K.Y.: Phys. Lett. A **97**, 45 (1983)
29. Koepke, M.E.: Phys. Plasmas **9**, 2420 (2002)
30. Lonngren, K.E.: Plasma Phys. **25**, 943 (1983)
31. Lui, A.T.Y., Eastman, D.T.E., Williamsa, D.J., Frank, L.A.: J. Geophys. Res. **88**, 7753 (1983)
32. Mace, R.L., Baboolal, S., Bharuthram, R., Hellberg, M.A.: J. Plasma Phys. **45**, 323 (1991)
33. Marklund, M., Shukla, P.K.: Rev. Mod. Phys. **78**, 591 (2006)
34. Markowich, P.A., Ringhofer, C.A., Schmeiser, C.: Semiconductor equations. Springer, New York (1990)
35. McKenzie, J.F., Verheest, F., Doyle, T.B., Hellberg, M.A.: Phys. Plasmas **11**, 1762 (2004)
36. McKenzie, J.F., Verheest, F., Doyle, T.B., Hellberg, M.A.: Phys. Plasmas **12**, 102305 (2005)
37. Mishra, K., Chhabra, R.S.: Phys. Plasmas **3**, 4446 (1996)
38. Misra, A.P.: Phys. Plasmas **16**, 033702 (2009)
39. Nakamura, N., Nakamura, M., Itoh, T.: Phys. Rev. Lett. **37**, 209 (1976)
40. Nakamura, Y.: IEEE Trans. Plasma Sci. **PS-7**, 232 (1982)
41. Onsager, T.G., Thomsen, M.F., Elphic, R.C., Gosling, J.T., Anderson, R.R., Kettmann, G.: J. Geophys. Res. **98**, 15509 (1993)
42. Parks, G.K.: J. Geophys. Res. **89**, 8885 (1984)
43. Ren, H., Wu, Z., Chu, P.: Phys. Plasmas **14**, 062102 (2007)
44. Rosenberg, M., Merlino, R.L.: Planet. Space Sci. **55**, 1464 (2007)
45. Rosenberg, M., Merlino, R.L.: J. Plasma Phys. **79**, 949 (2013)
46. Roy, K., Mishra, A.P., Chatterjee, P.: Phys. Plasmas **15**, 032310 (2008)
47. Sabetkar, A., Dorranian, D.: J. Plasma Phys. **80**, 565 (2014)
48. Sabetkar, A., Dorranian, D.: J. Theor. Appl. Phys. **9**, 141 (2015)
49. Sabetkar, A., Dorranian, D.: Phys. Scr. **90**, 035603 (2015)
50. Sabry, R., Moslem, W.M., Shukla, P.K.: Phys. Plasmas **16**, 032302 (2009)
51. Rehman, S.U.: Phys. Plasmas **17**, 062303 (2010)
52. Sahu, B., Singha, P.: Earth Moon Planets **110**, 165 (2013)
53. Saleem, H., Ahmed, A., Khan, S.A.: Phys. Plasmas **15**, 014503 (2008)
54. Shahmansouri, M.: Chinese Phys. Lett. **29**, 105201 (2012)
55. Shahmansouri, M.: Phys. Plasmas **20**, 102104 (2013)
56. Shahmansouri, M., Alinejad, H.: Astrophys. Space Sci. **344**, 463 (2013)
57. Shahmansouri, M., Alinejad, H.: Phys. Plasmas **20**, 082130 (2013)
58. Shahmansouri, M., Alinejad, H.: Phys. Plasmas **20**, 033704 (2013)
59. Shahmansouri, M., Mamun, A.A.: Phys. Plasmas **20**, 082122 (2013)
60. Shahmansouri, M., Shahmansouri, B., Darabi, D.: Ind. J. Phys. **87**, 711 (2013)
61. Shahmansouri, M., Tribrche, M.: Astrophys. Space Sci. **350**, 623 (2014)
62. Shahmansouri, M., Tribrche, M.: Astrophys. Space Sci. **349**, 781 (2014)
63. Shahmansouri, M., Astaraki, E.: J. Theor. Appl. Phys. **8**, 189 (2014)
64. Shahmoradi, N., Dorranian, D.: Phys. Scr. **89**, 065602 (2014)
65. Sharp, R.D., Carr, D.L., Peterson, W.K., Shelley, E.G.: J. Geophys. Res. **86**, 4639 (1981)
66. Sharp, R.D., Lennartsson, W., Peterson, W.K., Shelley, E.G.: J. Geophys. Res. **87**, 10420 (1982)
67. Shpatakovskaya, G.V.: J. Exp. Theor. Phys. **102**, 466 (2006)
68. Shukla, P.K., Stenflo, L.: Phys. Lett. A **355**, 378 (2006)
69. Shukla, P.K., Stenflo, L.: Phys. Lett. A **357**, 229 (2006)
70. Singh, D.K., Narayan, D., Singh, R.P.: Earth Moon Planets **77**, 75 (1996)
71. Singh, A.K., Narayan, D., Singh, R.P.: Earth Moon Planets **91**, 161 (2002)
72. Song, B., D'Angelo, N., Merlino, R.L.: Phys. Fluids B **3**, 284 (1991)
73. Stenflo, L., Shukla, P.K., Marklund, M.: Europhys. Lett. **74**, 844 (2006)
74. Stix, T.H.: Waves in plasmas. AIP, New York (1992)
75. Taniuti, T., Yajima, M.: J. Math. Phys. **10**, 1369 (1969)
76. Taniuti, T.: Prog. Theor. Phys. Suppl. **55**, 1 (1974)
77. Tran, M.Q., Coquerand, S.: Phys. Rev. A **14**, 2301 (1976)
78. Tran, M.Q.: Phys. Scr. **20**, 317 (1979)
79. Tribeche, M., Boubakour, N.: Phys. Plasmas **16**, 084502 (2009)
80. Tribeche, M., Mayout, S., Amour, R.: Phys. Plasmas **16**, 043706 (2009)
81. Verheest, F.: Waves in dusty space plasmas. Kluwer Academic, Dordrecht (2000)
82. Washimi, H., Taniuti, T.: T. Phys. Rev. Lett. **17**, 996 (1966)
83. Wang, Y., Dong, Y., Eliasson, B.: Phys. Lett. A **377**, 2604 (2013)
84. Wei, L., Wang, Y.N.: Phys. Rev. B **75**, 193407 (2007)
85. White, R.B., Fried, B.D., Coroniti, F.V.: Phys. Fluids **15**, 1484 (1972)
86. Yadav, L.L., Sharma, S.R.: Phys. Scr. **43**, 106 (1991)

25

Glass surface modification using Nd:YAG laser in SF_6 atmospheres

H. R. Dehghanpour · P. Parvin

Abstract Glass surface modification is one of the processes which changes its physical properties. Changing optical property is a significant effect of surface modification. Among the surface modification methods, laser surface modification is attractive for researchers. Here, SiO_2 glass target was irradiated by Q-switched Nd:YAG laser (1064 nm) in SF_6 atmospheres as well as vacuum. In this process, an Nd:YAG laser beam was focused near the glass target while the target was inserted in a chamber containing SF_6 atmosphere or at vacuum condition. The surface morphology changes were investigated by scanning electron microscopy (SEM). Noticeable surface changes were observed due to laser irradiation in SF_6 atmosphere. Although the absorption lines of SF_6 molecule are not near the Nd:YAG laser wavelength, laser-induced breakdown spectroscopy (LIBS) showed decomposition of SF_6 molecules near a glass target. The ablated fragments from the target impacted on gas molecules and decomposed them. In this work, LIBS showed that there were several fluorine ions when the target was irradiated in an SF_6 atmosphere. This demonstrates SF_6 molecule decomposition near the glass target by laser irradiation. The plasma temperature was $\sim$2000 K. This temperature is much less than metallic or other gaseous materials' plasma temperatures.

Keywords Surface modification · SF_6 decomposition · Fluorine penetration · Glass · Laser

H. R. Dehghanpour (✉)
Physics Department, Tafresh University, Tafresh, Iran
e-mail: h.dehghanpour@aut.ac.ir

P. Parvin
Physics Department, Amirkabir University of Technology, Tehran, Iran

Introduction

By increasing glass microdevices' applications, the necessary, low-priced, simple and flexible microstructuring techniques have vastly increased. Making precise surface microstructuring of the materials to fabricate highly integrated microdevices is one of the essential technologies of photonics. Moreover, UV-transparent materials such as amorphous SiO_2 glass, quartz, MgF_2, CaF_2 and LiF are important materials for optical device fabrication. Microfabrication of SiO_2 glass as a hard and brittle optical material is difficult. In practice, surface microstructures on glass are created by approaches based on photoresistant protection layer coating, lithography, etching and cleaning steps. Finding new techniques to decrease the time-consuming stages of the microfabrication is necessary [1].

An efficient and popular technique with attractive applications in many areas is laser-induced breakdown spectroscopy (LIBS). The main advantages of LIBS are: multi-elemental capability, low cost, the simplicity of working in wide range pressure environments and detection of various components in a great variety of matrices.

In LIBS, at first the pulsed laser is focused on the target. For the nanosecond pulses the energies are in the GW cm^{-2} range, which is high enough for plasma creation through the vaporization, atomization and ionization processes in a single setup. A high-temperature dense plasma is the result of the process which could be spectrally resolved and detected. It contains the characteristic peaks with significant information on the nature and concentration of the elements. The number density of the corresponding emitting species in the plume is proportional to integrated emission of the individual spectral lines.

Pulsed laser irradiation on the SiO_2 surface in an SF_6 atmosphere induces gas decomposition [2]. It produces reactive species to restructure the glass surface too. The irradiations on amorphous SiO_2 glass were performed at different wavelengths [1–3]. During Nd:YAG laser irradiation, the possibility for energy absorption by free electrons via the inverse Bremsstrahlung (IB) process leads to increase in the electron–molecule collision rate and the generation of weak plasma. It rarely creates molecular formation in accordance with a larger amount of fluorine trace in the solid [1]. Furthermore, fluorine penetration in the silicon content under plasma treatment and the diffusion of ion implanted-F in Si have been previously studied by various spectroscopic techniques [4]. Though theoretical studies emphasize that F diffusion involves normal interstitial motion in the perfect Si crystal, anomalous diffusion causes the depletion of F to only occur at temperatures above 550 °C, suggesting a thermally activated process which is strongly temperature dependent. Direct observation of volatile Si oxy-fluoride and Si fluoride moieties between F and SiO_2 (or Si) precede desorption of F from Si. The driving force for the observed anomalous F migration remains unknown [4]. In this work, we investigate surface modification on the glass surface by Q-switched Nd:YAG laser at different atmospheres. Furthermore, using LIBS the SF_6 decomposition near the glass target as well as the temperature of the irradiated glass surface were calculated.

Experimental apparatus

The experimental setup consists of a stainless steel irradiation chamber, $\sim 100\ cm^3$, with cross-type quartet windows, high vacuum-mixing system, Q-switched Nd:YAG laser (1064 nm, 150 mJ/pulse, 10 ns duration, 1–20 Hz) and conducting and focusing optics as shown in Fig. 1. An AR-coated MgF_2 window and a glass slab were inserted opposite to each other as coupler and target. A semiconductor detector (PIN, EG&G, FNT100), a Tektronix 7844 400 MHz oscilloscope and a Coherent^{TM} J m (Field Master, LM-P10 and LM-P5 LP heads) were used for the relative and absolute power measurements. The laser energy delivered to the cell was varied using several NDF aluminized attenuators with the appropriate optical densities. LIBS arrangement consists of a CCD-coupled transmission grating spectrometer and the corresponding optics and data processing. The light emission of plasma was collected by a fiber bundle (UV 600/660 type with SMA-905 fiber connector and 1 m length) using a quartz lens (25 mm diameter, 50 mm focal length) placed 80 mm away from the sample. The fiber output was coupled to the entrance slit of the compact wide range spectrometer (200–1100 nm) model S150 Solar laser system (50 mm focal length, transmission diffraction grating with 200 Grooves/mm with 0.02×3.0 mm of entrance slit and 0.5 nm spectral resolutions). A charged coupled device (CCD) detector array model Toshiba TCD1304AP with 3648 pixels was used to detect the dispersed light subsequently. The CCD camera was triggered $\sim 2\ \mu s$ after the onset of the laser shot using a suitable delay generator to reduce the continuum Bremsstrahlung radiation. The LIBS setup is shown in Fig. 2.

Experiments

The experiments were done using different cell content SF_6 gas at 800 mbar pressure as well as vacuum to investigate the surface morphology modification due to laser irradiation. At first, LIBS of amorphous SiO_2 glass in SF_6 atmosphere (800 mbar) was performed using a focused Nd:YAG laser beam at 1064 nm. Then, the surface morphology of the treated surfaces was studied. SEM micrographs of irradiated surfaces in the vacuum and chamber filled with SF_6 were obtained.

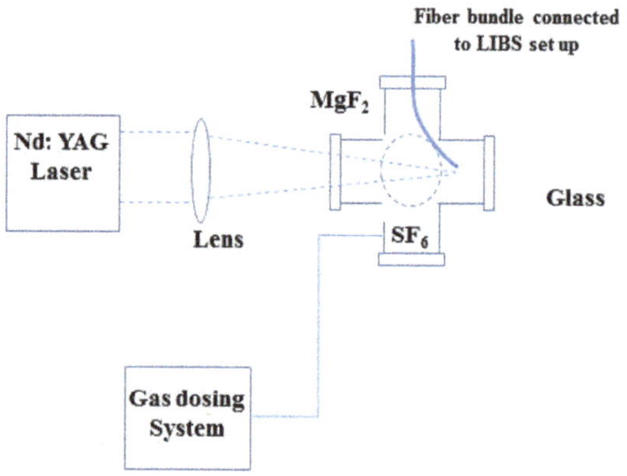

Fig. 1 Experimental arrangement for the laser irradiation of amorphous SiO_2 glass

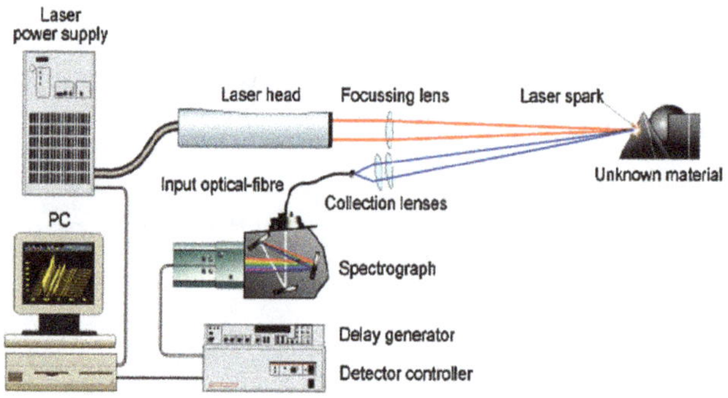

Fig. 2 A schematic of the LIBS setup

Theory

LIBS

The energy level population of the plasma species is proportional to the intensity of plasma's spectral line emission. Local thermodynamic equilibrium (LTE) and being optically thin are the basic assumptions about the plasma. The LTE condition is given by the relation [5]:

$$N_e \geq 1.6 \times 10^{12} T^{1/2} \Delta E^3, \tag{1}$$

where N_e is the electron density, $T(K)$ shows the plasma temperature and ΔE (eV) denotes the largest energy transition for which the condition holds. Electron density is known as an important plasma parameter that gives indications about the thermal equilibrium. The experimental condition satisfies the LTE condition and as a result Eq. 1 is valid. So, the Boltzmann equation is used to express the population of an excited level based on the number density ns of the species within the plasma [5]:

$$I_{ij} = \frac{A_{ij} g_i}{U_S(T)} n_S \exp\left(-\frac{E_i}{kT_e}\right), \tag{2}$$

where A_{ij}, E_i, g_i, k and U_S are the transition probability, the excitation energy of the level, the statistical weight for the upper level, the Boltzmann constant and the partition function of the species at electronic temperature T_e, respectively. Therefore, the plasma temperature is determined according to the Boltzmann plot:

$$\ln\left(\frac{I_{ij}}{A_{ij} g_i}\right) = \ln\left(\frac{n_S}{U_S(T)}\right) - \frac{E_i}{kT_e}. \tag{3}$$

Fluences higher than the threshold (~ 2 J cm^{-2}) usually lead to nanosecond laser ablation, so that the plasma is always created above the target surface. Plasma self-regulating absorption via inverse Bremsstrahlung (IB) is the reason for the laser ablation process. At higher laser irradiation, highly dense ionized plasma mediates the laser–target interactions. Via the IB process, free electrons absorb photons by which energy is gained from the laser beam during collisions with neutral and ionized atoms. Thus due to electron collisions with the excited and ground-state neutrals, vapor ionization and excitation take place. In the IR laser ablation process, vaporization creates primary electrons which have laser photon absorption ability at low electron density. Due to the high absorption of infrared radiation, its result is strong electron heating and breakdown ionization at very low laser intensities. On the other hand, in UV laser ablation, the photon energies are almost near the typical ionization energies of the excited atoms. Moreover, due to higher temperature of the electronic gas, plasma etching with UV irradiation might be possible. The higher density and temperature of the UV-induced plume lead to a large number of excited atoms, which could be photo-ionized by the several photon impacts. The presence of the highly excited fluorine ions in plume significantly increases the fragment combination rate to produce new clusters and molecules after UV photoablation. The effect of laser wavelengths causes different photoablation regimes, i.e., UV photoablation and plasma-induced ablation for excimer and Nd:YAG lasers, respectively.

IB is a wavelength-dependent effect proportional to λ^2; therefore it reduces four times using SHG of Nd:YAG at 532 nm with respect to the fundamental wavelength at 1064 nm. As a result, the contribution of IB becomes much weaker at shorter wavelengths of UV range, while the electron collisions in the plume notably contribute to the decomposition where the free electrons produced in the plume have gained energy in the electric field of the laser accordingly.

Fragment density diffused into the glass surface

Fluorine or fluorine content fragments can diffuse into glass surface so that we can write:

$$\frac{\partial N}{\partial t} = D \frac{\partial^2 N}{\partial x^2}, \tag{4}$$

where N, t, x and D are fragment density, time, depth of penetration and diffusivity, respectively. We suppose that fragment density in the glass surface before diffusion is negligible. The boundary conditions are at $x = x_0$, $N = N_0$ and at $x = \infty$, $N = 0$. The latter boundary condition is valid where there are not any fragments. The solution of (14) under those conditions is [6]:

$$N(x,t) = N_0 \operatorname{erfc} \frac{x}{2\sqrt{Dt}}. \tag{5}$$

The total number of fragments in unit area is:

$$Q(t) = \int_0^\infty N(x,t) \mathrm{d}x = \frac{2}{\sqrt{\pi}} \sqrt{Dt} N_0. \tag{6}$$

Then, we can estimate N_0 as:

$$N_0 = \frac{2Q(t)\sqrt{\pi}}{\sqrt{Dt}}. \tag{7}$$

Depth of penetration of fragments into a solid

We can use the following formula for approximating plasma fragments' penetration depth into glass surface [7]:

$$\frac{\mathrm{d}E}{\mathrm{d}x} = \frac{4\pi z^2 q^4 NZ \times (3 \times 10^9)^4}{Mv^2 \times 1.6 \times 10^{-6}} \left[\ln \frac{2Mv^2}{I} - \ln\left(1 - \frac{v^2}{c^2}\right) - \frac{v^2}{c^2}\right] \left(\frac{\text{MeV}}{\text{cm}}\right), \tag{8}$$

where z is the atomic number of ionized particles, M rthe est mass of ionized particles, v the velocity of ionized particles, N the number of absorber materials contained in unit volume, Z the atomic number of absorber material, c the velocity of light in vacuum and I the mean potential of ionization and excitation of absorber atoms ($I = 2.16 \times 10^{-11}$ Z).

Results and discussion

Figure 3 shows the whole irradiated sample. Laser-irradiated regions are in the central part of the sample. There are four craters on the sample. The first two craters on top of the sample surface correspond to laser irradiation at 800 mbar SF_6 ambient gas and the next two craters concern laser shooting at vacuum condition. The lack of any crack on the sample after irradiation by Nd:YAG laser is a noticeable result.

The morphology of the irradiated samples was studied subsequently. Figure **4a–c** illustrates the SEM micrograph of amorphous SiO_2 glass surface irradiated by 9 kJ/cm^2 of fundamental harmonic Nd:YAG laser dose at 800 mbar SF_6 pressure.

Figure **5a–c** illustrates the SEM micrograph of amorphous SiO_2 glass surface irradiated by 9 kJ/cm^2 of fundamental harmonic Nd:YAG laser dose in vacuum.

A comparison between the two irradiated regions shows that in SF_6 atmosphere, we have more significant morphology changes due to laser irradiation. The difference could be visible even on the boundary region.

LIBS of SiO_2 glass in a typical SF_6 atmosphere (800 mbar) was performed using a focused laser beam at 1064 nm to investigate the SF_6 decomposed components. Figure 6 shows the emission characteristic lines due to significant fluorine radicals during laser irradiation. Excited fluorine F(I) characteristic lines at 634.850 and 690.247 nm are detectable. It indicates that SF_6 molecules are decomposed due to the energetic electrons and excited atoms such as the reactive oxygen. The reactive oxygen O(I) and O(II) lines were also detected due to SiO_2 dissociation after the ejection into gaseous SF_6 atmosphere to contribute further SF_6 decomposition.

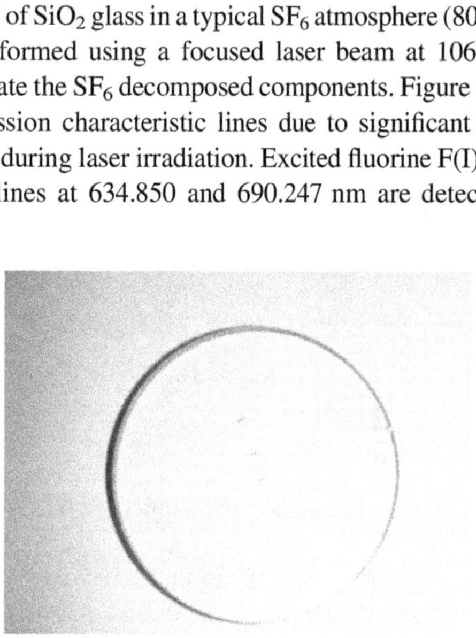

Fig. 3 Glass target after laser irradiation. The first two craters on top of the sample surface correspond to laser irradiation at 800 mbar SF_6 ambient gas and the next two craters correspond to laser shooting at vacuum condition

As mentioned in Eq. (3) due to linearity of that equation, we just need two points for obtaining the Boltzmann plot. The left-hand side of Eq. (3) versus E_i contains the slope of kT_e^{-1}. The coefficients A_{ij} and g_i are taken from the National Institute for Standards and Technology (NIST) atomic spectra database [8]. From F(I) lines in spectral region of 630– 695 nm, the plasma temperature was obtained as $\sim$2000 K. This temperature is much lesser than the metallic or other gaseous materials' plasma temperatures. The main reason for this may be the more electronegativity of fluorine with respect to metallic or other materials. Considering the excitation energy of the level of F(I), comparison with a metallic ion shows that the fluorine ion excitation energy is nearly twice greater than that of the metallic ion. So, a noticeable part of the laser energy is spent in ionizing the fluorine atoms and the remaining part of this energy belongs to the plasma temperature. In the case of metallic or other materials, the ionizer portion of the laser energy is lower than that of fluorine, so the plasma temperature is higher than that of fluorine.

From data of our previous work [1] and Eq. (3), we have: $Q_{\mathrm{Nd:YAG}} = 320{,}000$ mm^{-2}, $Q_{\mathrm{SHG\ \ Nd:YAG}} = 370{,}000$ -mm^{-2} and $Q_{\mathrm{ArF}} = 34{,}000$ mm^{-2}. So we conclude: $N_{0,\mathrm{Nd:YAG}}/N_{0,\mathrm{ArF}} = 9.4$, $N_{0,\mathrm{Nd:YAG}}/N_{0,\mathrm{SHG\ \ Nd:YAG}} = 0.9$ and $N_{0,\mathrm{SHG\ \ Nd:YAG}}/N_{0,\mathrm{ArF}} = 10.9$. These results show that although N_0 (solid solubility) is not a linear function of temperature, we can conclude that the temperature of the irradiated area of glass surface in the Nd:YAG laser exposure is several times greater in comparison with ArF laser irradiation. In addition, the temperature of the irradiated glass surface in SHG of Nd:Yag exposure is slightly greater than in Nd:YAG laser irradiation. Those results showed in another manner the plasma-induced ablation and dissociation of SF_6 molecules near the glass surface in Nd:YAG laser irradiation as well as photoablation and SF_6 molecule decomposition near the glass surface in ArF laser exposure.

Using Eq. (8) and ignoring the terms which contain (v/ c) ratio, considering the glass as the absorber material and F^- as the ionized particle and replacing their atomic number into the equation, we conclude that the penetration depth is very small (nearly zero) so that fluorine would be merely on the glass surface.

Conclusion

Glass surface modification occurred using Nd:YAG laser irradiation in SF_6 atmosphere. This kind of surface

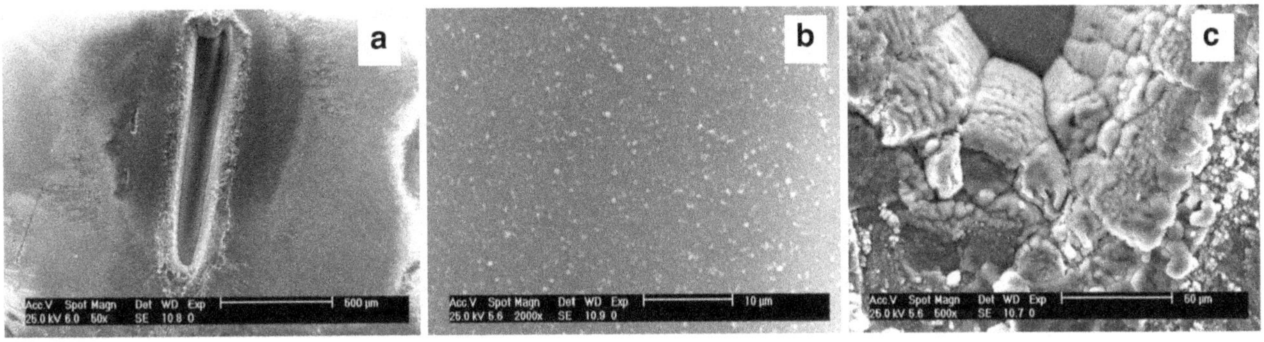

Fig. 4 **a** Crater created by laser irradiation on the glass target, **b** the crater's internal region surface morphology, **c** morphology of the crater boundary (fundamental harmonic of Nd:YAG laser, 9 kJ/cm^2, 800 mbar SF_6 atmosphere)

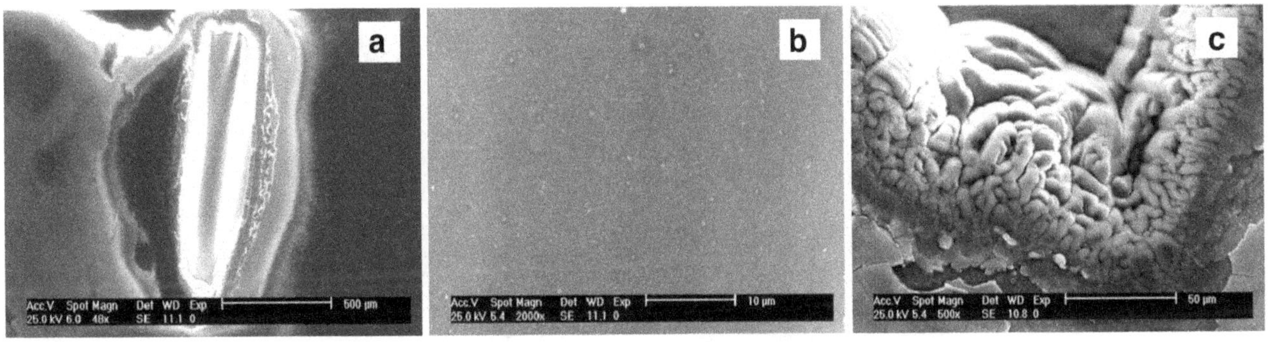

Fig. 5 **a** Crater created by laser irradiation on the glass target, **b** the crater's internal region surface morphology, **c** morphology of the crater boundary (fundamental harmonic of Nd:YAG laser, 9 kJ/cm^2, vacuum)

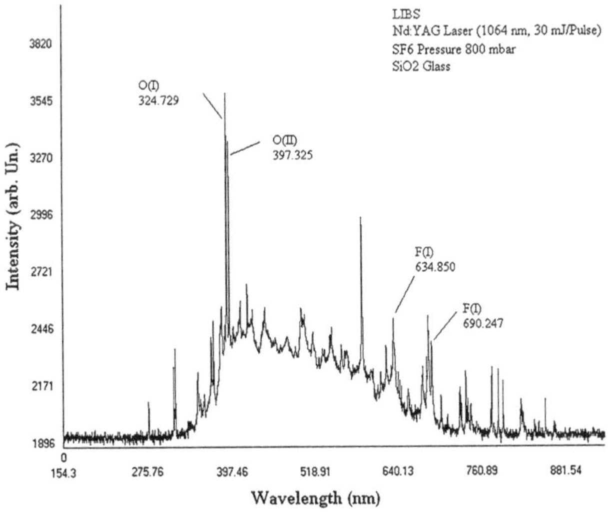

Fig. 6 Emission lines obtained by LIBS from amorphous SiO_2 glass surface within a cell filled with SF_6 at 800 mbar due to Nd:YAG laser exposure (10 ns, 1064 nm, 30 mJ/pulse)

modification embedded physical and chemical changes on the glass surface, which probably led to changes of the optical properties of irradiated glass. The laser irradiation on the SiO_2 glass surface in SF_6 ambient gas at 1064 nm wavelength was investigated. In fact, SF_6 decomposition by laser leads to the subsequent fluorine penetration into the SiO_2 layer. Laser breakdown spectroscopy (LIBS) shows that pulsed laser irradiation on SiO_2 surface in an SF_6 atmosphere induces gas decomposition. It produces the plasma containing reactive species to restructure the glass surface also. The plasma temperature creates at SF_6 atmosphere was lower than air ones. The main reason may be the more electronegativity of the fluorine comparison to other materials. The diffusion equation was used and it was shown that there were differences between irradiated zone temperatures at various laser wavelengths. Moreover, the use of penetration depth of ionized particle into a target showed that fluorine ions merely located on the glass surface and changed its surface chemical composition.

References

1. Dehghanpour, H.R., Parvin, P.: Jpn. J. Appl. Phys. **49**, 075803 (2010)
2. Sajad, B., Parvin, P., Bassam, M.A.: J. Phys. D: Appl. Phys. **37**, 3402 (2004)

3. Dehghanpour, H.R., Parvin, P.: Appl. Phys. B. **101**, 611 (2010)
4. Loper, G.L., Tabat, M.D.: Appl. Phys. Lett. **46**, 654 (1985)
5. Miziolek, A.W., Palleschi, V., Schechter, I.: Laser Breakdown Spectroscopy (LIBS) Fundamental and Applications. Cambridge University Press, New York (2006)
6. Hamilton, D.J., Howard, W.G.: Basic Integrated Circuit Engineering. McGraw-Hill, New York (1975)
7. Cember, H.: Introduction to Health Physics. Pergamon Press, USA (1983)
8. NIST electronic data base. http://physics.nist.gov/PhysRefData/ASD/Html/lineshelp.html

Self-focusing of a high-intensity laser pulse by a magnetized plasma lens in sub-relativistic regime

Mehdi Etehadi Abari[1] · Mahsa Sedaghat[1] · Mohammad Taghi Hosseinnejad[1]

Abstract Interaction of high power circularly polarized short laser pulses with a cold underdense magnetized thin plasma lens is analyzed in the sub-relativistic regime. The magnetic field is applied along the direction of the laser field propagation. The evolution equation of the beam spot size is derived and solved by making use of the variational principle approach method. The theoretical investigations reveal that not only the magnetized plasma lens more sufficiently decreases the laser spot size, but also the left-handed circularly polarized beam is more effectively focused by a magnetized plasma lens compared to the right-handed circularly polarized beam.

Keywords Laser plasma interaction · Magnetized underdense plasma · Variational principal approach method · Laser spot size

Introduction

Studying the evolution of high-intensity laser pulses propagating through underdense plasmas is an active area of research due to its importance to plasma accelerators [1], radiation schemes [2], X-ray lasers [2, 3], and etc. The problem of self-focusing of laser light in collisionless plasmas has been the subject of intense study in last years [4, 6]. However; laser pulse interaction with magnetized plasmas is, respectively, new and of interest to researchers and therefore the related physical processes have been studied. It has been seen that intense magnetic fields are generated in the laser plasma interaction [5]. These self-generated (or externally applied) transverse and axial magnetic fields affect the propagation of laser pulses in plasmas since the canonical momentum in a magnetized plasma is not conserved [6, 7]. Furthermore, laser pulse interaction with magnetized plasmas finds various applications for different branches such as: nonlinear interaction [8], wakefield excitation [9], modulation instability [10, 11], laser fusion schemas [12, 13], and fast ignition schemes in inertial confinement fusion (ICF) [14]. Moreover, self-generated magnetic fields in the corona region have been studied for intense laser pulse interaction with magnetized plasmas.

Self-focusing in plasma can occur through thermal, relativistic and ponderomotive effects [15]. Thermal self-focusing is due to collisional heating of plasma exposed to electromagnetic radiation and the rise in temperature induces a hydrodynamic expansion, which leads to an increase in the index of refraction and further heating [18]. Relativistic self-focusing is caused by the mass increase in electrons traveling at speed approaching the speed of light which modifies the plasma refraction index [4, 6, 19]. Ponderomotive self-focusing is caused by the ponderomotive force, which pushes electrons away from the region where the laser beam is more intense, therefore increasing the refraction index and inducing the focusing effect [20].

In the present paper we analyze this problem with the assumption that the plasma is cold so that before the passage of the laser radiation, the plasma electrons are at rest and the magnetic field does not affect them [21, 22]. Clearly, the ponderomotive force [23] tries to expel electrons; however, since the ions are assumed motionless (because of their inertia and the shortness of the pulse), expelled electrons lead to a charge imbalance, creating an electrostatic field which attracts the electrons back into this

✉ Mehdi Etehadi Abari
ettehadimehdii@gmail.com

[1] Young Researchers and Elites Club, Science and Research Branch, Islamic Azad University, Tehran, Iran

channel. The balance between the radial electrostatic and ponderomotive forces determines the electron density within the channel. In our study it will be assumed that the electrons are cold. This is justified if the quiver velocity of the electron motion in the wave is large compared to the electron thermal speed [24]. In additional, we consider the effect of nonlinear electric field in the dielectric function of plasma. It arises on account of redistribution of electron density in the transverse direction. Also because of the time duration of the laser pulse is shorter than the electrons plasma relaxation time ($t \prec t_{\varepsilon}$), we only consider the relativistic self-focusing nonlinearity. To investigate nonlinear interplay instabilities in the laser–plasma interaction it is desirable to obtain differential equations for the evolution of the macroscopic quantities that characterize the laser beam profile [25, 26], including the amplitude, spot size, phase, curvature radius, and centroid. There are various methods for attempting to obtain such envelope equations, including the variation method [27], the moment method [28], and the source dependent expansion (SDE) technique [29]. We can use these mentioned techniques to study the nonlinear relativistic self-focusing phenomena with taking into account the relativistic mass corrections. Relativistic self-focusing is caused by the mass increase of the electrons traveling at speed and approaching the speed of light which modifies the plasma refractive index [4, 6, 30–38]. As we know, the propagation of a laser beam without diffraction is an essential factor to get better application [39]. According to this reason, in this work we use a thin magnetized plasma lens to compress and focus a laser beam outside of the plasma in the vacuum region. The plasma lens is a thin plasma layer used to focus particle beams and laser pulses. By making use of a thin plasma lens, it is possible to compress and focus a laser beam outside of the plasma in the vacuum region. Plasma lens is not damaged by an intense laser beam which is an advantage of the plasma lens compared with an optical lens [40]. Here, we show that for high-intensity laser pulses propagating in the underdense magnetized plasma lens, in the sub-relativistic regime, the magnetic field sufficiently decreases the minimum spot size of the radiation field and increases the focal length of the laser beam for the circular polarization. The focal length is defined as the distance between the plasma lens and the laser beam focus point.

Using the magnetized plasma as a thin lens for focusing high-intensity lasers, we extend the variation principal technique to solve the nonlinear wave equation and analyze the self-focusing phenomenon in the presence of the static magnetic field. The focusing of intense laser beam in magnetized plasma lens has been studied by Jha et al. in [35]. Here, the propagation of an intense circularly polarized laser beam in axially magnetized plasma is analyzed. In this reference plasma is assumed to be cold so that before the passage of the laser radiation, the plasma electrons are at rest and the magnetic field does not affect them. Here, the effect of an axial magnetic field on the self-focusing property of the laser beam is discussed and an expression for the circular power for self-focusing is obtained. To compare our current work with this reference it should be noted that we consider the dependence of a laser pulse trial function on the group velocity dispersion in our calculation and by considering the circular polarization for the laser beam propagating through a magnetized plasma and taking into account the sub-relativistic effects, we obtain the modified laser spot size, minimum laser spot size, and the focal length parameters with the variations of the external magnetic field for right- and left-handed circular polarizations. It is important to note that in Jha et al. [35] only the variations of the normalized spot size versus normalized propagation distance for a constant laser pulse strength in right- and left-handed polarizations were presented. As a comparison, in our current work by considering the group velocity dispersion effect in our calculation the variations of the normalized beam spot size as a function of the normalized cyclotron frequency for different laser pulse strength in right- and left-handed circular polarizations are obtained. It is clear that in left-hand circular polarization, the normalized beam spot size decreased more with increasing the laser pulse strength. It should be noted that in our current work and in the Jha et al. [35], the pulse dimension is in the order of plasma wavelength and it assure us to neglect the instability effect [41].

This paper is arranged as follows: in "Analytical analysis", the current density generated in the plasma lens due to the interaction of the intense laser beam with an axially magnetized plasma is obtained. In "Spot size evolution", by making use of the variation principal approach, the envelope equation for laser spot size is solved, and the minimum spot size and focal length are obtained. Finally a summery and conclusion is presented in "Summery and conclusion".

Analytical analysis

We assume the laser beam propagating through magnetized plasma has circular polarization. In this case, the electric radiation field is given by:

$$\boldsymbol{E} = (\boldsymbol{e}_x \pm i\boldsymbol{e}_y)\left(E(r,z,t)e^{i(kz-\omega t)} + \text{c.c.}\right), \tag{1}$$

where $\boldsymbol{e}_x$ and $\boldsymbol{e}_y$ refer to the unit vectors in x and y directions, $E(r,z,t), k$ and ω are amplitude, wave number, and laser field frequency, respectively. Furthermore, the sings

$+$ and $-$ are used to define the left and right-handed circularly polarized beams. In addition, c.c. denotes the complex conjugate. The equations of electrons motion in the presence of the external magnetic field B_0 in $\mathbf{x}$ and $\mathbf{y}$ directions are:

$$\frac{dv_x}{dt} = -\frac{eE_x}{m_e} - \frac{eB_0}{m_e c}v_y, \tag{2}$$

$$\frac{dv_y}{dt} = -\frac{eE_y}{m_e} + \frac{eB_0}{m_e c}v_x. \tag{3}$$

Here, the direction of the external magnetic field is along the $\mathbf{z}$ axis ($\boldsymbol{B} = B_0\hat{z}$). Also, m, and c are the electron mass and the light speed, respectively. Assuming that the growth rate is slow compared to the pulse frequency, we find:

$$V_{\perp x} = \frac{ie}{m_e\omega}\left(E_x + i\left(\frac{\omega_{ce}}{\omega}\right)E_y\right)\left(1 - \left(\frac{\omega_{ce}}{\omega}\right)^2\right)^{-1}, \tag{4}$$

$$V_{\perp y} = \frac{ie}{m_e\omega}\left(E_y - i\left(\frac{\omega_{ce}}{\omega}\right)E_x\right)\left(1 - \left(\frac{\omega_{ce}}{\omega}\right)^2\right)^{-1}, \tag{5}$$

where $\omega_{ce} = \frac{eB_0}{m_e}$ is the electron cyclotron frequency. Now, using the Maxwell equation, the wave equation of the electric field is obtained as follows:

$$\left(\nabla^2 - \frac{1}{c^2}\frac{\partial^2}{\partial t^2}\right)\mathbf{E}_\perp = \frac{4\pi}{c^2}\frac{\partial \mathbf{J}_\perp}{\partial t}, \tag{6}$$

where $\mathbf{J}_\perp = -ne\mathbf{v}_\perp$ is the plasma transverse current density with respect to $\mathbf{E}$, $\mathbf{v}$ and n are the electron velocity and density.

Here, by defining the normalized laser pulse strength as $\mathbf{A}_\perp = \frac{eE_\perp}{m_e\omega c}$ and by considering the transverse electrons velocity as $\mathbf{v}_\perp = \boldsymbol{e}_x v_x + \boldsymbol{e}_y v_y$, $\left(v_\perp = \sqrt{v_x^2 + v_y^2}\right)$ in Eq. (6), then the equation of the laser vector potential is obtained as:

$$\nabla^2\boldsymbol{A}_\perp - \frac{1}{c^2}\frac{\partial^2}{\partial t^2}\boldsymbol{A}_\perp = \frac{\omega_{pe}^2}{c^2}\frac{1}{\gamma}\left(\frac{\omega}{\omega \pm \omega_{ce}}\right)\boldsymbol{A}_\perp, \tag{7}$$

where $\omega_{pe} = (4\pi ne^2/m)^{1/2}$ is the electron plasma frequency and $\gamma = \left(1 + \frac{p_\perp^2}{m_e^2c^2}\right)^{\frac{1}{2}}$ is the relativistic factor. It should be noted that in this equation the laser vector potential ($\boldsymbol{A}_\perp$) is normalized to $m_e c^2/e$.

It is obvious when the laser pulse strength magnitude is rather near to unity, the sub-relativistic effects should be taken into account and the plasma particles will experience a sub-relativistic nonlinear force. Here, the strength of wake field can exceed to $10^9 \frac{v}{m}$ and we have:

$$\frac{1}{\gamma} \cong 1 - \frac{v_\perp^2}{c^2} \cong 1 - \frac{|A_\perp|^2}{2}\left(\frac{\omega}{\omega \pm \omega_{ce}}\right)^2. \tag{8}$$

Then the final equation will be:

$$\nabla^2\boldsymbol{A}_\perp - \frac{1}{c^2}\frac{\partial^2}{\partial t^2}\boldsymbol{A}_\perp = \frac{\omega_{pe}^2}{c^2}\left(1 - \frac{|A_\perp|^2}{2}\left(\frac{\omega}{\omega \pm \omega_c}\right)^2\right)\left(\frac{\omega}{\omega \pm \omega_{ce}}\right)\boldsymbol{A}_\perp. \tag{9}$$

Now, we can write the transverse normalized potential vector $\boldsymbol{A}_\perp$ as the product of an envelope and a phase,

$$\boldsymbol{A} = (\boldsymbol{e}_x \pm i\boldsymbol{e}_y)\left(\frac{a}{2}e^{i(kz-\omega t)} + \text{c.c.}\right), \tag{10}$$

where $a = \frac{e|E_\perp|}{m_e\omega c}$ is the magnitude of laser pulse strength; ω, k satisfy the circularly polarized wave dispersion relation as:

$$\frac{\omega^2}{c^2} - k^2 = \frac{\omega_{pe}^2/c^2}{1 \pm \omega_{ce}/\omega}. \tag{11}$$

Using a mathematical transformation from the coordinate variables (t, z) to variables (ψ, τ) in which $\tau = z$ and $\psi = t - z/v_g$, where v_g is the group velocity of the radiation field, the envelope equation of $a(\psi, \tau)$ reduces to:

$$\begin{aligned} &2ik\frac{\partial a}{\partial \tau} + \nabla_\perp^2 a - \frac{\omega_{pe}^2/c^2}{(1 \pm \frac{\omega_{ce}}{\omega})}\left(1 - \frac{|a|^2/4}{(1 \pm \frac{\omega_{ce}}{\omega})^2}\right)a \\ &+ \left(\frac{1}{v_g^2} - \frac{1}{c^2}\right)\frac{\partial^2 a}{\partial \psi^2} - \frac{2}{v_g}\frac{\partial^2 a}{\partial \psi \partial \tau} + \left(\frac{\omega^2}{c^2} - k^2\right)a \\ &+ \left(\frac{2i\omega}{c^2} - \frac{2ik}{v_g}\right)\frac{\partial a}{\partial \psi} + \frac{\partial^2 a}{\partial \tau^2} \\ &= 0. \end{aligned} \tag{12}$$

In this equation, both the first and second terms are related to the diffraction of the pulse. The third term represents the reaction of the plasma and the features of the medium. This term shows the relativistic effects in the dispersion relation of the plasma. The other terms denote the group velocity dispersion. In this equation, the group velocity is obtained by:

$$v_g = c\frac{\sqrt{\omega^2 + (\omega/\omega_{ce})\omega_{pe}^2}}{\omega + \omega_{pe}^2/2\omega_{ce}} \cong c\frac{\sqrt{1 + \omega_{pe}^2/\omega\omega_{ce}}}{1 + \omega_{pe}^2/2\omega\omega_{ce}}. \tag{13}$$

Note that in Eq. (12), in the term $\frac{\omega_{pe}^2/c^2}{(1 \pm \frac{\omega_{ce}}{\omega})}\left(1 - \frac{|a|^2/4}{(1 \pm \frac{\omega_{ce}}{\omega})^2}\right)a$, the second part has large value and we can neglect the first part. Therefore, the simplest equation including self-focusing in the magnetized plasma is:

$$\begin{aligned} &2ik\frac{\partial a}{\partial \tau} + \frac{\varepsilon^2}{c^2}\frac{\partial^2 a}{\partial \psi^2} + \nabla_\perp^2 a + \left(\frac{2i\omega}{c^2} - \frac{2ik}{v_g}\right)\frac{\partial a}{\partial \psi} \\ &+ \frac{\omega_{pe}^2}{c^2}\frac{|a|^2/4}{(1 \pm \frac{\omega_{ce}}{\omega})^3}a = 0, \end{aligned} \tag{14}$$

where $\varepsilon = \frac{1}{c^2}\frac{\{\omega_{pe}^4/4\omega\omega_{ce}\}}{(\omega_{pe}^2 + \omega\omega_{ce})}$. This equation can be approximately solved by using the variational principle [24]. It is done by writing down the Lagrangian density (l) for Eq. 14 as follows:

$$l = ik\left(a\frac{\partial a^*}{\partial \tau} - a^*\frac{\partial a}{\partial \tau}\right) + \frac{\varepsilon^2}{c^2}\frac{\partial a^*}{\partial \psi}\frac{\partial a}{\partial \psi} + \nabla_\perp a^*.\nabla_\perp a - i\left(\frac{\omega}{c^2} - \frac{k}{v_g}\right)\left(a\frac{\partial a^*}{\partial \psi} - a^*\frac{\partial a}{\partial \psi}\right) - \frac{\omega_{pe}^2}{8c^2}\frac{a^2 a^{*2}}{\left(1 \pm \frac{\omega_{ce}}{\omega}\right)^3}, \tag{15}$$

i.e., Eq. (12) is the Euler–Lagrange equation. With this equation we can minimize the action integral $S = \int l d\tau dx_\perp d\psi$ for variations of a^* and a. Next we use the following trial function:

$$a = a(\tau)e^{-i\varphi(\tau)}e^{i\psi^2/\eta(\tau)}e^{-\psi^2/L^2}e^{ik_0 r^2/2R(\tau)}e^{-r^2/W^2}. \tag{16}$$

Here, L, W, a, ϕ, η, and R, being real quantities, are the pulse length, spot size, amplitude, phase, chirp, and radius of curvature of the Gaussian laser beam, respectively. Substituting this trial function into Eq. (15) and integrating over ψ and $x_\perp (dx_\perp = rdr)$, we obtain the reduced Lagrangian density:

$$l \equiv \frac{4}{\sqrt{\pi}}\int_0^{+\infty} rdr \int_{-\infty}^{+\infty} l d\psi = -\sqrt{2}a^2 k\left(\frac{d\varphi}{d\tau} + \frac{L^2}{4\eta^2}\frac{d\eta}{d\tau} + \frac{kW^2}{4R^2}\frac{dR}{d\tau}\right) + \frac{\varepsilon^2}{c^2}\frac{a^2}{\sqrt{2}}\left(\frac{1}{L^2} + \frac{L^2}{\eta^2}\right) + \frac{a^2}{\sqrt{2}}\left(\frac{2}{W^2} + \frac{k^2W^2}{2R^2}\right) - \frac{a^4}{32W^2L}\frac{\omega_{pe}^2/c^2}{\left(1 \pm \frac{\omega_{ce}}{\omega}\right)^3}. \tag{17}$$

Spot size evolution

We use the Euler–Lagrange equations of the reduced Lagrangian density to drive the evolution of W in τ. We consider the ideal case where a thin magnetized plasma lens with a thickness Δ is placed at the vacuum waist ($\tau = 0$) of a Gaussian laser pulse with the spot size W_0. The laser exits the plasma at $\tau = \Delta$ and reaches to its focus, where it has the minimum spot size $W_{\min}$, at $\tau = f$. If the pulse is assumed to remain Gaussian and finite pulse length effects are neglected, i.e., L, and η are assumed to remain fixed, then the evolution equation for spot size W inside the plasma can be derived from the reduced Lagrangian given by Eq. (17). Varying φ leads to power conservation,

$$\frac{da^2}{d\tau} = 0. \tag{18}$$

Furthermore, varying R relates R and $dW/d\tau$ as:

$$\frac{W}{R} = \frac{dW}{d\tau}, \tag{19}$$

and varying W gives another equation for R and W as:

$$-\frac{k^2}{\sqrt{2}}\frac{a^2 W}{R^2}\frac{dR}{d\tau} + \frac{a^2}{\sqrt{2}}\left(\frac{-4}{W^3} + \frac{k^2W}{R^2}\right) + \frac{\omega_{pe}^2/c^2}{\left(1 \pm \frac{\omega_{ce}}{\omega}\right)^3}\frac{a^4}{16W^3L} = 0. \tag{20}$$

Combining Eqs. (19) and (20), we can obtain the evolution equation for the laser beam spot size W as:

$$\frac{d^2W}{d\tau^2} = -\frac{4}{k^2W^3}\left(\frac{P/P_c}{\left(1 \pm \frac{\omega_{ce}}{\omega}\right)^3} - 1\right), \tag{21}$$

where $P/P_c = \sqrt{2}a^2\omega_{pe}^2 W_0^2/(64c^2)$ is independent of τ. The solution of Eq. (21) with the initial condition of $W = W_0$ and $dW/d\tau = 0$ at $\tau = 0$, gives laser beam spot size inside the magnetized plasma lens as:

$$W = W_0\sqrt{1 - \left(\frac{P/P_c}{\left(1 \pm \frac{\omega_{ce}}{\omega}\right)^3} - 1\right)\left(\frac{\tau^2}{\tau_R^2}\right)}, \tag{22}$$

where $\tau_R \equiv kW_0^2/2$ is the vacuum Raleigh length of the laser pulse. It is found that spot size will eventually go to zero if $P/Pc > 1$. When the laser beam exits the plasma, spot size evolves as:

$$\frac{d^2W}{d\tau^2} = \frac{4}{k^2W^3}. \tag{23}$$

If the laser beam is focused in the focal length f, and the minimum spot size $W_{\min}$, we can obtain laser spot size outside of the plasma lens in the vacuum region by solving Eq. (23) as:

$$W = W_{\min}\sqrt{1 + \left(\frac{\tau - f}{\tau_R}\right)^2 \frac{W_0^4}{W_{\min}^4}}. \tag{24}$$

By matching W and $dW/d\tau$ from Eqs. (22) and (24), we can obtain the minimum spot size $W_{\min}$ and focal length f as follows:

$$W_{\min} = W_0\sqrt{\frac{1 - \delta^2}{1 + \left(\frac{P/P_c}{\left(1 \pm \frac{\omega_{ce}}{\omega}\right)^3} - 1\right)\delta^2}}, \tag{25}$$

$$f = \Delta\frac{\frac{P/P_c}{(1 \pm \omega_c/\omega)^3}}{1 + \left(\frac{P/P_c}{\left(1 \pm \frac{\omega_{ce}}{\omega}\right)^3} - 1\right)\delta^2}, \tag{26}$$

where $\delta = (\Delta/\tau_R)(P/P_c - 1)^{1/2}$ is the normalized lens thickness.

Although some methods have been already studied to obtain the variations of laser spot size with plasma distance such as SDE (source dependence expansion) [35, 36], but these methods present only the variation of laser spot size with distance. In the present method (Lagrangian density), as we see, other parameters such as the variation of laser spot size with the variations of the plasma distance and the external magnetic field for right and left-handed circular polarizations are obtained. Furthermore, the variations of the minimum laser spot size and laser focal length versus the variations of the external magnetic field and laser pulse strength for right and left-handed circular polarizations are presented. Also, it is noticeable to note that we can study evaluation of spot size versus other parameter such as external magnetic field, propagation length and plasma density through SDE method.

Figure 1 shows the plot of the normalized beam spot size $\frac{W}{W_0}$ as a dimensionless parameters ω_{ce}/ω_0 for right and left circular polarizations at the fixed point $\frac{\tau}{\tau_R} = 0.03$ inside the plasma lens. The laser source is Nd:Yag with frequency $f_0 = 1.88 \times 10^{15}\text{s}^{-1}$, wavelength $\lambda = 1$ μm, and initial beam spot size $W_0 = 100$ μm. This figure predicts that the self-focusing property is enhanced by increasing the external magnetic field for left-handed circular polarization. In this figure we see that the maximum reduction for the left-handed circularly polarized beam propagating through the magnetized pure electron plasma lens is smaller than $0.15W_0$. On the other hand, this figure shows that the right-handed circularly polarized beam diffraction is increased slightly by increasing the magnetic field.

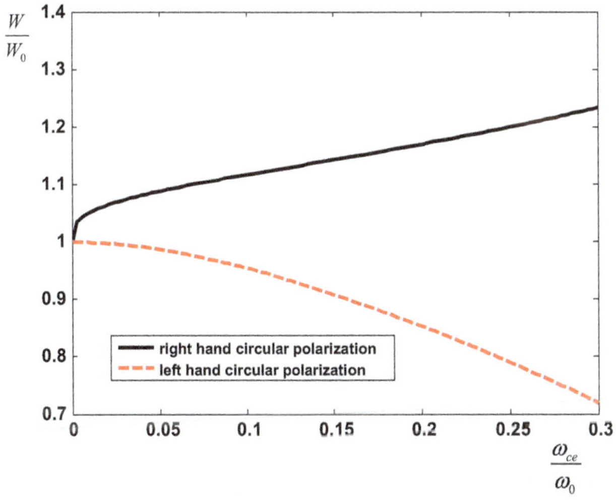

Fig. 1 Variation of the normalized spot size $\frac{W}{W_0}$ as a dimensionless parameter ω_{ce}/ω_0 for right (*solid line*) and *left circular* polarized (*dashed line*) laser beams. The parameters are chosen as $\omega_p/\omega_0 = 0.09$, $W_0 = 100$ μm, $\omega_0 = 1.88 \times 10^{15}\,\text{s}^{-1}$, laser wavelength $\lambda = 1$ μm and $a = 0.7$, for a fix point $\frac{\tau}{\tau_R} = 0.03$ inside the plasma lens with thickness $\Delta = 1$ mm

Figure 2 illustrates the normalized beam spot size $\frac{W}{W_0}$ as a function of the dimensionless parameter $\frac{\omega_{ce}}{\omega_0}$ for different laser pulse strength with the right-handed circular polarization at the fixed point $\frac{\tau}{\tau_R} = 0.03$ inside the plasma lens. It is clear that the diffraction property of the laser beam is enhanced strongly with increasing the strength of the external magnetic field. In Fig. 3 the normalized spot size $\frac{W}{W_0}$ is plotted as a function of the normalized parameter $\frac{\omega_{ce}}{\omega_0}$ for different laser pulse strength with left-handed circular polarization. It is seen, in contrast to the right-handed circularly polarized beam, the normalized beam spot size decreases with increasing the strength of the external magnetic field. In the above figures, it should be noted that the application of external magnetic field enhances the quality of self-focusing for left-hand polarization and reduces it for right-hand polarization. Physically, the left-hand wave drives electrons in the direction of their cyclotron motion. In this polarization the increase in magnetic field causes the increase in transverse velocity of electrons and it leads to increase in nonlinear current density or nonlinearity of plasma. Furthermore, the rotation sense of the right-hand polarization is opposite to the electrons cyclotron motion and the velocity of the electrons or consequently, the nonlinearity of plasma medium decreases with increase in external magnetic field. Finally, we conclude that, self-focusing for left-handed circularly laser light in magnetized plasma is stronger toward the right-handed circular polarization.

Figures 4 and 5 show the normalized minimum spot size $\frac{W_{\min}}{W_0}$ of a laser beam as a function of the dimensionless parameter $\frac{\omega_{ce}}{\omega_0}$ for different laser pulse strength with right and left-handed circular polarizations, respectively. It is shown that by increasing laser pulse strength the initial

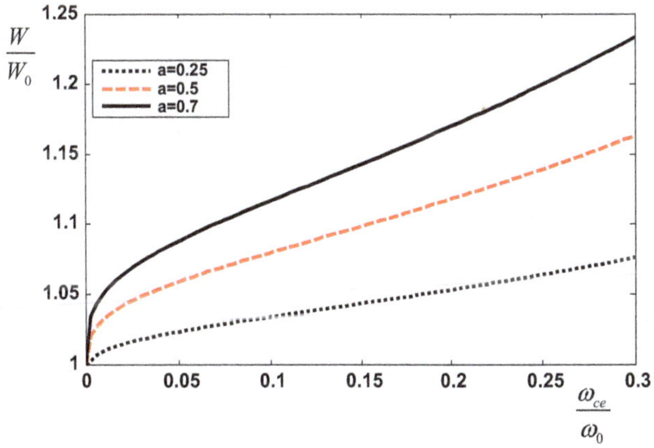

Fig. 2 Variation of normalized spot size $\frac{W}{W_0}$ as a dimensionless parameter ω_{ce}/ω_0 for *right circularly* polarized laser beam with different laser pulses strength ($a = 0.25$ *dotted line*), ($a = 0.5$ *dashed line*) and ($a = 0.7$ *solid line*), in the fix point $\frac{\tau}{\tau_R} = 0.03$ inside the plasma lens. The other parameters are chosen as Fig. 1

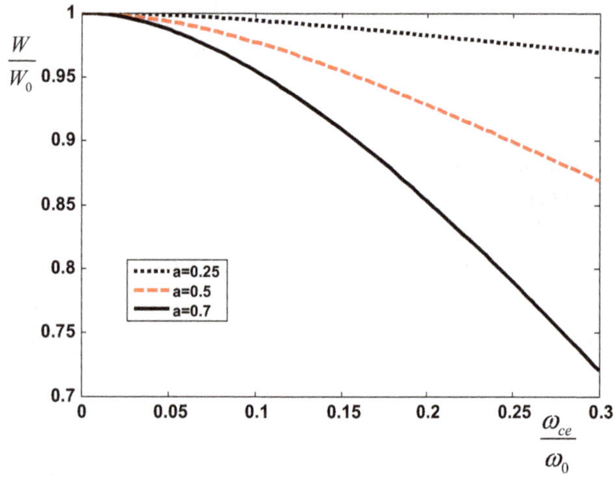

Fig. 3 Variation of normalized spot size $\frac{W}{W_0}$ as a dimensionless parameter ω_{ce}/ω_0 for *left circularly* polarized laser beam with different laser pulses strength ($a = 0.25$ *dotted line*), ($a = 0.5$ *dashed line*) and ($a = 0.7$ *solid line*), in the fix point $\frac{\tau}{\tau_R} = 0.03$ inside plasma lens. The other parameters are chosen as Fig. 1

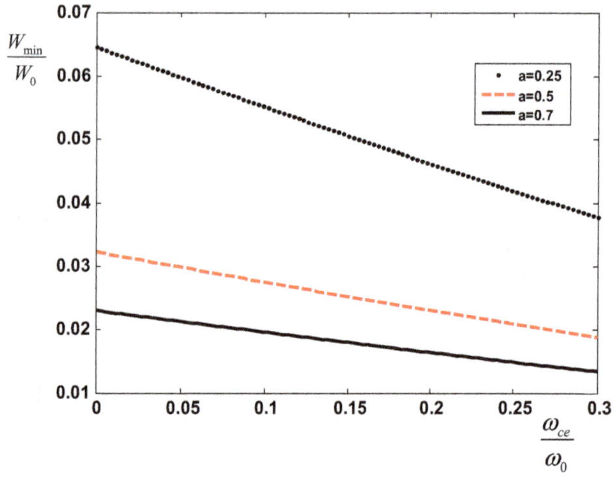

Fig. 4 Variation of the minimum normalized spot size $\frac{W_{\min}}{W_0}$ as a dimensionless parameter ω_{ce}/ω_0 for *right circularly* polarized laser beam with different strengths ($a = 0.25$ *dotted line*), ($a = 0.5$ *dashed line*) and ($a = 0.7$ *solid line*) with plasma lens thickness $\Delta = 1$ mm. The other parameters are chosen as Fig. 1

minimum spot size is decreased for the left-handed circular polarization, while it is increased for the right-handed circular polarization. Figures 6 and 7 show the plot of the normalized focal length variations as a function of the normalized parameter $\frac{\omega_{ce}}{\omega_0}$ for different laser pulse strength with right and left-handed circular polarizations, respectively. These figures show that we can focus a weakly relativistic ($a < 1$) right-handed circularly polarized laser beam to a long focal length by applying an external magnetic field, while it is plausible to focus a rather strong ($a \approx 1$) left-handed circularly polarized beam to a long focal length by making use of a magnetized plasma lens.

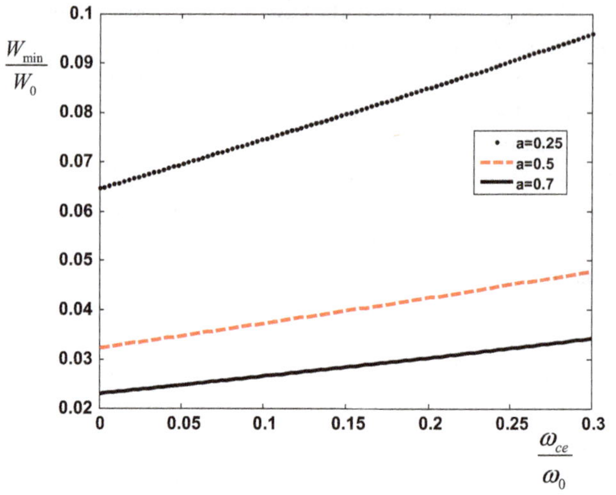

Fig. 5 Variation of the minimum normalized spot size $\frac{W_{\min}}{W_0}$ as a dimensionless parameter ω_{ce}/ω_0 for *left circularly* polarized laser beam with different strengths ($a = 0.25$ *dotted line*), ($a = 0.5$ *dashed line*) and ($a = 0.7$ *solid line*) with plasma lens thickness $\Delta = 1$mm. The other parameters are chosen as Fig. 1

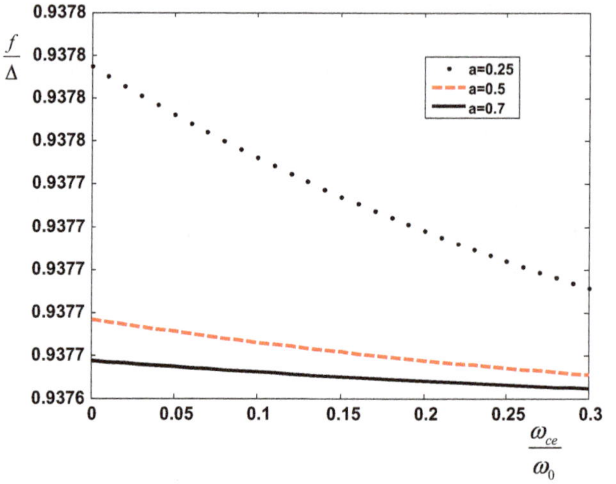

Fig. 6 Variation of the normalized focal length $\frac{f}{\Delta}$ as a dimensionless parameter ω_{ce}/ω_0 for right circularly polarized laser beam with different strengths ($a = 0.25$ *dotted line*), ($a = 0.5$ *dashed line*) and ($a = 0.7$ *solid line*) with plasma lens thickness $\Delta = 1$mm. The other parameters are chosen as Fig. 1

Summary and conclusion

We have derived analytical formulas for the spot size evolution of a high-intensity laser pulse propagating in a thin magnetized plasma lens in the sub-relativistic regime. Our results show that how $W_{\min}$ and f depend on plasma lens thickness and the normalized laser power P/P_c. Clearly, the magnetic field modifies the plasma density and intensifies the nonlinear effects. In such a case, the self-focusing of the left-handed circularly polarized beam is

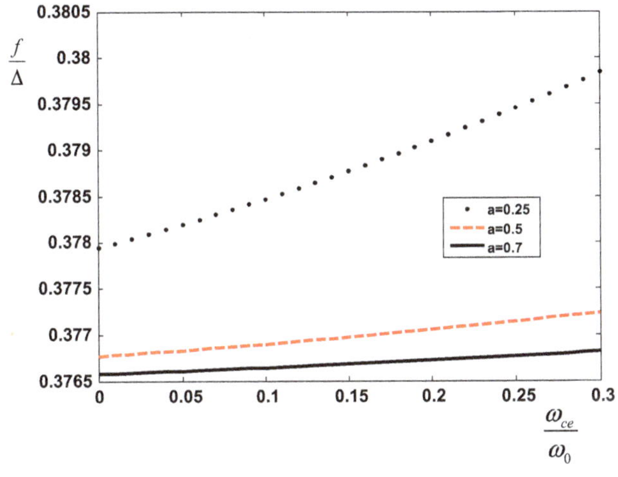

Fig. 7 Variation of the normalized focal length $\frac{f}{\Delta}$ as a dimensionless parameter ω_{ce}/ω_0 for *left circularly* polarization laser beam with different strengths ($a = 0.25$ *dotted line*), ($a = 0.5$ *dashed line*) and ($a = 0.7$ *solid line*) with plasma lens thickness $\Delta = 1$ mm. The other parameters are chosen as Fig. 1

enhanced by increasing the external magnetic field, while this property is diminished for the right-handed circularly polarized beam. The analytical investigation predicts the increase of the laser beam focal length and the decrease of the minimum spot size by increasing the external magnetic field for a left-handed circularly polarized case. As a final conclusion, we can focus a strong left-handed circularly polarized laser beam to a long focal length by applying the external magnetic field on the plasma lens.

Authors' contributions The plasma physics of this article (basic plasma physics, achieving equations) and writing the manuscript were obtained by MEA (Corr author), the analytical simulations was done by MTH (second author), and finally the valuable help in responding reviewers process was done by NP (third author).

References

1. Tajima, T., Dawson, J.M.: Laser electron accelerator. Phys. Rev. Lett. **43**, 267 (1979)
2. Amendt, P., Eder, D.C., Wilks, S.C.: X-ray lasing by optical-field-induced ionization. Phys. Rev. Lett. **66**, 2589 (1991)
3. Lemoff, B.E., Yin, G.Y., Gordan III, C.L., Harris, S.E.: Demonstration of a 10-Hz femtosecond-pulse-driven XUV Laser at 41.8 nm in Xe IX. Phys. Rev. Lett. **74**, 1574 (1995)
4. Litvak, A.G.: Finite-amplitude wave beams in a magnetoactive plasma. Zh. Eksp. Teor. Fiz. **57**, 629–636 (1968)
5. Najmudin, Z., Tatarakis, M., Pukhov, A., et al.: Measurements of the inverse faraday effect from relativistic laser interactions with an underdense plasma. Phys. Rev. Lett. **87**, 215004 (2001)
6. Sprangle, P., Esarey, E.: Generation of stimulated backscattered harmonic radiation from intense-laser interactions with beams and plasmas. Phys. Rev. A. **45**, 5872 (1976)
7. Deutch, C., Furukawa, H., Mima, K., Murakami, M., Nishihara, K.: Interaction physics of the fast ignitor concept. Phys. Rev. Lett. **77**, 2483 (1996)
8. Wadhwani, N., Kumar, P., Jha, P.: Nonlinear theory of propagation of intense laser pulses in magnetized plasma. Phys. Plasmas **9**, 263 (2002)
9. Ren, C., Mori, W.B.: Nonlinear and three-dimensional theory for cross-magnetic field propagation of short-pulse lasers in underdense plasmas. Phys. Plasmas. **11**, 1978 (2004)
10. Jha, P., Kumar, P., Paj, G., Upadhyaya, A.K.: Modulation instability of laser pulse in magnetized plasma. Phys. Plasmas **12**, 123104 (2005)
11. Rios, L.A., Shukla, P.K.: Modulational instabilities of electromagnetic electron cyclotron waves in a dense magnetized plasma. Phys. Plasmas. **15**, 074501 (2008)
12. Burnett, H., Corkum, P.B.: Cold-plasma production for recombination extreme-ultraviolet lasers by optical-field-induced ionization. J. Opt. Soc. Am. B **6**, 1195 (1989)
13. Tabak, M., Hammer, J., Glinsky, M.E., Kruer, W.L., Koson, R.J.: Ignition and high gain with ultra-powerful lasers. Phys. Plasmas **1**, 1626 (1994)
14. Borghesi, M., Mackinnon, A.J., Gailard, R., Willi, O., Pukhov, A., Meyerter-Vehn, J.: Large quasistatic magnetic fields generated by a relativistically Intense laser pulse propagating in a preionized Plasma. Phys. Rev. Lett. **80**, 5137 (1998)
15. Mori, W.B., Joshi, C., Dawson, J.M., Forslund, D.W., Kindel, J.M.: Evolution of self-focusing of intense electromagnetic waves in plasma. Phys. Rev. Lett. **60**, 1298 (1988)
16. Etehadi Abari, M., Shokri, B.: Study of nonlinear ohmic heating and ponderomotive force on the self-focusing and defocusing of a laser pulse in underdense plasmas. Phys. Plasmas. **19**, 113107 (2012)
17. Etehadi Abari, M., Sedaghat, M., Shokri, B.: Self-focusing and de-focusing of Gaussian laser beams in collisional underdense magnetized plasmas with considering the nonlinear ohmic heating and ponderomotive force effects. Phys. Plasmas. **22**, 103112 (2015)
18. Perkins, F.N., Valeo, E.J.: Thermal self-focusing of electromagnetic waves in plasmas. Phys. Rev. Lett. **32**, 1234 (1974)
19. Max, C., Arons, J., Langdon, A.B.: Self-modulation and self-focusing of electromagnetic waves in plasmas. Phys. Rev. Lett. **33**, 209 (1974)
20. Gupta, D.N., Sharma, A.K.: Transient self-focusing of an intense short pulse laser in magnetized plasma. Phys. Scr. **66**, 262 (2002)
21. Ghorbanalilu, M.: Reduction of laser pulse diffraction by magnetized electron–positron Plasma lens. IEEE Trans. Plasma Sci. **40**, 12 (2012)
22. Jha, P., Mishra, Rohit K., et al.: Spot-size evolution of laser beam propagating in plasma embedded in axial magnetic field. Phys. Plasmas **14**, 114504 (2007)
23. Gupta, D.N., Sharma, A.K.: Transient self-focusing of an intense short pulse laser in magnetized plasma. Phys. Scr. **66**, 262 (2002)
24. Schmit, G., Horton, W.: Self-focusing of laser beams in the beat-wave accelerator. Comments Plasma Phys. Controlled Fusion **9**, 8 (1985)
25. Mackinstrie, C.J., Bingham, R.: Stimulated Raman forward scattering and the relativistic modulational instability of light waves in rarefied plasma. Phys. Fluids B **4**, 2626 (1992)
26. Antonsen, T.M., Mora, P.: Self-focusing and Raman scattering of laser pulses in tenuous plasmas. Phys. Rev. Lett. **69**, 2204 (1992)
27. Anderson, D., Bonnedal, M.: Variational approach to nonlinear self-focusing of Gaussian laser beams. Phys. Fluids **22**, 105 (1979)
28. Lam, J.F., et al.: Moment theory of self-trapped laser beams with nonlinear saturation. Opt. Commun. **15**, 419 (1975)
29. Sprangle, P., et al.: Analysis of radiation focusing and steering in the free-electron laser by use of a source-dependent expansion technique. Phys. Rev. A **36**, 2773 (1987)

30. Max, C., Arons, J., Langdon, A.B.: Self-modulation and self-focusing of electromagnetic waves in plasmas. Phys. Rev. Lett. **33**, 209 (1974)
31. Sprangle, P., Tang, C.M., Esarey, E.: Relativistic self-focusing of short-pulse radiation beams in plasmas. IEEE Trans. Plasma Sci. **15**, 145 (1987)
32. Sun, G.Z., Ott, E., Lee, Y.C., Guzdar, P.: Self-focusing of short intense pulses in plasmas. Phys. Fluids **30**, 526 (1987)
33. Chen, X.L., Sudan, R.N.: Necessary and sufficient conditions for self-focusing of short ultraintense laser pulse in underdense plasma. Phys. Rev. Lett. **70**, 2082 (1993)
34. Ghorbanalilu, M.: Focusing of intense laser beam by a thin axially magnetized plasma lens. Phys. Plasmas **17**, 023111 (2010)
35. Jha, P., Mishra, R.K., Upadhyay, A.K., Raj, G.: Spot-size evolution of laser beam propagating in plasma embedded in axial magnetic field. Phys. Plasmas **14**, 114504 (2007)
36. Jha, P., Mishra, R.K., Upadhyay, A.K., Raj, G.: Self-focusing of intense laser beam in magnetized plasma. Phys. Plasmas **13**, 103102 (2006)
37. Malekshahi, M., Dorranian, D., Ranjbar Askari, H.: Self-focusing of the high intensity ultra short laser pulse propagating through relativistic magnetized plasma. Opt. Commun. **332**, 227 (2014)
38. Gupta, D., Suk, H., Hur, M.: Realistic laser focusing effect on electron acceleration in the presence of a pulsed magnetic field. Appl. Phys. Lett **91**, 211101 (2007)
39. Shvests, G., Tushentsov, M., Tokman, M.D., Kryachko, A.: Propagation of electromagnetic waves in the plasma near electron cyclotron resonance: undulator-induced transparency. Phys. Plasmas **12**, 056701 (2005)
40. Ren, C., Duda, B.J., Hemker, R.G., Mori, W.B., Katsouleas, T.: Compressing and focusing a short laser pulse by a thin plasma lens. Phys. Rev. E. **63**(2), 026411-1–026411-8 (2001)
41. Bulanov, S.V., Sakharov, A.S.: Induced focusing of electromagnetic wave in a wake plasma wave. JETP Lett. **54**, 203–207 (1991)

27

Proof of factorization of χ_{cJ} production in non-equilibrium QCD at RHIC and LHC in color singlet mechanism

Gouranga C Nayak[1]

Abstract Recently we have proved the factorization of NRQCD S-wave heavy quarkonium production at all orders in coupling constant. In this paper we extend this to prove the factorization of infrared divergences in χ_{cJ} production from color singlet $c\bar{c}$ pair in non-equilibrium QCD at RHIC and LHC at all orders in coupling constant. This can be relevant to study the quark–gluon plasma at RHIC and LHC.

Keywords Heavy quarkonium · Quark-gluon plasma · RHIC · LHC · Factorization · Non-equilibrium QCD

Introduction

The factorization of infrared divergences in nonrelativistic QCD (NRQCD) color octet S-wave heavy quarkonium production at high energy colliders at all orders in coupling constant is recently proved in [1]. In this paper we extend this formalism to non-equilibrium QCD by using the closed-time path integral formulation to prove the factorization of infrared divergences in χ_{cJ} production from the color singlet $c\bar{c}$ pair in non-equilibrium QCD at all orders in coupling constant at RHIC and LHC. We also predict the correct definition of the non-perturbative matrix element of the χ_{cJ} production from color singlet $c\bar{c}$ pair in non-equilibrium QCD at RHIC and LHC. This can be relevant to study the quark-gluon plasma (QGP) at RHIC and LHC.

At very high temperature ($\geq$ 200 MeV) the normal hadronic matter becomes a new state of matter known as the QGP. About 10^{-12} s after the big bang our universe was filled with the QGP which makes it important to produce it in the laboratory at RHIC and LHC by colliding two heavy nuclei at very high energy [2]. Since the confinement in QCD prevents us to detect the QGP directly at RHIC and LHC, various indirect signatures (such as the heavy quarkonium production/suppression [3]) are proposed for its detection.

Since the center of mass energy $\sqrt{s}$ = 200 GeV (5.5 TeV) of Au–Au (Pb–Pb) collisions at RHIC (LHC) is very high, the two nuclei at RHIC (LHC) travel almost at the speed of light creating the non-equilibrium quark–gluon plasma just after the heavy-ion collisions. Because of the very small hadronization time scale in QCD ($\sim 10^{-24}$ s) there may not be enough secondary partonic collisions to bring this non-equilibrium QGP to equilibrium. Hence the QGP at RHIC (LHC) may be in non-equilibrium where one cannot define a temperature.

The hard (high p_T) parton production at RHIC and LHC can be calculated by using pQCD but the soft parton production calculation needs non-perturbative QCD which is not solved yet. This implies that there remains uncertainty in determining the soft partons production at RHIC and LHC. Note that the soft partons play an important role in determining the bulk properties of the QGP at RHIC and LHC.

It should be mentioned here that the study of hadronization from non-equilibrium QGP at RHIC and LHC is one of the most difficult and important problems because the confinement problem in QCD is not solved yet due to the lack of our understanding of non-perturbative QCD. This implies that the first principle calculation of hadron production from non-equilibrium partons at RHIC and LHC is not known.

G. C. Nayak was affiliated with C. N. Yang Institute for Theoretical Physics in 2004–2007.

✉ Gouranga C Nayak
nayakg138@gmail.com

[1] C. N. Yang Institute for Theoretical Physics, Stony Brook University, Stony Brook, NY 11794-3840, USA

Because of these reasons one finds that in order to detect the QGP at RHIC and LHC by using the first principle calculation one needs to study the nonequilibrium–nonperturbative QCD by using the closed-time path integral formalism which is not easy [4–7]. If one does not perform the exact first principle nonequilibrium–nonperturbative QCD calculation then the comparison of the theoretical calculation with the experimental data at RHIC and LHC becomes questionable. For example, some of the limitations of the present theoretical approaches are listed below.

The lattice QCD at finite temperature [8] is a common tool to study the properties of the QGP. However, for the reasons explained above, the actual QGP at RHIC and LHC may be in non-equilibrium where one cannot define a temperature. Hence the lattice QCD at finite temperature has no application in non-equilibrium QGP at RHIC and LHC.

Similarly the hydrodynamics [9–14] is not applicable in non-equilibrium QGP at RHIC and LHC. Another limitation of the hydrodynamics [9–14] is that it does not answer the question how the partons become hadrons from first principle. As shown in [15] the parton to hadron fragmentation function in QCD in vacuum cannot be used to study the hadrons production from partons from the quark–gluon plasma at RHIC and LHC. It is important to observe that even if the experimental data at RHIC and LHC is explained by using the hydrodynamics [9–14] it does not prove that the QGP is in equilibrium. In order to make sure that the QGP is in equilibrium at RHIC and LHC one has to prove that the same experimental data cannot be explained by using the non-equilibrium QGP for which one has to study the nonequilibrium-nonperturbative QCD by using the closed-time path integral formalism.

As far as the actual physics at RHIC and LHC heavy-ion collisions is concerned the AdS/CFT based studies [16, 17] and the supersymmetric Yang–Mills plasma-based studies [18] have nothing to do with it because of the lack of experimental verification of the string theory and the supersymmetry.

Regarding the initial condition for the QGP formation and the color glass condensate (CGC) [19, 20], as discussed above, the hard (high p_T) parton production at RHIC and LHC can be calculated by using the pQCD but the soft parton production can only be correctly calculated from the first principle by using the non-perturbative QCD which is yet to be solved.

The jet quenching study, see for example [21, 22, and references therein], directly/indirectly uses the parton to hadron fragmentation function in QCD in vacuum. This is not possible because unlike the leading order perturbative gluon propagator in non-equilibrium QCD the non-perturbative fragmentation function in non-equilibrium QCD cannot be decomposed into the vacuum part and the medium part [15].

Hence from the above discussions one finds that, although a lot of experimental data are available at RHIC and LHC heavy-ion colliders, there exists no exact first principle theoretical calculation to explain these experimental data. It is almost impossible to make an exact first principle theoretical calculation at RHIC and LHC without studying the nonequilibrium–nonperturbative QCD by using the closed-time path integral formalism.

The first principle way to study non-equilibrium quantum field theory is the Schwinger–Keldysh closed-time path (CTP) formalism [4, 5]. Although the non-equilibrium QED is usually studied by using the canonical quantization formalism, the closed-time path integral formalism is useful to study the nonequilibrium–nonperturbative QCD due to the self-gluon interactions and the hadronization.

As mentioned earlier, the heavy quarkonium is one of the indirect signatures for the detection of QGP [3]. Both j/ψ and χ_{cJ} are measured by various collaborations at the RHIC and LHC heavy-ion collider experiments. In order to study heavy quarkonium production from the QGP at RHIC and LHC one needs to prove factorization of infrared divergences; otherwise, one will predict infinite cross section for the heavy quarkonium production.

The infrared divergences issue in the case of P-wave heavy quarkonium production is more complicated than that of the j/ψ production. This is because there are no uncanceled infrared divergences due to eikonal gluons exchange in the case of S-wave heavy quarkonium (j/ψ) production in the color singlet mechanism, whereas there are uncanceled infrared divergences due to eikonal gluons exchange in case of P-wave heavy quarkonium (χ_{cJ}) production in the color singlet mechanism [23–28].

Recently, we have shown that these uncanceled infrared divergences can be factored into the correct definition of the color singlet P-wave heavy quarkonium non-perturbative matrix element by supplying the eikonal lines or the gauge links [29]. In this paper we will extend this to the non-equilibrium QCD by using the closed-time path integral formalism. We will prove the factorization of infrared divergences in the χ_{cJ} production from the color singlet $c\bar{c}$ pair in non-equilibrium QCD at RHIC and LHC at all orders in coupling constant. We will predict the correct definition of the non-perturbative matrix element of the χ_{cJ} production from the color singlet $c\bar{c}$ pair in non-equilibrium QCD at RHIC and LHC. This can be relevant to detect the QGP at RHIC and LHC.

The paper is organized as follows: In "Closed-time path integral formalism and the generating functional in non-equilibrium QCD" section a brief discussion on the generating functional in non-equilibrium QCD is presented. In "Infrared divergences in χ_{cJ} production from color singlet $C\bar{C}$ pair" section we discuss the non-canceling infrared divergences in color singlet χ_{cJ} production. In "Infrared divergence due to eikonal gluon and the SU(3) pure gauge background field" section we show that the infrared divergences due to eikonal gluons exchange can be studied by using the SU(3) pure gauge. In "Proof of factorization of χ_{cJ} production in non-equilibrium QCD at RHIC and LHC in color singlet mechanism" section we prove the factorization of infrared divergences in the χ_{cJ} production from color singlet $c\bar{c}$ pair in non-equilibrium QCD at RHIC and LHC at all orders in coupling constant. In "Correct definition of χ_{cJ} production in non-equilibrium QCD at RHIC and LHC in color singlet mechanism" section we predict the correct definition of the non-perturbative matrix element of the χ_{cJ} production from color singlet $c\bar{c}$ pair in non-equilibrium QCD at RHIC and LHC. We conclude in "Conclusion" section.

Closed-time path integral formalism and the generating functional in non-equilibrium QCD

Since we will use the background field method of QCD in this paper we denote the gluon field by $Q^{\lambda d}(x)$ and the background field by $A^{\lambda d}(x)$ where $\lambda = 0,1,2,3$ and $d = 1,\ldots,8$. The generating functional in non-equilibrium QCD (without the background field) in the closed-time path integral formalism is given by [6, 7]

$$\begin{aligned}
&Z[\rho,J_+,J_-,\eta_{1+},\bar{\eta}_{1+},\eta_{1-},\bar{\eta}_{1-},\eta_{2+},\bar{\eta}_{2+},\eta_{2-},\bar{\eta}_{2-},\\
&\eta_{3+},\bar{\eta}_{3+},\eta_{3-},\bar{\eta}_{3-},\eta_{l+},\bar{\eta}_{l+},\eta_{l-},\bar{\eta}_{l-}]\\
&= \int [dQ_+][dQ_-]\Pi_{k=1}^{3}[d\bar{\psi}_{k+}][d\bar{\psi}_{k-}][d\psi_{k+}][d\psi_{k-}][d\bar{\Psi}_+]\\
&\times [d\bar{\Psi}_-][d\Psi_+][d\Psi_-] \times \det\left(\frac{\delta\partial_\lambda Q_+^{\lambda d}}{\delta\omega_+^e}\right) \times \det\left(\frac{\delta\partial_\lambda Q_-^{\lambda d}}{\delta\omega_-^e}\right)\\
&\times \exp[i\int d^4x\{-\frac{1}{4}F^{d^2}_{\lambda\delta}[Q_+] + \frac{1}{4}F^{d^2}_{\lambda\delta}[Q_-]\\
&-\frac{1}{2\alpha}(\partial_\lambda Q_+^{\lambda d})^2 + \frac{1}{2\alpha}(\partial_\lambda Q_-^{\lambda d})^2\\
&+\sum_{k=1}^{3}\bar{\psi}_{k+}[i\gamma^\lambda\partial_\lambda - m_k + gT^d\gamma^\lambda Q^d_{\lambda+}]\psi_{k+}\\
&-\sum_{k=1}^{3}\bar{\psi}_{k-}[i\gamma^\lambda\partial_\lambda - m_k + gT^d\gamma^\lambda Q^d_{\lambda-}]\psi_{k-}\\
&+\bar{\Psi}_+[i\gamma^\lambda\partial_\lambda - M + gT^d\gamma^\lambda Q^d_{\lambda+}]\Psi_+\\
&-\bar{\Psi}_-[i\gamma^\lambda\partial_\lambda - M + gT^d\gamma^\lambda Q^d_{\lambda-}]\Psi_- + J_+Q_+ - J_-Q_-\\
&+\sum_{k=1}^{3}[\bar{\psi}_{k+}\eta_{k+} - \bar{\psi}_k\ \eta_{k-} + \bar{\eta}_{k+}\psi_{k+} - \bar{\eta}_{k-}\psi_{k-}]\\
&+\bar{\Psi}_+\eta_{l+} - \bar{\Psi}_-\eta_{l-} + \bar{\eta}_{l+}\Psi_+ - \bar{\eta}_{l-}\Psi_-\}]\\
&\times <Q_+,\psi_{1+},\bar{\psi}_{1+},\psi_{2+},\bar{\psi}_{2+},\psi_{3+},\bar{\psi}_{3+},\Psi_+,\bar{\Psi}_+,0|\rho|0,\\
&\times \bar{\psi}_{1-},\psi_{1-},\bar{\psi}_{2-},\psi_{2-},\bar{\psi}_{3-},\psi_{3-},\\
&\times \bar{\Psi}_-,\Psi_-,Q_->,
\end{aligned} \tag{1}$$

where $\delta = 0,1,2,3$ and we have included the heavy quark. In Eq. (1) the symbol $k = 1,2,3 = u,d,s$ stands for up, down and strange quark with mass m_k and field ψ_k. The heavy quark field is Ψ and the heavy quark mass is M. The initial density of states is denoted by ρ, the arbitrary gauge fixing parameter is α, the $|0,\bar{\psi}_{1-},\psi_{1-},\bar{\psi}_{2-},\psi_{2-},\bar{\psi}_{3-},\psi_{3-},\bar{\Psi}_-,\Psi_-,Q_->$ corresponds to the state at the initial time and

$$\begin{aligned}
F^{d^2}_{\lambda\delta}[Q_+] =& [\partial_\lambda Q^d_{\delta+} - \partial_\delta Q^d_{\lambda+} + gf^{dba}Q^b_{\lambda+}Q^a_{\delta+}]\\
&\times [\partial^\lambda Q_+^{\delta d} - \partial^\delta Q_+^{\lambda d} + gf^{dce}Q_+^{\lambda c}Q_+^{\delta e}],
\end{aligned} \tag{2}$$

and similarly for the $-$ index where $+,-$ stand for the closed-time path indices. Note that we do not introduce ghost fields as we directly work with the ghost determinant $\det(\frac{\delta\partial_\lambda Q_+^{\lambda d}}{\delta\omega_+^e})$ in Eq. (1).

The corresponding non-equilibrium QCD generating functional in the closed-time path integral formalism of the background field method of QCD is given by [6, 7, 30–32]

$$\begin{aligned}
&Z[A,\rho,J_+,J_-,\eta_{1+},\bar{\eta}_{1+},\eta_{1-},\bar{\eta}_{1-},\eta_{2+},\bar{\eta}_{2+},\eta_{2-},\bar{\eta}_{2-},\\
&\eta_{3+},\bar{\eta}_{3+},\eta_{3-},\bar{\eta}_{3-},\eta_{l+},\bar{\eta}_{l+},\eta_{l-},\bar{\eta}_{l-}]\\
&= \int [dQ_+][dQ_-]\Pi_{k=1}^{3}[d\bar{\psi}_{k+}][d\bar{\psi}_{k-}][d\psi_{k+}][d\psi_{k-}][d\bar{\Psi}_+]\\
&[d\bar{\Psi}_-][d\Psi_+][d\Psi_-]\\
&\times \det\left(\frac{\delta G^d(Q_+)}{\delta\omega_+^e}\right) \times \det\left(\frac{\delta G^d(Q_-)}{\delta\omega_-^e}\right)\\
&\times \exp[i\int d^4x\{-\frac{1}{4}F^{d^2}_{\lambda\delta}[Q_+ + A_+] + \frac{1}{4}F^{d^2}_{\lambda\delta}[Q_- + A_-]\\
&-\frac{1}{2\alpha}(G^d(Q_+))^2 + \frac{1}{2\alpha}(G^d(Q_-))^2\\
&+\sum_{k=1}^{3}\bar{\psi}_{k+}[i\gamma^\lambda\partial_\lambda - m_k + gT^d\gamma^\lambda(Q+A)^d_{\lambda+}]\psi_{k+}\\
&-\sum_{k=1}^{3}\bar{\psi}_{k-}[i\gamma^\lambda\partial_\lambda - m_k + gT^d\gamma^\lambda(Q+A)^d_{\lambda-}]\psi_{k-}\\
&+\bar{\Psi}_+[i\gamma^\lambda\partial_\lambda - M + gT^d\gamma^\lambda(Q+A)^d_{\lambda+}]\Psi_+ - \bar{\Psi}_-[i\gamma^\lambda\partial_\lambda\\
&- M + gT^d\gamma^\lambda(Q+A)^d_{\lambda-}]\Psi_- + \sum_{k=1}^{3}[\bar{\psi}_{k+}\eta_{k+}\\
&-\bar{\psi}_{k-}\eta_{k-} + \bar{\eta}_{k+}\psi_{k+} - \bar{\eta}_{k-}\psi_{k-}] + \bar{\Psi}_+\eta_{l+} - \bar{\Psi}_-\eta_{l-}\\
&+\bar{\eta}_{l+}\Psi_+ - \bar{\eta}_{l-}\Psi_- + J_+Q_+ - J_-Q_-\}]\\
&\times <Q_+ + A_+,\psi_{1+},\bar{\psi}_{1+},\psi_{2+},\bar{\psi}_{2+},\psi_{3+},\bar{\psi}_{3+},\Psi_+,\bar{\Psi}_+,\\
&0|\rho|0,\bar{\psi}_{1-},\psi_{1-},\bar{\psi}_{2-},\psi_{2-},\bar{\psi}_{3-},\psi_{3-},\bar{\Psi}_-,\Psi_-,Q_- + A_->
\end{aligned} \tag{3}$$

where the background gauge fixing

$$G^d(Q_+) = \partial_\lambda Q_+^{\lambda d} + gf^{dba}A^b_{\lambda+}Q_+^{\lambda a} \tag{4}$$

depends on the background field $A^{\lambda d}(x)$. In Eq. (3)

$$F^{d2}_{\lambda\delta}[Q_+ + A_+] = [\partial_\lambda [A^d_{\delta+} + Q^d_{\delta+}] - \partial_\delta [A^d_{\lambda+} + Q^d_{\lambda+}] + gf^{dba} [A^b_{\lambda+} + Q^b_{\lambda+}] [A^a_{\delta+} + Q^a_{\delta+}]] \times [\partial^\lambda [A^{\delta d}_+ + Q^{\delta d}_+] - \partial^\delta [A^{\lambda d}_+ + Q^{\lambda d}_+] + gf^{dce} [A^{\lambda c}_+ + Q^{\lambda c}_+] [A^{\delta e}_+ + Q^{\delta e}_+]] \tag{5}$$

and we do not have any ghost fields because we directly work with the ghost determinant $\det(\frac{\delta G^d(Q_+)}{\delta \omega^e_+})$ in Eq. (3).

For the type I gauge transformation we have [31, 32]

$$T^d A'^{\lambda d}_+ = \Phi_+ T^d A^{\lambda d}_+ \Phi_+^{-1} + \frac{1}{ig}(\partial^\lambda \Phi_+)\Phi_+^{-1}, \quad T^d Q'^{\lambda d}_+ = \Phi_+ T^d Q^{\lambda d}_+ \Phi_+^{-1}, \tag{6}$$

where the light-like gauge link or the light-like eikonal line in the fundamental representation of SU(3) is given by [1, 33, 34]

$$\Phi_+(x) = e^{igT^d \omega^d_+(x)} = \mathcal{P} e^{-igT^d \int_0^\infty d\tau l \cdot A^d_+(x+\tau l)}, \quad l^2 = 0, \tag{7}$$

where l^λ is the light-like four-velocity.

In this paper we will use the generating functionals from Eqs. (1) and (3) in the path integral formulation to prove the factorization of infrared divergences in the χ_{cJ} production from the color singlet $c\bar{c}$ pair in non-equilibrium QCD at RHIC and LHC at all orders of coupling constant.

Infrared divergences in χ_{cJ} production from color singlet $C\bar{C}$ pair

The non-canceling infrared divergences were found in the higher order pQCD calculation of the annihilation of heavy quark–antiquark pair to light partons in the hadronic decay of the color singlet P-wave heavy quarkonium [23–28]. For example, in the partonic processes [23–28]

$$\chi_{cJ} \to q\bar{q}g, \quad h_c \to ggg \tag{8}$$

of the hadronic decay of χ_{cJ} and h_c, respectively, one finds the non-canceling infrared divergences due to real soft gluons (eikonal gluons) emission/absorption [23–28, 35, 36].

Now let us discuss the hadroproduction of χ_{cJ} from color singlet $c\bar{c}$ pair at high-energy colliders. If the factorization theorem is valid [1, 33, 34, 37–46] then the χ_{cJ} production from the color singlet $c\bar{c}$ pair at high energy colliders is given by

$$d\sigma_{pp\to\chi_{cJ}+X(p_T)} = \sum_{k,j} \int dx_1 dx_2 f_{k/p}(x_1, Q) f_{j/p}(x_2, Q) d\hat{\sigma}_{kj\to C\bar{C}[^3P_J]+X(p_T)} <0|\mathcal{O}_{\chi_{cJ}}|0>, \tag{9}$$

where $d\hat{\sigma}_{kj\to C\bar{C}[^3P_J]+X(p_T)}$ is the partonic level cross section for the $c\bar{c}$ production in ^3P_J state. This partonic level cross section can be calculated by using pQCD where $k, j = q, \bar{q}, g$. The parton distribution function $f_{k/p}(x, Q)$ of the parton k inside the proton p is a non-perturbative quantity in QCD. The non-perturbative matrix element of χ_{cJ} production from the color singlet $c\bar{c}$ pair is denoted by $<0|\mathcal{O}_{\chi_{cJ}}|0>$.

As mentioned above the non-canceling infrared divergences were found in the hadronic decay of the color singlet P-wave heavy quarkonium [23–28, 35, 36]. Similarly, the non-canceling infrared divergences were also found in the hadroproduction of the color singlet P-wave heavy quarkonium [36].

Note that for S-wave and P-wave color singlet heavy quarkonium the infrared divergences occur due to coulomb gluon and eikonal gluon exchanges. The infrared divergence due to Coulomb gluon exchange is analogous to the infrared divergence due to the Coulomb photon exchange in QED, see [47]. This Coulomb gluon infrared divergence is also known as the $\frac{1}{v} \to \infty$ divergence where v is the relative velocity of the heavy quark–antiquark which is normally absorbed into the normalization of the bound state wave function [23–28] similar to that in QED [47].

In case of j/ψ production the infrared divergences due to the eikonal gluons interacting with charm quark exactly cancel with the corresponding infrared divergences associated with the charm antiquark [23–28]. Hence there is no uncanceled infrared divergences due to eikonal gluons exchange in case of j/ψ production. That is why there are no gauge links in the definition of the j/ψ wave function [29].

However, in case of χ_{cJ} production the non-canceling infrared divergences occur due to the eikonal gluons [23–28]. At NLO in coupling constant the non-canceling infrared divergence due to the eikonal gluons exchange is found in the quark–antiquark fusion process [36]

$$q\bar{q} \to \chi_{cJ} g. \tag{10}$$

Because of the existence of these non-canceling infrared divergences, we have shown in [29] that the gauge links are necessary in the definition of the color singlet P-wave non-perturbative matrix element of the heavy quarkonium production. These gauge links make the non-perturbative matrix element gauge invariant and cancel these non-canceling infrared divergences.

Hence the correct definition of the non-perturbative matrix element of the χ_{c0} production from color singlet $c\bar{c}$ pair at high-energy colliders which is consistent with the factorization of infrared divergences at all orders in coupling constant in QCD is given by [29]

$$<0|\mathcal{O}_{\chi_{c0}}|0> \;=\; <0|\zeta^\dagger \Phi \bar{\nabla} \Phi^\dagger \xi a^\dagger_{\chi_{c0}} \cdot a_{\chi_{c0}} \xi^\dagger \Phi \bar{\nabla} \Phi^\dagger \zeta|0>, \tag{11}$$

where ζ (ξ) is the two-component Dirac spinor field that creates (annihilates) a heavy quark and

$$\zeta^\dagger \Phi \bar{\nabla} \Phi^\dagger \xi = \zeta^\dagger \Phi (\nabla \Phi^\dagger \xi) - (\nabla \Phi^\dagger \zeta)^\dagger \Phi^\dagger \xi. \tag{12}$$

In Eq. (11), the $a^\dagger_{\chi_{c0}}$ is the creation operator of the χ_{c0}, the $<0|\mathcal{O}_{\chi_{c0}}|0>$ is evaluated at the origin and

$$\Phi(x) = \mathcal{P} e^{-igT^d \int_0^\infty d\tau l \cdot A^d(x+\tau l)}, \quad l^2 = 0 \tag{13}$$

is the light-like gauge link or the light-like eikonal line in the fundamental representation of SU(3).

Infrared divergence due to eikonal gluon and the SU(3) pure gauge background field

As mentioned earlier the real gluon emission/absorption is the source of the non-canceling infrared divergences in case of P-wave heavy quarkonium production/deacy [23–28, 35, 36]. In this section we will briefly discuss the infrared divergence due to real gluon emission/absorption which can be described by eikonal Feynman rules in QCD. Let us first discuss the eikonal Feynman rules in QED before proceeding to QCD as the eikonal Feynman rules in QCD is similar to that in QED.

In QED the Feynman diagram contribution for an electron emitting a real photon is given by [48]

$$\frac{1}{\not{r} - \not{k} - m} \not{\epsilon}(k) u(r) = -\frac{r \cdot \epsilon(k)}{r \cdot k} u(r) + \frac{\not{k} \not{\epsilon}(k)}{2r \cdot k} u(r), \tag{14}$$

where r^λ (k^λ) is the momentum of electron (photon). Eq. (14) has both eikonal part

$$\frac{r \cdot \epsilon(k)}{r \cdot k} u(r) \to \infty \quad \text{when} \quad k^\lambda \to 0 \tag{15}$$

and the non-eikonal part

$$\frac{\not{k} \not{\epsilon}(k)}{2r \cdot k} u(r) \to \text{finite} \quad \text{when} \quad k^\lambda \to 0. \tag{16}$$

The eikonal part is the source of the infrared divergence as Eq. (15) diverges in the infrared limit $k^\lambda \to 0$. The non-eikonal part in Eq. (16) does not diverge in the infrared limit $k^\lambda \to 0$. This implies that the infrared divergence due to the emission of real photon from the electron can be studied by using only the eikonal term $\frac{r \cdot \epsilon(k)}{r \cdot k} u(r)$ without taking into account the non-eikonal term $\frac{\not{k} \not{\epsilon}(k)}{2r \cdot k} u(r)$ in the Feynman diagram contribution in Eq. (14).

Now we will show that the study of the infrared divergences due to the eikonal photons at all order in coupling constant in QED can be enormously simplified when the electron is light-like ($r^2 = 0$). The effective lagrangian density of the photon in the presence of current density $K^\lambda(x)$ in quantum field theory is given by [1]

$$\begin{aligned} \int d^4x \mathcal{L}_{eff}(x) &= -i \ln <0|0>_K = -i \ln [\frac{Z[K]}{Z[0]}] \\ &= -\frac{1}{2} \int d^4x K^\lambda(x) \frac{1}{\partial^2} K_\lambda(x), \end{aligned} \tag{17}$$

where the generating functional $Z[K]$ in the path integral formulation involving the photon field $Q^\lambda(x)$ is given by

$$Z[K] = \int [dQ] e^{i \int d^4x [-\frac{1}{4}[\partial_\delta Q_\lambda(x) - \partial_\lambda Q_\delta(x)][\partial^\delta Q^\lambda(x) - \partial^\lambda Q^\delta(x)] - \frac{1}{2\alpha}(\partial_\lambda Q^\lambda)^2 + K_\lambda(x) Q^\lambda(x)]}. \tag{18}$$

From Eq. (15) the eikonal contribution

$$e \int \frac{d^4k}{(2\pi)^4} \frac{l_\lambda Q^\lambda(k)}{l \cdot k + i\epsilon} = -i \int d^4x Q^\lambda(x) K_\lambda(x) \tag{19}$$

gives the eikonal current density

$$K^\lambda(x) = e \int_0^\infty d\tau l^\lambda \delta^{(4)}(x - l\tau), \tag{20}$$

where l^λ is the light-like four-velocity ($l^2 = 0$) of the electron.

Using eq. (20) in (17) we find that

$$\mathcal{L}_{eff}(x) = \frac{[el^2]^2}{[\sqrt{2}(l \cdot x)^2]^2} = 0, \quad \text{when} \quad l \cdot x \neq 0, \quad l^2 = 0. \tag{21}$$

From Eq. (21) we find that the light-like eikonal current produces pure gauge field in quantum field theory at all space–time points except at the positions perpendicular to the direction of motion of the charge at the time of closest approach, a result which agrees with the classical mechanics [41, 49, 50].

Hence we find from Eq. (21) that the calculation of infrared divergences due to the real photons' emission from the light-like electron can be simplified by using the pure gauge field in QED. This can also be seen from Grammer–Yennie approximation [48] as follows: We write the photon polarization as the sum of the transverse (physical) polarization plus the longitudinal (pure gauge) polarization to find [48]

$$\epsilon^{\lambda}(k) = \epsilon^{\lambda}_{\text{physical}}(k) + \epsilon^{\lambda}_{\text{pure gauge}}(k), \tag{22}$$

where

$$\epsilon^{\lambda}_{\text{physical}}(k) = \left[\epsilon^{\lambda}(k) - k^{\lambda}\frac{r\cdot\epsilon(k)}{r\cdot k}\right], \tag{23}$$

which contributes to the physical (finite) cross section and

$$\epsilon^{\lambda}_{\text{pure gauge}}(k) = k^{\lambda}\frac{r\cdot\epsilon(k)}{r\cdot k}, \tag{24}$$

which does not contribute to the physical (finite) cross section but contributes to the infrared divergence. This can be explicitly seen by using Eq. (22) in the eikonal part in Eq. (14) to find

$$\frac{r\cdot\epsilon(k)}{r\cdot k}u(r) = \frac{r\cdot\epsilon_{\text{pure gauge}}(k)}{r\cdot k}u(r) \to \infty \quad \text{when} \quad k^{\lambda}\to 0, \tag{25}$$

$$\frac{r\cdot\epsilon_{\text{physical}}(k)}{r\cdot k}u(r) = 0, \tag{26}$$

and in the non-eikonal part in Eq. (14) to find

$$\frac{\not{k}\not{\epsilon}(k)}{2r\cdot k}u(r) = \frac{\not{k}\not{\epsilon}_{\text{physical}}(k)}{2r\cdot k}u(r) \to \text{finite} \quad \text{when} \quad k^{\lambda}\to 0 \tag{27}$$

and

$$\frac{\not{k}\not{\epsilon}_{\text{pure gauge}}(k)}{2r\cdot k}u(r) = 0. \tag{28}$$

From Eq. (16) the non-eikonal contribution

$$e\int \frac{d^4k}{(2\pi)^4}\frac{\not{k}\not{Q}(k)}{2r\cdot k + i\epsilon} = \int d^4x K(x)\cdot Q(x) \tag{29}$$

gives the non-eikonal current density

$$K^{\lambda}(x) = \frac{e}{2}\gamma^{\delta}\gamma^{\lambda}\int dw \frac{\partial}{\partial x^{\delta}}\delta^{(4)}(x - rw), \tag{30}$$

where r^{λ} is light-like ($r^2 = 0$) or non-light-like ($r^2 \neq 0$) momentum of the electron. Using Eqs. (20) and (30) in Eq. (17) we find that the interaction between the (light-like or non-light-like) non-eikonal line with four-momentum r^{λ} and the gauge field generated by the light-like eikonal line with four-velocity l^{λ} ($l^2 = 0$) gives the interaction (effective) lagrangian density

$$\mathcal{L}^{\text{interaction}}_{\text{eff}}(x) = \frac{l^2e^2[(r\cdot l)(r\cdot x) - r^2 l\cdot x]}{2(l\cdot x)^3[(r\cdot x)^2 - r^2x^2]^{\frac{3}{2}}} = 0, \quad \text{when} \quad l\cdot x \neq 0, \quad r\cdot x\neq 0. \tag{31}$$

From Eq. (31) we find that, in quantum field theory, the interaction between the non-eikonal line and the gauge field generated by the light-like eikonal line does not contribute to the interaction (effective) lagrangian density. Since the light-like eikonal line produces pure gauge field in quantum field theory (see Eq. (21)) we find from Eqs. (31) and (28) that the light-like eikonal line does not modify the finite physical cross section.

Hence we find from Eqs. (21), (31), (25), (26), (27) and (28) that the study of infrared divergences in QED due to real photon emission from the light-like electron can be enormously simplified by using the pure gauge field without modifying the finite value of the cross section.

We have shown in Eqs. (21) and (31) that the light-like electron produces pure gauge field in QED. This result in QED agrees with classical mechanics [41, 49, 50]. Hence we find that the infrared divergences at all orders in coupling constant due to the real photons" emission from the light-like electron in quantum field theory can be studied by using the path integral formulation of the background field method of quantum field theory in the presence of pure gauge background field [1, 33, 34, 51, 52].

In QED the U(1) pure gauge field $A^{\lambda}(x)$ is given by $A^{\lambda}(x) = \partial^{\lambda}\omega(x)$ and in QCD the SU(3) pure gauge field $A^{\lambda d}(x)$ is given by [1, 33, 34]

$$T^d A^{\lambda d}(x) = \frac{1}{ig}[\partial^{\lambda}\Phi(x)]\Phi^{-1}(x), \tag{32}$$

where $\Phi(x)$ is the light-like gauge link or the light-like eikonal line in the fundamental representation of SU(3) given by Eq. (13).

Proof of factorization of χ_{cJ} production in non-equilibrium QCD at RHIC and LHC in color singlet mechanism

As discussed in "Infrared divergence due to eikonal gluon and the SU(3) pure gauge background field" section the infrared divergences due to the exchange of eikonal gluons with the light-like parton in QCD can be studied by using the path integral formulation of the background field method of QCD in the presence of SU(3) pure gauge background field as given by Eq. (32) [1, 33, 34]. Note that the path integral technique is suitable to study the properties of the non-perturbative quantities in QCD. It should be mentioned here that the properties of a non-perturbative function may not always be correctly studied by using the perturbative method no matter how many orders of perturbation theory is used. Take, for example, a non-perturbative function

$$f(g) = e^{-\frac{1}{g^2}}. \tag{33}$$

The Taylor series at $g = 0$ gives $f(g) = 0$ to all orders in perturbation theory but $f(g) \neq 0$ for $g \neq 0$.

Having considered the points mentioned above, one should note that perturbative QCD entered a new phase when the cancelation of the leading-order (LO) renormalons between the QCD potential and the pole masses of quark and antiquark was discovered (see for example [53, 54]). Convergence of the perturbative series improved dramatically and much more accurate perturbative predictions became available. Hence, in some later works (see, for example, [55, 56]) it was shown that perturbative predictions in QCD agree well with phenomenological QCD results (determined from heavy quarkonium spectroscopy) and lattice QCD calculations. For recent developments on color potential produced by the color charge of the quark, see [49, 50].

In this paper we will use the path integral formulation of the background field method of QCD to predict the correct definition of the non-perturbative matrix element of the χ_{cJ} production from color singlet $c\bar{c}$ pair in non-equilibrium QCD which is gauge invariant and is consistent with the factorization of infrared divergences at all orders in coupling constant.

In the closed-time path integral formulation the generating functional in non-equilibrium QCD is given by Eq. (1). Hence from Eq. (1) we find that the heavy quark–antiquark non-perturbative correlation function of the type $<in|\bar{\Psi}_r(x')\Psi_r(x')\bar{\Psi}_s(x'')\Psi_s(x'')|in>$ in non-equilibrium QCD is given by [6, 7, 32, 57]

$$\begin{aligned}
&<in|\bar{\Psi}_r(x')\bar{\nabla}_{x'}\Psi_r(x')\cdot\bar{\Psi}_s(x'')\bar{\nabla}_{x''}\Psi_s(x'')|in> \\
&= \int [dQ_+][dQ_-]\Pi_{k=1}^3[d\bar{\psi}_{k+}][d\bar{\psi}_{k-}][d\psi_{k+}][d\psi_{k-}] \\
&[d\bar{\Psi}_+][d\bar{\Psi}_-][d\Psi_+][d\Psi_-] \\
&\times \bar{\Psi}_r(x')\bar{\nabla}_{x'}\Psi_r(x')\cdot\bar{\Psi}_s(x'')\bar{\nabla}_{x''}\Psi_s(x'')\times\det\left(\frac{\delta\partial_\lambda Q_+^{\lambda d}}{\delta\omega_+^e}\right)\times\det\left(\frac{\delta\partial_\lambda Q_-^{\lambda d}}{\delta\omega_-^e}\right) \\
&\exp[i\int d^4x\{-\frac{1}{4}F^{d2}_{\lambda\delta}[Q_+]+\frac{1}{4}F^{d2}_{\lambda\delta}[Q_-]-\frac{1}{2\alpha}(\partial_\lambda Q_+^{\lambda d})^2+\frac{1}{2\alpha}(\partial_\lambda Q_-^{\lambda d})^2 \\
&+\sum_{k=1}^3\bar{\psi}_{k+}[i\gamma^\lambda\partial_\lambda - m_k + gT^d\gamma^\lambda Q_{\lambda+}^d]\psi_{k+} \\
&-\sum_{k=1}^3\bar{\psi}_{k-}[i\gamma^\lambda\partial_\lambda - m_k + gT^d\gamma^\lambda Q_{\lambda-}^d]\psi_{k-} \\
&+\bar{\Psi}_+[i\gamma^\lambda\partial_\lambda - M + gT^d\gamma^\lambda Q_{\lambda+}^d]\Psi_+ \\
&-\bar{\Psi}_-[i\gamma^\lambda\partial_\lambda - M + gT^d\gamma^\lambda Q_{\lambda-}^d]\Psi_-\}] \\
&\times <Q_+,\psi_{1+},\bar{\psi}_{1+},\psi_{2+},\bar{\psi}_{2+},\psi_{3+},\bar{\psi}_{3+},\Psi_+,\bar{\Psi}_+, \\
&\times 0|\rho|0,\bar{\psi}_{1-},\psi_{1-},\bar{\psi}_{2-},\psi_{2-},\bar{\psi}_{3-},\psi_{3-},\bar{\Psi}_-,\Psi_-,Q_->,
\end{aligned} \tag{34}$$

where $r, s = +, -$ are the closed-time path indices in non-equilibrium QCD (the repeated closed-time path indices r, s in Eq. (34) are not summed) and $|in>$ is the ground state in non-equilibrium QCD.

In the closed-time path integral formulation in non-equilibrium the generating functional in the background field method of QCD is given by Eq. (3). Hence from Eq. (3) we find that the heavy quark–antiquark nonequilibrium–nonperturbative correlation function of the type $<in|\bar{\Psi}_r(x')\Psi_r(x')\bar{\Psi}_s(x'')\Psi_s(x'')|in>_A$ in the background field method of QCD is given by [6, 7, 30–32]

$$\begin{aligned}
&<in|\bar{\Psi}_r(x')\bar{\nabla}_{x'}\Psi_r(x')\cdot\bar{\Psi}_s(x'')\bar{\nabla}_{x''}\Psi_s(x'')|in>_A \\
&= \int [dQ_+][dQ_-]\Pi_{k=1}^3[d\bar{\psi}_{k+}][d\bar{\psi}_{k-}][d\psi_{k+}][d\psi_{k-}][d\bar{\Psi}_+] \\
&[d\bar{\Psi}_-][d\Psi_+][d\Psi_-] \\
&\times\bar{\Psi}_r(x')\bar{\nabla}_{x'}\Psi_r(x')\cdot\bar{\Psi}_s(x'')\bar{\nabla}_{x''}\Psi_s(x'')\times\det\left(\frac{\delta G^d(Q_+)}{\delta\omega_+^e}\right) \\
&\times\det\left(\frac{\delta G^d(Q_-)}{\delta\omega_-^e}\right)\exp[i\int d^4x\{-\frac{1}{4}F^{d2}_{\lambda\delta}[Q_++A_+] \\
&+\frac{1}{4}F^{d2}_{\lambda\delta}[Q_-+A_-]-\frac{1}{2\alpha}(G^d(Q_+))^2+\frac{1}{2\alpha}(G^d(Q_-))^2 \\
&+\sum_{k=1}^3\bar{\psi}_{k+}[i\gamma^\lambda\partial_\lambda - m_k + gT^d\gamma^\lambda(Q+A)_{\lambda+}^d]\psi_{k+} \\
&-\sum_{k=1}^3\bar{\psi}_{k-}[i\gamma^\lambda\partial_\lambda - m_k + gT^d\gamma^\lambda(Q+A)_{\lambda-}^d]\psi_{k-} \\
&+\bar{\Psi}_+[i\gamma^\lambda\partial_\lambda - M + gT^d\gamma^\lambda(Q+A)_{\lambda+}^d]\Psi_+ - \bar{\Psi}_-[i\gamma^\lambda\partial_\lambda \\
&-M+gT^d\gamma^\lambda(Q+A)_{\lambda-}^d]\Psi_-\}] \\
&<Q_++A_+,\psi_{1+},\bar{\psi}_{1+},\psi_{2+},\bar{\psi}_{2+},\psi_{3+},\bar{\psi}_{3+},\Psi_+,\bar{\Psi}_+,0| \\
&\rho|0,\bar{\psi}_{1-},\psi_{1-},\bar{\psi}_{2-},\psi_{2-},\bar{\psi}_{3-},\psi_{3-},\bar{\Psi}_-,\Psi_-,Q_-+A_->.
\end{aligned} \tag{35}$$

From Eq. (35) we find

$$\begin{aligned}
&<in|\bar{\Psi}_r(x')\Phi_r(x')\bar{\nabla}_{x'}\Phi_r^\dagger(x')\Psi_r(x')\cdot\bar{\Psi}_s(x'')\Phi_s(x'')\bar{\nabla}_{x''}\Phi_s^\dagger(x'') \\
&\Psi_s(x'')|in>_A \\
&= \int [dQ_+][dQ_-]\Pi_{k=1}^3[d\bar{\psi}_{k+}][d\bar{\psi}_{k-}][d\psi_{k+}][d\psi_{k-}][d\bar{\Psi}_+] \\
&[d\bar{\Psi}_-][d\Psi_+][\Psi_-] \\
&\times\bar{\Psi}_r(x')\Phi_r(x')\bar{\nabla}_{x'}\Phi_r^\dagger(x')\Psi_r(x')\cdot\bar{\Psi}_s(x'')\Phi_s(x'')\bar{\nabla}_{x''}\Phi_s^\dagger(x'')\Psi_s(x'') \\
&\times\det\left(\frac{\delta G^d(Q_+)}{\delta\omega_+^e}\right)\times\det\left(\frac{\delta G^d(Q_-)}{\delta\omega_-^e}\right) \\
&\times\exp[i\int d^4x\{-\frac{1}{4}F^{d2}_{\lambda\delta}[Q_++A_+]+\frac{1}{4}F^{d2}_{\lambda\delta}[Q_-+A_-] \\
&-\frac{1}{2\alpha}(G^d(Q_+))^2+\frac{1}{2\alpha}(G^d(Q_-))^2 \\
&+\sum_{k-1}^3\bar{\psi}_{k+}[i\gamma^\lambda\partial_\lambda - m_k + gT^d\gamma^\lambda(Q+A)_{\lambda+}^d]\psi_{k+} \\
&-\sum_{k=1}^3\bar{\psi}_{k-}[i\gamma^\lambda\partial_\lambda - m_k + gT^d\gamma^\lambda(Q+A)_{\lambda-}^d]\psi_{k-} \\
&+\bar{\Psi}_+[i\gamma^\lambda\partial_\lambda - M + gT^d\gamma^\lambda(Q+A)_{\lambda+}^d]\Psi_+ \\
&-\bar{\Psi}_-[i\gamma^\lambda\partial_\lambda - M + gT^d\gamma^\lambda(Q+A)_{\lambda-}^d]\Psi_-\}] \\
&\times <Q_++A_+,\psi_{1+},\bar{\psi}_{1+},\psi_{2+},\bar{\psi}_{2+},\psi_{3+},\bar{\psi}_{3+},\Psi_+,\bar{\Psi}_+, \\
&0|\rho|0,\bar{\psi}_{1-},\psi_{1-},\bar{\psi}_{2-},\psi_{2-},\bar{\psi}_{3-},\psi_{3-},\bar{\Psi}_-,\Psi_-,Q_-+A_->,
\end{aligned} \tag{36}$$

where $\Phi(x)$ is the light-like gauge link or the light-like eikonal line in the fundamental representation of SU(3) given by Eq. (13).

Since Q is the integration variable inside the path integration we change the integration variable $Q \to Q - A$ in Eq. (36) to find

$$\begin{aligned}
&< in|\bar{\Psi}_r(x')\Phi_r(x')\bar{\nabla}_{x'}\Phi_r^\dagger(x')\Psi_r(x')\cdot\bar{\Psi}_s(x'')\Phi_s(x'')\\
&\bar{\nabla}_{x''}\Phi_s^\dagger(x'')\Psi_s(x'')|in>_A\\
&=\int[dQ_+][dQ_-]\Pi_{k=1}^3[d\bar{\psi}_{k+}][d\bar{\psi}_{k-}][d\psi_{k+}][d\psi_{k-}][d\bar{\Psi}_+]\\
&\times[d\bar{\Psi}_-][d\Psi_+][d\Psi_-]\\
&\times\bar{\Psi}_r(x')\Phi_r(x')\bar{\nabla}_{x'}\Phi_r^\dagger(x')\Psi_r(x')\cdot\bar{\Psi}_s(x'')\Phi_s(x'')\\
&\bar{\nabla}_{x''}\Phi_s^\dagger(x'')\Psi_s(x'')\times\det\left(\frac{\delta G_f^d(Q_+)}{\delta\omega_+^e}\right)\\
&\times\det\left(\frac{\delta G_f^d(Q_-)}{\delta\omega_-^e}\right)\times\exp[i\int d^4x\{-\frac{1}{4}F^{d^2}_{\lambda\delta}[Q_+]+\frac{1}{4}F^{d^2}_{\lambda\delta}[Q_-]\\
&-\frac{1}{2\alpha}(G_f^d(Q_+))^2+\frac{1}{2\alpha}(G_f^d(Q_-))^2+\sum_{k=1}^3\bar{\psi}_{k+}[i\gamma^\lambda\partial_\lambda\\
&-m_k+gT^d\gamma^\lambda Q^d_{\lambda+}]\psi_{k+}-\sum_{k=1}^3\bar{\psi}_{k-}[i\gamma^\lambda\partial_\lambda-m_k+gT^d\gamma^\lambda Q^d_{\lambda-}]\psi_{k-}\\
&+\bar{\Psi}_+[i\gamma^\lambda\partial_\lambda-M+gT^d\gamma^\lambda Q^d_{\lambda+}]\Psi_+-\bar{\Psi}_-[i\gamma^\lambda\partial_\lambda-M\\
&+gT^d\gamma^\lambda Q^d_{\lambda-}]\Psi_-\}]\\
&\times<Q_+,\psi_{1+},\bar{\psi}_{1+},\psi_{2+},\bar{\psi}_{2+},\psi_{3+},\bar{\psi}_{3+},\Psi_+,\bar{\Psi}_+,\\
&0|\rho|0,\bar{\psi}_{1-},\psi_{1-},\bar{\psi}_{2-},\psi_{2-},\bar{\psi}_{3-},\psi_{3-},\bar{\Psi}_-,\Psi_-,Q_->,
\end{aligned} \tag{37}$$

where from Eqs. (4) and (6) we have

$$\begin{aligned}
&G_f^d(Q_+)=\partial_\lambda Q_+^{\lambda d}+gf^{dba}A^b_{\lambda+}Q_+^{\lambda a}-\partial_\lambda A_+^{\lambda d},\\
&T^dQ_+^{\prime\lambda d}=\Phi_+T^dQ_+^{\lambda d}\Phi_+^{-1}+\frac{1}{ig}(\partial^\lambda\Phi_+)\Phi_+^{-1}.
\end{aligned} \tag{38}$$

Since Q, ψ, $\bar{\psi}$, Ψ and $\bar{\Psi}$ are integration variables inside the path integration we can change the unprimed integration variables to primed integration variables in Eq. (37) to find

$$\begin{aligned}
&< in|\bar{\Psi}_r(x')\Phi_r(x')\bar{\nabla}_{x'}\Phi_r^\dagger(x')\Psi_r(x')\cdot\bar{\Psi}_s(x'')\Phi_s(x'')\\
&\bar{\nabla}_{x''}\Phi_s^\dagger(x'')\Psi_s(x'')|in>_A\\
&=\int[dQ'_+][dQ'_-]\Pi_{k=1}^3[d\bar{\psi}'_{k+}][d\bar{\psi}'_{k-}][d\psi'_{k+}][d\psi'_{k-}][d\bar{\Psi}'_+]\\
&\times[d\bar{\Psi}'_-][d\Psi'_+][d\Psi'_-]\bar{\Psi}'_r(x')\Phi_r(x')\bar{\nabla}_{x'}\Phi_r^\dagger(x')\Psi'_r(x')\cdot\\
&\bar{\Psi}'_s(x'')\Phi_s(x'')\bar{\nabla}_{x''}\Phi_s^\dagger(x'')\Psi'_s(x'')\times\det(\frac{\delta G_f^d(Q'_+)}{\delta\omega_+^e})\\
&\times\det(\frac{\delta G_f^d(Q'_-)}{\delta\omega_-^e})\times\exp[i\int d^4x\{-\frac{1}{4}F^{d^2}_{\lambda\delta}[Q'_+]+\frac{1}{4}F^{d^2}_{\lambda\delta}[Q'_-]\\
&-\frac{1}{2\alpha}(G_f^d(Q'_+))^2+\frac{1}{2\alpha}(G_f^d(Q'_-))^2\\
&+\sum_{k=1}^3\bar{\psi}'_{k+}[i\gamma^\lambda\partial_\lambda-m_k+gT^d\gamma^\lambda Q'^d_{\lambda+}]\psi'_{k+}\\
&-\sum_{k=1}^3\bar{\psi}'_{k-}[i\gamma^\lambda\partial_\lambda-m_k+gT^d\gamma^\lambda Q'^d_{\lambda-}]\psi'_{k-}\\
&+\bar{\Psi}'_+[i\gamma^\lambda\partial_\lambda-M+gT^d\gamma^\lambda Q'^d_{\lambda+}]\Psi'_+\\
&-\bar{\Psi}'_-[i\gamma^\lambda\partial_\lambda-M+gT^d\gamma^\lambda Q'^d_{\lambda-}]\Psi'_-\}]\\
&\times<Q'_+,\psi'_{1+},\bar{\psi}'_{1+},\psi'_{2+},\bar{\psi}'_{2+},\psi'_{3+},\bar{\psi}'_{3+},\Psi'_+,\\
&\bar{\Psi}'_+,0|\rho|0,\bar{\psi}'_{1-},\psi'_{1-},\bar{\psi}'_{2-},\psi'_{2-},\bar{\psi}'_{3-},\psi'_{3-},\bar{\Psi}'_-,\Psi'_-,Q'_->.
\end{aligned} \tag{39}$$

The SU(3) pure gauge background field $A^{\lambda d}(x)$ given by Eq. (32). Using the background field $A^{\lambda d}(x)$ as the SU(3) pure gauge background field given by Eq. (32) we find from

$$\psi'_+(x)=\Phi_+(x)\psi_+(x) \tag{40}$$

and from Eq. (38) that [1, 33, 34]

$$\begin{aligned}
&[d\bar{\psi}'_{k+}][d\psi'_{k+}]=[d\bar{\psi}_{k+}][d\psi_{k+}],\quad [dQ'_+]=[dQ_+],\\
&[d\bar{\Psi}'_+][d\Psi'_+]=[d\bar{\Psi}_+][d\Psi_+],\\
&(G_f^d(Q'_+))^2=(\partial_\lambda Q_+^{\lambda d}(x))^2,\quad \det\left[\frac{\delta G_f^d(Q'_+)}{\delta\omega_+^e}\right]=\det\left[\frac{\delta(\partial_\lambda Q_+^{\lambda d}(x))}{\delta\omega_+^e}\right]\\
&\bar{\psi}'_{k+}[i\gamma^\lambda\partial_\lambda-m_k+gT^d\gamma^\lambda Q'^d_{\lambda+}]\psi'_{k+}=\bar{\psi}_{k+}[i\gamma^\lambda\partial_\lambda-m_k+gT^d\gamma^\lambda Q^d_{\lambda+}]\psi_{k+},\\
&\bar{\Psi}'_+[i\gamma^\lambda\partial_\lambda-M+gT^d\gamma^\lambda Q'^d_{\lambda+}]\Psi'_+=\bar{\Psi}_+[i\gamma^\lambda\partial_\lambda-M+gT^d\gamma^\lambda Q^d_{\lambda+}]\Psi_+.
\end{aligned} \tag{41}$$

At the initial time we are working in the frozen ghost formalism for the non-equilibrium QCD at the initial time [6, 7]. This implies from Eqs. (38) and (40) that at the initial time the $<Q_+,\psi_{1+},\bar{\psi}_{1+},\psi_{2+},\bar{\psi}_{2+},\psi_{3+},\bar{\psi}_{3+},\Psi_+,\bar{\Psi}_+,0|\rho|0,\bar{\psi}_{1-},\psi_{1-},\bar{\psi}_{2-},\psi_{2-},\bar{\psi}_{3-},\psi_{3-},\bar{\Psi}_-,\Psi_-,Q_->$ in non-equilibrium QCD at the initial time is gauge invariant by definition, *i. e.*, [34]

$$\begin{aligned}
&<Q'_+,\psi'_{1+},\bar{\psi}'_{1+},\psi'_{2+},\bar{\psi}'_{2+},\psi'_{3+},\bar{\psi}'_{3+},\Psi'_+,\\
&\times\bar{\Psi}'_+,0|\rho|0,\bar{\psi}'_{1-},\psi'_{1-},\bar{\psi}'_{2-},\psi'_{2-},\bar{\psi}'_{3-},\psi'_{3-},\\
&\bar{\Psi}'_-,\Psi'_-,Q'_->\\
&=<Q_+,\psi_{1+},\bar{\psi}_{1+},\psi_{2+},\bar{\psi}_{2+},\psi_{3+},\bar{\psi}_{3+},\Psi_+,\\
&\times\bar{\Psi}_+,0|\rho|0,\bar{\psi}_{1-},\psi_{1-},\bar{\psi}_{2-},\psi_{2-},\bar{\psi}_{3-},\psi_{3-},\\
&\bar{\Psi}_-,\Psi_-,Q_->.
\end{aligned} \tag{42}$$

From Eqs. (41), (40), (42), (39) and (34) we finally obtain

$$\begin{aligned}
&<in|\bar{\Psi}_r(x')\bar{\nabla}_{x'}\Psi_r(x')a_H^\dagger\cdot a_H\bar{\Psi}_s(x)\bar{\nabla}_x\Psi_s(x)|in>\\
&=<in|\bar{\Psi}_r(x')\Phi_r(x')\bar{\nabla}_{x'}\Phi_r^\dagger(x')\Psi_r(x')a_H^\dagger\cdot a_H\bar{\Psi}_s(x)\\
&\times\Phi_s(x)\bar{\nabla}_x\Phi_s^\dagger(x)\Psi_s(x)|in>_A,
\end{aligned} \tag{43}$$

which proves the factorization of infrared divergences in χ_{cJ} production from color singlet $c\bar{c}$ pair in non-equilibrium QCD at all order in coupling constant where the light-like gauge link or the light-like eikonal line $\Phi_+(x)$ in the fundamental representation of SU(3) is given by

$$\Phi_+(x) = \mathcal{P}e^{-igT^d \int_0^\infty d\tau l \cdot A_+^d(x+\tau l)}. \tag{44}$$

Correct definition of χ_{cJ} production in non-equilibrium QCD at RHIC and LHC in color singlet mechanism

From Eq. (43) we find that the correct definition of the gauge invariant non-perturbative matrix element of the χ_{c0} production from the color singlet $c\bar{c}$ pair in non-equilibrium QCD which is consistent with factorization of infrared divergences at all orders in coupling constant is given by

$$<in|\mathcal{O}_{\chi_{c0}}|in> \;=\; <in|\zeta^\dagger \Phi \bar{\nabla} \Phi^\dagger \xi a^\dagger_{\chi_{c0}} \cdot a_{\chi_{c0}} \xi^\dagger \Phi \bar{\nabla} \Phi^\dagger \zeta|in>. \tag{45}$$

Since the left-hand side of Eq. (43) is independent of the light-like four-velocity l^λ we find that the long-distance behavior of the χ_{c0} non-perturbative matrix element $<in|\mathcal{O}_{\chi_{c0}}|in> \;=\; <in|\zeta^\dagger \Phi \bar{\nabla} \Phi^\dagger \xi a^\dagger_{\chi_{c0}} \cdot a_{\chi_{c0}} \xi^\dagger \Phi \bar{\nabla} \Phi^\dagger \zeta|in>$ in Eq. (45) in non-equilibrium QCD is independent of the light-like vector l^λ used to define the light-like gauge link or the light-like eikonal line in Eq. (44) at all orders in coupling constant.

Conclusions

Recently we have proved the factorization of NRQCD S-wave heavy quarkonium production at all orders in coupling constant. In this paper we have extended this to prove the factorization of infrared divergences in χ_{cJ} production from color singlet $c\bar{c}$ pair in non-equilibrium QCD at RHIC and LHC at all orders in coupling constant. This can be relevant to study the quark–gluon plasma at RHIC and LHC.

References

1. Nayak, G.C.: Proof of NRQCD factorization at all order in coupling constant in heavy quarkonium production. Eur. Phys. J. C **76**, 448 (2016)
2. Cooper, F., Mottola, E., Nayak, G.C.: Minijet initial conditions for non-equilibrium parton evolution at RHIC and LHC. Phys. Lett. B **555**, 181 (2003)
3. Matsui, T., Satz, H.: J/ψ suppression by quark-gluon plasma formation. Phys. Lett. B **178**, 416 (1986)
4. Schwinger, J.: Brownian motion of a quantum oscillator. J. Math. Phys. **2**, 407 (1961)
5. Keldysh, L.V.: Diagram technique for nonequilibrium processes. JETP **20**, 1018 (1965)
6. Kao, C.-W., Nayak, G.C., Greiner, W.: Closed-time path integral formalism and medium effects of non-equilibrium QCD matter. Phys. Rev. D **66**, 034017 (2002)
7. Cooper, F., Kao, C.-W., Nayak, G.C.: Infrared behaviour of the gluon propagator in non-equilibrium situations. Phys. Rev. D **66**, 114016 (2002)
8. Bazavov, A., et al.: Equation of state and QCD transition at finite temperature. Phys. Rev. D **80**, 014504 (2009)
9. Teaney, D., Lauret, J., Shuryak, E.V.: A hydrodynamic description of heavy ion collisions at the SPS and RHIC. arXiv:nucl-th/0110037
10. Teaney, D., Lauret, J., Shuryak, E.V.: Flow at the SPS and RHIC as a quark gluon plasma signature. Phys. Rev. Lett. **86**, 4783 (2001)
11. Luzum, M., Romatschke, P.: Conformal relativistic viscous hydrodynamics: applications to RHIC results at $\sqrt{s_{NN}}$ = 200-GeV. Phys. Rev. C **78**, 034915 (2008) Erratum: Phys. Rev. C **79**, 039903 (2009)
12. Kolb, P.F., Heinz, U.W.: Hydrodynamic description of ultrarelativistic heavy ion collisions. In: Hwa, R.C. (ed.) et al.: Quark gluon plasma, pp. 634–714. arXiv:nucl-th/0305084
13. Muller, B.: From quark-gluon plasma to the perfect liquid. Acta Phys. Polon. B **38**, 3705 (2007)
14. Shuryak, E.V.: Why does the quark gluon plasma at RHIC behave as a nearly ideal fluid? Progr. Part Nucl. Phys. **53**, 273 (2004)
15. Nayak, G.C.: Jet quenching at RHIC and LHC and the fragmentation function in vacuum. arXiv:1705.04878 [hep-ph]
16. Liu, H., Rajagopal, K., Wiedemann, U.A.: Wilson loops in heavy ion collisions and their calculation in AdS/CFT. JHEP **0703**, 066 (2007)
17. Albacete, J.L., Kovchegov, Y.V., Taliotis, A.: Modeling heavy ion collisions in AdS/CFT. JHEP **0807**, 100 (2008)
18. Herzog, C.P., et al.: Energy loss of a heavy quark moving through N = 4 supersymmetric Yang–Mills plasma. JHEP **0607**, 013 (2006)
19. McLerran, L.D., Venugopalan, R.: Computing quark and gluon distribution functions for very large nuclei. Phys. Rev. D **49**, 2233 (1994)
20. McLerran, L.D., Venugopalan, R.: Gluon distribution functions for very large nuclei at small transverse momentum. Phys. Rev. D **49**, 3352 (1994)
21. Gyulassy, M., Vitev, I., Wang, X.-N., Zhang, B.-W.: Jet quenching and radiative energy loss in dense nuclear matter. In: Hwa, R.C. (ed.) et al.: Quark gluon plasma, pp. 123–191. arXiv: nucl-th/0302077
22. Chien, Y.-T., Emerman, A., Kang, Z.-B., Ovanesyan, G., Vitev, I.: Jet quenching from QCD evolution. Phys. Rev. D **93**, 074030 (2016). arXiv:1509.02936 [hep-ph]
23. Barbieri, R., et al.: Strong radiative corrections to annihilations of quarkonia in QCD. Nucl. Phys. B **154**, 535 (1979)
24. Barbieri, R., et al.: Singular binding dependence in the hadronic widths of 1^{++} and 1^{+-} heavy quark anti-quark bound states. Phys. Lett. **61B**, 465 (1976)
25. Barbieri, R., et al.: Gluon jets from quarkonia. Nucl. Phys. B **162**, 220 (1980)
26. Barbieri, R., et al.: Strong QCD corrections to p-wave quarkonium decays. Phys. Lett. **95B**, 93 (1980)
27. Barbieri, R., et al.: QCD corrections to P-wave quarkonium decays. Nucl. Phys. B **192**, 61 (1981)
28. Kwong, W., et al.: Quarkonium annihilation rates. Phys. Rev. D **37**, 3210 (1988)

29. Nayak, G.C.: Correct definition of color singlet p-wave non-perturbative matrix element of heavy quarkonium production. JHEP **2017**, 90 (2017)arXiv:1704.07449v2 [hep-ph]
30. Hooft, G.: An algorithm for the poles at dimension four in the dimensional regularization procedure. Nucl. Phys. B **62**, 444 (1973)
31. Klueberg-Stern, H., Zuber, J.B.: Renormalization of nonabelian gauge theories in a background field gauge. 1. Green functions. Phys. Rev. D **12**, 482 (1975)
32. Abbott, L.F.: The background field method beyond one loop. Nucl. Phys. B **185**, 189 (1981)
33. Nayak, G.C.: Light-like Wilson line in QCD without path ordering. Phys. Part Nucl. Lett. **13**, 417 (2016)
34. Nayak, G.C.: Jet quenching and gluon to hadron fragmentation function in non-equilibrium QCD at RHIC and LHC. Phys. Part Nucl. Lett. **14**, 18 (2017)
35. Braaten, E., Chen, Y.-Q.: Calculation of P wave charmonium decay rates using dimensional regularization. Phys. Rev. D **55**, 7152 (1997). arXiv:hep-ph/9701242
36. Petrelli, A., et al.: NLO production and decay of quarkonium. Nucl. Phys. B **514**, 245 (1998). arXiv:hep-ph/9707223
37. Collins, J.C., Soper, D.E., Sterman, G.: Factorization for one loop corrections in the Drell–Yan process. Nucl. Phys. B **223**, 381 (1983)
38. Collins, J., Soper, D.E., Sterman, G.: Does the Drell–Yan cross-section factorize? Phys. Lett. **109B**, 388 (1982)
39. Collins, J., Soper, D.E., Sterman, G.: Relation of parton distribution functions in Drell-Yan process to deeply inelastic scattering. Phys. Lett. **126B**, 275 (1983)
40. Collins, J., Soper, D.E., Sterman, G.: All order factorization for Drell-Yan cross-sections. Phys. Lett. **134B**, 263 (1984)
41. Collins, J.C., Soper, D.E., Sterman, G.: Factorization for short distance hadron–hadron scattering. Nucl. Phys. B **261**, 104 (1985)
42. Nayak, G.C., Qiu, J.-W., Sterman, G.: Fragmentation, factorization and infrared poles in heavy quarkonium production. Phys. Lett. B **613**, 45 (2005)
43. Nayak, G.C., Qiu, J.-W., Sterman, G.: Fragmentation, NRQCD and NNLO factorization analysis in heavy quarkonium production. Phys. Rev. D **72**, 114012 (2005)
44. Nayak, G.C., Qiu, J.-W., Sterman, G.: NRQCD factorization and velocity-dependence of NNLO poles in heavy quarkonium production. Phys. Rev. D **74**, 074007 (2006)
45. Nayak, G.C., Qiu, J.-W., Sterman, G.: Color transfer in associated heavy-quarkonium production. Phys. Rev. Lett. **99**, 212001 (2007)
46. Nayak, G.C., Qiu, J.-W., Sterman, G.: Color transfer enhancement for heavy quarkonium production. Phys. Rev. D **77**, 034022 (2008)
47. Harris, I., Brown, L.M.: Radiative corrections to pair annihilation. Phys. Rev. **105**, 1656 (1957)
48. Grammer, G., Yennie, D.R.: Improved treatment for the infrared divergence problem in quantum electrodynamics. Phys. Rev. D **8**, 4332 (1973)
49. Nayak, G.C.: General form of the color potential produced by color charges of the quark. JHEP **1303**, 001 (2013)
50. Nayak, G.C.: General form of color charge of the quark. Eur. Phys. J. C **73**, 2442 (2013)
51. Tucci, R.: Factorization of soft and collinear divergences in QCD in Feynman gauge via background field gauge. Phys. Rev. D **32**, 945 (1985)
52. Nayak, G.C.: Factorization of soft and collinear divergences in non-equilibrium quantum field theory. Ann. Phys. **324**, 2579 (2009)
53. Beneke, M.: A quark mass definition adequate for threshold problems. Phys. Lett. B **434**, 115 (1998)
54. Hoang, A., Smith, M., Stelzer, T., Willenbrock, S.: Quarkonia and the pole mass. Phys. Rev. D **59**, 114014 (1999)
55. Sumino, Y.: A connection between the perturbative QCD potential and phenomenological potentials. Phys. Rev. D **65**, 054003 (2002)
56. Sumino, Y.: QCD potential as a 'Coulomb plus linear' potential. Phys. Lett. B **571**, 173 (2003)
57. Muta, T.: Foundations of quantum chromodynamics. World Scientific lecture notes in physics, vol. 5 (1987)

28

Effects of Landau damping on finite amplitude low-frequency nonlinear waves in a dusty plasma

Arnab Sikdar[1] · Manoranjan Khan[2]

Abstract The effect of linear ion Landau damping on weakly nonlinear as well as weakly dispersive low-frequency waves in a dusty plasma is investigated. The standard perturbative approach leads to the Korteweg–de Vries (KdV) equation with a linear Landau damping term for the dynamics of the low-frequency nonlinear wave. Landau damping causes the wave amplitude to decay with time and the dust charge variation enhances the damping rate.

Keywords Landau damping · Dusty plasma · Korteweg–de Vries equation · Dust acoustic wave

Introduction

The Landau damping is a physical phenomenon which is related to the resonant particles (the particles whose velocity is nearly equal to the wave phase velocity) [1, 2]. The resonant particles may include both trapped and un-trapped particles. The usual ion acoustic wave in electron-ion plasma suffers Landau damping due to these resonant particles [1–3]. However, the presence of charged dust grains in a plasma gives rise to very low-frequency new mode ($\sim$10–15 Hz), called dust acoustic wave (DAW) [4–8], where the inertia is provided by the charged and massive dust grains. In the linear theory, it has already been seen that this mode also suffers Landau damping due to the resonant wave–particle interactions [9, 10]. Another well known non-Landau damping mechanism in a dusty plasma is due to the dust charge variations in the presence of waves [11, 20]. Actually, dust grains immersed in a plasma can exhibit self-consistent charge variations in response to the surrounding plasma oscillations and thus become a time-dependent dynamical variable which causes an anomalous dissipation in a dusty plasma.

In the nonlinear theory, the linear electron Landau damping effects on ion acoustic solitary wave have been investigated in an electron-ion plasma neglecting the particle trapping effect under the assumption that the particle trapping time is much larger than the Landau damping time [1, 13]. It has been shown that the solitary wave amplitude decays with time due to the linear electron Landau damping [13]. Later, theoretical [14, 15] and experimental [16] results show similar behavior. The wave–particle interactions also cause the oscillations in the tail of the solitary waves in which the shape of the tail depends on the strength of the Landau damping [17]. Recent experimental observation also predicts the formation of ion acoustic shock wave due to the Landau damping induced dissipation [18]. However, no study of nonlinear DAW is carried out including Landau damping in a dusty plasma. In this paper, the effect of linear ion Landau damping on dust acoustic solitary wave has been investigated neglecting the particle trapping effect. The instantaneous dust charge variation effects are also incorporated.

The manuscript is organized in the following manner. Formulation of the problem including the physical assumptions and basic equations is described in Sect. 2. The Korteweg–de Vries (KdV) equation with linear damping is derived using the reductive perturbation technique in Sect. 3. The analytical solution and the effect of

✉ Arnab Sikdar
arnabs.ju@gmail.com

Manoranjan Khan
mkhan_ju@yahoo.com

1 Department of Mathematics, Calcutta Institute of Engineering and Management, 24/1A Chandi Ghosh Road, Kolkata 700040, India

2 Department of Instrumentation Science, Jadavpur University, Kolkata 700032, India

Landau damping on the solitary wave solution are investigated in Sect. 4. The results of the present investigation are summarized in Sect. 5.

Formulation of the problem and the basic equations

A fully ionized, un-magnetized plasma consisting of electrons, ions, and negatively charged dust grains are considered. The plasma is assumed to be in its equilibrium state at $-\infty$, where electrostatic potential $\phi = 0$, electron number density $n_e = n_{e0}$, ion density $n_i = n_{i0}$, dust density $n_d = n_{d0}$, and dust charge $q_d = -z_{de}$, so that the quasi-neutrality condition $n_{e0} + z_d n_{d0} = n_{i0}$ is satisfied, where z_d is the number of electrons residing on the dust grains, n_j is the number density of the jth ($e =$ electron , $i =$ ion and $d =$ dust grain) species, and n_{j0} be its equilibrium value.

The charge on the dust grain varies continuously in space (x) and time (t). The temperature of dust grain is very low compared to that of electrons (T_e) and ions (T_i). Therefore, the dust grains are effectively cold with respect to the electrons and ions. The dust grains are moving with fluid velocity U. It is convenient to express all the variables in non-dimensional form before going to the details of the basic formalism of the problem. For this purpose, let us introduce the following normalization: $\Phi = e\phi/T_i$, $N = n_d/n_{d0}$, $N_e = n_e/n_{e0}$, $N_i = n_i/n_{i0}$, $\bar{x} = x/\lambda_D$, $\bar{t} = \omega_{pd}t$, $\bar{U} = U/C_s$ and $\bar{q_d} = q_d/z_{de} = -1 + \Delta Q$, and ΔQ is the fluctuating dust charge. Here, $\omega_{pd} = (z_d^2 e^2 n_{d0}/\epsilon_0 m_d)^{1/2}$ is the dust plasma frequency, $\lambda_D = (\epsilon_0 T_i/n_{i0}e^2)^{1/2}$ is the plasma Debye length, $C_s = (z_d T_i/m_d)^{1/2}$ is the dust acoustic speed, $\delta_i = n_{i0}/n_{e0}$, $\sigma = T_i/T_e$, and $z = z_{de}^2/4\pi\epsilon_0 r_d T_e$ are the dimensionless dusty plasma parameters (the ratio of the electrostatic energy of a dust grain of radius r_d to the electron thermal energy). Hereafter, we will be using these new variables and remove all the bars for simplicity of notations.

We assume that $\delta_i(m_i/m_d)^{1/2} \gg \sigma^{3/2}(m_e/m_i)^{1/2}$, so that the electron Landau damping effect is neglected. The dust Landau damping effect is also neglected as the dust thermal velocity is much smaller than the wave phase velocity. Moreover, we are interested to study the low-frequency nonlinear DAW and, therefore, we neglect the inertia of the electrons compared to the dust grains. On this slow time scale, the electrons are in local thermodynamic equilibrium and their densities are modeled by the Boltzmann distribution: $n_e = n_{e0}\exp(\sigma\Phi)$.

The ions are treated kinetically, so that their number densities are given by

$$N_i = \int_{-\infty}^{+\infty} f dV. \tag{1}$$

The velocity distribution function of ion f (normalized) satisfies the following Vlasov–Boltzmann equation:

$$M\frac{\partial f}{\partial t} + V\frac{\partial f}{\partial x} - \frac{\partial \Phi}{\partial x}\frac{\partial f}{\partial V} = 0. \tag{2}$$

Here, the parameter $M = (z_d m_i/m_d)^{1/2}$ represents the effect of finite ion inertia on propagation characteristics of DAW. The ion velocity V is normalized in units of ion thermal velocity $V_{ti} = (T_i/m_i)^{1/2}$ and velocity distribution function f is normalized by V_{ti}/n_{i0}. It is to be noted that when the plasma is in thermodynamical equilibrium, the velocity distribution of the ion is given by the following Maxwellian distribution:

$$f^{(0)} = \frac{1}{\sqrt{2\pi}}\exp(-V^2/2) \tag{3}$$

which is also the solution of Eq. (2) in the absence of external force ($\Phi = 0$).

Finally, the nonlinear propagation of low phase velocity (in comparison with the electron and ion thermal velocities) DAW is governed by the following normalized basic equations:

$$\frac{\partial N}{\partial t} + \frac{\partial (NU)}{\partial x} = 0 \tag{4}$$

$$\frac{\partial U}{\partial t} + U\frac{\partial U}{\partial x} = -(\Delta Q - 1)\frac{\partial \Phi}{\partial x} \tag{5}$$

$$\frac{\partial^2 \Phi}{\partial x^2} = \left[\frac{1}{\delta_i - 1}\exp(\sigma\Phi) - \frac{\delta_i}{\delta_i - 1}N_i - (\Delta Q - 1)N\right]. \tag{6}$$

The normalized dust grain charging equation becomes

$$\left(\frac{\omega_{pd}}{v_d}\right)\frac{d\Delta Q}{dt} = \frac{I_e + I_i}{v_d z_d e} \tag{7}$$

where $v_d = r_d\omega_{pi}^2(1 + z + \sigma)/\sqrt{2\pi}V_{ti}$ is the dust charging frequency and ω_{pi} is the ion plasma frequency. The expressions for electron current (I_e) and ion current (I_i) for negatively charged dust grains are given by

$$I_e = -\pi r_d^2 e\sqrt{\frac{8T_e}{\pi m_e}}n_{e0}\exp(\sigma\Phi)\exp[z(\Delta Q - 1)] \tag{8}$$

and

$$I_i = \pi r_d^2 e\sqrt{\frac{8T_i}{\pi m_i}}n_{i0}N_i\left[1 - \frac{z}{\sigma}(\Delta Q - 1)\right]. \tag{9}$$

Korteweg–de Vries equation with Landau damping

To study the propagation characteristics of finite amplitude nonlinear DAW, the reductive perturbation technique is adopted. Accordingly, the stretched co-ordinates and power series expansion (in powers of ϵ) of dependent variables are as follows:

$$\xi = \sqrt{\epsilon}(x - \Lambda t),\ \tau = \epsilon^{3/2} t \tag{10}$$

$$h = h^{(0)} + \epsilon h^{(1)} + \epsilon^2 h^{(2)} + \cdots \tag{11}$$

where ϵ is the smallness parameter that indicates the magnitude of the rate of change and Λ is the normalized wave phase velocity. It is to be noted that $h^{(0)} = 1(0)$, $h \equiv N_j(\Phi, U)$, and $h = f^{(0)}$ for $h \equiv f$. To incorporate the effects of ion inertia on finite amplitude nonlinear DAW, the following scaling is assumed which is consistent with the perturbation:

$$M \sim O(\epsilon) \Rightarrow M = \mu\epsilon. \tag{12}$$

Substituting Eqs. (10)–(12) in Eqs. (1)–(8) and then equating different powers of ϵ on both sides of these equations, the following relations are obtained. In the lowest order of ϵ, Eqs. (1)—(8) reduce to

$$N_i^{(1)} = \int_{-\infty}^{+\infty} f^{(1)} \mathrm{d}V,\ \Lambda N^{(1)} = U^{(1)} \tag{13}$$

$$\Lambda U^{(1)} = -\Phi^{(1)},\ \Delta Q^{(1)} = \frac{\sigma}{\delta_i - 1}\Phi^{(1)} - \frac{\delta_i}{\delta_i - 1}N_i^{(1)} + N^{(1)}. \tag{14}$$

The Vlasov–Boltzmann equation (2) for ions at the order of $\epsilon^{3/2}$ yields

$$V\frac{\partial f^{(1)}}{\partial \xi} + Vf^{(0)}\frac{\partial \Phi^{(1)}}{\partial \xi} = 0. \tag{15}$$

This equation does not have unique solution [13]. However, the non-uniqueness can be removed by including a $\partial f^{(1)}/\partial \tau$ term [13] in Eq. (15), and thus, we have the following equation:

$$\mu\epsilon^2\frac{\partial f_\epsilon^{(1)}}{\partial \tau} + V\frac{\partial f_\epsilon^{(1)}}{\partial \xi} = -Vf^{(0)}\frac{\partial \Phi^{(1)}}{\partial \xi}. \tag{16}$$

Then, $f^{(1)}$ is uniquely determined from the solution of the Eq. (16) by taking $f^{(1)} = \lim_{\epsilon \to 0} f_\epsilon^{(1)}$ [13]. Finally, we get

$$f^{(1)} = -f^{(0)}\Phi^{(1)}. \tag{17}$$

It is well known that the non-steady dust charge variations produce an anomalous dissipation which leads to collisionless, non-Landau wave damping in a dusty plasma [19–22]. However, for a typical laboratory dusty plasma [6], dust oscillation frequency $\omega_{\mathrm{pd}} \approx 10^2\ \mathrm{s}^{-1}$ and dust charging frequency $\nu_{\mathrm{d}} \approx 10^8\ s^{-1}$ imply $\omega_{\mathrm{pd}}/\nu_{\mathrm{d}} \approx 10^{-6}$, and thus, the charging equation (7) can be approximated as

$$I_{\mathrm{e}} + I_{\mathrm{i}} \approx 0, \tag{18}$$

so that the charge on the dust grains instantaneously reaches its equilibrium value, which is known as "adiabatic variation" of dust charge. In this adiabatic approximation, Eq. (18) together with Eqs. (8) and (9), at the order of ϵ gives the following relation:

$$\Delta Q^{(1)} = \beta_{\mathrm{d}}\left(N_i^{(1)} - \sigma\Phi^{(1)}\right),\ \beta_{\mathrm{d}} = \frac{(z+\sigma)}{z(1+z+\sigma)}. \tag{19}$$

This equation together with Eqs. (13)–(15), (19) self-consistently determined the (normalized) phase velocity of DAW

$$\Lambda^2 = \frac{(\delta_{\mathrm{i}} - 1)}{(\sigma + \delta_{\mathrm{i}}) + \beta_{\mathrm{d}}(1+\sigma)(\delta_{\mathrm{i}} - 1)}. \tag{20}$$

In the absence of dust charge variations ($\beta_{\mathrm{d}} = 0$), this Eq. (20) can be written in dimensional form, as

$$\omega^2 = \frac{k^2 C_{\mathrm{s}}^2(\delta_{\mathrm{i}} - 1)}{(\delta_{\mathrm{i}} + \sigma)},$$

which is the phase velocity of the usual DAW in the long wavelength limit [4].

In the next higher order of ϵ, Eqs. (1), (2) and (4)–(6) yield the following relations:

$$\frac{\partial N^{(1)}}{\partial \tau} + \frac{\partial (N^{(1)}U^{(1)})}{\partial \xi} = \Lambda\frac{\partial N^{(2)}}{\partial \xi} - \frac{\partial U^{(2)}}{\partial \xi} \tag{21}$$

$$\frac{\partial U^{(1)}}{\partial \tau} + \Delta Q^{(1)}\frac{\partial \Phi^{(1)}}{\partial \xi} + U^{(1)}\frac{\partial U^{(1)}}{\partial \xi} = \Lambda\frac{\partial U^{(2)}}{\partial \xi} + \frac{\partial \Phi^{(2)}}{\partial \xi} \tag{22}$$

$$\begin{aligned}\frac{\partial^2 \Phi^{(1)}}{\partial \xi^2} = {} & \frac{\sigma}{\delta_i - 1}\Phi^{(2)} + \frac{\sigma^2}{2(\delta_i - 1)}{\Phi^{(1)}}^2 - \frac{\delta_i}{\delta_i - 1}N_i^{(2)} \\ & - \Delta Q^{(1)}N^{(1)} - \Delta Q^{(2)} + N^{(2)}\end{aligned} \tag{23}$$

$$N_i^{(2)} = \int_{-\infty}^{+\infty} f^{(2)}\mathrm{d}V. \tag{24}$$

To get the unique solution for $f^{(2)}$ of Eq. (2), proceeding as before, a term containing time derivative of $f^{(2)}$ (as in the lowest order case) is included in the equation of order of $\epsilon^{5/2}$ of Eq. (2), and thus, we obtain

$$\mu\epsilon^2\frac{\partial f_\epsilon^{(2)}}{\partial \tau} + V\frac{\partial f_\epsilon^{(2)}}{\partial \xi} = -\left[\Lambda\mu f^{(0)}\frac{\partial \Phi^{(1)}}{\partial \xi} + Vf^{(0)}\frac{\partial \Phi^{(2)}}{\partial \xi} - Vf^{(0)}\Phi^{(1)}\frac{\partial \Phi^{(1)}}{\partial \xi}\right]. \tag{25}$$

As before, $f^{(2)}$ is uniquely determined from the solution of Eq. (25) by taking the limit $f^{(2)} = \lim_{\epsilon \to 0} f_\epsilon^{(2)}$ and thus finally using the relation (24), we obtain the following relation:

$$\frac{\partial N_{\mathrm{i}}^{(2)}}{\partial \xi} = -\frac{\partial \Phi^{(2)}}{\partial \xi} + \Phi^{(1)}\frac{\partial \Phi^{(1)}}{\partial \xi} + \left(\frac{\Lambda\mu}{\sqrt{2\pi}}\right)\wp\int_{-\infty}^{+\infty}\frac{\partial \Phi^{(1)}}{\partial \acute{\xi}}\frac{d\acute{\xi}}{\xi - \acute{\xi}} \tag{26}$$

where $\wp$ represents Cauchy's principal value. Note that in the derivation of the above relation (26), we use the properties of generalized functions [25]: $1/(\eta - i0) = \wp(1/\eta) + i\pi\delta(\eta)$, $\eta\wp(1/\eta) = 1$, $\eta\delta(\eta) = 0$ and $\delta(k\eta) = (\mathrm{sgn}(k)/k)\delta(\eta)$.

The Eq. (18) together with Eqs. (8) and (9) at the order of ϵ^2 gives

$$\Delta Q^{(2)} = \beta_{\mathrm{d}}\Big(N_i^{(2)} - \sigma\Phi^{(2)} - z\sigma\Delta Q^{(1)}\Phi^{(1)} - \frac{\sigma^2}{2}\Phi^{(1)^2} - \frac{z^2}{2}\Delta Q^{(1)^2} - \frac{z}{\sigma+z}N_i^{(1)}\Delta Q^{(1)}\Big). \quad (27)$$

Finally, the usual perturbation analysis (the elimination of all the second-order terms) yields the following modified form of Korteweg–de Vries (KdV) equation modified by Landau damping:

$$\frac{\partial\Phi^{(1)}}{\partial\tau} - \alpha\Phi^{(1)}\frac{\partial\Phi^{(1)}}{\partial\xi} + \beta\frac{\partial^3\Phi^{(1)}}{\partial\xi^3} + \gamma_L\wp\int_{-\infty}^{+\infty}\frac{\partial\Phi^{(1)}}{\partial\acute{\xi}}\frac{d\acute{\xi}}{\xi-\acute{\xi}} = 0 \quad (28)$$

where

$$\alpha = \beta\left[\frac{3}{\Lambda^4} + \frac{\sigma^2-\delta_i}{\delta_i-1} - \beta_{\mathrm{d}}\left(\frac{3}{\Lambda^2}(1+\sigma) + 1 - \frac{z^2\beta_{\mathrm{d}}^2}{(\sigma+z)^2}\left((1+\sigma+z)^2 + (1+\sigma)^2\right)\right)\right], \quad (29)$$

$$\beta = \frac{\Lambda^3}{2} \quad (30)$$

and

$$\gamma_{\mathrm{L}} = \frac{\Lambda^4\mu}{2\sqrt{2\pi}}\left(\beta_{\mathrm{d}} + \frac{\delta_i}{\delta_i-1}\right). \quad (31)$$

The variations of γ_{L} with ion-electron density ratio for different ion-electron temperature ratio are shown graphically in Fig. 3. Note that $\mu = 0 \Rightarrow \gamma_{\mathrm{L}} = 0$ and then from Eq. (28), we recover the usual KdV equation for the finite amplitude nonlinear DAW. The term β_{d} present in the expression for γ_{L} is responsible for the instantaneous dust charge variations. It is also to be noted that one can easily obtain the regular Landau damping of DAW from Eq. (28) for $\alpha = \beta = 0$. Let us discuss it briefly: Taking the Fourier transform of Eq. (28) with $\alpha = \beta = 0$ with respect to ξ and τ [according to the formula, $\tilde{g}(\omega,k) = \int_{-\infty}^{\infty}\int_{-\infty}^{\infty} g(\xi,\tau)\exp\{i(k\xi-\omega\tau)\}\mathrm{d}\xi\mathrm{d}\tau$] and then treating the integral as a convolution with the help of the result that the inverse transform of $[i\mathrm{sgn}(k)] = -(1/\pi)\wp(1/\xi)$ [25], the following equation is obtained:

$$\omega = -i\pi k\gamma_{\mathrm{L}} = -ik\mu\sqrt{\frac{\pi}{8}}\Lambda^4\left(\beta_{\mathrm{d}} + \frac{\delta_{\mathrm{i}}}{\delta_{\mathrm{i}}-1}\right). \quad (32)$$

This clearly shows that the wave becomes damped due to finite ion inertia effects as $\mu \propto (z_{\mathrm{d}}m_{\mathrm{i}}/m_{\mathrm{d}})^{1/2}$ with the damping decrement (normalized) $|\gamma| = \pi\gamma_{\mathrm{L}}$. More precisely, in the absence of dust charge variations ($\beta_{\mathrm{d}} = 0$), we obtain the following Landau damping decrement (normalized):

$$|\gamma| = \pi\gamma_{\mathrm{L}}(\beta_{\mathrm{d}} = 0) = \sqrt{\frac{\pi}{8}}\frac{\sqrt{z_{\mathrm{d}}(\delta_{\mathrm{i}}-1)}\,\delta_{\mathrm{i}}}{(1+\delta_{\mathrm{i}}T_{\mathrm{e}}/T_{\mathrm{i}})^{3/2}}\left(\frac{T_{\mathrm{e}}}{T_{\mathrm{i}}}\right)^{3/2}\sqrt{\frac{m_{\mathrm{i}}}{m_{\mathrm{d}}}} \quad (33)$$

of DAW in usual dusty plasma [9]. These discussions clearly show that the term γ_{L} arises only due to the Landau damping, which is the consequence of the presence of μ and the scaling (Eq. 12). This expression also shows that the dust charge variations enhance the Landau damping rate.

Landau damping effect on dust acoustic solitary wave

The KdV equation (28) without the Landau damping ($\gamma_{\mathrm{L}} = 0$) represents a completely integrable Hamiltonian system which has an infinite set of conservation laws. We consider the following energy conservation equation:

$$\frac{1}{2}\frac{\partial}{\partial\tau}\int_{-\infty}^{+\infty}\Phi^{(1)^2}(\xi,\tau)\mathrm{d}\xi = 0. \quad (34)$$

This shows that in the absence of Landau damping, the wave energy is conserved and possesses the following single soliton solution:

$$\Phi^{(1)}(\xi,\tau) = A\,\mathrm{sech}^2\left[\sqrt{\frac{\alpha A}{12\beta}}\left(\xi - \frac{\alpha}{3}A\tau\right)\right] \quad (35)$$

where A is the amplitude of the solitary wave, $3A/\alpha$ is the solitary wave velocity, and $(12\beta/A\alpha)^{1/2}$ is the spatial width of the solitary wave.

However, in the presence of Landau damping ($\gamma_{\mathrm{L}} \neq 0$), the KdV equation (28) does not represent a completely integrable Hamiltonian system, and in this case, the above energy equation (34) becomes

$$\frac{1}{2}\frac{\partial}{\partial\tau}\int_{-\infty}^{+\infty}\Phi^{(1)^2}(\xi,\tau)\mathrm{d}\xi = -\gamma_{\mathrm{L}}\int_{-\infty}^{+\infty}\Phi^{(1)}(\xi,\tau)\left[\wp\int_{-\infty}^{+\infty}\frac{\mathrm{d}\Phi^{(1)}(\xi,\tau)}{\mathrm{d}\acute{\xi}}\frac{\mathrm{d}\acute{\xi}}{\xi-\acute{\xi}}\right]\mathrm{d}\xi. \quad (36)$$

Now, following the procedures of Refs. [15, 23, 24], in the presence of Landau damping, a slow time-dependent form of the solution of Eq. (28) is considered

$$\Phi^{(1)}(\xi,\tau) = A(\tau)\,\mathrm{sech}^2\left[\sqrt{\frac{\alpha A(\tau)}{12\beta}}\left(\xi - \frac{\alpha A(\tau)}{3}\tau\right)\right]. \quad (37)$$

Finally, substitution of Eq. (37) in Eq. (36) yields the following solution:

$$A(\tau) = A_0\left(1 + \frac{\tau}{\tau_0}\right)^{-2} \quad (38)$$

where

$$\tau_0^{-1} = \frac{6\gamma_L \zeta(3)}{\pi^2}\sqrt{\frac{\alpha A_0}{3\beta}} \quad (39)$$

where $A_0 = A(\tau = 0) > 0$ is the initial amplitude and $\zeta(3)$ be the Riemann Zeta function [26] defined by

$$\zeta(3) = \frac{\pi^2}{24}\wp\int_{-\infty}^{+\infty} d\xi \int_{-\infty}^{+\infty} \frac{\sec h^2\xi d\xi'}{\xi - \xi'}\frac{d}{d\xi}(\sec h^2\xi).$$

The above solution (38) shows that the linear ion Landau damping causes the dust acoustic solitary wave amplitude $A(\tau)$ to decay algebraically with time (τ) and the decay rate is proportional to γ_L (Landau damping). In the presence of Landau damping, the amplitude modulations with time τ for different ion-electron density ratio (δ_i) and temperature ratio (σ) are shown in Figs. 1 and 2.

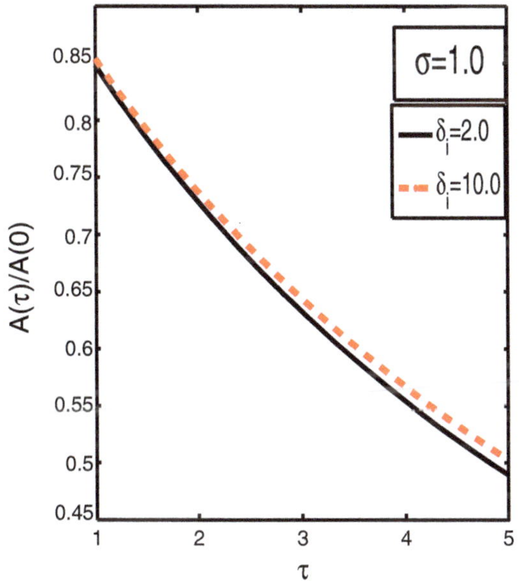

Fig. 1 (Color online) Time variations of wave amplitude with $\sigma(= T_i/T_e) = 1$ for different $\delta_i(= n_{i0}/n_{e0})$

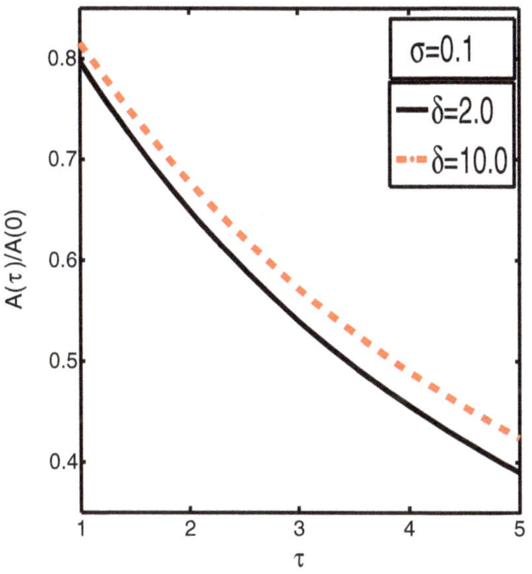

Fig. 2 (Color online) Time variations of wave amplitude with $\sigma = 0.1$ for different δ_i

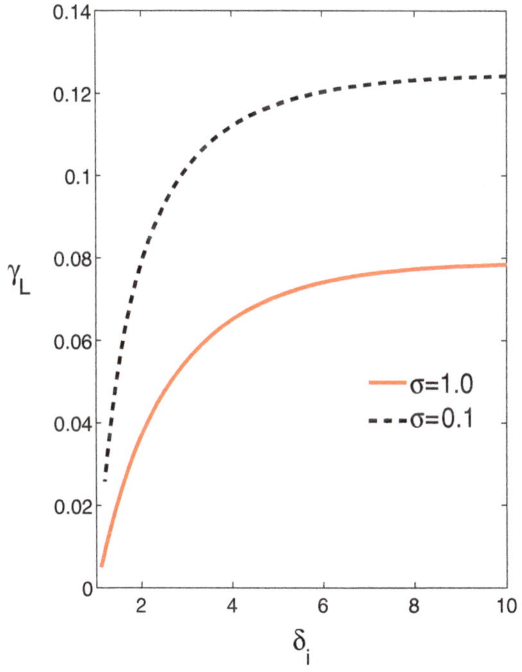

Fig. 3 (Color online) Variations of damping rate γ_L with δ_i for different σ

Conclusions

In this paper, we have investigated the effects of ion Landau damping on nonlinear dust acoustic wave. It is shown that the nonlinear wave is governed by a modified form of KdV equation [see Eq. (28)]. In the presence of Landau damping, approximate analytical solutions reveal that the wave amplitude decays algebraically with time. To understand the feature, the wave amplitude modulations with time τ [see Eqs. (38) and (39)] are shown graphically for different σ and δ_i in Figs. 1 and 2 for K^+ ion and electron plasma. Figures 1 and 2 show that wave amplitude decreases with time τ for any fixed value of δ_i and σ. However for any fixed time τ, the amplitude increases with the increase of ion-electron temperature ratio (σ) and ion-electron density ratio (δ_i). In addition, the Landau damping rate (γ_L) decreases with the increase of ion-electron temperature ratio (σ), as shown in Fig. 3.

Acknowledgements The author (A.S) would like to thank Prof. Samiran Ghosh of Department of Applied Mathematics, University of Calcutta for fruitful discussion.

References

1. Davidson, R.C.: Methods in nonlinear plasma theory. Academic, New York (1972)
2. Chen, F.F.: Introduction to plasma physics and controlled fusion. Springer, Berlin (2006)
3. Fried, B.D., Gould, R.W.: Longitudinal ion oscillations in a hot plasma. Phys. Fluids **4**, 139 (1961)
4. Rao, N.N., Shukla, P.K., Yu, M.Y.: Dust-acoustic waves in dusty plasmas. Planet. Space Sci. **38**, 543 (1990)
5. D'Angelo, N.: Low-frequency electrostatic waves in dusty plasmas. Planet. Space Sci. **38**, 1143 (1990)
6. Barkan, A., Merlino, R.L., D'Angelo, N.: Laboratory observation of the dust-acoustic wave mode. Phys. Plasmas **2**, 3563 (1995)
7. D'Angelo, N.: Coulomb solids and low-frequency fluctuations in RF dusty plasmas. J. Phys. D **28**, 1009 (1995)
8. Merlino, R.L., Heinrich, J.R., Kim, S.H., Meyer, J.K.: Dusty plasmas: experiments on nonlinear dust acoustic waves, shocks and structures. Plasma Phys. Control. Fusion **54**, 124014 (2012)
9. Rosenberg, M.: Ion and dust-acoustic instabilities in dusty plasmas. Planet. Space Sci. **41**, 229 (1993)
10. Lee, M.J.: Landau damping of dust acoustic waves in a Lorentzian plasma. Phys. Plasmas **14**, 032112 (2007)
11. Melandso, F., Aslaksen, T., Havnes, O.: A new damping effect for the dust-acoustic wave. Planet. Space Sci. **41**, 321 (1993)
12. Varma, R.K., Shukla, P.K., Krishan, V.: Electrostatic oscillations in the presence of grain-charge perturbations in dusty plasmas. Phys. Rev. E **47**, 3612 (1993)
13. Ott, E., Sudan, R.N.: Nonlinear theory of ion acoustic waves with Landau damping. Phys. Fluids **12**, 2388 (1969)
14. Hirose, A., Alexeff, I., Jones, W.D., Krall, N.A., Montgomery, D.: Landau damping of electrostatic ion waves in a uniform magnetic field. Phys. Lett. A **29**, 31 (1969)
15. Ghosh, S., Bharuthram, R.: Ion acoustic solitary wave in electron-positron-ion plasma: effect of Landau damping. Astrophys. Space Sci. **331**, 163 (2011)
16. Saitou, Y., Nakamura, Y.: Ion-acoustic soliton-like waves undergoing Landau damping. Phys. Lett. A **343**, 397 (2005)
17. Karpman, V.I., Lynov, J.P., Michelsen, P., Pecseli, H.L., Rasmussen, J.J.: Modification of plasma solitons by resonant particles. Phys. Fluids **23**, 1782 (1980)
18. Nakamura, Y., Bailung, H., Saitou, Y.: Observation of ion-acoustic solitary waves in a dusty plasma. Phys. Plasmas **11**, 3925 (2003)
19. Jana, M.R., Sen, A., Kaw, P.K.: Collective effects due to charge-fluctuation dynamics in a dusty plasma. Phys. Rev. E **48**, 3930 (1993)
20. Varma, R.K., Shukla, P.K., Krishan, V.: Electrostatic oscillation in the presence of grain-charge perturbation in dusty plasmas. Phys. Rev. E **47**, 3612 (1993)
21. Gupta, M.R., Sarkar, S., Ghosh, S., Debnath, M., Khan, M.: Effect of nonadiabaticity of dust charge variation on dust acoustic waves: generation of dust acoustic shock waves. Phys. Rev. E **63**, 046406 (2001)
22. Gupta, M.R., Sarkar, S., Khan, M., Ghosh, S.: Dust acoustic shock wave generation due to dust charge variation in a dusty plasma. Pramana J. Phys. **61**, 1197 (2003)
23. Karpman, V.I., Maslov, E.M.: Perturbation theory for solitons. Sov. Phys. JETP **46**, 281 (1977)
24. Herman, R.L.: Conservation laws and the perturbed KdV equation. J. Phys. A **23**, 4719 (1990)
25. Lighthill, M.J.: Fourier analysis and generalized functions. Cambridge University Press, London (1964)
26. Abramowitz, M., Stegun, I.A.: Handbook of mathematical functions, p. 256. Dover, New York (1970)

Permissions

All chapters in this book were first published in JTAP, by Springer International Publishing AG.; hereby published with permission under the Creative Commons Attribution License or equivalent. Every chapter published in this book has been scrutinized by our experts. Their significance has been extensively debated. The topics covered herein carry significant findings which will fuel the growth of the discipline. They may even be implemented as practical applications or may be referred to as a beginning point for another development.

The contributors of this book come from diverse backgrounds, making this book a truly international effort. This book will bring forth new frontiers with its revolutionizing research information and detailed analysis of the nascent developments around the world.

We would like to thank all the contributing authors for lending their expertise to make the book truly unique. They have played a crucial role in the development of this book. Without their invaluable contributions this book wouldn't have been possible. They have made vital efforts to compile up to date information on the varied aspects of this subject to make this book a valuable addition to the collection of many professionals and students.

This book was conceptualized with the vision of imparting up-to-date information and advanced data in this field. To ensure the same, a matchless editorial board was set up. Every individual on the board went through rigorous rounds of assessment to prove their worth. After which they invested a large part of their time researching and compiling the most relevant data for our readers.

The editorial board has been involved in producing this book since its inception. They have spent rigorous hours researching and exploring the diverse topics which have resulted in the successful publishing of this book. They have passed on their knowledge of decades through this book. To expedite this challenging task, the publisher supported the team at every step. A small team of assistant editors was also appointed to further simplify the editing procedure and attain best results for the readers.

Apart from the editorial board, the designing team has also invested a significant amount of their time in understanding the subject and creating the most relevant covers. They scrutinized every image to scout for the most suitable representation of the subject and create an appropriate cover for the book.

The publishing team has been an ardent support to the editorial, designing and production team. Their endless efforts to recruit the best for this project, has resulted in the accomplishment of this book. They are a veteran in the field of academics and their pool of knowledge is as vast as their experience in printing. Their expertise and guidance has proved useful at every step. Their uncompromising quality standards have made this book an exceptional effort. Their encouragement from time to time has been an inspiration for everyone.

The publisher and the editorial board hope that this book will prove to be a valuable piece of knowledge for researchers, students, practitioners and scholars across the globe.

List of Contributors

S. Meydanloo and S. Saviz
Plasma Physics Research Center, Science and Research Branch, Islamic Azad University, Tehran, Iran

Ziba Matinzadeh, Mahmood Ghoranneviss and Mohammad Kazem Salem
Plasma Physics Research Center, Science and Research Branch, Islamic Azad University, Tehran, Iran

Farhad Shahgoli
Department of Energy Engineering and Physics, Amirkabir University of Technology, Tehran, Iran

Hamed Abbasi
Biomedical Laser and Optics Group, Department of Biomedical Engineering, University of Basel, Allschwil, Switzerland

N. Sepehri Javan
Department of Physics, University of Mohaghegh Ardabili, PO Box 179, Ardabil, Iran

Asit Saha and Tapash Saha
Department of Mathematics, Sikkim Manipal Institute of Technology, Sikkim Manipal University, Majitar, Rangpo, East Sikkim 737136, India

Nikhil Pal and Prasanta Chatterjee
Department of Mathematics, Siksha-Bhavana, Visva-Bharati University, Santiniketan 731235, India

M. K. Ghorui
Department of Mathematics, B.B. College, Ushagram, Asansol 713303, India

M. Nikpour
Atomic and Molecular Physics Department, Faculty of Basic Sciences, University of Mazandaran, Babolsar, Iran

F. Sohbatzadeh and S. Mirzanejhad
Atomic and Molecular Physics Department, Faculty of Basic Sciences, University of Mazandaran, Babolsar, Iran
Nano and Biotechnology Research Group, Faculty of Basic Sciences, University of Mazandaran, Babolsar, Iran

H. Shokri
Faculty of Veterinary Medicine, Amol University of Special Modern Technologies, Amol, Iran

B. S. Chahal, Yashika Ghai and N. S. Saini
Department of Physics, Guru Nanak Dev University, Amritsar, India

Masoomeh Mahmoodi-Darian
Department of Physics, Karaj Branch, Islamic Azad University, Karaj, Iran

Mehdi Ettehadi-Abari and Mahsa Sedaghat
Physics Department and Laser Research Institute of Beheshti University, G.C., Evin, 19839 Tehran, Iran

Mansour Khoram
Department of Physics, Borujerd Branch, Islamic Azad University, Borujerd, Iran

Hamid Ghomi
Laser and Plasma Research Institute, Shahid Beheshti University, Evin, Tehran 1983963113, Iran

Kiomars Yasserian
Department of Physics, Karaj Branch, Islamic Azad University, Karaj, Iran

Morteza Aslaninejad
Institute for Research in Fundamental Sciences (IPM), School of Particles and Accelerators, P.O. Box 19395-5531, Tehran, Iran

O. P. Malik
Department of ECE, Al-Falah University, Dhauj, Faridabad, Haryana, India

Sukhmander Singh
Motilal Nehru College, South Campus, Delhi University, New Delhi 110 021, India

Hitendra K. Malik
PWAPA Laboratory, Department of Physics, Indian Institute of Technology Delhi, New Delhi 110 016, India

A. Kumar
Department of Applied Sciences, Al-Falah University, Dhauj, Faridabad, Haryana, India

Zahra Javadi
Central Tehran Branch, Islamic Azad University, Tehran, Iran

Shahrooz Saviz
Plasma Physics Research Center, Science and Research Branch, Islamic Azad University, Tehran, Iran

Y. Golian and D. Dorranian
Laser Laboratory, Plasma Physics Research Center, Science and Research Branch, Islamic Azad University, Tehran, Iran

Mehrnaz Gharagozalian, Davoud Dorranian and Mahmood Ghoranneviss
Biotechnology Lab., Plasma Physics Research Center, Science and Research Branch, Islamic Azad University, Tehran, Iran

Samina Dehghanizadeh
Central Tehran Branch, Islamic Azad University, Tehran, Iran

Shahrooz Saviz
Science and Research Branch, Plasma Physics Research Center, Islamic Azad University, Tehran, Iran

Akbar Sabetkar and Davoud Dorranian
Laser Laboratory, Plasma Physics Research Center, Science and Research Branch, Islamic Azad University, Tehran, Iran

Manesh Michael, Neethu Jayakumar, Sijo Sebastian, G. Sreekala and Chandu Venugopal
School of Pure and Applied Physics, Mahatma Gandhi University, Priyadarshini Hills, Kottayam 686 560, Kerala, India

Neethu T. Willington
Department of Physics, C. M. S. College, Kottayam 686001, Kerala, India

Shalini and N. S. Saini
Department of Physics, Guru Nanak Dev University, Amritsar, India

A. P. Misra
Department of Mathematics, Siksha Bhavana, Visva-Bharati University, Santiniketan, India

Uday Narayan Ghosh, Nikhil Pal and Prasanta Chatterjee
Department of Mathematics, Siksha Bhavana, Visva Bharati University, Santiniketan 731235, India

Asit Saha
Department of Mathematics, Siksha Bhavana, Visva Bharati University, Santiniketan 731235, India
Department of Mathematics, Sikkim Manipal Institute of Technology, Majitar, Rangpo, East-Sikkim 737136, India

A. M. El-Hanbaly, M. Sallah and H. F. Darweesh
Theoretical Physics Research Group, Physics Department, Faculty of Science, Mansoura University, P.O. Box 35516, Mansoura, Egypt

E. K. El-Shewy
Department of Physics, Taibah University, Al-Madinah Al-Munawarrah, Kingdom of Saudi Arabia

Reenu Gill and Hitendra K. Malik
Plasma Waves and Particle Acceleration (PWAPA) Laboratory, Department of Physics, Indian Institute of Technology Delhi, New Delhi 110016, India

Divya Singh
Rajdhani College, Delhi University, New Delhi, India

Hakima Ababsa, Med Tayeb Meftah and Thouria Chohra
LRPPS Laboratory, Department of Physics, Faculty of Mathematics, Kasdi Merbah University, 30000 Ouargla, Algeria

M. Momenei and Z. Khodabakhshei
Faculty of Physics, University of Shahrood, Shahrood, Iran

N. Panahi
Department of Physics, Bandar Abbas Branch, Islamic Azad University, Bandar Abbas, Iran

M. A. Mohammadi
Department of Atomic and Molecular Physics, Faculty of Physics, University of Tabriz, Tabriz, Iran

Mehran Shahmansouri
Department of Physics, Faculty of Science, Arak University, Arak 38156-8 8349, Iran

Mouloud Tribeche
Theoretical Physics Laboratory, Faculty of Physics, Plasma Physics Group, University of Bab-Ezzouar, USTHB, B.P. 32, El Alia, Algiers 16111, Algeria

H. R. Dehghanpour
Physics Department, Tafresh University, Tafresh, Iran

P. Parvin
Physics Department, Amirkabir University of Technology, Tehran, Iran

Mehdi Etehadi Abari, Mahsa Sedaghat and Mohammad Taghi Hosseinnejad
Young Researchers and Elites Club, Science and Research Branch, Islamic Azad University, Tehran, Iran

Gouranga C Nayak.
C. N. Yang Institute for Theoretical Physics, Stony Brook University, Stony Brook, NY 11794-3840, USA

Arnab Sikdar
Department of Mathematics, Calcutta Institute of Engineering and Management, 24/1A Chandi Ghosh Road, Kolkata 700040, India

Manoranjan Khan
Department of Instrumentation Science, Jadavpur University, Kolkata 700032, India

Index

www.ingramcontent.com/pod-product-compliance
Lightning Source LLC
Chambersburg PA
CBHW082057190326
41458CB00010B/3514
* 9 7 8 1 6 3 2 3 8 6 5 0 2 *